Selected Titles in This Series

IAS/PARK CITY
MATHEMATICS SERIES

Volume 5

Hyperbolic Equations and Frequency Interactions

Luis Caffarelli

Weinan E

Editors

American Mathematical Society
Institute for Advanced Study

IAS/Park City Mathematics Institute runs mathematics education programs that bring together high school mathematics teachers, researchers in mathematics and mathematics education, undergraduate mathematics faculty, graduate students, and undergraduates to participate in distinct but overlapping programs of research and education. This volume contains the lecture notes from the Graduate Summer School program on Nonlinear Wave Phenomena, which took place in Park City, Utah, July 9–29, 1995.

Supported by the National Science Foundation.

1991 *Mathematics Subject Classification.* Primary 35Lxx, 35Qxx, 42-XX, 73Dxx, 76-XX.

Library of Congress Cataloging-in-Publication Data

Hyperbolic equations and frequency interactions / Luis Caffarelli, Weinan E, editors.
  p. cm. – (IAS/Park City mathematics series, ISSN 1079-5634 ; v. 5)
  Includes bibliographical references.
  ISBN 0-8218-0592-4 (alk. paper)
  1. Geometrical optics–Mathematics. 2. Differential equations, Partial. 3. Schrödinger equation. 4. Wavelets (Mathematics) I. Caffarelli, Luis A. II. E, Weinan, 1963- . III. Series.
QC381.H96 1998   98-30060
515′.353–dc21   CIP

Contents

Jeffrey Rauch with the assistance of Markus Keel, Lectures on Geometric Optics

Preface

The Institute for Advanced Study/Park City Mathematics Institute (PCMI) was founded in 1991 as part of the "Regional Geometry Institute" initiative of the National Science Foundation. In mid 1993 the program found an institutional home at the Institute for Advanced Study (IAS) in Princeton, New Jersey. The PCMI will continue to hold summer programs alternately in Park City and in Princeton.

The IAS/Park City Mathematics Institute encourages both research and education in mathematics and fosters interaction between the two. The three-week Summer Session offers programs for researchers and postdoctoral scholars, graduate students, undergraduate students, high school teachers and undergraduate faculty. One of PCMI's most important goals is to make all of the participants aware of the total spectrum of activities that occur in both mathematics education and research: we wish to involve professional mathematicians in education and to bring modern concepts in mathematics to the attention of educators. To that end the Summer Session also features general seminars designed to encourage interaction among the various groups.

Beginning in 1997 the PCMI Summer Session also includes an Undergraduate Faculty Program. In-year activities at sites around the country form an integral part of the Program for High School Teachers.

Each summer a different topic is chosen as the focus of the Research Program and Graduate Summer School. Activities in the Undergraduate Program deal with this topic as well. Lecture notes from the Graduate Summer School are being published each year in this series. The first five volumes are:

Volume 1: *Geometry and Quantum Field Theory* (1991)
Volume 2: *Nonlinear Partial Differential Equations in Differential Geometry* (1992)
Volume 3: *Complex Algebraic Geometry* (1993)
Volume 4: *Gauge Theory and the Topology of Four-Manifolds* (1994)
Volume 5: *Nonlinear Wave Phenomena* (1995)

Future volumes from the 1996 Summer School *Probability* and the 1997 Summer School *Symplectic Geometry and Topology* are in preparation. The 1998 Research Program and Graduate Summer School topic is *Representation Theory of Lie Groups* and the 1999 Research Program and Graduate Summer School topic is *Number Theory*.

We plan to publish material from other parts of the IAS/Park City Mathematics Institute in the future. This will include material from the Undergraduate Program, the High School Teachers Program and publications documenting the interactive activities which are a primary focus of the PCMI. At the Summer Session late afternoons are devoted to programs of common interest to all participants. Many deal with current issues in education; others treat mathematical topics at a level which encourages broad participation. The PCMI has also spawned interactions between universities and high schools at a local level. We hope to share these activities with a wider audience in future volumes.

The IAS/Park City Mathematics Institute is sponsored by the Institute for Advanced Study and receives major funding from the National Science Foundation.

Dan Freed, Series Editor
May, 1998

Introduction

The second Institute for Advanced Study/Park City Mathematics Institute Summer Session, a continuation of the Utah Regional Geometry Institute, took place in the summer of 1995. The research topic was Nonlinear Wave Phenomena, the first major topic not connected with geometry.

Our idea was to bring together those mathematicians from the more theoretical areas of partial differential equations with those involved in applications to exchange ideas, knowledge, and perspectives in the field. How waves, or "frequencies," interact in nonlinear phenomena has been a central issue in many of the recent developments in pure and applied analysis. It is present in the discussion of singularity formation for hyperbolic equations, in the new functional spaces introduced by Bourgain in the treatment of Schrödinger and related equations, in the study of geometric optics as a limit of high frequency phenomena, and in many other areas. It was our impression that wavelet theory—with its simultaneous localization in both physical and frequency space and its lacunarity—is and will be a fundamental new tool in the treatment of the phenomena.

Thus the two "general methods and tools" courses were the ones by Jeff Rauch and Ingrid Daubechies. Rauch's Lectures on Geometric Optics study geometric optics as an asymptotic limit of high frequency phenomena. He shows how nonlinear effects are reflected in the asymptotic theory. In Daubechies' Harmonic Analysis, Wavelets and Applications the main structure of the by now fundamental wavelet theory is presented.

The more "specialized" course (this is not a sharp distinction since all the courses were both general and specialized in some way) were Nonlinear Schrödinger Equations by Jean Bourgain; Waves and Transport by George Papanicolaou; and Nonlinear Waves: Patterns, Oscillations, Singularities, and Stochasticity by David McLaughlin, who unfortunately was unable to prepare a written version of his beautiful series of lectures.

Finally, Susan Friedlander kindly agreed to prepare a written version of her series of lectures Stability and Instability of an Ideal Fluid given at the Mentoring Program for Women in Mathematics in May 1995, a preliminary program to the Summer Session, directed by Karen Uhlenbeck.

Of course, the Summer Session did not limit itself to the courses discussed above. There were several guest lectures that offered insight and challenging problems, and the participants themselves organized several discussion groups in subareas: fluid dynamics, blowup of parabolic and hyperbolic equations, combustion, etc.

In closing, we would like to stress the importance of a summer school like IAS/Park City Mathematics Institute. In the course of three weeks, about ninety graduate students, postdoctoral fellows, and established mathematicians from all over the United States and the world came together, had a chance to follow an array of courses in their field that are very unlikely to be repeated, and exchanged ideas in a very exciting atmosphere. We would like to thank in this context the lecturers who put so much effort and dedication in their courses; the different boards and committees that compose the IAS/Park City Mathematics Institute for choosing us to organize this school, and also for helping us in every aspect of the organization (Herb Clemens, Dan Freed, John Polking, Karen Uhlenbeck); and Anne Humes and the wonderful staff at the Institute for Advanced Study that, as usual, made everything perfect for us to enjoy our work.

Luis Caffarelli and Weinan E, Volume Editors

May, 1998

Nonlinear Schrödinger Equations

Jean Bourgain

IAS/Park City Mathematics Series
Volume 5, 1999

Nonlinear Schrödinger Equations

Jean Bourgain

Introduction

The main theme is the behavior of the flow of Schrödinger equations in one or more space variables, on compact domains (periodic boundary conditions for instance). A large part of the methods discussed apply, however, to other Hamiltonian PDE as well and occasionally results for nonlinear wave equations will be mentioned.

Some lectures will deal with the Cauchy problem aspect (thus mainly the behavior of the flow local in time). This topic is well understood on the line (and R^n) but was only recently studied under periodic boundary conditions. We prove periodic Strichartz type inequalities and exploit them in certain new function spaces with an appropriately defined space-time norm. These space-time norms may be defined naturally for a variety of equations (including wave equations and KdV) and turned out to lead to essential progress on wellposedness problems, also in the nonperiodic setting. Our discussion here aims mainly to provide results with 'rough' (i.e. nonsmooth) initial data, as needed later on in the invariant measure discussions. This material appears in my papers published in Geometric and Functional Analysis 3 (1993), 107-156 and 209-262. There is also a recent exposition of this material in Sem. Bourbaki (1994-95, N 796) by J. Ginibre.

The main part of the course will deal with longtime behavior of the flow, more precisely applications of methods of statistical mechanics and dynamical systems in PDE context. The first topic is the invariance of certain (normalized) Gibbs measures under the flow of the equation. This program was initiated in a paper of Lebowitz, Rose and Speer (Comm. in Stat. Phys., 88). In $1D$, as shown in this paper, the measure may be normalized also in the focusing case, as long as the nonlinearity does not exceed the critical blow-up exponent. In $2D$, only the cubic defocusing case is considered and normalization obtained by Wick ordering. The main new aspect is the construction of a well defined dynamics on the support of these measures.

---

[1]School of Mathematics, Institute for Advanced Study, Princeton, NJ 08540

E-mail: bourgain@math.ias.edu

The second topic is the application of the KAM technology to construct quasi periodic solutions. If one tries to apply the standard technique in the PDE-case, several problems appear. First, the phase space is infinite dimensional (and the interactions not finite range). This requires substantial adjustments of the method and the most relevant work in this aspect is S. Kuksin's book (Springer LNM, 1556). The next problem is the appearance of resonant normal frequencies, especially in higher space dimension, the conventional KAM scheme does not seem able to deal with. Subsequent to some work of W. Craig, E. Wayne (Comm. Pure and Applied Math. 93), I developed a scheme to treat normally resonant 'Melnikov' type problems. This method gives new persistency results, also in finite dimensional phase space. [(Intern. Math. Res. Notices, Vol. 11, 94 and various preprints).] It permits to progress considerably on PDE problems and construct for instance (non-trivial) quasi-periodic solutions for $2D$ Schrödinger equations. This is today an active research area. A closely related subject is 'Nekhoroshev' type stability. Some results in the non-resonant regime are described.

Thus several significantly different issues are considered with a varying level of understanding. As a consequence, the presentation style in these notes may differ from section to section and some are more expository than others.

LECTURE 1
Generalities and Initial Value Problems

We will consider as basic model of Hamiltonian evolution equation a nonlinear Schrödinger equation (NLS) of the form

$$iu_t + \Delta u + \frac{\partial F}{\partial \overline{u}} = 0 \tag{1}$$

where F is a real sufficiently smooth function $F = F(u, \overline{u})$.

One may typically take $F(u) = \lambda |u|^p$, $\lambda \in \mathbb{R}$, $p > 2$, in which case (1) becomes

$$iu_t + \Delta u + \frac{p}{2}\lambda |u|^{p-2}u = 0 \tag{2}$$

and hence essentially

$$iu_t + \Delta u \pm |u|^{p-2}u = 0 \quad \begin{pmatrix} + & : & \text{focusing} \\ - & : & \text{defocusing} \end{pmatrix}. \tag{3}$$

In (1), $u = u(x, t)$ is a complex valued function. We will be mostly concerned with periodic boundary values in space, i.e. u is 1-periodic in each of the x-variables. Hence u is a function on $\mathbb{T}^d \times I$. $\mathbb{T}^d = d$-torus and I some time interval. Equation (1) is Hamiltonian, since it may be written as

$$iu_t = \frac{\partial H}{\partial \overline{u}} \tag{4}$$

where

$$H(\phi) = \int_{\mathbb{T}^d} |\nabla \phi|^2 dx - \int_{\mathbb{T}^d} F(\phi). \tag{5}$$

Hence (5) yields in case (3)

$$\int_{\mathbb{T}^d} |\nabla \phi|^2 dx \mp \int |\phi|^p \quad \begin{pmatrix} - & : & \text{focusing} \\ + & : & \text{defocusing} \end{pmatrix}. \tag{6}$$

The "phase space" is formally obtained considering $(\operatorname{Re}\phi, \operatorname{Im}\phi)$ as pairs of conjugate variables and (4) is equivalent to

$$\begin{cases} (\operatorname{Re} u)^\bullet = \frac{1}{2}\frac{\partial H}{\partial(\operatorname{Im} u)} \\[2mm] (\operatorname{Im} u)^\bullet = -\frac{1}{2}\frac{\partial H}{\partial(\operatorname{Re} u)} \end{cases} \tag{7}$$

which is the standard format for a Hamiltonian system.

7

Since ϕ is a function on $\mathbb{T}^d$, one may expand it as Fourier series (assume ϕ at least in L^2 say)

$$\phi = \sum_{n \in \mathbb{Z}^d} \widehat{\phi}(n) e^{i\langle n, x\rangle}. \tag{8}$$

Identifying ϕ to its Fourier transform

$$\phi \leftrightarrow \{\widehat{\phi}(n)\}_{n \in \mathbb{Z}^d}$$

one may replace the phase space by the sequence space $\{(\operatorname{Re}\widehat{\phi}(n), \operatorname{Im}\widehat{\phi}(n))\}$ with $(\operatorname{Re}\widehat{\phi}(n), \operatorname{Im}\widehat{\phi}(n))$ as pairs of canonical variables. Considering H as a function of $\{\widehat{\phi}(n)\}$, (4) is indeed equivalent to

$$i\dot{a}_n = \frac{\partial H}{\partial \bar{a}_n} \quad (a_n = \widehat{\phi}(n)). \tag{9}$$

This description of phase space will be mostly used to make formal considerations rigorous, since it easily permits reduction to the finite dimensional situation (by projection on the N first Fourier modes). This discussion will be pursued more in the context of invariant Gibbs measures. Observe at this stage that the symplectic topology induced by (9) on the phase space is

$$\left(\sum_{n \in \mathbb{Z}^d} |a_n|^2\right)^{1/2} = \|\phi\|_{L^2(\mathbb{T}^d)}. \tag{10}$$

The Hamiltonian (5) is preserved under the flow of (4). In particular, the Hamiltonian

$$\frac{1}{2} \int |\nabla \phi|^2 \mp \frac{1}{p} \int |\phi|^p \tag{11}$$

is invariant under the flow of the respective equation

$$iu_t + \Delta u \pm |u|^{p-2}u = 0. \tag{12}$$

In the particular case $F(u, \bar{u})$ is of the form $f(|u|^2)$, there is also conservation of the L^2-norm

$$\int_{\mathbb{T}^2} |\phi|^2 dx. \tag{13}$$

Remark. The case of the $1D$ NLS with nonlinearity $\lambda u|u|^2$ ($1D$ cubic NLS) is special in the sense that besides (13), (11) there are other invariants of motion. In fact this equation is integrable in the sense that there is a "complete set" of conserved quantities controlling the flow. This property is very special and the so-called "integrable-equations" form a distinguished list. Our aim here is to study more general PDE and Hamiltonian aspects and we will not go into the theory of these integrable cases.

Observe that for equation (12), there is an apriori bound of the L^2-norm $\|u(t)\|_2$ for all time t.

In the defocusing case, the Hamiltonian

$$\frac{1}{2} \int \nabla \phi|^2 + \frac{1}{p} \int |\phi|^p \tag{14}$$

controls in particular the H^1-norm

$$\|u(t)\|_{H^1(\mathbb{T}^d)} = \left(\int_{\mathbb{T}^d} |\nabla_x u(t)|^2\right)^{1/2} \tag{15}$$

for all time t.

In the focusing case, the Hamiltonian is given by

$$\frac{1}{2}\int |\nabla\phi|^2 - \frac{1}{p}\int |\phi|^p \tag{16}$$

(not necessarily bounded from below) and previous issue depends on dimension d and exponent p in the nonlinearity. More precisely, from standard Sobolev inequality, one has if

$$\frac{1}{p} \geq \frac{1}{2} - \frac{1}{d} \tag{17}$$

$$\|\phi\|_p \leq c_p \, \|\phi\|_{H^1}^\theta \, \|\phi\|_2^{1-\theta}; \quad \theta = d\left(\frac{1}{2} - \frac{1}{p}\right). \tag{18}$$

Hence

$$H(\phi) \equiv \frac{1}{2}\int |\nabla\phi|^2 - \frac{1}{p}\int |\phi|^p \geq \frac{1}{2}\|\phi\|_{H^1}^2 - c_p \, \|\phi\|_{H^1}^{\theta p} \, \|\phi\|_2^{(1-\theta)p} \tag{19}$$

and we distinguish the cases

  (i) $p < 2 + \frac{4}{d}$ Then $\theta p < 2$ and $\|\phi\|_{H^1}$ is controlled by $H(\phi)$, $\|\phi\|_2$ which are conserved quantities

  (ii) $p = 2 + \frac{4}{d}$ Then $\theta p = 2$ and $\|\phi\|_{H_1}$ is controlled by $H(\phi)$, provided $\|\phi\|_2$ is small enough

  (iii) $p > 2 + \frac{4}{d}, (17)$ Then $\theta p > 2$ and $\|\phi\|_{H^1}$ is controlled by $H(\phi)$ for $H(\phi)$ sufficiently small

Proof of (18). Since $1 < p < \infty$, and has from the Littlewood-Paley square function theorem

$$\|\phi\|_p \sim \|S(\phi)\|_p \tag{20}$$

$$S(\phi) = \left(|\widehat{\phi}(0)|^2 + \sum_{k=1}^\infty \Big| \underbrace{\sum_{2^{k-1}\leq|n|<2^k} \widehat{\phi}(n) \, e^{\langle n,x\rangle}}_{\text{\tiny III} \; \Delta_k\phi}\Big|^2\right)^{1/2} \tag{21}$$

and hence by convexity

$$\|\phi\|_p \leq c_p\|S(\phi)\|_p \leq c_p\left(\sum_k \|\Delta_k\phi\|_p^2\right)^{1/2}. \tag{22}$$

Define

$$\|\phi\|_{H^s} = \left(\sum(1+|n|^2)^s|\widehat{\phi}(n)|^2\right)^{1/2}. \tag{23}$$

Since

$$\|\Delta_k\phi\|_\infty \leq 2^{k\frac{d}{2}} \, \|\Delta_k\phi\|_2 \sim \|\Delta_k\phi\|_{H^{d/2}} \tag{24}$$

it follows from interpolation for $2 < p < \infty$ between $(H^{d/2}, L^\infty)$. and (H^0, L^2)

$$\frac{1}{p} = \frac{1-\theta}{\infty} + \frac{\theta}{2}, \quad \theta = \frac{2}{p}$$

$$\|\Delta_k \phi\|_p \lesssim \|\Delta_k \phi\|_{H^{\frac{d}{2}(1-\theta)}} = \|\Delta_k \phi\|_{H^{\frac{d}{2}\left(1-\frac{2}{p}\right)}}. \tag{25}$$

Thus, from (22), also

$$\|\phi\|_p \leq c_p \, \|\phi\|_{H^{\frac{d}{2}\left(1-\frac{2}{p}\right)}}. \tag{26}$$

If $\frac{d}{2}\left(1 - \frac{2}{p}\right) \leq 1$, i.e. (17)

$$\frac{1}{p} \geq \frac{1}{2} - \frac{1}{d}$$

one gets by interpolation again (or Hölder's inequality)

$$\|\phi\|_{H^{\frac{d}{2}\left(1-\frac{2}{p}\right)}} \leq \|\phi\|_{H^1}^{\theta} \, \|\phi\|_{H^0}^{1-\theta}, \quad \theta = d\left(\frac{1}{2} - \frac{1}{p}\right)$$

which is (18).

As a corollary of the preceding, one may thus conclude the following.
Let u be a (sufficiently smooth) solution of (1.10)

$$iu_t + \Delta u \mp u|u|^p = 0 \quad \begin{pmatrix} - & : & \text{defocusing} \\ + & : & \text{focusing} \end{pmatrix}.$$

There is the L^2-conservation

$$\|u(t)\|_2 = \|u(0)\|_2.$$

In the defocusing case, there is an apriori bound of $\|u(t)\|_{H^1}$ in terms of $\|u(0)\|_{H^1}$.
In the focusing case, we distinguish the cases

$$p < 2 + \frac{4}{d} : \text{ a priori bound of } \|u(t)\|_{H^1} \text{ from } \|u(0)\|_{H^1}$$

$$p = 2 + \frac{4}{d} : \;\dotfill\; \text{ for sufficiently small } \|u(0)\|_2$$

$$\frac{1}{p} > \frac{1}{2} - \frac{1}{d}, p > 2 + \frac{4}{d} : \;\dotfill\; \text{ for sufficiently small } \|u(0)\|_{H^1}.$$

Remarks.
(1) In the focusing case, when $p > 2 + \frac{4}{d}$ or $p = 2 + \frac{4}{d}$ and the initial data has
   sufficiently large L^2-norm, solutions with smooth initial data may develop
   singularities in finite time.
Consider the case $p = 2 + \frac{4}{d}$ and the equation

$$iu_t + \Delta u + u|u|^{4/d} = 0 \tag{27}$$

on $\mathbb{R}^d$. This equation has certain conformal properties. If $u = u(x,t)$ is a solution
of (27), then so is

$$v(x,t) = \frac{1}{|T-t|^{d/2}} \, e^{i|x|^2/4(t-T)} \, \overline{u}\left(\frac{x}{T-t}, \frac{1}{T-t}\right). \tag{28}$$

This invariance permits to construct explicit blow-up solutions of given (sufficiently large) L^2-norm. One starts with a solution of the form

$$u(x,t) = e^{i\omega^2 t}\omega^{d/2}\, Q(\omega x) \tag{29}$$

where Q solves the problem

$$\begin{cases} \Delta Q + |Q|^{4/d}\, Q = Q & \text{in } \mathbb{R}^d \\ Q > 0 & \text{in } \mathbb{R}^d \end{cases} \tag{30}$$

(there is a radial solution, achieving minimum L^2-norm for solutions $Q' \neq 0$ of the equation $\Delta Q' + |Q'|^{4/d}\, Q' = \omega Q'$ $\omega > 0$, called a ground state).

Then v given by (28) yields a blow-up solution of (27) at time $t = T$. More precisely, for $t \to T$, $|v(t)|^2$ will converge to the Dirac measure at $x = 0$. There is blow-up of the solution in any norm H^s for $s > 0$.[*]

Because of its local character, the previous example may be carried over to the periodic setting.

Besides the previous explicit construction of blow-up solutions, there are general conditions (on p and the initial data ϕ) that ensure blow-up (at least on $\mathbb{R}^d$) (R. Glassey; $p > 2 + \frac{4}{d}, H(\phi) < 0$). Many important open problems remain here (as to the blow-up speed for instance) but will not be pursued here.

(2) Omitting the special case of the $1D$ cubic NLS (which is integrable - Zakharov-Shabat), the L^2-norm and H^1-norm only (modulo preceding discussion) are subject to an a priori bound. In general, estimates on smooth solutions in higher Sobolev spaces grow when $|t| \to \infty$. Some estimates on this matter for NLS will be outlined in the next section and for wave equations in the Appendix. This possible growth of higher Sobolev norms (related to weak turbulence) due to a lack of additional constants of motion is a phenomenon comparable to Arnold diffusion in a dynamical system (see the Appendix). The "true" speed of growth is essentially on open problem.

As mentioned earlier, we will need (for various reasons) to consider "generalized" solutions for "rough" data. We introduce the notion of *wellposedness* of the Cauchy problem on a time interval I

$$\begin{cases} iu_t + \Delta u + \frac{\partial F}{\partial \overline{u}} = 0 \\ u(0) = \phi \end{cases} \tag{31}$$

for data ϕ in a certain function space B (this definition is of course not restricted to NLS). We will assume that the smooth functions are dense in B. Sometimes B will be a space of distributions.

We say that (31) is wellposed on I for data $\phi \in B$

$$[\text{in a space } X \text{ contained in } C_B(I)]$$

provided there is a well defined continuous map

$$\phi \mapsto u_\phi$$

---

[*]This is a general feature, at least on $\mathbb{R}^d$. For $p = 2 + 4/2$, the blow-up time (if $< \infty$) is the same in all H^s-spaces, $s > 0$.

of B into X such that if ϕ is sufficiently smooth, u_ϕ coincides with the classical solution.

Hence, given any $\phi \in B$ and regularizing sequence $\{\phi_n\}, \phi_n \to \phi$ in B of smooth functions, u_ϕ is the unique limit point in X of the sequence $\{u_{\phi_n}\}$. In what follows the space B will be mostly an $H^s(\mathbb{T}^d)$-space, $s \geq 0$. Spaces of distributions will appear for instance in the context of invariant Gibbs measures for $d > 1$. The space X will be certain space-time spaces often defined in terms of the Fourier transform of u (and are essentially Besov spaces). Their introduction is in fact the main novelty arising from the study of the periodic case, as mentioned earlier.

As a general philosophy, wellposedness for data in H^s will go together with wellposedness in $H^{s_1}, s_1 \geq s$.

We further distinguish between *local* and *global* wellposedness according to whether the time interval I is a neighborhood of 0 or $\mathbb{R}$.

Wellposedness problems for the NLS are much better understood on $\mathbb{R}^d$ than in the periodic case (the periodic case was only recently looked at). The reason for the difference will be further made clear in the next section.

On $\mathbb{R}^d$, there is the following local wellposedness theorem for the equation

$$\begin{cases} iu_t + \Delta u + \lambda u|u|^{p-2} = 0 \\ u(0) = \phi. \end{cases} \tag{32}$$

Theorem 33. *(Casanave-Weissler). The Cauchy problem (32) is locally wellposed for data $\phi \in H^s, s \geq 0$, provided $p - 2 > [s]$ and if $p > 2 + \frac{4}{d}$*

$$s \geq s_*, \quad s_* \text{ defined by } p - 2 = \frac{4}{d - 2s_*}. \tag{34}$$

In this case, the spacetime spaces X are classical mixed $L^p - L^q$ (or Besov) spaces. In the context of Theorem 33, there is moreover global wellposedness provided $\|\phi\|_{H^{s_*}}$ is sufficiently small.

On the other hand, global wellposedness may be obtained combining Theorem 33 with the conservation laws (the global solution is gotten piecing together local solutions).

Theorem 35. *The Cauchy problem (32) is globally (i.e. on any finite time interval) wellposed*
- *For data $\phi \in H^s, s \geq 0$ provided $p < 2 + \frac{4}{d}$ (from L^2-conservation)*
- *For data $\phi \in H^s, s \geq 1$ provided $p < 2 + \frac{4}{d-2}$ and the Hamiltonian controls the H^1-norm*

Remarks.
(1) The issue when the H^1-norm is controlled by the Hamiltonian was discussed earlier.
(2) The time interval of local wellposedness in Theorem 33 depends on $\|\phi\|_{H^s}$ provided $s > s_*$. In the case $s = s_*$, it also depends on ϕ (hence there is no contradiction with the singularity formations when $s_* = 0$, $p = 2 + \frac{4}{d}$ discussed earlier).
(3) In Theorem 35, there is no a priori bound on $\|u(t)\|_{H^s}$ for $s > 1$ (assuming $\phi \in H^s$). An exponential estimate $C^{|t|}$ is immediately deduced from the

method. The argument given at the end of the next lecture in the periodic case would improve this to a powerlike bound in time.

We now return to the periodic case. It would be reasonable to expect the analogue of Theorem 33, except that in (34) one should replace $s \geq s_*$ by $s > s_*$ (for reasons that will be clear later and relates to certain failure of Strichartz's inequality in the periodic case). Our knowledge is however at the present far less complete. We state next some model results, which will be proved in the next section.

Theorem 36. *The Cauchy problem in 1D*

$$\begin{cases} iu_t + u_{xx} + \lambda u|u|^2 = 0 \\ u(0) = \phi \end{cases}$$

is globally wellposed for data $\phi \in H^s(\mathbb{T})$, $s \geq 0$.

Theorem 37. *The Cauchy problem in 1D*

$$\begin{cases} iu_t + u_{xx} + \lambda u|u|^{p-2} = 0 \qquad (p > 2) \\ u(0) = \phi \end{cases}$$

is locally wellposed for data $\phi \in H^s(\mathbb{T})$ when

$$s > 0 \quad \text{and} \quad p \leq 6 \tag{38}$$

$$s > s_*, \quad p - 2 = \frac{4}{1 - 2s_*}, \qquad p > 6. \tag{39}$$

Theorem 40. *The Cauchy problem in 2D*

(2D)

$$\begin{cases} iu_t + \Delta u + \lambda u|u|^2 = 0 \\ u(0) = \phi \end{cases}$$

is locally wellposed for $\phi \in H^s(\mathbb{T}^2)$, $s > 0$ and hence globally for $\phi \in H^s, s \geq 1$, provided the Hamiltonian controls H^1-norm.

Theorem 41. *The Cauchy problem in 3D*

(3D)

$$\begin{cases} iu_t + \Delta u + \lambda u|u|^2 = 0 \\ u(0) = \phi \end{cases}$$

is globally wellposed for $\phi \in H^s, s \geq 1$, provided Hamiltonian controls the H^1-norm.

Theorem 42. *Consider the-Cauchy problem in 4D*

(4D)

$$\begin{cases} iu_t + \Delta u + ug(|u|^2) = 0 \\ u(0) = \phi \end{cases} \tag{43}$$

where $g \in C^2(\mathbb{R}^+)$, $|g(s)| < cs^{1/2}, |g'(s)| < cs^{-1/2}, |g''(s)| < cs^{-3/2}$. Then (43) is globally wellposed for data $\phi \in H^s(\mathbb{T}^4), s \geq 2$, provided the H^1-norm is controlled by the energy (=Hamiltonian), which is the case for $g < 0$ or $\|\phi\|_2$ small enough.

Comments and references related to Lecture 1

Concerning the Cauchy problem for NLS on $\mathbb{R}^d$, these are some standard references.

[C-W]   T. Cazenave, F. Weissler, *The Cauchy problem for the critical non-linear Schrödinger equation in H^s*, Nonlinear Analysis, Theory Methods and Applications 14 (10) (1990), 807-936.

[G-V]   J. Ginibre, G. Velo, *The global Cauchy problem for the nonlinear Schrödinger equation, H. Poincaré*, Analyse Non Linéaire 2 (1985), 309-327.

References in the case of periodic boundary conditions

[B]   J. Bourgain, *Restriction phenomena for lattice subsets and applications to nonlinear evolution equations I, Schrödinger equations, Geometric and Functional Analysis*, (GAFA), Vol. 3, N2 (1993), 107-156.

[B1]   J. Bourgain, *Exponential sums and nonlinear Schrödinger equations*, GAFA, Vol. 3, N2 (1992), 157-178.

References concerning blowup solutions of NLS and NLS with critical nonlinearity

[G1]   R.T. Glassey, *On the blow-up of solutions to the Cauchy problem for the nonlinear Schrödinger equation*, J. Math. Phys. 18 (1977), 1794-1797.

[Mer1] F. Merle, *Determination of blow-up solutions with minimal mass for Schrödinger equation with critical power*, Duke J. Math. 69, N2 (1993), 427-454.

[Mer2] F. Merle, *On uniqueness and continuation properties after blow-up time of self similar solutions of nonlinear Schrödinger equation with critical exponent and critical mass*, Commun. Pure and Appl. Math. 45 (1992), 203-254.

LECTURE 2
The Initial Value Problem (continued)

In this lecture, we will mainly concentrate on proving the wellposedness theorems mentioned above. Consider the equation

$$iu_t + \Delta u + \frac{\partial F}{\partial \overline{u}} = 0 \tag{1}$$

with main example $F(u) = \pm|u|^p$.

In studying the Cauchy problem in the periodic case, we basically follow the same scheme as on the line. Thus the first step is to study the linear problem,

$$\begin{cases} iu_t + \Delta u = 0 \\ u(0) = \phi. \end{cases} \tag{2}$$

The solution of (3) is given by an explicit formula. In the $\mathbb{R}^d$-case, one gets a Fourier integral

$$u(x,t) = S(t)\phi(x) = \int_{\mathbb{R}^d} \hat{\phi}(\xi)\ e^{i(\langle x,\xi\rangle + |\xi|^2 t)} d\xi \tag{3}$$

$$\hat{\phi}(\xi) = \int_{\mathbb{R}^d} \phi(x)\ e^{-i\langle x,\xi\rangle} dx$$

and in the case of periodic boundary conditions, as a Fourier series

$$u(x,t) = S(t)\phi(x) = \sum_{n\in\mathbb{Z}^d} \hat{\phi}(n)\ e^{i(\langle n,x\rangle + |n|^2 t)} \tag{4}$$

$$\hat{\phi}(n) = \int_{\mathbb{T}^d} \phi(x)\ e^{-2\pi i\langle n,x\rangle} dx. \tag{5}$$

Observe that $S(t)$ forms a unitary group.

To study (1), we replace the PDE by the equivalent integral equation

$$u(t) = S(t)\phi + i\int_0^t S(t-\tau)\left(\frac{\partial F}{\partial \overline{u}}\right)(\tau) d\tau \tag{6}$$

(in the case of classical solutions, this equivalence is clear). We will study (6) instead of (1). In particular, local wellposedness will be established by showing a

15

contractive property of the map in an appropriate space X

$$u \mapsto S(t)\phi + i \int_0^t S(t-\tau)\left(\frac{\partial F}{\partial \overline{u}}\right)(\tau)d\tau \tag{7}$$

on a sufficiently small time interval. The main ingredients in this discussion are the following
- Estimates on $S(t)$ (i.e. Strichartz inequalities)
- The definition of appropriate spaces X.

In the $\mathbb{R}^d$-case, the relevant Strichartz inequality is following (global) space-time inequality

$$\|S(t)\phi(x)\|_{L^p(dxdt)} \leq c\|\phi\|_{L_x^2} \qquad p = \frac{2(d+2)}{d}. \tag{8}$$

This result is equivalent with the "restriction theorem" for the paraboloid. In the space periodic case, one may not expect global estimates in time. The natural analogue of (8) would be

$$``\|S(t)\phi(x)\|_{L_{x,t}^p(\mathbb{T}^{d+1})} \leq c\|\phi\|_{L_x^2(\mathbb{T}^d)}; \qquad p = \frac{2(d+2)}{d}". \tag{9}$$

Estimate (9) is false. For instance, for $d = 1$, one may show that

$$\left\|\frac{1}{N^{1/2}} \sum_{|n|\leq N} e^{i(nx+n^2 t)}\right\|_{L^6(\mathbb{T}^2)} \geq (\log N)^{1/6}. \tag{10}$$

It would be reasonable to conjecture that (9) holds for $p < \frac{2(d+2)}{d}$ and

$$\left\|\sum_{|n|\leq N} a_n\, e^{i(\langle n,x\rangle+|n|^2 t)}\right\|_{L^{\frac{2(d+2)}{d}}(\mathbb{T}^{d+1})} \ll N^\varepsilon \left(\sum |a_n|^2\right)^{1/2} \tag{11}$$

for all $\varepsilon > 0$.

This conjecture is open in any dimension. The validity of (11) is shown for $d = 1, 2$.

To estimate "exponential sums" of the form

$$\sum_{|n|\leq N} a_n\, e^{i(\langle n,x\rangle+|n|^2 t)} \tag{12}$$

we will rely on two methods
- arithmetic methods (for p on even integers)   (13)
- methods based on the circle method   (14)

The following results are obtained for instance using (13).

Proposition 15. *There is in 1D the L^4-inequality*

$$\left\|\sum a_n\, e^{i(nx+n^2 t)}\right\|_{L_{x,t}^4} \leq c\left(\sum |a_n|^2\right)^{1/2} \tag{16}$$

and, more generally,

$$\left\|\sum a_{n,m}\, e^{i(nx+mt)}\right\|_{L^4_{x,t}} \le c\left(\sum (1+|n^2-m|)^{3/4}|a_{nm}|^2\right)^{1/2} \tag{17}$$

(the 3/4-exponent in (17) is optimal)

Proposition 18. *In 1D, there is inequality (11), in the form*

$$\left\|\sum_{|n|\le N} a_n\, e^{i(nx+n^2 t)}\right\|_{L^6_{x,t}} \le c\left(\exp \frac{\log N}{\log\log N}\right)\left(\sum |a_n|^2\right)^{1/2}. \tag{19}$$

Proposition 20. *In 2D, there is inequality (11) in the form*

$$\left\|\sum_{|n|\le N} a_n\, e^{i(\langle n,x\rangle+|n|^2 t)}\right\|_{L^4_{x,t}} \le c\left(\exp \frac{\log N}{\log\log N}\right)\left(\sum |a_n|^2\right)^{1/2}. \tag{21}$$

These facts may be proved by direct calculation, since we have an even integer as moment p. More precisely, for p an even integer, write by Parseval's identity

$$\left\|\sum_{|n|\le N} a_n\, e^{i(\langle n,x\rangle+|n|^2 t)}\right\|_{L^p_{x,t}}^p = \sum_{n,m} |b_{n,m}|^2 \tag{22}$$

where the $b_{n,m}$ are defined by

$$\left[\sum a_n\, e^{i(\langle n,x\rangle+|n|^2 t)}\right]^{p/2} = \sum_{n,m} b_{n,m}\, e^{i(\langle n,x\rangle+mt)}. \tag{23}$$

Hence

$$b_{n,m} = \sum_{\substack{n=n_1+\cdots+n_k \\ m=|n_1|^2+\cdots+|n_k|^2}} a_{n_1} a_{n_2}\ldots a_{n_k} \quad \left(k=\frac{p}{2}\right) \tag{24}$$

and it follows that

$$\left\|\sum_{|n|\le N} a_n e^{i(\langle n,x\rangle-|n|^2 t)}\right\|_p \le c(N)\left(\sum |a_n|^2\right)^{1/2} \tag{25}$$

where $c(N)$ is a uniform bound on the number of solutions of the system of equations

$$\begin{cases} n_1+n_2+\ldots+n_k = n,\ |n_i|\le N \\ |n_1|^2+|n_2|^2+\cdots+|n_k|^2 = m. \end{cases} \tag{26}$$

Hence the problem reduces to counting the number of lattice points $(n_1,\ldots,n_{k-1})$ on the quadratic surface

$$|n_1|^2+\ldots+|n_{k-1}|^2+|n-n_1-\ldots-n_{k-1}|^2 = m. \tag{27}$$

In (16), $d=1, p=4, k=1$. The equation $n_1^2+(n-n_1)^2 = m$ has at most 2 solutions, implying the inequality.

In (19), $d=1, p=6, k=2$ and we have the equation

$$n_1^2+n_2^2+(n-n_1-n_2)^2 = m \tag{28}$$

hence

$$3(n_1 + n_2)^2 + (n_1 - n_2)^2 - 4n(n_1 + n_2) = 2(m - n^2).$$

Putting $m_1 = n_1 + n_2$, $m_2 = n_1 - n_2$, this gives

$$(3m_1 - 2n)^2 + 3m_2^2 = 6m - 2n^2. \tag{29}$$

Consider the equation

$$X^2 + 3Y^2 = A \quad (X, Y, A \in \mathbb{Z}). \tag{30}$$

Denoting $\rho = e^{\frac{2\pi i}{3}} = \frac{1 + i\sqrt{3}}{2}$, the ring $\mathbb{Z} + \rho\mathbb{Z}$ is known to be an euclidean division domain (implying unique prime factorization), similar to the Gaussian integers $\mathbb{Z} + i\mathbb{Z}$. Observe that if $X, Y \in \mathbb{Z}$ satisfy $X^2 + 3Y^2 = A$, then $X + i\sqrt{3}Y$ is a divisor of A in $\mathbb{Z} + \rho\mathbb{Z}$ and that X, Y determine uniquely (n_1, n_2). Hence the factor in (19) follows simply from an estimate on the divisor function.

In (21), $d = 2, p = 4, k = 1$. We get the equation $|n_1|^2 + |n - n_1|^2 = m, n_1 \in \mathbb{Z}^2$ and invoke the bound on the maximum number of lattice points on a circle in $\mathbb{R}^2$ of radius R (which is $\exp \frac{\log R}{\log \log R}$ from Gaussian integer factorization).

Remark. In view of (10), it would be of interest to determine whether (19) is the right behaviour or bounds are possibly logarithmic. An estimate with factor $(\log N)^\alpha$, $\alpha > 0$ sufficiently small, rather than $\exp c \frac{\log N}{\log \log N}$ in (19) would have significant PDE applications.

A related formally simpler problem is to determine the best constant c_N in the inequality

$$\left\| \sum_{|n| \leq N} a_n e^{in^2 t} \right\|_{L^4(\mathbb{T})} \leq c_N \left(\sum |a_n|^2 \right)^{1/2} \tag{31}$$

(the "Λ_4-constants" for the sequence of squares $\{n^2\}$). One has again

$$(\log N)^{1/4} < c_N < \exp \frac{\log N}{\log \log N}.$$

An alternative method for proving Strichartz' type inequalities in the periodic case consists in combining the method of proof in the $\mathbb{R}^d$-case (mainly interpolation between $L^2 \to L^2$ and $L^1 \to L^\infty$) with elements of the Hardy-Littlewood circle method in analytic number theory. A result obtained this way is a partial analogue of (8) in the form of the following distributional inequality.

Proposition 32. *In dimension d, letting*

$$F(x, t) = \sum_{|n| \leq N} a_n e^{i(\langle n, x \rangle + |n|^2 t)}, \quad \sum |a_n|^2 \leq 1$$

one has for $\lambda > N^{d/4}$

$$\mathrm{meas}\{(x, t) \in \mathbb{T}^{d+1} \big| |F(x, t)| > \lambda\} \ll N^\varepsilon \lambda^{-\frac{2(d+2)}{d}}. \tag{33}$$

Details of the proof are more complicated and will not be elaborated on here. Proposition 32 plays a role in the proof of the wellposedness theorem 42 ($d = 4$) stated in Lecture 1.

Next, we will discuss certain space-time norms $X^{s,b}$ used in the study of the local wellposedness problem. Fix a time interval $I \subset \mathbb{R}$. Define $X^{s,b}(I)$ as the space of functions u on $\mathbb{T}^d \times I$, which are represented as

$$\sum_{n \in \mathbb{Z}^d} \int_{\mathbb{R}} d\lambda \, \tilde{u}(n, \lambda) \, e^{i(\langle n, x \rangle + \lambda t)}|_{\mathbb{T}^d \times I} \tag{34}$$

where $\hat{u}$ satisfies

$$\left(\sum_{n \in \mathbb{Z}^d} (1 + |n|^2)^s \int_{\mathbb{R}} d\lambda (1 + |\lambda - |n|^2|)^{2b} |\tilde{u}(n, \lambda)|^2 \right)^{1/2} < \infty. \tag{35}$$

We define then $\|u\|_{X^{s,b}(I)}$ as the infimum of (35) for all representations (34) (i.e. as restriction norm).

Remarks.
   (1) It is clear from preceding definition that if $I' \subset I$, then the restriction map $u \to u|_{I'}$ maps $X^{s,b}(I)$ in $X^{s,b}(I')$.
   (2) We will usually denote (abusively) $\tilde{u}(n, \lambda)$ in (34) by $\hat{u}(n, \lambda)$. Denoting for $u \in X^{s,b}(I)$

$$U(x, t) = \sum_{n \in \mathbb{Z}^d} \int_{\mathbb{R}} d\lambda \tilde{u}(n, \lambda) \, e^{i(\langle n, x \rangle + \lambda t)} \tag{36}$$

one has indeed that $\hat{U} = \tilde{u}$ satisfies (35) but the Fourier transform of the restriction $u = U|_{\mathbb{T}^d \times I}$ may fail.

On the other hand, in certain arguments we will aim to decrease a given interval say $I = [0, T]$ to a smaller interval $I' = [0, \delta]$ and use for $u|_{I'}$ the representation

$$U_1(x, t) = U(x, t)\psi(t) \tag{37}$$

where U is given by (36) and ψ is a smooth localizing bumpfunction

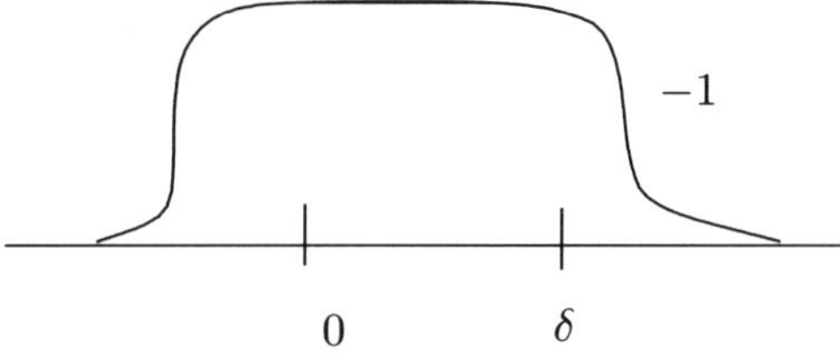

$\psi = 1$ on $[-\delta, \delta]$. One then has the estimate

$$\|(1 + |n|)^s (1 + |\lambda - |n|^2|)^b \hat{U}_1\|_2 \leq$$

$$\begin{cases} c\|(1 + |n|^2)(1 + |\lambda - |n|^2|)^b \hat{U}\| = c\|u\|_{X^{s,b}} & \text{if } b < \tfrac{1}{2} \\ c(\log \tfrac{1}{\delta})^{1/2}\|u\|_{X^{s,b}} & \text{if } b = \tfrac{1}{2} \\ c(\tfrac{1}{\delta})^{b - \tfrac{1}{2}}\|u\|_{X^{s,b}} & \text{if } b > \tfrac{1}{2}. \end{cases} \tag{38}$$

For $b < \tfrac{1}{2}$, the same is true when ψ is taken to be $\chi_{[0,\delta]}$.

Replacement of U by U_1 in estimating the nonlinear term in $X^{s,b}([0,\delta])$ may however lead to savings when $\delta \to 0$ beating the $\frac{1}{\delta}$-factors in (38).

(3) It is clear that for $b > \frac{1}{2}$

$$\|u\|_{L^\infty_{H^s(\mathbb{T}^d)}(I)} \le c_b \, \|u\|_{X^{s,b}(I)} \tag{39}$$

as a consequence of the inclusion $H^b(\mathbb{R}) \subset L^\infty(\mathbb{R})$ for $b > \frac{1}{2}$.

Proof of Theorem 1.36

Consider the Cauchy problem in $1D$ with cubic nonlinearity

$$\begin{cases} iu_t + u_{xx} + \lambda u|u|^2 = 0 \\ u(0) = \phi \in H^s(\mathbb{T}), s \ge 0 \end{cases} \tag{40}$$

and the equivalent integral formulation (6)

$$u(t) = S(t)\phi + i\lambda \int_0^t S(t-\tau)(u|u|^2)(\tau)d\tau. \tag{41}$$

We will show that the corresponding map satisfies the contraction principle in the space $X^{s,b}([0,\delta])$ for $\frac{3}{8} < b < \frac{5}{8}$ and $\delta > 0$ small enough (depending on the size of ϕ.) This will yield local wellposedness in the corresponding space. In order to get global wellposedness, we will make a further discussion of the size of δ.

The main ingredient in what follows is Strichartz inequality (17), implying the following inequality for functions on $\mathbb{T} \times \mathbb{R}$

$$\|F\|_{L^4(\mathbb{T}\times[0,1])} \le c \, \|F\|_{X^{0,3/8}([0,1])}. \tag{42}$$

(The same is true for any bounded interval I).

Let $u = u|_{\mathbb{T}\times[0,\delta]}$. Recall that then, for $b_1 < \frac{1}{2}$

$$\left(\sum_n \int d\lambda(1 + |n|^{2s})(1 + |\lambda - n^2|)^{2b_1} \, |\hat{u}(n,\lambda)|^2\right)^{1/2} \le c\|u\|_{X^{s,b_1}}. \tag{43}$$

Denote $w = u|u|^2$. The nonlinear term in (41) is then given by

$$\sum_n \int d\lambda \, \hat{w}(n,\lambda) \, e^{i(nx+n^2t)} \, \frac{e^{i(\lambda-n^2)t} - 1}{\lambda - n^2} \tag{44}$$

which we express as the sum of the following 3 terms

$$\left[\sum_n \int_{|\lambda-n^2|\le 1} d\lambda \, \hat{w}(n,\lambda)e^{i(nx+n^2t)} \, \frac{e^{i(\lambda-n^2)t} - 1}{\lambda - n^2}\right]\psi(t) \tag{45}$$

$$\sum_n \int_{|\lambda-n^2|>1} d\lambda \, \frac{\hat{w}(n,\lambda)}{\lambda - n^2} \, e^{i(nx+\lambda t)} \tag{46}$$

$$\left[\sum_n \left(\int_{|\lambda-n^2|>1} d\lambda \, \frac{\hat{w}(n,\lambda)}{\lambda - n^2}\right)e^{i(nx+n^2t)}\right]\psi(t) \tag{47}$$

with ψ a localizing bumpfunction as above.

One has clearly

$$\|(46)\|_{X^{s,b}([0,\delta])} \leq \left[\sum_n \int d\lambda (1+|n|^s)^2(1+|\lambda-n^2|)^{2(b-1)}\, |\hat{w}(n,\lambda)|^2\right]^{1/2} \qquad (48)$$

and

$$\|(47)\|_{X^{s,b}([0,\delta])} \leq c\left[\sum_n (1+|n|^s)^2 \left|\int_{|\lambda-n^2|>1} d\lambda\, \frac{\hat{w}(n,\lambda)}{\lambda-n^2}\right|^2\right]^{1/2} \qquad (49)$$

and invoking Cauchy-Schwartz, both terms are clearly bounded by

$$\left[\sum_n \int d\lambda(1+|n|^s)^2(1+|\lambda-n^2|)^{-2(1-b')}\, |\hat{w}(n,\lambda)|^2\right]^{1/2} \qquad (50)$$

where

$$b' > \frac{1}{2}, b' \geq b. \qquad (51)$$

Estimate (50) by duality. Thus consider a system

$$\{c_{n,\lambda}\}_{n\in\mathbb{Z},\lambda\in\mathbb{R}} \qquad (52)$$

of complex numbers satisfying

$$\|c\|_2 = \left(\sum_n \int d\lambda |c_{n,\lambda}|^2\right)^{1/2} \leq 1 \qquad (53)$$

and estimate

$$\sum_n \int d\lambda(1+|n|^s)(1+|\lambda-n^2|)^{-(1-b')}c_{n,\lambda}\, \hat{w}(n,\lambda) \qquad (54)$$

which is from the definition of w

$$\sum_{n_1,n_2,n_3} \int d\lambda_1 d\lambda_2 d\lambda_3\, c_{n,\lambda}\, \frac{1+|n|^s}{(1+|\lambda-n^2|)^{1-b'}}\, \hat{u}(n_1,\lambda_1)\cdot\overline{\hat{u}(n_2,\lambda_2)}\cdot\hat{u}(n_3,\lambda_3) \qquad (55)$$

where $n = n_1 - n_2 + n_3, \lambda = \lambda_1 - \lambda_2 + \lambda_3$.

Observe that

$$|n|^s \leq c\max(|n_1|^s, |n_2|^s, |n_3|^s). \qquad (56)$$

Assume $|n|^s \leq c|n_1|^s$ say. Then (55) is bounded by

$$\sum_{n_1,n_2,n_3} \int d\lambda_1 d\lambda_2 d\lambda_3\, \frac{|c_{n,\lambda}|}{(1+|\lambda-n^2|)^{1-b'}} \qquad (57)$$

$$(1+|n_1|^s)|\hat{u}(n_1,\lambda_1)|\cdot|\hat{u}(n_2,\lambda_2)|\cdot|\hat{u}(n_3,\lambda_3)|.$$

Take a localizing bumpfunction $\psi_\delta, 0 \leq \psi_\delta \leq 1, \hat{\psi}_\delta \geq 0$, $\psi_\delta = 1$ on $[-\delta,\delta]$, supp $\psi_\delta \subset [-2\delta, 2\delta]$. Then

$$u = u\psi_\delta, \hat{u} = \hat{u} * \hat{\psi}_\delta \qquad (58)$$

$$|\hat{u}| \leq |\hat{u}| * \hat{\psi}_\delta = (|\hat{u}|^\vee \psi_\delta)^\wedge. \qquad (59)$$

Define the functions

$$F(x,t) = \sum_n \int d\lambda \frac{|c_{n\lambda}|}{(1 + |\lambda - n^2|)^{1-b'}}\, e^{i(nx+\lambda t)}) \tag{60}$$

$$G(x,t) = \sum_n \int d\lambda\, |\hat{u}(n,\lambda)| e^{i(nx+\lambda t)} \tag{61}$$

$$H(x,t) = \sum_n \int d\lambda (1 + |n|^s)\, |\hat{u}(n,\lambda)|\, e^{i(nx+\lambda t)}. \tag{62}$$

Thus from (57), (59)

$$(57) \le \int G(x,t)^2 H(x,t) F(x,t)\psi_\delta(t)^3 dx dt$$

which by Hölder's inequality is bounded by

$$\|G\|_{L^4}^2\, \|H\|_{L^4}\, \|F\psi_\delta\|_{L^4}. \tag{63}$$

By (42), (43)

$$\|G\|_4 \le c\, \|G\|_{X^{0,3/8}} \le \left(\sum_n \int d\lambda (1+|\lambda-n^2|)^{3/4}\, |\hat{u}(n,\lambda)|^2 \right)^{1/2} \le c\|u\|_{X^{0,3/8}} \tag{64}$$

and similarly

$$\|H\|_4 \le c\, \|H\|_{X^{0,3/8}} \le c\, \|u\|_{X^{s,3/8}}. \tag{65}$$

Estimate further by (42)

$$\|F\psi_\delta\|_4 \le c\, \|F\psi_\delta\|_{X^{0,3/8}} \le \left[\sum_n \int d\lambda (1 + |\lambda - n^2|)^{3/4} |\widehat{F\psi_\delta}(n,\lambda)|^2 \right]^{1/2}. \tag{66}$$

Since $b' < \frac{5}{8}$, cf. (51), $1 - b' > \frac{3}{8}$. By Hölder's inequality (interpolation between $X^{0,0}, X^{0,1-b'}$), we get thus

$$(66) \le \|F\psi_\delta\|_2^\theta \left(\sum_n \int d\lambda (1 + |\lambda - n^2|)^{2(1-b')} |\widehat{F\psi_\delta}(n,\lambda)|^2 \right)^{\frac{1-\theta}{2}} \tag{67}$$

for some $\theta > 0$. One has by Hölder's inequality (38), (60), (53)

$$\|F\psi_\delta\|_2 \le \delta^{1/4}\, \|F\psi_\delta\|_4 \le \delta^{1/4}\, \|F\psi_\delta\|_{X^{0,3/8}}$$

$$\le \delta^{1/4} \left[\sum_n \int d\lambda (1 + |\lambda - n^2|)^{3/4}\, \frac{|c_{n\lambda}|^2}{(1 + |\lambda - n^2|)^{2(1-b')}} \right]^{1/2} \tag{68}$$

$$\le c\delta^{1/4}$$

and similarly, since $1 - b' < \frac{1}{2}$, the second factor in (67) is also bounded.

Collecting estimates (64), (65), (68), it follows that the $X^{s,b}$, norm of (46), (47) is at most

$$c\delta^{\theta/4}\, \|u\|^2_{X^{0,3/8}[0,\delta]}\, \|u\|_{X^{s,3/8}[0,\delta]}. \tag{69}$$

To estimate (45), use a series expansion

$$\frac{e^{iut}-1}{\mu} = \sum_{j=1}^{\infty} \frac{i^j}{j!}\mu^{j-1}t^j$$

which yields

$$(45) = \sum_{j\geq 1}\frac{i^j}{j!}\sum_{n}\left[\int_{|\lambda-n^2|\leq 1}\hat{w}(n,\lambda)(\lambda-n^2)^{j-1}d\lambda\right]e^{i(nx+n^2 t)}\,\psi(t)t^j. \tag{70}$$

The $X^{s,b}$-norm may be bounded by

$$c\left[\sum_{n}(1+|n|^{2s})\left(\int_{|\lambda-n^2|\leq 1}|\hat{w}(n,\lambda)|d\lambda\right)^2\right]^{1/2}. \tag{71}$$

Since (71) is in particular bounded by (50), the same estimate (69) is valid.

As a conclusion of the preceding, we get that

$$\|u - S(t)\phi\|_{X^{s,b}(0,\delta)} \leq c\delta^{\alpha}\,\|u\|^2_{X^{0,3/8}([0,\delta])}\,\|u\|_{X^{s,3/8}[0,\delta]} \tag{72}$$

for some $\alpha > 0$.

Denote (cf. (41))

$$Tu = S(t)\phi + i\lambda\int_0^t S(t-\tau)(u|u|^2)(\tau)d\tau. \tag{73}$$

Obviously

$$\|S(t)\phi\|_{X^{s,b}} \leq c\,\|\phi\|_{H^s(\mathbb{T})}. \tag{74}$$

Hence, from (72), the map T satisfies in particular ($b \geq 3/8$)

$$\|Tu\|_{X^{s,b}[0,\delta]} \leq c\,\|\phi\|_{H^s} + \delta^{\alpha}\,\|u\|^3_{X^{s,b}[0,\delta]}. \tag{75}$$

Hence for δ sufficiently small (depending on $\|\phi\|_s$), T will map a ball in $X^{s,b}[0,\delta]$ space into itself. Moreover, writing

$$u|u|^2 - v|v|^2 = (u-v)|u|^2 + v\overline{u}(u-v) + v^2(\overline{u}-\overline{v}) \tag{76}$$

and repeating previous estimates on each of the (76)-factors, one gets

$$\|Tu - Tv\|_{X^{s,b}[0,\delta]} \leq \delta^{\alpha}(\|u\|_{X^{s,b}} + \|v\|_{X^{sb}})^2\,\|u-v\|_{X^{s,b}[0,\delta]}. \tag{77}$$

Hence, for $\delta > 0$ small enough, T is a contraction and Picard's fixpoint theorem yields a unique solution in the space $X^{s,b}_{[0,\delta]}$.

The drawback of previous argument (to establish global wellposedness) is that the size δ of our time interval depends on $\|\varphi\|_{H^s}$ on which we do not have apriori bounds when $s > 0$. We therefore reorganize the preceding as follows. First take $s = 0$. We get a time interval of size $\delta_0 = \delta_0(\|\phi\|_{L^2})$ and a solution u in the space $X^{0,b}[0,\delta_0]$. Assuming $\phi \in H^s$, write then from (72), (74)

$$\|u\|_{X^{s,b}[0,\delta_0]} \leq c\,\|\phi\|_{H^s} + c\delta_0^{\alpha}\,\|u\|^2_{X^{0,b}[0,\delta_0]}\,\|u\|_{X^{s,b}[0,\delta_0]}. \tag{78}$$

Hence, since from the choice of δ_0, the factor

$$c\delta_0^\alpha \, \|u\|_{X^{a,b}[0,\delta_0]}^2 < \frac{1}{10} \tag{79}$$

say, we conclude that $u \in X^{s,b}[0,\delta_0]$.

The map

$$\phi \mapsto u_\phi \tag{80}$$

is Lipschitz continuous from $H^s(\mathbb{T})$ to $X^{s,b}[0,\delta_0]$.

Writing by (41)

$$u_\phi - u_\psi = S(t)(\phi - \psi) + i \int_0^t S(t-\tau)(u_\phi|u_\phi|^2 - u_\psi|u_\psi|^2)(\tau)d\tau \tag{81}$$

it follows from the preceding that

$$\|u_\phi - u_\psi\|_{X^{s,b}[0,\delta]} \leq \|\phi - \psi\|_{H^s} + \delta^\alpha (\|u_\phi\|_{X^{s,b}} + \|u_\psi\|_{X^{s,b}})^2 \, \|u_\phi - u_\psi\|_{X^{s,b}[0,\delta]} \tag{82}$$

and hence, if δ is small enough

$$\|u_\phi - u_\psi\|_{X^{s,b}[0,\delta]} \leq 2 \, \|\phi - \psi\|_{H^s} \tag{83}$$

(here δ depends again on $\|\phi\|_{H^s}$, $\|\psi\|_{H^s}$). In order to prove the property for the space $X^{s,b}[0,\delta_0]$, consider a covering of $[0,\delta_0]$ with overlapping intervals $\{I_\alpha\}$ of size δ. For each α, one has (83), i.e.

$$\|u_\phi - u_\psi\|_{X^{s,b}(I_\alpha)} \leq 2 \, \|\phi - \psi\|_{H^s} \tag{84}$$

$$\|F\|_{X^{s,b}[0,\delta_0]} \text{ may be evaluated as } c\max_\alpha \, \|F|_{I_\alpha}\|_{X^{s,b}(I_\alpha)}$$

considering a standard partition of unity argument.

Summarizing the preceding, we proved wellposedness of (40) on a time interval $[0,\delta_0]$, δ_0 depending on $\|\phi\|_2$, with Lipschitz dependence of the solution $u_\phi \in X^{s,b}[0,\delta_0]$ $(b < \frac{5}{8})$ on the data $\phi \in H^s(\mathbb{T}), s \geq 0$.

Assume $b > \frac{1}{2}$. Then $u_\phi \in C_{H^s}[0,\delta_0]$. The preceding statement is valid for any bounded time interval $[0,T]$, since we may cover by overlapping intervals of size δ_0. It is of course essential here that the local wellposedness result is known to hold on a time-interval which size is only dependent on a conserved quantity (the L^2-norm $\|u_\phi(t)\|_2 = \|\phi\|_2$).

This proves Theorem 36. Observe that from the preceding, for $\phi, \psi \in H^s$

$$\|u_\phi(t)\|_{H^s} \leq C^{|t|} \, \|\phi\|_{H^s} \tag{$*$} \tag{85}$$

and

$$\|u_\phi(t) - u_\psi(t)\|_{H^s} = C^{|t|} \, \|\phi - \psi\|_{H^s} \tag{86}$$

where $C = C(\|\phi\|_2, \|\psi\|_2)$. Moreover $\partial_x^s u_\phi \in L^4(\mathbb{T} \times [0,T])$ for all T.

Remark. In the case of NLS (40) with polynomial nonlinearity, the solution u_ϕ obtained above has in fact a real analytic dependence on ϕ.

$(*)$ This exponential estimates may be improved to bounds of the form $|t|^A$ for some $A = A(s)$. The method for achieving this will be outlined at the end of this section.

Comments on the proofs of Theorems 1.37, 1.40, 1.41

There are the following sources of difficulties. First, the failure (or partial knowledge) of the exact periodic analogue of the linear Strichartz inequality (cf. Propositions 18, 20) makes the estimate of the nonlinear term more delicate and requires some new ideas. Secondly (this is also the case in the nonperiodic case), dealing with nonlinearities $u|u|^{p-2}$ for general p (not necessarily an even integer) leads to certain technicalities. More precisely, required differentiability properties of the function x^{p-1} forces p to be sufficiently large. We only discuss the first aspect. This discussion is again technically easiest when p is an even integer. We restrict ourself to Theorem 40, i.e. a cubic NLS in $2D$. The relevant Strichartz inequality here is given by Proposition 20, i.e.

$$\left\| \sum_{|n| \leq N} a_n \, e^{i(\langle n, x \rangle + |n|^2 t)} \right\|_{L^4(\mathbb{T}^3)} \ll N^\varepsilon \left(\sum_{|n| \leq N} |a_n|^2 \right)^{1/2}. \tag{87}$$

We first derive some consequences of (87) in terms of the norms $X^{s,b}$ defined earlier. Consider a ball Q in $\mathbb{Z}^2$ of size N entered at a point n_0. Writing $n = n_0 + m, |n| \leq N$ we have

$$|n|^2 = |n_0|^2 + 2\langle n_0, m \rangle + |m|^2$$

and

$$\langle n, x \rangle + |n|^2 t = (\langle n_0, x \rangle + |n_0|^2 t) + \langle m, x + 2n_0 t \rangle + |m|^2 t.$$

Making a change of variable $x' = x + 2n_0 t$, it therefore follows from (87) that

$$\left\| \sum_{n \in Q} a_n \, e^{i\langle n, x \rangle + |n|^2 t)} \right\|_4 \ll |Q|^\varepsilon \left(\sum_{n \in Q} |a_n|^2 \right)^{1/2}. \tag{88}$$

Consider now more generally

$$u(x, t) = \sum_{n \in Q} \int d\lambda \, \hat{u}(n, \lambda) \, e^{i(\langle n, x \rangle + \lambda t)} \tag{89}$$

$$= \int_{\mathbb{R}} \left[\sum_{n \in Q} \hat{u}(n, |n|^2 + \mu) \, e^{i(\langle n, x \rangle + |n|^2 t)} \right] e^{i\mu t} d\mu. \tag{90}$$

Estimate using (88) and Hölder's inequality

$$\|u\|_{L^4_{\text{loc}}} \leq \int_{\mathbb{R}} \left\| \sum_{n \in Q} \hat{u}(n, |n|^2 + \mu) \, e^{i\langle n, x \rangle + |n|^2 t)} \right\|_{L^4_{\text{loc}}} d\mu \tag{91}$$

$$\ll |Q|^\varepsilon \int_{\mathbb{R}} \left(\sum_{n \in Q} |\hat{u}(n, |n|^2 + \mu)|^2 \right)^{1/2} d\mu \tag{92}$$

$$\ll |Q|^\varepsilon \left(\sum_{n \in Q} \int d\lambda (1 + |\lambda - |n|^2|)^{2b} |\hat{u}(n, \lambda)|^2 \right)^{1/2} \tag{93}$$

for $b > \frac{1}{2}$.

In particular, one has thus

$$\|u\|_{L^4_{\mathrm{loc}}} \le c \, \|u\|_{X^{s,b}} \ \text{if } s > 0, b > \frac{1}{2}. \tag{94}$$

One may also estimate by Hausdorff-Young and Hölder's inequality

$$\|u\|_4 \le \left(\sum_{n \in Q} \int d\lambda |\hat{u}(n, \lambda)|^{4/3} \right)^{3/4} \tag{95}$$

$$\le \left(\sum_{n \in Q} \left(\int d\lambda (1 + |\lambda - |n|^2|)^{2b'} |\hat{u}(n, \lambda)|^2 \right)^{2/3} \right)^{3/4}$$

$$\le N^{1/2} \left(\sum_{n \in Q} \int d\lambda (1 + |\lambda - |n|^2|)^{2b'} |\hat{u}(n, \lambda)|^2 \right)^{1/2} \tag{96}$$

where N is the size of Q and $b' > \frac{1}{4}$.

Interpolation between (93) and (96) implies that for u defined by (89)

$$\|u\|_{L^4_{\mathrm{loc}}} < N^{s_1} \left(\sum_{n \in Q} \int d\lambda (1 + |\lambda - |n|^2|)^{2b_1} |\hat{u}(n, \lambda)|^2 \right)^{1/2} \tag{97}$$

for

$$b_1 > \frac{1 - (s_1 \wedge \frac{1}{2})}{2}. \tag{98}$$

This inequality will replace (42).

Proceeding for the cubic $2D$ NLS

$$\begin{cases} iu_t + \Delta u + \lambda u |u|^2 = 0 \\ u(0) = \phi \end{cases} \tag{99}$$

as in $1D$, estimate the $X^{s,b}[0, \delta]$-norm of the nonlinear term

$$\int_0^t S(t - \tau)(u|u|^2)(\tau)d\tau \tag{100}$$

by the expression (57), i.e.

$$\sum_{n_1, n_2, n_3} \int d\lambda_1 d\lambda_2 d\lambda_3 \, \frac{|c_{n,\lambda}|}{(1 + |\lambda - |n|^2|)^{1-b'}} \, (1 + |n|^s)|\hat{u}(n_1, \lambda_1)| \, |\hat{u}(n_2, \lambda_2)| \, |\hat{u}(n_3, \lambda_3)| \tag{101}$$

where $n = n_1 - n_2 + n_3, \lambda = \lambda_1 - \lambda_2 + \lambda_3$ and

$$b' > \frac{1}{2}, \ b' \ge b. \tag{102}$$

For each n_i-index $(i = 1, 2, 3)$, we subdivide the $\mathbb{Z}^2$-index set in dyadic regions (a standard dyadic partitioning as in Littlewood-Paley theory). Thus

$$\mathbb{Z}^2 = \cup_{k=n}^\infty D_k \tag{103}$$

and $D_k = [n \in \mathbb{Z}^2 | \ |n| \sim 2^k]$. Write

$$\sum_{n_1,n_2,n_3 \in \mathbb{Z}^2} = \sum_{k_1 \geq k_2 \geq k_3} \ \sum_{\substack{n_i \in D_{k_i} \\ i=1,2,3}} . \tag{104}$$

Fix $k_1 \geq k_2 \geq k_3$. Consider a further partition of $D_{k_1} = \cup_\alpha Q_\alpha$ in balls of size 2^{k_2}. One may thus essentially write

$$\sum_{n_i \in D_{k_i}} \int d\lambda_1 \lambda_2 d\lambda_3 \ \frac{|c_{n\lambda}|}{(1+|\lambda - |n|^2|)^{1-b'}} \ (1+|n|^s)|\hat{u}(n_1,\lambda_1)| \ |\hat{u}(n_2,\lambda_2)| \ |\hat{u}(n_3,\lambda_3)|$$

$$\sim 2^{k_1 s} \sum_\alpha \ \sum_{\substack{n,n_1 \in Q_\alpha \\ n_2 \in D_{k_2}, n_3 \in D_{k_3}}} \int d\lambda_1 d\lambda_2 d\lambda_3 \ \frac{|c_{n,\lambda}|}{(1+|\lambda - |n|^2|)^{1-b'}}$$

$$|\hat{u}(n_1,\lambda_1)| \ |\hat{u}(n_2,\lambda_2)| \ |\hat{u}(n_3,\lambda_3)|. \tag{105}$$

Define the functions

$$F_\alpha(x,t) = \sum_{n \in Q_\alpha} \int d\lambda \ \frac{|c_{n,\lambda}|}{(1+|\lambda - |n|^2|)^{1-b'}} \ e^{i(\langle n,x \rangle + \lambda t)} \tag{106}$$

$$G_\alpha(x,t) = \sum_{n \in Q_\alpha} \int d\lambda \ |\hat{u}(n,\lambda)| \ e^{i(\langle n,x \rangle + \lambda t)} \tag{107}$$

$$H_i(x,t) = \sum_{n \in D_{k_i}} \int d\lambda \ |\hat{u}(n,\lambda)| \ e^{i(\langle n,x \rangle + \lambda t)} \ (i = 2,3) \tag{108}$$

(cf. (60)-(62)) and bound (105) by

$$2^{k_1 s} \sum_\alpha \int F_\alpha \cdot G_\alpha \cdot H_2 \cdot H_3 \cdot \psi_\delta^3 \, dx dt$$

$$\leq 2^{k_1 s} \sum_\alpha \|H_2\|_4 \ \|H_3\|_4 \ \|G_\alpha\|_4 \ \|F_\alpha \psi_\delta\|_4. \tag{109}$$

Choose $s_1 < \frac{1}{2}$, $\frac{1-s_1}{2} < b_1 < \frac{1}{2}$ to satisfy (97), (98).
 Thus, for $i = 2,3$

$$\|H_i\|_4 \leq 2^{k_i s_1} \left(\sum_{n \in D_{k_i}} \int d\lambda \ (1+|\lambda - |n|^2|)^{2b_1} |\hat{u}(n,\lambda)|^2 \right)^{1/2} \tag{110}$$

and since each Q_α is of size 2^{k_2}, also

$$\|G_\alpha\|_4 \leq 2^{k_2 s_1} \left(\sum_{n \in Q_\alpha} \int d\lambda \ (1+|\lambda - |n|^2|)^{2b_1} |\hat{u}(n,\lambda)|^2 \right)^{1/2} \tag{111}$$

$$\|F_\alpha \psi_\delta\|_4 \leq 2^{k_2 s_1} \left(\sum_n \int d\lambda \ (1+|\lambda - |n|^2|)^{2b_1} |\widehat{F_\alpha \psi_\delta}(n,\lambda)|^2 \right)^{1/2}$$
$$\tag{112}$$

Thus, summing over α, it follows that (105), (109) are bounded by

$$4^{k_2 s_1} \Pi_{i=2,3} \left(\sum_{n \in D_{k_i}} \int d\lambda \, (1 + |\lambda - |n|^2|)^{2b_1} (1 + |n|)^{2s_1} |\hat{u}(n,\lambda)|^2 \right)^{1/2}.$$

$$\left(\sum_{n \in D_{k_1}} \int d\lambda \, (1 + |\lambda - |n|^2|)^{2b_1} (1 + |n|)^{2s} |\hat{u}(n,\lambda)|^2 \right)^{1/2}.$$

$$\left(\sum_{\alpha} \sum_{n} \int d\lambda \, (1 + |\lambda - |n|^2|)^{2b_1} |\widehat{F_\alpha \psi_\delta}(n,\lambda)|^2 \right)^{1/2}. \tag{113}$$

Assume

$$1 - b' > b_1. \tag{114}$$

Proceeding as in the $1D$ case, we may for the last factor in (113) interpolate between $0, 1 - b'$ and reach the estimate $\delta^\theta 2^{k_2 s_1} \|c|_{D_{k_1}}\|_2$ for some $\theta > 0$ (cf. (67), (68)). This yields for (113) the estimate

$$8^{k_2 s_1} \|c|_{D_{k_1}}\|_2 \left(\sum_{n \in D_{k_1}} \int d\lambda \, (1 + |\lambda - |n|^2|)^{2b_1} (1 + |n|^{2s}) |\hat{u}(n,\lambda)|^2 \right)^{1/2}.$$

$$\Pi_{i=2,3} \left(\sum_{n \in D_{k_i}} \int d\lambda \, (1 + |\lambda - |n|^2|)^{2b_1} (1 + |n|^{2s_1}) |\hat{u}(n,\lambda)|^2 \right)^{1/2} \tag{115}$$

$$\leq \delta^\theta \|c|_{D_{k_1}}\|_2 \cdot \left(\sum_{n \in D_{k_1}} \int d\lambda \, (1 + |\lambda - |n|^2|)^{2b_1} (1 + |n|^{2s}) |\hat{u}(n,\lambda)|^2 \right)^{1/2}.$$

$$\left(\sum_{n \in D_{k_2}} \int d\lambda \, (1 + |\lambda - |n|^2|)^{2b_1} (1 + |n|^{8s_1}) |\hat{u}(n,\lambda)|^2 \right)^{1/2}$$

$$\left(\sum_{n \in D_{k_3}} \int d\lambda \, (1 + |\lambda - |n|^2|)^{2b_1} (1 + |n|)^{2s_1} |\hat{u}(n,\lambda)|^2 \right)^{1/2}. \tag{116}$$

Hence, summation on $k_1 \geq k_2 \geq k_3$ permits to bound (101) by

$$\delta^\theta \left(\sum_n \int d\lambda (1 + |n|)^{2s} (1 + |\lambda - |n|^2|)^{2b_1} |\hat{u}(n,\lambda)|^2 \right)^{1/2}$$

$$\left(\sum_n \int d\lambda (1 + |n|)^{10s_1} (1 + |\lambda - |n|^2|)^{2b_1} |\hat{u}(n,\lambda)|^2 \right). \tag{117}$$

Fix $s > 0$ and let $s_1 = \min(\frac{s}{5}, \frac{1}{4})$. One may take $b = b' = b(s) > \frac{1}{2}$ such that $1 - b(s) > \frac{1 - s_1}{2}$, hence (114) holds with $\frac{1 - s_1}{2} < b_1 < \frac{1}{2}$.

From (117),

$$\|(100)\|_{X^{s,b}[0,\delta]} = \|u - S(t)\phi\|_{X^{s,b}[0,\delta]} \leq c\delta^\theta \, \|u\|^3_{X^{s,b_1}[0,\delta]} \leq c\delta^\theta \, \|u\|^3_{X^{s,b}[0,\delta]} \tag{118}$$

and similarly, for T the map defined by (73)

$$\|Tu - Tv\|_{X^{s,b}[0,\delta]} \leq c\delta^\theta (\|u\|_{X^{s,b}[0,\delta]} + \|v\|_{X^{s,b}[0,\delta]})^2 \|u - v\|_{X^{s,b}[0,\delta]}. \tag{119}$$

Thus, for $\delta = \delta(\|\phi\|_{H^s(\mathbb{T}^2)})$, the contraction principle applies again to prove local wellposedness in the space $X^{s,b}[0, \delta]$ for data $\phi \in H^s(\mathbb{T}^2)$.

For $s = 1$, assuming the Hamiltonian yields an apriori bound on the H^1-norm of $u(t)$ (see the discussion in section 1), one has global wellposedness for H^1-data. For $s \geq 1$, choose $s_1 = \frac{1}{5}$ in the second factor of (117) so that one obtains, for $\phi \in H^s$

$$\|u\|_{X^{s,b}[0,\delta]} = \|Tu\|_{X^{s,b}[0,\delta]} \leq \|\phi\|_{H^s} + \delta^\theta \, \|u\|^2_{X^{1,b_1}[0,\delta]} \, \|u\|_{X^{s,b_1}[0,\delta]} \tag{120}$$

and hence a bound on $\|u\|_{X^{s,b}[0,\delta]}$ for a time interval of size δ only dependent on $\|\phi\|_{H^1}$. The situation is analogous to the 1D-case discussed earlier with L^2-norm replaced by the H^1-norm.

This yields the proof of Theorem 40.

Addition to Lecture 2

Remarks on the growth of higher Sobolev norms

We wish to outline a refinement of the scheme described above leading to an improvement of the exponential bounds

$$\|u(t)\|_{H^s} < C^{|t|} \tag{121}$$

on higher Sobolev norms when $|t| \to \infty$. In fact, we prove estimates of the form

$$\|u(t)\|_{H^s} \lesssim (1 + |t|)^A \qquad A = A(s) \sim s \tag{122}$$

provided one may treat locally the Cauchy problem in H^ρ with $\rho < 1$ and dispose of an apriori control of the H^1-norm from the Hamiltonian. The type of argument described below in the NLS-case is fairly general and applies to other equations, for instance nonlinear wave equations (NLW). In the NLW-case, the procedure is actually simpler due to smoothing properties of $\square^{-1}$ and details will be given in the Appendix of this text.

Recall that (127) is gotten from iterating the estimate

$$\|u(t)\|_{H^s} \leq 2\|u(0)\|_{H^s} \text{ for } |t| < \tau \tag{123}$$

obtained from the local in time analysis. We wish to improve on this, by deriving an inequality of the form

$$\|u(t)\|_{H^s} \leq \|u(0)\|_{H_s} + \|u(0)\|_{H^s}^{1-\delta} \qquad (|t| < \tau) \tag{124}$$

for some $\delta > 0$. The polynomial bound (122) immediately follows.

Consider the Cauchy problem (local in time)

$$\begin{cases} i\dot{u} + \Delta u + \frac{\partial H}{\partial \bar{u}} = 0 \\ u(0) = \phi \end{cases} \tag{125}$$

assuming the situation described above. An interesting example would be that of the 2D cubic NLS

$$i\dot{u} \pm \Delta u + u|u|^2 = 0 \tag{126}$$

(considering data of sufficiently small L^2-norm in the focusing case).

Write $\phi = \phi_1 + \phi_2$ when $\phi_1 = P_{K_1}\phi = \sum_{|n|\leq K_1} \hat{\phi}(n)\, e^{in.x}$ and K_1 will be specified depending on $\|\phi\|_{H^s}$. Denote u_1 the solution of the Cauchy problem

$$\begin{cases} i\dot{u}_1 + \Delta u_1 + \frac{\partial H}{\partial \overline{u}_1} = 0 \\ u_1(0) = \phi_1. \end{cases} \tag{127}$$

Hence, from the wellposedness theory

$$\|u_1(t)\|_{H^s} < 2\|\phi_1\|_{H^s} \lesssim K_1^{s-1}\|u(0)\|_{H^1} \tag{128}$$

and it remains to control $v = u - u_1$, solving

$$\begin{cases} i\dot{v} + \Delta v + \left(\frac{\partial^2 H}{\partial u_1 \partial \overline{u}_1}\right) v + \left(\frac{\partial^2 H}{\partial \overline{u}_1^2}\right)\overline{v} + F(v,\overline{v}) = 0 \\ v(0) = \phi_2 \end{cases} \tag{129}$$

with F at least quadratic in $v, \overline{v}$. Observe that by construction, for $\rho < 1$

$$\|\phi_2\|_{H^s} \lesssim K_1^{\rho-1}\|\phi\|_{H^1} \tag{130}$$

hence, since we assume local wellposedness of (125) in a Sobolev norm below H^1, the effect of the nonlinear term $F(v,\overline{v})$ in (129) will be small for $K_1 \to \infty$. Consider next the term $\frac{\partial^2 H}{\partial \overline{u}_1^2}.\overline{v}$. When treating (129) perturbatively, we consider again the norms $X^{s,b}$, thus essentially

$$\|v\|_{X^{s,b}} = \left(\sum |n|^{2s} |\lambda - |n|^2|^b\, |\hat{v}(n,\lambda)|^2\right)^{1/2}. \tag{131}$$

Let $\Omega = \frac{\partial^2 H}{\partial \overline{u}_1^2}$, depending on u_1 (hence K_1) and $S(t) = e^{it\Delta}$. We have essentially

$$\left\|\int_0^t S(t-\tau)(\Omega\overline{v})(\tau)d\tau\right\|_{s,b} \lesssim \left(\sum_n \int d\lambda\, |n|^{2s}\, \frac{|\widehat{\Omega\overline{v}}(n,\lambda)|^2}{|\lambda - |n|^2|^{2(1-b)}}\right)^{1/2} \tag{132}$$

and linearizing

$$\sim \sum_{\substack{n=n_1+n_2 \\ \lambda=\lambda_1+\lambda_2}} \int |n|^s\, \frac{d(n,\lambda)}{|\lambda - |n|^2|^{1-b}}\, \overline{\hat{v}(-n_1,-\lambda_1)}\widehat{\Omega}(n_2,\lambda_2)$$

$$\lesssim \sum_{\substack{n=n_1+n_2 \\ \lambda=\lambda_1+\lambda_2}} \int \frac{d(n,\lambda)}{|\lambda - |n|^2|^{1-b}}\, \frac{c(n_1,\lambda_1)}{|\lambda_1 + |n_1|^2|^b}\widehat{\Omega}(n_2,\lambda_2) \tag{133}$$

where

$$\left(\sum_n \int d\lambda |c(n,\lambda)|^2\right)^{1/2} \leq \|v\|_{s,b} \quad \text{and} \quad \left(\sum_n \int d\lambda |d(n,\lambda)|^2\right)^{1/2} \leq 1. \tag{134}$$

Since Ω is controlled by K_1, it may for large $|n|$ essentially be ignored in (133). Considering $n_1 \approx n$, $\lambda_1 \approx \lambda$, the main point is that

$$\max(|\lambda - |n|^2|, |\lambda_1 + |n_1|^2|) \gtrsim |n|^2 \tag{135}$$

hence smoothing occurs. (In the $2D$-model (126), there would be a smoothing of order $1-$). As a consequence of this smoothing, also the term $\frac{\partial^2 H}{\partial \overline{u}_1^2}.\overline{v}$ may be considered as a lower order perturbation in estimating $\|v(t)\|_{H^s}$.

The preceding does not apply to the term

$$\left(\frac{\partial^2 H}{\partial u_1 \partial \overline{u}_1}\right) v \tag{136}$$

and we absorb it in the linear part, replacing the linear operator $i\dot{v} + \Delta v$ by

$$i\dot{v} + \Delta v + \left(\frac{\partial^2 H}{\partial u_1 \partial \overline{u}_1}\right) v. \tag{137}$$

The main fact at this stage is that the solution operator of

$$\begin{cases} i\dot{v} + \Delta v + \left(\frac{\partial^2 H}{\partial u_1, \partial \overline{u}_1}\right) v = 0 \\ v(0) = \phi_2 \end{cases} \tag{138}$$

has essentially a "unitary behaviour" for large frequencies. More precisely (in $2D$), there is an inequality

$$\|v(t)\|_{H_s} \leq \left(1 + \frac{K_1^{1+}}{\|\phi_2\|_{H^s}^{\frac{1}{s-1}}}\right) \|\phi_2\|_{H^s} + K_1^{s+} \tag{139}$$

(assume $s \geq 2$ an even integer and $\|\phi\|_{H^1} \leq 1$ say). One has indeed since $\Omega = \frac{\partial^2 H}{\partial u_1 \partial \overline{u}_1}$ is real from the equation (138)

$$\begin{aligned} \frac{d}{dt}\|v(t)\|_s^2 &= \frac{d}{dt}\langle B^s v, B^s v\rangle \qquad (B = \sqrt{-\Delta}) \\ &= 2Re\langle B^s \dot{v}, B^s v\rangle \\ &= 2Im\langle B^s(\Omega v), B^s v\rangle \\ &= 2\underbrace{Im\langle \Omega B^s v, B^s v\rangle}_{\substack{\| \\ 0}} + 0\left(\sum_{\substack{|\alpha|+|\beta|=s \\ |\beta|\geq 1}} \|D^\alpha v\|_2 \|D^\beta \Omega\|_\infty \|B^s v\|_2\right) \end{aligned} \tag{140}$$

$$= 0\left(\sum_{j=1}^{s} \|\phi_2\|_{s-j}\|\Omega\|_{j+1+}\right)\|\phi_2\|_s. \tag{141}$$

By interpolation between H^1 and H^s, for $j \leq s - 1$

$$\|\phi_2\|_{s-j} \lesssim \|\phi_2\|_s^{\frac{s-j-1}{s-1}} \tag{142}$$

and since $\Omega = \frac{\partial^2 H}{\partial u_1 \partial \overline{u}_1}$

$$\|\Omega\|_{s'} \ll K_1^+\|u_1\|_{s'} < K_1^{s'-1+}. \tag{143}$$

Thus, from (141) - (143)

$$\|v(t)\|_s^2 \leq \|\phi_2\|_s^2 + c\|\phi_2\|_s\{K_1^{1+}\|\phi_2\|_s^{\frac{s-2}{s-1}} + K_1^{s+}\} \tag{144}$$

implying (139).

Coming back to (129), construct v by successive corrections

$$v = v_0 + v_1 + v_2 + \cdots \tag{145}$$

defining

$$\begin{cases} i\dot{v}_0 + \Delta v_0 + \frac{\partial^2 H}{\partial u_1 \partial \overline{u}_1} \, v_0 = 0 \\ v_0(0) = \phi_2 \end{cases} \tag{146}$$

$$\begin{cases} i\dot{v}_1 + \Delta v_1 + \frac{\partial^2 H}{\partial \overline{u}_1^2} \, \overline{v}_0 + F(v_0, \overline{v}_0) = 0 \\ v_1(0) = 0 \end{cases} \tag{147}$$

$$\begin{cases} i\dot{v}_2 + \Delta v_2 + \frac{\partial^2 H}{\partial u_1 \partial \overline{u}_1} \, v_1 + \frac{\partial^2 H}{\partial \overline{u}_1^2} \, \overline{v}_1 + F(v_0 + v_1, \overline{v}_0 + \overline{v}_1) - F(v_0, \overline{v}_0) = 0 \\ v_2(0) = 0 \end{cases} \tag{148}$$

and in general for $k \geq 2$

$$\begin{cases} i\dot{v}_{k+1} + \Delta v_{k+1} + \frac{\partial^2 H}{\partial u_1 \partial \overline{u}_1} \, v_k + \frac{\partial^2 H}{\partial \overline{u}_1^2} \, \overline{v}_k + F(\sum_{s \leq k} v_s) - F(\sum_{s \leq k-1} v_s) = 0 \\ v_{k+1}(0) = 0. \end{cases} \tag{149}$$

The estimate on $v_0(t)$, $|t| \leq \tau$ is given by (139). The solution of (147) is given by

$$v_1(t) = i \int_0^t S(t - \tau) \left[\frac{\partial^2 H}{\partial \overline{u}_1^2} \, \overline{v}_0 + F(v_0, \overline{v}_0) \right] (\tau) d\tau \tag{150}$$

for which, according to the earlier discussion, one has for some $\gamma > 0$

$$\|v_1\|_{s,b} < K_1^{-\gamma} \|u(0)\|_s \tag{151}$$

$$\|v_1(t)\|_s < K_1^{-\gamma} \|u(0)\|_s. \tag{152}$$

In particular, for the $2D$-model (126), there is local wellposedness in H^ρ for any $\rho > 0$ and in (151), (152), we may take any $\gamma < 1$.

Considering (148), (149) the iterative process yields estimates for $k \geq 2$

$$\|v_k\|_s \leq \frac{1}{10} \big(\|v_{k-1}\|_s + \|\phi\|_{H^s} \|v_{k-1}\|_\rho \big) < C 10^{-k} \|v_1\|_s + C 10^{-k} K_1^{-\gamma} \|\phi\|_{H^s} \tag{153}$$

$$\|v_k(t)\|_s < C 10^{-k} \|v_1\|_s + C 10^{-k} K_1^{-\gamma} \|\phi\|_{H^s}$$

provided the interval $[0, \tau]$ is taken small enough. Hence one gets as conclusion a bound on the H^s-norm of the form

$$\|u(t)\|_s \leq \|u_1(t)\|_s + \|v(t)\|_s \leq K_1^{s-1} + (139) + K_1^{-\gamma} \|u(0)\|_s \tag{154}$$

and hence for $|t| \leq \tau$

$$\|u(t)\|_s \leq \left(1 + K_1^{-\gamma} + \frac{K_1^{1+}}{\|u(0)\|_s^{\frac{1}{s-1}}} \right) \|u(0)\|_s + K_1^{s+}. \tag{155}$$

Thus an inequality of the form (124) is derived choosing K_1 appropriately in (155). In the case (126) in $2D$, we get $\delta < \frac{1}{2(s-1)}$ in (124), yielding $A = 2(s-1)+$ in (122).

Comments and references related to Lecture 2.

Most of its content appears in [**B**] (see lecture 1).

The results at the end concerning growth of higher Sobolev norms are new. Details here were discussed in the context of periodic boundary conditions but the method works on the line as well. The main ingredient for this approach is to dispose of a local wellposedness theory which is subcritical wrt the energy norm. At this stage, I did not try to optimize the procedure and for instance the estimate $T^{2(1-s)+}$, $|t| \leq T$ valid for the 2D NLS $iu_t + \Delta u - u|u|^2 = 0$ with data $\dot{u}(0) \in H^s(\mathbb{T}^2)$, $s > 1$, may be improvable. In the Appendix, the example of the wave equation is discussed. A more systematic study appears in [**B2**] and was further pursued by G. Staffiliani.

[B2]   J. Bourgain, *On the growth in time of higher Sobolev norms of smooth solutions of Hamiltonian PDE*, IMRN, 1996, N6, 277-304.

[Sta]   G. Staffilani, Duke Math. J, Vol. 86, N1 (1997), 79-142.

LECTURE 3
A Digression: The Initial Value Problem for the KdV Equation

We wish here to make some comments on further applications of the type of norms introduced in previous section. We mainly want to show how, considering say the Kortweg de Vries equation (KdV)

$$u_t + u_{xxx} + uu_x = 0 \tag{1}$$

instead of the NLS, derivative loss in the nonlinear term may be recovered from "regularizing" properties of the linear solution operator. In our approach, this mechanism is based on certain simple arithmetics of n^3. It depends on the structure of both the linear and nonlinear part of (3.1) and not just a smoothing property of the linear solution operator (here Kato's smoothing inequality, see below), not valid in the periodic case.

Theorem 2. *The Cauchy problem (with periodic boundary conditions)*

$$\begin{cases} u_t + u_{xxx} + uu_x = 0 \\ u(0) = \phi \end{cases} \tag{3}$$

is globally wellposed for data $\phi \in H^s(\mathbb{T})$, $s \geq 0$ (ϕ real).

Remarks.
  (1) The requirement described below yields also local wellposedness for data $\phi \in H^s(\mathbb{T}), s > -\frac{1}{2}$.
  (2) Theorem 2 is also valid on the line; the local wellposedness result is known to hold here for $\phi \in H^s(\mathbb{R}), s > -\frac{3}{4}$. Results on the line may be gotten combining the technique described below for the periodic case together with the more "traditional" approach for *KdV* based mainly on Kato's smoothing properties of the linear solution group $S(t)$, given here by

$$S(t)\phi = \int_{-\infty}^{\infty} \hat{\phi}(\lambda) \, e^{i(\lambda x + \lambda^3 t)} d\lambda. \tag{4}$$

More precisely, there is so-called Kato's local smoothing inequality

$$\|\partial_x S(t)\phi\|_{L_x^\infty L_t^2} \leq c \, \|\phi\|_{L^2(\mathbb{R})}. \tag{5}$$

In the periodic case, the solution $S(t)\phi$ of the Cauchy problem

$$\begin{cases} u_t + u_{xxx} = 0 \\ u(0) = \phi, u(x,t) = u(x+1,t) \end{cases} \tag{6}$$

is given by the exponential sum

$$S(t)\phi = \sum \hat{\phi}(n)\, e^{i(nx+n^3 t)}. \tag{7}$$

First consider Strichartz' inequality in L^4-norm. There is the following fact generalizing inequality (2.17) from previous section.

Proposition 8. *Let $p(x) = x^k + p_{k-1}(x)$ $(k \geq 2)$ be a polynomial with integer coefficients.*
Define

$$\delta_k = \frac{k+1}{4k}. \tag{9}$$

Then one has the following inequality for all coefficient systems $\{a_{mn}\}$

$$\left\| \sum_{m,n} a_{mn}\, e^{i(nx+nt)} \right\|_{L^4(\mathbb{T}^2)} \leq c\left(\sum_{m,n} (1 + |n - p(m)|)^{2\delta_k} |a_{mn}|^2 \right)^{1/2}. \tag{10}$$

Hence, for $p(x) = x^3, k = 3$

$$\left\| \sum_{m,n} a_{mn}\, e^{i(mx+nt)} \right\|_{L^4(\mathbb{T}^2)} \leq c\left(\sum_{m,n} (1 + |n - m^3|)^{2/3} |a_{mn}|^2 \right)^{1/2}. \tag{11}$$

The analogue of the NLS spaces $X^{s,b}$ are here obtained replacing the multiplier $|\lambda - |n|^2|^b$ by $|\lambda - n^3|^b$. For reasons clear from what will follow, only the value $b = \frac{1}{2}$ is considered.

As for NLS, we consider the corresponding integral equation

$$u(t) = S(t)\phi - \int_0^t S(t-\tau)(uu_x)(\tau)d\tau \tag{12}$$

with $S(t)$ the group defined by (7). For the local wellposedness result, we show that for δ sufficiently small, depending on $\|\phi\|_2$, the map

$$Tu = S(t)\phi - \int_0^t S(t-\tau)(uu_x)(\tau)d\tau \tag{13}$$

is a contraction on $X^{s,\frac{1}{2}}[0,\delta]$.

Strictly speaking, we will assume

$$\hat{\phi}(0) = \int_{\mathbb{T}} \phi(x)dx = 0 \tag{14}$$

(observe that $\int \phi$ is a conserved quantity for the KdV flow).

If $\hat{\phi}(0) \neq 0$, one replaces the equation

$$u_t + u_{xxx} + uu_x = 0$$

by

$$v_t + v_{xxx} + \hat{\phi}(0)v_x + vv_x = 0 \tag{15}$$

with linear part $v_t + v_{xxx} + \hat{\phi}(0)v_x$. The polynomial in Proposition 8 becomes now $x^3 + \hat{\phi}(0)x$. For the considerations below, the additional term $\hat{\phi}(0)v_x$ will be harmless.

We take $s = 0$. The modifications for $s > 0$ are inessential.

The control of the nonlinear term in (13) is obtained from considering

$$\left\{ \sum_{n \in \mathbb{Z}\backslash\{0\}} \int_{-\infty}^{\infty} \frac{|\hat{w}(n, \lambda)|^2}{1 + |\lambda - n^3|} d\lambda \right\}^{1/2} \tag{16}$$

and

$$\left\{ \sum_{n \in \mathbb{Z}\backslash\{0\}} \left(\int_{-\infty}^{\infty} \frac{|\hat{w}(n, \lambda)|}{1 + |\lambda - n^3|} d\lambda \right)^2 \right\}^{1/2} \tag{17}$$

where $w = uu_x$; $\hat{w} = \frac{n}{2}(\hat{u} * \hat{u})$.

Observe that (17) yields a bound on $\| \int_0^t S(t - \tau)(uu_x)(\tau)d\tau \|_{L_t^\infty L_x^2}$ so that a posteriori the solution $u_\phi \in X^{0,\frac{1}{2}}$ of (3) will also satisfy $u_\phi \in C_{L_x^2}[0, T]$ with Lipschitz dependence on ϕ.

We only treat (16). The expression (17) is done similarly.

Define

$$\begin{cases} c(n, \lambda) = (1 + |\lambda - n^3|)^{1/2}|\hat{u}(n, \lambda)| \\ = 0 \, (\text{for } n = 0) \end{cases} \tag{18}$$

so that

$$\|u\|_{X^{0,\frac{1}{2}}} = \|c\|_2 \equiv \left(\sum_n \int d\lambda \, c(n, \lambda)^2 \right)^{1/2}. \tag{19}$$

Write

$$|\hat{w}(n, \lambda)| \leq \sum_{n_1 + n_2 = n} \int d\lambda_1 \frac{|n|c(n_1, \lambda_1)c(n_2, \lambda - \lambda_1)}{(1 + |\lambda_1 - n_1^3|)^{1/2}(1 + |\lambda - \lambda_1 - n_2^3|)^{1/2}}. \tag{20}$$

Estimate again (16) by dualization, i.e. consider $d = \{d(n, \lambda) \geq 0\}$ such that

$$\|d\|_2 = \left(\sum_n \int d\lambda \, d(n, \lambda)^2 \right)^{1/2} \leq 1 \tag{21}$$

and

$$\sum_n \int d\lambda \frac{|\hat{w}(n, \lambda)|d(n, \lambda)}{(1 + |\lambda - n^3|)^{1/2}} \tag{22}$$

which, by (20) is bounded by the expression

$$\sum_{n_1, n_2} \int d\lambda \, d\lambda_1 \frac{|n|c(n_1, \lambda_1)c(n_2, \lambda - \lambda_1)d(n, \lambda)}{(1 + |\lambda - n^3|)^{1/2}(1 + |\lambda_1 - n_1^3|)^{1/2}(1 + |\lambda - \lambda_1 - n_2^3|)^{1/2}}. \tag{23}$$

The main point is to observe that since $n = n_1 + n_2$

$$n^3 - n_1^3 - n_2^3 = 3nn_1n_2 \tag{24}$$

and therefore

$$\max\{|\lambda - n^3|^{1/2}, |\lambda_1 - n_1^3|^{1/2}, |\lambda_2 - n_2^3|^{1/2}\} > (|n|\,|n_1|\,|n_2|)^{1/2}. \qquad (25)$$

Assume $|n_1| \geq |n_2|$. Then the right member of (25) is at least $|n|\,|n_2|^{1/2}$. One can distinguish 2 cases, which are treated similarly

$$\begin{cases} |\lambda - n^3|^{1/2} > |n|\,|n_2|^{1/2} & (26) \\[2ex] |\lambda_j - n_j^3|^{1/2} > |n|\,|n_2|^{1/2} \text{ for } j = 1 \text{ or } j = 2. & (27) \end{cases}$$

Assume (26). Since $|n_2| \geq 1$, the corresponding contribution to (23) is at most

$$\sum_{n,n_2} \int d\lambda d\lambda_1 \; d(n,\lambda) \; \frac{c(n_1,\lambda_1)}{(1+|\lambda_1 - n_1^3|)^{1/2}} \; \frac{c(n_2, \lambda - \lambda_1)}{(1+|\lambda - \lambda_1 - n_2^3|)^{1/2}}. \qquad (28)$$

Define the following functions

$$G(x,t) = \sum_n \int d\lambda \; d(n,\lambda) \; e^{i(nx+\lambda t)} \qquad (29)$$

$$H(x,t) = \sum_n \int d\lambda \frac{c(n,\lambda)}{(1+|\lambda - n^3|)^{1/2}} \; e^{i(nx+\lambda t)} \qquad (30)$$

so that

$$(28) = \langle \hat{G}, \hat{H} * \hat{H} \rangle = \langle G, H^2 \rangle \qquad (31)$$

which by Hölder's inequality, (21), is bounded by

$$\|G\|_2 \|H\|_4^2 \leq \|H\|_4^2. \qquad (32)$$

More precisely, as in the NLS-case, consider a localizing function $0 \leq \psi_\delta \leq 1$, $\hat{\psi}_\delta \geq 0$, $\psi_\delta = 1$ on $[-\delta, \delta]$, supp $\psi_\delta \subset [-2\delta, 2\delta]$. Then $u = u\psi_\delta, |\hat{u}| \leq |\hat{u}| * \hat{\psi}_\delta$ and (32) yields in fact $\|H\psi_\delta\|_4^2$.

Since from (11)

$$\|F\|_{L^4_{\mathrm{loc}}} \leq c \, \|F\|_{X^{0,\frac{1}{3}}} \qquad (33)$$

we get from the same chain of inequalities as for $1D$ NLS, an estimate

$$\|H\psi_\delta\|_4 \leq c\delta^\theta \, \|H\|_{X^{0,\frac{1}{2}}} \leq c\delta^\theta \, \|u\|_{X^{0,\frac{1}{2}}[0,\delta]} \qquad (34)$$

for some $\theta > 0$.

This yields that

$$\|Tu - S(t)\phi\|_{X^{0,\frac{1}{2}}[0,\delta]} \leq c\delta^\theta \, \|u\|_{X^{0,\frac{1}{2}}[0,\delta]}^2. \qquad (35)$$

The subsequent discussion is completely analogous to the $1D$ cubic NLS case treated in the preceding section and in particular global results follow from the L^2-conservation.

Remark.

(1) In fact, the KdV equation considered above

$$u_t + u_{xxx} + uu_x = 0$$

is integrable and a full set of invariants of notion are available. The preceding analysis is clearly independent of this integrability aspects however. It may be carried out (modulo a substantial amount of technicalities) to prove the following result for generalized KdV type equations of the form

$$u_t + u_{xxx} + F'(u)u_x = 0 \tag{36}$$

with $F(u)$ a polynomial or sufficiently smooth function.

Theorem 37. *The Cauchy problem with periodic boundary conditions*

$$\begin{cases} u_t + u_{xxx} + F'(u)u_x = 0 \\ u(0) = \phi \end{cases} \tag{38}$$

is globally wellposed for sufficiently smooth data ϕ, provided the H^1-norm is controlled by the Hamiltonian

$$\frac{1}{2} \int |\nabla \phi|^2 - \int G(\phi) \quad (G' = F) \tag{39}$$

(which is the case for small data).

Observe that in the general case $(F(u) = u(= KdV)$ and $F(u) = \lambda u^2$ $(=$ modified $KdV)$ are the integrable exceptions), one only disposes of

$$\int \phi \qquad \|\phi\|_2 = \left(\int |\phi(x)|^2 \right)^{1/2} \qquad H(\phi) = \int |\nabla \phi|^2 - \int G(\phi) \tag{40}$$

as conserved quantities. Thus the main point in proving Theorem 37 is to get the local result on a time interval only depending on the H_1-norm.

(2) On the other hand, there is a rich algebra-geometric structure tied to the KdV equation $u_t + u_{xxx} + uu_x = 0$. If $\phi = u(t)$, then the periodic spectrum $\lambda_0 < \lambda_1 \leq \lambda_2 < \cdots < \lambda_{2n-1} \leq \lambda_{2n} < \cdots$ of the Sturm-Liouville operator $Q = -\frac{d^2}{dx^2} + u(t)$ is invariant under the flow of the equation. The "gaps" $\{\lambda_{2n} - \lambda_{2n-1}\}$ measure essentially the size of the Fourier-coefficients $\{\hat{\phi}(n)\}$. This fact explains the presence of other conserved integrals of the form (40). Note however that the spectral invariants $\{\lambda_n\}$ are already defined for any L^2-potential. Theorem 2 establishes a well-defined continuous flow, acting on the isospectral manifold

$$\{q \in L^2(\mathbb{T}) | \lambda_n(q) = \lambda_n(\phi) \text{ for all } n\}. \tag{41}$$

These manifolds are compact and should be viewed as infinite dimensional tori. It can be shown that the solution u of (3) for any $q \in L^2$ is an almost periodic function of time. Historically, Lax and Novikov independently proved quasi-periodicity of the solutions corresponding to "finite gap" potentials, i.e. $\lambda_{2n-1}(q) = \lambda_{2n}(q)$ for n sufficiently large. These solutions are given explicitly in terms of θ-functions on the Riemann surface (here of finite genus) of the hyperelliptic irrationality $[-(\lambda-\lambda_0)(\lambda-\lambda_1)(\lambda-\lambda_2)\cdots(\lambda-\lambda_{2n-1})(\lambda-\lambda_{2n})]^{1/2}$. This theory was extended to the case of smooth potentials (data) by H. McKean, E. Trubowitz. Here the Riemann surface is infinite genus (compact) determined by $\sqrt{\Delta^2(\lambda)-4}$ where $\Delta(\lambda) = y_1(1,\lambda) + y_2'(1,\lambda)$ is the usual discriminant

$$\left(-\frac{d^2}{dx^2}+q\right) y_i = \lambda y_i \quad \begin{cases} y_1(0) = y_2'(0) = 1 \\ y_1'(0) = y_2(0) = 0. \end{cases}$$

Their work carries over to the L^2-setting.

Comments and references related to Lecture 3.

Papers of the author related to the Cauchy problem for KdV-type equations (especially under periodic boundary conditions) are

[B3]   J. Bourgain, *Restriction phenomena for lattice subsets and applications to nonlinear evolution equations II, KdV-equation*, GAFA, Vol. 3, N3 (1993), 209-262.

[B4]   J. Bourgain, *On the Cauchy problem for periodic KdV type equations*, Fourier Anal. and Appl., Proc. Conf. J.P Kahane (special Issue), 1995, 17-86.

There is an extensive literature for problems on the line. We mention the following papers dealing with initial value in negative Sobolev spaces.

[KPV]   C. Kenig, G. Ponce, L. Vega, *The Cauchy problem for the Kortweg-De Vries equation in Sobolev spaces of negative indices*, Duke Math. J. 71, N1 (1994), 1-21.

Idem: *A bilinear estimate with applications to the KdV equation*, TAMS 348 (1996), No 8, 3323-3355.

A survey exposition may be found in

[Gin]   J. Ginibre, Sem. Bourbaki (1994-95), N796.

The main reference[*] concerning almost periodicity of solutions of the periodic KdV equation is

[McK, Tr] H.P. McKean, E. Trubowitz, *Hill's operator and hyperelliptic function theory in the presence of infinitely many branch points*, Comm. Pure Appl. Math. 29 (1976), 142-226.

This subject also has an extensive history and we refer to [**McK, Tr**] for the discussion of part of it and additional references (it is not the subject of these lectures). The result of [**McK, Tr**] yields almost periodic behaviour in time of any solution of the KdV-equation $u_t + u_{xxx} + uu_x = 0$ with smooth initial data $u(0)$. In [**B3**], we extended this fact to L^2-data for which we established a wellposedness theory.

---

[*]not in the historical sense

LECTURE 4
1D Invariant Gibbs Measures

The Lebesgue measure on $\mathbb{R}^{2n}$ is invariant under the Hamiltonian flow

$$\begin{cases} \dot{p}_i = \frac{\partial H}{\partial q_i} \\ \dot{q}_i = -\frac{\partial H}{\partial p_i} \end{cases} \tag{1}$$

with Hamiltonian $H(p_1, \ldots, p_n, q_1, \ldots q_n)$, since (1) defines a divergence free vector-field. Consequently, from the preservation of H, also the measures (Gibbs measure)

$$e^{-\beta H} \Pi_{i=1}^n dp_i dq_i \tag{2}$$

are invariant.

Considering the Schrödinger equation

$$iu_t = \frac{\partial H}{\partial \overline{u}} \tag{3}$$

with

$$H(\phi) = \frac{1}{2} \int |\nabla\phi|^2 \pm \frac{1}{p} \int |\phi|^p \tag{4}$$

we get a Hamiltonian system with infinite dimensional phase space; the canonical coordinates $\{p_n, q_n\}$ are taken to be

$$p_n = Re\ \hat{\phi}(n), q_n = I_m \hat{\phi}(n) \tag{5}$$

for $n \in \mathbb{Z}^d$.

The interest of (2) is that this expression may be given a sense (possibly after an appropriate normalization) in this infinite dimensional context. The factor β ($\frac{1}{\beta}$ = temperature in statistical physics) will be taken 1.

Rewrite (2) as

$$e^{\mp \frac{1}{p} \int |\phi|^p} \underbrace{e^{-\frac{1}{2} \int |\nabla\phi|^2} \Pi d^2\phi}_{\overset{\|}{d\mu}}. \tag{6}$$

The factor $d\mu$ corresponds to (unnormalized) Wiener measure. Identifying ϕ and its Fourier transform $\hat{\phi} = \{\hat{\phi}(n) | n \in \mathbb{Z}^d\}$, we have

$$\int |\nabla\phi|^2 = \sum |n|^2 |\hat{\phi}(n)|^2 \tag{7}$$

41

and thus, writing $a_n = \hat{\phi}(n)$, the normalized measure $d\mu$ on $\{a_n | n \in \mathbb{Z}^d\}$ is given by

$$d\mu = \frac{e^{-\frac{1}{2}\sum |n|^2 |a_n|^2}\Pi d^2 a_n}{\int e^{-\frac{1}{2}\sum |n|^2 |a_n|^2}\Pi d^2 a_n} = \Pi_{n\in\mathbb{Z}^d} \frac{e^{-\frac{1}{2}|n|^2 |a_n|^2} d^2 a_n}{\int_{\mathbb{C}} e^{-\frac{1}{2}|n|^2 |a_n|^2} d^2 a_n}. \tag{8}$$

Thus this measure corresponds to Gaussian distribution for each $|n|\hat{\phi}(n)$. Equivalently, this yields for ϕ the distribution of a random Fourier series

$$\phi = \phi_\omega = \sum_{n\in\mathbb{Z}^d} \frac{g_n(\omega)}{|n|} e^{i\langle n,x\rangle} \tag{9}$$

where the $\{g_n\}$ are identically distributed complex Gaussian random variables.

Remark. (i) In the preceding, there is a problem with the 0-Fourier mode $n = 0$. This may be solved, replacing $-\Delta$ by $-\Delta + \rho, \rho > 0$, so that (9) gets replaced by

$$\phi_\omega = \sum_{n\in\mathbb{Z}^d} \frac{g_n(\omega)}{\sqrt{|n|^2 + \rho}} e^{i\langle n,x\rangle}. \tag{10}$$

The series (10) defines in the $1D$-case almost surely in ω a function $\phi_\omega \in H^s(\Pi)$, $s < \frac{1}{2}$. In higher dimension, one gets a distribution.

The discussion in this section will be $1D$. Our goal is to establish the invariance of (6) under the flow of the corresponding $1D$ NLS

$$iu_t + \Delta u \mp u|u|^{p-2} = 0 \tag{11}$$

(which is formally the case).

To give (6) a meaning, we consider it as a "weighted Wiener measure"

$$e^{\mp \int |\phi|^p} d\rho(\phi) \tag{12}$$

with density $e^{\mp \int |\phi|^p}$. Thus the questions arising here are the following

$$\text{- Existence of } \int |\phi_\omega|^p \text{ a.s.} \tag{13}$$

$$\text{- Integrability of the factor } e^{\mp \int |\phi_\omega|^p} \text{ in } \omega \tag{14}$$

In $1D$, since $\phi_\omega \in H^s(\Pi), s < \frac{1}{2}$ a.s., (13) is fulfilled.
In the defocusing case, one gets the factor

$$e^{-\int |\phi_\omega|^p} \tag{15}$$

which is clearly bounded.
In the focusing case, the factor

$$e^{\int |\phi_\omega|^p} \tag{16}$$

is not integrable for $p > 2$. This problem was studied by Lebowitz, Rose, Speer and solved as follows. Observe that since $\|\phi\|_2$ (the L^2-norm) is a conserved quantity,

$$d\mu = \chi_{[\|\phi\|_2 \leq B]} \, e^{\int |\phi|^p} d\rho \tag{17}$$

obtained by restriction of (6), (12) to a ball in L^2 of radius B, is still formally invariant.

Theorem 18. *(L-R-S)*

  (1) *If $p < 6$, the factor $\chi_{[\|\phi\|_2 \leq B]}\, e^{\int |\phi|^p}$ belongs to $L^r(d\mu)$ for all $r < \infty, B < \infty$. In particular, (17) defines a measure absolutely continuous wrt Wiener measure.*

  (2) *If $p = 6$, the preceding is true provided B is sufficiently small; $B < B_r$.*

Remark.

  (i) Instead of an L^2-cutoff in Theorem 18 (1), one may multiply with a decay function $e^{-\|\phi\|_2^c}$ for large enough c.

  (ii) The role of $p = 6$ is related to blowup phenomena ($p = 6$ is $1D$ critical).

Proof of Theorem 18. We have to estimate the measure

$$\mathbb{P}_\omega\left[\left\|\sum \frac{g_n(\omega)}{|n|}\, e^{inx}\right\|_p > \lambda, \ \left(\sum \frac{|g_n(\omega)|^2}{n^2}\right)^{1/2} < B\right]. \tag{19}$$

Split the trigonometric system in dyadic intervals. One has for $2 \leq p \leq \infty$

$$\left\|\sum_{n \sim M} a_n e^{inx}\right\|_p \leq \left\|\sum_{|n| < 2M} a_n e^{inx}\right\|_p \leq c M^{\frac{1}{2} - \frac{1}{p}}\left(\sum |a_n|^2\right)^{1/2}. \tag{20}$$

Hence, in estimating (19), we let

$$M > M_0 \sim \left(\frac{\lambda}{B}\right)^{\frac{1}{\frac{1}{2} - \frac{1}{p}}}. \tag{21}$$

Let $(\sigma_M)_{\substack{M > M_0 \\ M \text{dyadic}}}$ be positive numbers such that

$$\sum \sigma_M < 1. \tag{22}$$

We estimate for each $M < M_0$

$$\mathbb{P}_\omega\left[\left\|\sum_{n \sim M} g_n(\omega)e^{inx}\right\|_p > \sigma_M M \lambda\right]. \tag{23}$$

Again by (20), if

$$\left\|\sum_{n \sim M} g_n(\omega)e^{inx}\right\|_p > K \tag{24}$$

then

$$\left(\sum_{n \sim M} |g_n(\omega)|^2\right)^{1/2} > K M^{\frac{1}{p} - \frac{1}{2}} \tag{25}$$

and (by duality), the probability of event (25) is at most

$$e^{CM - cK^2 M^{\frac{2}{p} - 1}}. \tag{26}$$

Hence

$$(23) < e^{-c\sigma_M^2 M^{\frac{2}{p} + 1} \lambda^2} \tag{27}$$

and

$$(19) < \sum_{M > M_0} e^{-c\sigma_M^2 M^{\frac{2}{p} + 1} \lambda^2} < e^{-c M_o^{\frac{2}{p} + 1} \lambda^2} \tag{28}$$

for an appropriate choice of (22).

Thus for $\chi_{[\|\phi\|_2 \leq B]} e^{\int |\phi|^p}$ to be in $L^r(d\omega)$, we require by (21)

$$r\lambda^p < c\left(\frac{\lambda}{B}\right)^{\frac{\frac{2}{p}+1}{\frac{1}{2}-\frac{1}{p}}} \lambda^2 \tag{29}$$

for λ sufficiently large. The result follows.

Next we discuss the invariance of the measure $d\mu$ obtained above for the flow of the corresponding equation

$$iu_t + u_{xx} \mp u|u|^{p-2} = 0. \tag{30}$$

The role of the sign (defocusing or focusing) played a role in the construction of previous measure, i.e.

$$\begin{cases} \text{defocusing : all } p \text{ allowed} \\ \text{focusing : } p \leq 6 \text{ with appropriate } L^2\text{-cutoff} \end{cases}$$

but will be unimportant in our subsequent discussion.

The fact that $d\mu$ is invariant under the flow of (30) is formally correct. More precisely, take a large N and consider instead of (30) the truncated equation (ODE)

$$iu_t^N + u_{xx}^N \mp P_N(u^N|u^N|^{p-2}) = 0 \tag{31}$$

defining

$$P_N\phi = \sum_{|n| \leq N} \hat{\phi}(n)\, e^{inx}. \tag{32}$$

Then the measure μ_N corresponding to the measure μ above, i.e.

$$d\mu_N = \chi_{[\|\phi_N\|_2 \leq B]}\, e^{\mp \int |\phi_N|^p}\, d\rho_N \tag{33}$$

$$\phi_N = \sum_{|n| \leq N} a_n\, e^{inx} \tag{34}$$

on the phase space $\mathbb{C}^{2N+1}$ is invariant under the flow of (31) (which exists globally because of the L^2-conservation). Now ρ (resp. μ) are the limit of ρ_N (resp. μ_N) for $N \to \infty$. Denoting

$$E_N = [|e^{2\pi inx}; |n| \leq N] \tag{35}$$

one may indeed verify (cf. Theorem 18).

Lemma 36. *Let U be an open set in $H^s(\mathbb{T}), s < \frac{1}{2}$. Then*

$$\rho(U) = \lim \rho_N(U \cap E_N) \tag{37}$$

$$\mu(U) = \lim \mu_N(U \cap E_N). \tag{38}$$

Our next task is to compare the flow of (31) and (30). We first discuss the case $p = 4$. Then for almost all ϕ, $\phi \in H^s(\mathbb{T}), s < \frac{1}{2}$ and Theorem 1.36 ensures global wellposedness of the Cauchy problem

$$\begin{cases} iu_t + u_{xx} \mp u|u|^2 = 0 \\ u(0) = \phi. \end{cases} \tag{39}$$

Let for each N, u_N be the solution of

$$\begin{cases} iu_t^N + u_{xx}^N \mp_N^P (u^N |u^N|^2) = 0 \\ u^N(0) = \phi_N = P_N \phi \end{cases} \tag{40}$$

then following approximation property holds for (39), (40).

Lemma 41. *Assume the data ϕ in (39) in H^s, $s > 0$. Then, for all $T < \infty$, $s_1 < s$*

$$\|u - u^N\|_{C_{H^{s_1}}[-T,T]} \overset{N \to \infty}{\longrightarrow} 0. \tag{42}$$

Proof. We will compare locally in time the solutions of (39) and

$$\begin{cases} iv_t + v_{xx} \mp P_N(v|v|^2) = 0 \\ v(0) = \psi = P_N\psi. \end{cases} \tag{43}$$

We show that

$$\|u - v\|_{C_{H^{s_1}}[-\delta,\delta]} \tag{44}$$

may be made small assuming N sufficiently large and $\|\phi - \psi\|_{H^{s_1}}$ sufficiently small. The time interval is of size $\delta = \delta(\|\phi\|_{H^s})$. The lemma will then follow from repeated applications of this approximation property. Observe that here ϕ will be taken $u(t)$ for $|t| < T$ and the solution u of (39) satisfies a bounded

$$\|u(t)\|_{H^s} \leq C^{|t|} < C^T \tag{45}$$

(cf. Theorem 1.36).

Rewriting (39), (43) as integral equation, we have

$$u(t) = \phi + i \int_0^t S(t - \tau)(u|u|^2)(\tau)d\tau \tag{46}$$

$$v(t) = \psi + i \int_0^t S(t - \tau)(P_N(v|v|^2))(\tau)d\tau \tag{47}$$

hence

$$u(t) - v(t) = \phi - \psi + i \int_0^t S(t - \tau)[u|u|^2 - P_N(v|v|^2)](\tau)d\tau. \tag{48}$$

Rewrite

$$u|u|^2 - P_N(v|v|^2) = u|u|^2 - P_N(u|u|^2) + P_N(u|u|^2 - v|v|^2) \tag{49}$$

$$u|u|^2 - P_N(u|u|^2) = u|u|^2 - P_{\frac{N}{3}}u|P_{\frac{N}{3}}u|^2 - P_N(u|u|^2 - P_{\frac{N}{3}}u|P_{\frac{N}{3}}u|^2) \tag{50}$$

and

$$u|u|^2 - u^1|u^1|^2 = (u - u^1)|u|^2 + \overline{u}u^1(u - u^1) + (u^1)^2(\overline{u} - \overline{u^1}). \tag{51}$$

From estimate (2.72) in the proof of Theorem 1.36, we get for $b < \frac{5}{8}$

$$\left\| \int_0^t S(t - \tau)[u|u|^2 - P_N(v|v|^2)](\tau)d\tau \right\|_{X^{s_1,b}[-\delta,\delta]} \leq$$

$$c.\delta^\alpha (\|u\|_{X^{s_1,\frac{3}{8}}[-\delta,\delta]} + \|v\|_{X^{s_1,\frac{3}{8}}[-\delta,\delta]})^2 (\|u - v\|_{X^{s_1,\frac{3}{8}}[-\delta,\delta]} + \|u - P_{\frac{N}{3}}u\|_{X^{s_1,\frac{3}{8}}[-\delta,\delta]}). \tag{52}$$

Write

$$\|v\|_{X^{s_1,\frac{3}{8}}} \leq \|u\|_{X^{s_1,b}} + \|u - v\|_{X^{s_1,b}} \tag{53}$$

by (45)

$$\|u\|_{X^{s_1,b}} \leq \|u\|_{X^{s,b}} < C_{s,T} \tag{54}$$

$$\|u - P_{\frac{N}{3}}u\|_{X^{s_1,3/8}} \leq cN^{s_1-s}\|u\|_{X^{s,3/8}} < c_{s,T}N^{s_1-s}. \tag{55}$$

Hence, from (48), (52) and the preceding

$$\|u - v\|_{X^{s_1,b}[-\delta,\delta]} \leq c\|\phi - \psi\|_{H^{s_1}} + c\delta^{\alpha}(\|u - v\|_{X^{s_1,b}[-\delta,\delta]} + N^{s_1-s}). \tag{56}$$

Thus, for δ sufficiently small (depending on (54))

$$\|u - v\|_{C_{H^{s_1}}[-\delta,\delta]} \leq \|u - v\|_{X^{s_1,b}} \tag{57}$$

$$\leq c\|\phi - \psi\|_{H^{s_1}} + N^{s_1-s}. \tag{58}$$

This proves Lemma 41.

Denote $S'(t)$ (resp. $S'_N(t)$) the time flow corresponding to the equation $iu_t + u_{xx} \mp u|u|^2 = 0$ (resp. $u_t^N + u_{xx}^N \mp P_N(u^N(u^N|^2) = 0)$). It follows from Lemma 41 that if $\phi \in H^s(\mathbb{T}), s > 0$, then

$$S'(t)\phi = \lim_{N \to \infty} S'_N(t)(P_N\phi) \text{ in } H^{s_1}, s_1 < s \tag{59}$$

uniformly on finite time interval and bounded subsets of H^s.

We now show the invariance of Gibbs-measure μ.

Fix time $t, 0 < s_1 < \frac{1}{2}$. Let K be a compact in $\cap_{s<\frac{1}{2}}H^s$ and denote B_ε the ball $B_{H^{s_1}}(0,\varepsilon)$. It follows from (38) that

$$\mu(S'(K) + B_\varepsilon) = \lim_{N \to \infty} \mu_N\big((S'(K) + B_\varepsilon) \cap E_N\big). \tag{60}$$

It follows from (59) that for $N > N_0$

$$S'_N(P_N K) \subset S'(K) + B_{\varepsilon/2} \tag{61}$$

and hence, for sufficiently small ε_1 (independent of N)

$$S'_N\big((K + B_{\varepsilon_1}) \cap E_N\big) \subset S'_N(P_N K) + B_{\varepsilon/2} \subset S'(K) + B_\varepsilon. \tag{62}$$

Since S'_N leaves μ_N invariant for each N, (62) implies for $N > N_0$

$$\mu_N\big((K + B_{\varepsilon_1}) \cap E_N\big) \leq \mu_N\big((S'(K) + B_\varepsilon) \cap E_N\big). \tag{63}$$

Passing to the limit for $N \to \infty$ yields from (38), (60)

$$\mu(K) \leq \mu(K + B_{\varepsilon_1}) \leq \mu\big(S'(K) + B_\varepsilon\big). \tag{64}$$

Hence, letting $\varepsilon \to 0$

$$u(K) \leq \mu(S'(K)) \tag{65}$$

and since the flow is reversible

$$\mu(K) = \mu\big(S'(K)\big). \tag{66}$$

This proves the invariance.

Remark. The preceding remains valid for $2 < p < 4$.

The problem for $p > 4$ is the fact that Theorem 1.37 only yields a local existence result. Thus consider the Cauchy problem

$$\begin{cases} iu_t + u_{xx} \mp u|u|^{p-2} = 0 \\ u(0) = \phi \in H^s, s > \frac{1}{2} - \frac{2}{p-2}. \end{cases} \tag{67}$$

Then there is local wellposedness. The time interval δ may be estimated as $\delta = (1+\|\phi\|_{H^s})^{-c}$ from below. This follows from the proof of Theorem 1.37. Also, there is approximation in $C_{H^{s_1}}[-\delta, \delta], s_1 < s$, by solutions of the truncated problem

$$\begin{cases} iu_t^N + u_{xx}^N \pm P_N(u^N|u^N|^{p-2}) = 0 \\ u^N(0) = P_N\phi \end{cases} \tag{68}$$

for $N \to \infty$.

The main idea at this stage is to exploit the invariance of the Gibbs measure as a substitute for a conservation law and show that for almost all $\phi \in \operatorname{supp}\mu$ (67) is in fact globally wellposed. Since this Gibbs measure invariance is only available for (68) at this stage, we consider the flow of the truncated equation (bounds are independent of the truncation N). Moreover in the focusing case, we need to assume $p \le 6$ (because of the measure existence).

Lemma 69. *Let $0 < s < \frac{1}{2}, p \le 6, T < \infty, \delta > 0$. There is a set $\Omega_\delta \subset \cap_{s < \frac{1}{2}} H^s$ such that $\mu(\Omega_\delta^c) < \delta$ and for $\phi \in \Omega_\delta$, the solution u of the IVP*

$$\begin{cases} iu_t + u_{xx} + P_N(u|u|^{p-2}) = 0 \\ u(0) = P_N\phi \end{cases} \tag{70}$$

satisfies for $|t| \le T$

$$\|u(t)\|_{H^s} \le c(\log \frac{T}{\delta})^{1/2}. \tag{71}$$

Having a bound on $\|u^N(t)\|_{H^s}$ for $|t| < T$ at our disposal, one may carry out the proof of Lemma 41 comparing (67), (68) with the same conclusion.

Remark. We considered the focusing case in Lemma 69. In the defocusing case, no restriction on p is necessary.

Proof of Lemma 69. Define for $K > 1$ the set

$$\Omega_{s,K} = \{\phi_N \in E_N| \ \|\phi_N\|_{H^s(\mathbb{T})} \le K\}. \tag{72}$$

Then, for $s < \frac{1}{2}$ (independently of N)

$$\rho_N(\Omega_{s,K}^c) < e^{-cK^2} \tag{73}$$

and hence also

$$\mu_N(\Omega_{s,K}^c) < e^{-cK^2}. \tag{74}$$

(73) results from the general Gaussian estimate $\|X_\omega\|_{L_B^{\psi_2}} \sim \|X_\omega\|_{L_B^2}$, if X_ω is a B-valued Gaussian process ($B=$ Banach space). Here $\psi_2(\lambda) = e^{c\lambda^2} - 1$.

Denote $S_N(t)$ the flow map corresponding to (70). Thus μ_N is $S_N(t)$-invariant for all t. Fix K (to be specified). If $\phi \in \Omega_{s,K}$, then

$$\|u_N(t)\|_{H^s} < 2K \quad \text{for } |t| < \delta \sim K^{-C} \tag{75}$$

(independently of N, as a consequence of (the proof of) the local wellposedness of (67)). Denote $S = S_N(\frac{\delta}{2})$ and consider the set

$$\Omega_\delta = \Omega_{s,K} \cap S^{-1}(\Omega_{s,K}) \cap S^{-2}(\Omega_{s,K}) \cdots \cap S^{-\left[\frac{2T}{\delta}\right]}(\Omega_{s,K}). \tag{76}$$

Since μ_N is S-invariant, it follows that

$$\mu_N(\Omega_\delta^c) \leq \frac{T}{\delta} \cdot \mu_N(\Omega_{s,K}^c) < TK^C e^{-cK^2} \tag{77}$$

by (74), (76). Hence, for $K \sim (\log \frac{T}{\delta})^{1/2}$, we ensure $\mu_N(C\Omega_\delta) < \delta$.

If $\phi_N \in \Omega_\delta$, it follows that for all $j = 0, 1, \ldots, \left[\frac{2T}{\delta}\right]$

$$\left\|S_N\left(j\frac{\delta}{2}\right)\phi_N\right\|_{H^s} \leq K. \tag{78}$$

Hence, by (75), also

$$\|S_N(t)\phi_N\|_{H^s} \leq 2K \tag{79}$$

for all $0 \leq t \leq T$. This gives the result.

Remark. The preceding reasoning has a general character. Essentially speaking, it shows that in order to establish the invariance of the Gibbs measure, only a local wellposedness of the Cauchy problem for typical data in the support of the measure is required. Moreover, the existence of such invariant measure has an interest from purely PDE point of view, since it implies that the local solution extends to a global one, for almost all data in the support of the measure.

Along previous lines, one may study the $1D$ Zakharov models (with periodic boundary conditions)

$$\begin{cases} iu_t = -u_{xx} + nu \\ n_{tt} - c^2 n_{xx} = c^2(|u|^2)_{xx}. \end{cases} \tag{80}$$

where c is a parameter. The physical meaning of u, n, c are respectively the electrostatic envelope field, the ion density fluctuation field and the ion sound speed. The cubic NLS $iu_t + u_{xx} + u|u|^2 = 0$ may be seen as a limit of (80) if $c \to \infty$ (80) is a Hamiltonian system (it is a coupling of an NLS and a wave equation). Defining an auxiliary field $V(x,t)$ by

$$\begin{cases} n_t = -c^2 V_x \\ V_t = -n_x - (|u|^2)_x. \end{cases} \tag{81}$$

The Hamiltonian is given by

$$H = \frac{1}{2}\int \left[|u_x|^2 + \frac{1}{2}(n^2 + c^2 V^2) + n|u|^2\right]dx \tag{82}$$

and $(\operatorname{Re} u, \operatorname{Im} u)$, $(\tilde{n}, \tilde{V})$ with $\tilde{n} = 2^{-1/2}n$, $\tilde{V} = 2^{-1/2}\int^x V$ are pairs of conjugate variables.

The formal associated Gibbs measure is (after L^2-truncation of u, as for focusing NLS)

$$d\mu = e^{-\beta H}\chi_{[\|u\|_2 \leq B]}\Pi_x d^2 u(x)d\tilde{n}(x)d\tilde{V}(x). \tag{83}$$

It may be given a rigorous meaning as in the NLS case. In fact one gets the $1D$ cubic NLS Gibbs measure constructed above as marginal distribution of the u field.

One may prove the following

Theorem 84. *The $1D$ (ZE) system (80) is globally wellposed for almost all data $(u_0, \tilde{n}_0, \tilde{V}_0)$ in the support of the Gibbs measure $d\mu$ which is invariant under the resulting flow.*

The proof of Theorem 84 follows the scheme described for the NLS. The first step consists in proving a local wellposedness property for the Cauchy problem with appropriate data. This result applies in particular to data

$$(u^0, n^0, n_t^0) \in H^1(\mathbb{T}) \times L^2(\mathbb{T}) \times H^{-1}(\mathbb{T}) \tag{85}$$

(u is complex, n is real).

From the conservation of $\|u(t)\|_{L^2}$ and H^1, this local result extends to a global one.

Comments and references related to Lecture 4.

The program discussed here and in the next lecture originates from the paper

[L-R-S]   J. Lebowitz, R. Rose, E. Speer, *Statistical mechanics of the nonlinear Schrödinger equation*, J.Stat. Phys. V. 50 (1988), 657-687.

The discussion of the $1D$ NLS follows the author's paper

[B5]   J. Bourgain, *Periodic Nonlinear Schrödinger equation and invariant measures*, Comm. Math. Phys. 166 (1994), 1-26.

See also for the construction of an invariant measure for the Zakharov system

[B6]   J. Bourgain, *On the Cauchy and invariant measure problem for the periodic Zakharov system*, Duke Math. J, Vol.76 (1994), 175-202.

There has been related work in the $1D$ case by other authors, including H. McKean, K. Vaninski, P. Zhidkov. We mention some of it.

[McK-V]   H. McKean, K. Vaninski, *Statistical mechanics of nonlinear wave equations and Brownian notin with restoring drift: The petit and microcanonical ensembles*, Comm. Math. Phys. 160 (1994), 615-630.

and various preprints.

This work has a more probabilistic flavor.

[Z]   P. Zhidkov, *On the invariant measure for the nonlinear Schrödinger equation*, Doklady Akad. Nauk. SSSR 317, N3 543, (1991).

Idem: An invariant measure for a nonlinear wave equation. Preprint 92.

LECTURE 5
Invariant Measures (2D)

We will discuss here some aspects of Gibbs measure normalization and invariance under the flow of the corresponding evolution equation in $2D$. Our main model will be that of the $2D$ defocusing cubic NLS

$$iu_t + \Delta u - u|u|^2 = 0. \tag{1}$$

For the formal Gibbs measure, we get

$$e^{-H(\phi)}\Pi d^2\phi(x) = e^{-\int |\phi|^4}\underbrace{e^{-\frac{1}{2}\int |\nabla\phi|^2}\Pi d^2\phi}_{\text{Wiener measure}} = e^{-\int |\phi|^4}d\rho. \tag{2}$$

As mentioned earlier, the distribution of ϕ given by the random Fourier series

$$\phi_\omega = \sum_{n\in\mathbb{Z}^2} \frac{g_n(\omega)}{(|n|^2+\rho)^{1/2}}\, e^{i\langle n,x\rangle} \tag{3}$$

yields distributions and not functions when $D > 1$. Hence for typical ϕ the expression $\int |\phi|^4$ is unbounded, i.e.

$$\lim_{N\to\infty}\int |P_N\phi_\omega|^4 = \infty \qquad \omega \quad a.s. \tag{4}$$

This problem is overcome in $2D$ by so-called *Wick-ordering*. The general process consists in associating to a monomial x^{2k} the corresponding Hermith polynomial $P_{2k}(x)$, where (in the real case)

$$P_n(x) = \sum_{j=0}^{[\frac{n}{2}]} (-1)^j\, c_{n,j}\, x^{n-2j} \tag{5}$$

$$c_{n,j} = \frac{n!}{(n-2j)!2^j j!} \tag{6}$$

obtained by orthogonalization of the monomials $\{x^n\}$ *wrt* Gaussian measure on $\mathbb{R}$. Thus one has

$$P_{2k}(x) = x^{2k} + \text{ lower degree terms.}$$

51

The complex case is similar (but the coefficients (5.6) are different). We denote

$$a_N = \int |\phi_\omega^N|^2 d\omega \sim \sum_{|n| < N} \frac{1}{|n|^2 + \rho} \quad (\sim \log N \text{ in } 2D) \tag{7}$$

and the Wick ordering of ϕ^{2k}

$$: \phi_N^{2k} := a_N^k P_{2k}\left(\frac{\phi_N}{\sqrt{a_N}}\right). \tag{8}$$

In the complex case, we have in our situation (2)

$$: |\phi_N|^4 := |\phi_N|^4 - 4a_N|\phi_N|^2 + 2a_N^2. \tag{9}$$

The next basic result is a general feature (we consider the case $k = 4$).

Proposition 10.
 (1) $\int : |\phi_N|^4 : dx$ *converges as to a finite limit for* $N \to \infty$.
 (2) *The measures* $d\mu_N = e^{-\int :|\phi_N|^4:} d\rho_N$ *converge to a weighted Wiener measure with density in* $\cap_{q<\infty} L^q(d\rho)$.

This "normalized" Gibbs measure $d\mu = e^{-\int :|\phi|^4:} d\rho$ is obtained by modification of the Hamiltonian according to (9). It will require to modify the original equation (1) slightly in order to get the corresponding evolution equation.

Proof of Proposition 10. Denoting $\phi = \phi_N$, one has first from (9)

$$- : |\phi|^4 := -|\phi|^4 + 4a_N|\phi|^2 - 2a_N^2 = -(|\phi|^2 - 2a_N)^2 + 2a_N^2 < C(\log N)^2. \tag{10}$$

Hence

$$- \int : |\phi_N|^4 :< C(\log N)^2. \tag{11}$$

Write expanding in Fourier, by (3)

$$- \int : |\phi|^4 := -4 \sum_{\substack{n_1 - n_2 + n_3 - n_4 = 0 \\ n_1 \neq n_2, n_4 \\ |n_i| < N}} \frac{g_{n_1}(\omega)}{|n_1|} \frac{\overline{g_{n_2}(\omega)}}{|n_2|} \frac{g_{n_3}(\omega)}{|n_3|} \frac{\overline{g_{n_4}(\omega)}}{|n_4|} \tag{12}$$

$$- 2\left(\sum_{|n|<N} \frac{|g_n(\omega)|^2}{|n|^2}\right)^2 + \sum_{|n|<N} \frac{|g_n(\omega)|^4}{|n|^4} \tag{13}$$

$$+ 4a_N \sum_{|n| \leq N} \frac{|g_n(\omega)|^2}{|n|^2} - 2a_N^2 \tag{14}$$

where (13) + (14) equals (assuming $\{g_n\}$ L^2-normalized)

$$- 2\left(\sum_{|n|<N} \frac{|g_n(\omega)|^2 - 1}{|n|^2}\right)^2 + \sum_{|n|<N} \frac{|g_n(\omega)|^4}{|n|^4} \tag{15}$$

$$= -2\left(\int : |\phi|^2 :\right)^2 + \sum_{|n|<N} \frac{|g_n(\omega)|^4}{|n|^4}. \tag{16}$$

The expression (12) + (15) converge a.s. for $N \to \infty$. In fact, we need some more precise estimates. Analyzing (12), consider for a given N_0

$$\sum_{\substack{n_1 - n_2 + n_3 - n_4 = 0 \\ n_1 \neq n_2, n_4 \\ \max |n_i| > N_0}} \frac{g_{n_1}(\omega)}{|n_1|} \frac{\overline{g_{n_2}(\omega)}}{|n_2|} \frac{g_{n_3}(\omega)}{|n_3|} \frac{\overline{g_{n_4}(\omega)}}{|n_4|}. \tag{17}$$

This is a sum of products of 4 Gaussians. Thus from hypercontractivity properties related to products of Gaussians, there is equivalence of the $L^2(d\omega)$-norm and the Orlicz norm $L^{\psi_{1/2}}(d\omega)$ with $\psi_{1/2}(\lambda) = e^{c\lambda^{1/2}} - 1$. (In general, for k-products, the corresponding Orliez function is $e^{c\lambda^{2/k}} - 1$).

Since the $\{g_n(\omega)\}$ are independent complex Gaussians, one gets for the $L^2(d\omega)$-norm essentially

$$\left\{ \int_{|x_1| > N_0} (1 + |x_1|)^{-2} (1 + |x_2|)^{-2} (1 + |x_3|)^{-2} (1 + |x_1 - x_2 + x_3|)^{-2} \, dx_1 dx_2 dx_3 \right\}^{1/2}$$
$$\leq N_0^{-1/2}. \tag{18}$$

Consequently

$$\mathbb{P}_\omega[|(17)| > \gamma] < e^{-c\gamma^{1/2} N_0^{1/4}}. \tag{19}$$

Write for the first term in (15)

$$\left(\sum_{|n| < N} \frac{|g_n(\omega)|^2 - 1}{|n|^2} \right)^2 - \left(\sum_{|n| < N_0} \frac{|g_n(\omega)|^2 - 1}{|n|^2} \right)^2 =$$

$$\left(2 \sum_{|n| \leq N_0} \frac{|g_n(\omega)|^2 - 1}{|n|^2} + \sum_{N > |n| > N_0} \frac{|g_n(\omega)|^2 - 1}{|n|^2} \right) \left(\sum_{N_0 < |n| < N} \frac{|g_n(\omega)|^2 - 1}{|n|^2} \right) \tag{20}$$

which norm is at most $C \left(\sum_{|n| > N_0} \frac{1}{|n|^4} \right)^{1/2} \lesssim N_0^{-1}$, since $\int |g_n(\omega)|^2 = 1$.

Hence also

$$\mathbb{P}_\omega[|(20)| > \gamma] < e^{-c\gamma^{1/2} N_0^{1/2}}. \tag{21}$$

Similarly, for the second term in (16),

$$\sum_{N_0 < |n| < N} \frac{|g_n(\omega)|^4}{|n|^4} \tag{22}$$

has norm $< \sum_{|n| > N_0} \frac{1}{|n|^4} < C N_0^{-2}$, thus

$$\mathbb{P}_\omega[|(22)| > \gamma] < e^{-c\gamma^{1/2} N_0}. \tag{23}$$

It follows from the preceding (19), (21), (23) that

$$\rho \left[\left| \int : |\phi_N|^4 : - \int : |\phi_{N_0}|^4 : \right| > \gamma \right] < e^{-c\gamma^{1/2} N_0^{1/4}}. \tag{24}$$

Thus in particular,

$$\int : |\phi_N|^4 : \text{ converges a.s. for } N \to \infty. \tag{25}$$

Moreover, for λ large

$$\rho\left[-\int : |\phi_N|^4 :> \lambda \right]$$

$$\leq \rho\left[-\int : |\phi_N|^4 : + \int : |\phi_{N_0}|^4 :> \lambda - c(\log N_0)^2 \right]$$

$$\leq \rho\left[\left| \int : |\phi_N|^4 : - \int : |\phi_{N_0}|^4 : \right| > 1 \right] < e^{-cN_0^{1/4}} < e^{-e^{c\sqrt{\lambda}}} \tag{26}$$

invoking (11) with $N = N_0$, $\log N_0 \sim \sqrt{\lambda}$ and (24).

Since $\frac{e^{c\sqrt{\lambda}}}{\lambda} \underset{\lambda \to \infty}{\to} \infty$, the second part of Proposition 10 is clear from (25), (26).

The Wick ordering of the nonlinearity leads to a modified Hamiltonian. For the truncated equation, one gets

$$i\dot{u}_N = \frac{\partial H_N}{\partial \overline{u}_N} \tag{27}$$

where

$$H_N(\phi_N) = \int |\nabla \phi_N|^2 + \frac{1}{2} \int : |\phi_N|^4 :$$

$$= \int |\nabla \phi_N|^2 + \frac{1}{2} \int |\phi_N|^4 - 2a_N \int |\phi_N|^2 + a_N^2. \tag{28}$$

Hence

$$\frac{\partial H_N}{\partial \overline{u}_N} = -\Delta u_N + P_N(|u_N|^2 u_N) - 2a_N u_N \tag{29}$$

and (29) becomes

$$i\dot{u}_N + \Delta u_N - P_N(|u_N|^2 u_N) + 2a_N u_N = 0 \tag{30}$$

or equivalently

$$i\dot{u}_N + \Delta u_N + 2(a_N - \int |u_N|^2)u_N - P_N(u_N|u_N|^2 - 2u_N \int |u_N|^2) = 0. \tag{31}$$

From the L^2-conservation, $\int |u_N|^2 = \int |\phi_N|^2 = \sum_{|n|<N} \frac{|g_n(\omega)|^2}{(|n|^2+\rho)^{1/2}}$, hence

$$\int |u_N|^2 - a_n = \int : |\phi_N|^2 := c_N(\phi) \underset{N \to \infty}{\to} c(\phi) \tag{32}$$

a.s. in ϕ. Now the third term in the equation

$$i\dot{u}_N + \Delta u_N - 2c_N u_N - P_N(u_N|u_N|^2 - 2u_N \int |u_N|^2) = 0 \tag{33}$$

may be removed by multiplying with an exponential.

Define

$$v_N = e^{2ic_N t} u_N \tag{34}$$

satisfying the equation

$$i\dot{v}_N + \Delta v_N - P_N\left(v_N |v_N|^2 - 2v_N \int |v_N|^2\right) = 0. \tag{35}$$

Equation (35) appears as truncated version of the NLS

$$i\dot{v} + \Delta v - \left(v|v|^2 - 2v \int |v|^2\right) = 0 \tag{36}$$

obtained by replacement of the $u|u^2$-nonlinearity in the original cubic NLS (1) by $u|u|^2 - 2u \int |u|^2$. We will call (36) the Wick ordered cubic NLS. The measure $e^{-H_N(\phi_N)}\Pi d^2\phi_N = d\mu_N$ is invariant under the flow of (27), (30). Observe that the map

$$\phi_N \rightarrow e^{2i(\int :|\phi_N|^2:)t}\phi_N \tag{37}$$

for each given t is ρ_N- and μ_N-measure preserving, since

$$\int :|\phi_N|^2: := \sum_{|n| \leq N}\left(|\hat{\phi}_N(n)|^2 - \frac{1}{|n|^2 + \rho}\right)$$

depends only on $\{|\hat{\phi}_N(n)|\}$. Thus μ_N is also invariant under the flow of (35).

Hence, one expects (36) to be the evolution equation of the Wick ordered Gibbs-measure $\mu = \lim \mu_N$ given by Proposition 10. The remaining problem is to establish convergence of the solutions of (35) with data $v_N(0) = P_N\phi$, $N \rightarrow \infty$ a.s. in ϕ. The main result concerning the Cauchy problem local in time is the following.

Proposition 38. *The Cauchy problem*

$$\begin{cases} iu_t + \Delta u - (u|u|^2 - 2u \int |u|^2) = 0 \\ u(0) = \phi_\omega = \sum \frac{g_n(\omega)}{|n|} e^{i\langle x, n\rangle} \end{cases} \tag{39}$$

is locally wellposed on a time interval $[0, \tau]$, except for ω in a set of measure at most e^{-1/τ^c} (for some $c > 0$). The solution u is the distributional limit of $\{u_N\}$, where u_N solves the truncated problem

$$\begin{cases} iu_t^N + \Delta u^N - P_N(u^N|u^N|^2 - 2u^N \int |u^N|^2) = 0 \\ u^N(0) = P_N(\phi_\omega). \end{cases} \tag{40}$$

In fact, the difference

$$u - \sum \frac{g_n(\omega)}{|n|} e^{i(\langle x, n\rangle + |n|^2 t)} \tag{41}$$

of the solution of (39) and the solution of the linear problem

$$\begin{cases} iu_t + \Delta u = 0 \\ u(0) = \phi_\omega \end{cases} \tag{42}$$

is in $C_{H^s(\mathbb{T}^2)}[0, \tau]$ for some $\underline{s > 0}$ and the limit in this space of

$$u_N - \sum_{|n| \leq N} \frac{g_n(\omega)}{|n|} e^{i(\langle n, x\rangle + |n|^2 t)}. \tag{43}$$

Once this local result is obtained, one proceeds as in the $1D$ case to get a result global in time, exploiting the invariance of μ_N under the flow of (40). This yields

Theorem 44. *For μ almost all ϕ, the sequence*

$$u_N - \sum_{|n|\leq N} \hat{\phi}(n)\, e^{i(\langle n,x\rangle+|n|^2 t)} \tag{45}$$

where u_N solves the ODE

$$\begin{cases} i\dot{u}_N + \Delta u_N - P_N(u_N|u_N|^2 - 2u_N \int |u_N|^2) = 0 \\ u_N(0) = P_N\phi \end{cases} \tag{46}$$

converges in $C_{H^s(\mathbb{T}^2)}[0,T]$ for all $T < \infty$ ($s > 0$ given by Proposition 38) to

$$u - \sum_{n\in\mathbb{Z}^2} \hat{\phi}(n)\, e^{i(\langle n,x\rangle+|n|^2 t)} \tag{47}$$

and the map $\phi \to u(t)$ preserves $d\mu$ (=Wick ordered Gibbs measure) for all time.

Remark. The invariance of μ means that if $F = F(\phi)$ is a bounded weakly continuous function (i.e. continues for the weak topology on $\{\hat{\phi}(n)\}$), then

$$\int F(\phi)\mu(d\phi) = \int F(S'\phi)\mu(d\phi) \tag{48}$$

where $S'\phi$ is the map $\phi \mapsto u(t)$ for some given t. One has indeed

$$\int F(\phi)\mu(d\phi) = \lim_N \int F(P_N\phi)\mu(d\phi)$$

$$= \lim_N \int F(\phi_N)\mu_N(d\phi_N) \qquad \text{(from Prop. 10)}$$

$$= \lim_N \int F(S'_N\phi_N)\mu_N(d\phi_N) \qquad (\mu_N \text{ is } S'_N \text{ invariant})$$

$$= \lim_N \int F(S'_N\phi_N)\mu(d\phi) \qquad \text{(from Prop. 10)}$$

$$= \int F(S'\phi)\mu(d\phi) \qquad \text{(from Th. 44)}$$

The proof of Proposition 38 is rather lengthy and we reproduce only a few key estimates. Theorem 1.40 as a general Cauchy result does not apply, since $\phi = \phi_\omega$ does not satisfy the required initial data condition for a typical ω. The method will be a mixture of techniques similar to those used in section 2 and probabilistic considerations.

We first set up a fixed point problem. Consider again the integral equation associated to (39)

$$u(t) = S(t)\phi + i \int_0^t S(t-\tau)[(u|u|^2 - 2u \int |u|^2)(\tau)]d\tau \tag{49}$$

where $\phi = \phi_\omega$ and $S(t) = e^{it\Delta}$.

We consider the space $X^{s,\frac{1}{2}}[0,\tau]$ with norm (see section 2 for precise definition)

$$|||u|||_s = \left(\sum_n \int d\lambda (1+|n|^{2s})(1+|\lambda-|n|^2|)|\hat{u}(n,\lambda)|^2 \right)^{1/2} \tag{50}$$

and denote $B = B(0,1)$ the unit ball in X^s.

We show that the map

$$u \mapsto S(t)\phi_\omega + i \int_0^t S(t-\tau)[u|u|^2 - 2u \int |u|^2](\tau)d\tau \tag{51}$$

defines a $||| \ |||_s$-contraction on the set $S(t)\phi_\omega + B$, except for a set of ω's of measure $< e^{-1/\tau^c}$.

Write

$$u|u|^2 - 2u \int |u|^2 = \sum_{n_2 \neq n_1, n_3} \hat{u}(n_1)\overline{\hat{u}(n_2)}\,\hat{u}(n_3)\,e^{i\langle n_1-n_2+n_3,x\rangle} - \sum_n \hat{u}(n)|\hat{u}(n)|^2\,e^{i\langle n,x\rangle} \tag{52}$$

and consider the first term as a trilinear expression

$$\sum_{n_2 \neq n_1, n_3} \hat{u}_1(n_1)\,\overline{\hat{u}_2(n_2)}\,\hat{u}_3(n_3)\,e^{i\langle n_1-n_2+n_3,x\rangle}. \tag{53}$$

(The second term in (52) is trivial to analyze.)

To make estimates on (53), divide the support of each $\hat{u}_i$ in dyadic regions $|n_i| \sim N_i$. Since $u_i \in S(t)\phi_\omega + B$, there are for each $i = 1,2,3$ two possibilities

$$u_i = \frac{1}{N_i} \sum_{|n|\sim N_i} g_n(\omega)\,e^{i(\langle n,x\rangle+t|n|^2)} \tag{54}$$

$$|||u_i|||_s \leq 1. \tag{55}$$

Observe that if $|||u_i|||_s \leq 1 (i = 1,2,3)$, then Theorem 1.40 permits to estimate the nonlinear term in $||| \ |||_s$-norm. It remains to analyze the remaining cases, distinguishing the various size ratios for N_1, N_2, N_3. It turns out that in case (55) we may essentially assume u_i of the form

$$\begin{cases} u_i = \sum_{|n|\sim N_i} a_i(n)\,e^{i(\langle n,x\rangle+|n|^2 t)} \\ \sum n^{2s}|a_i(n)|^2 \leq 1. \end{cases} \tag{56}$$

Denoting (53) by w

$$w(x,t) = \sum_{n_2 \neq n_1, n_3} \hat{u}_1(n_1)\overline{\hat{u}_2(n_2)}\,\hat{u}_3(n_3)\,e^{i\langle n_1-n_2+n_3,x\rangle}$$

we have to estimate

$$\left\| \int_0^t S(t-\tau)w(\tau)d\tau \right\|_s \tag{57}$$

hence (cf. section 2)

$$\left\{ \sum_n \int d\lambda\, \frac{|n|^{2s}|\hat{w}(n,\lambda)|^2}{1+|\lambda-|n|^2|} \right\}^{1/2}. \tag{58}$$

We will consider a few cases.

Assume first all u_i in (53) are of type (54). Thus (58) yields

$$\sum_{m,n} \frac{|n|^{2s}}{1+|m-|n|^2|} \left| \sum_{\substack{n=n_1-n_2+n_3, n_2\neq n_1,n_3 \\ m=|n_1|^2-|n_2|^2+|n_3|^2 \\ |\lambda|\sim N_i}} \frac{g_{n_1}(\omega)}{1+|n_1|} \frac{\overline{g_{n_2}(\omega)}}{1+|n_2|} \frac{g_{n_3}(\omega)}{1+|n_3|} \right|^2. \qquad (59)$$

Since $\{g_n\}$ are independent complex Gaussians, $n_2 \neq n_1, n_3$, integration of (59) in ω yields

$$\sum_{\substack{n=n_1-n_2+n_3, n_2\neq n_1,n_3 \\ |n_i|\sim N_i}} \frac{|n|^{2s}}{1+\left||n_1|^2-|n_2|^2+|n_3|^2-|n|^2\right|} \frac{1}{1+|n_1|^2} \frac{1}{1+|n_2|^2} \frac{1}{1+|n_3|^2}. \qquad (60)$$

Since

$$|n_1-n_2+n_3|^2 - |n_1|^2 + |n_2|^2 - |n_3|^2 = 2\langle n_2-n_1, n_2-n_3\rangle \qquad (61)$$

$$(60) \sim (N_1N_2N_3)^{-2} \sum_{\substack{n=n_1-n_2+n_3, n_2\neq n_1,n_3 \\ |n_i|\sim N_i}} \frac{|n|^{2s}}{|\langle n_2-n_1, n_2-n_3\rangle|+1}. \qquad (62)$$

Case 1. $N_1 \geq N_2, N_3$.

Since for each $\mu \in \mathbb{Z}$, the set

$$S_\mu = \{(n_1, n_2, n_3) \in \mathbb{Z}^2 \times \mathbb{Z}^2 \times \mathbb{Z}^2 \mid |n_i| \sim N_i, \langle n_2-n_1, n_2-n_3\rangle = \mu\} \qquad (63)$$

satisfies

$$\#S_\mu \leq N_2^2 N_3^2 N_1 \qquad (64)$$

(fixing n_2, n_3)

$$(62) < (N_1N_2N_3)^{-2} N_1^{2s} N_2^2 N_3^2 N_1 \sum_{|\mu|<N_1} \frac{1}{1+|\mu|} \ll N_1^{-1+2s+\varepsilon}. \qquad (65)$$

Case 2. $N_2 \geq N_1, N_3$.

Fixing N_1, N_3 and considering lattice points on a circle of radius N_2, we get

$$\#S_\mu \ll N_1^2 N_3^2 N_2^\varepsilon \qquad (66)$$

$$(62) \ll (N_1N_2N_3)^{-2} N_2^{2s} (N_1^2 N_3^2) N_2^\varepsilon \sum_{|\mu|<N_2} \frac{1}{1+|\mu|} \ll N_2^{-2+2s+\varepsilon}. \qquad (67)$$

From (65), (67), it follows that here $s < \frac{1}{2}$ is permitted.

Consider one more case, $n\ell$ $N_1 \geq N_2 \geq N_3$ and

$$\left\{ \begin{array}{ll} u_1 & (54) \\ u_2, u_3 & (56) \end{array} \right. \qquad (68)$$

Expression (58) gives

$$\left\{ \sum_{m,n} \frac{|n|^{2s}}{1+|m-|n|^2|} \left| \sum_{\substack{n=n_1-n_2+n_3, n_2\neq n_1,n_3 \\ m=|n_1|^2-|n_2|^2+|n_3|^2 \\ |n_i|\sim N_i}} \frac{g_{n_1}(\omega)}{1+|n_1|} \overline{a}_{n_2} a_{n_3} \right|^2 \right\}^{1/2}. \qquad (69)$$

Denote

$$b_n = (1 + |n|^s)a_n \tag{70}$$

hence, from (56)

$$\sum |b_n|^2 \le 1. \tag{71}$$

Writing as above $2\mu = -m + |n|^2$, $\{\cdots\}$ in (69) is bounded by (for some $\mu \in \mathbb{Z}$)

$$N_1^{2s-2+\varepsilon} N_2^{-2s} N_3^{-2s} \sum_n \left| \sum_{\substack{n=n_1-n_2+n_3,\, n_2\neq n_1, n_3 \\ \langle n_2-n_1,\, n_2-n_3\rangle=\mu \\ |n_i|\sim N_i}} g_{n_1}(\omega)\bar{b}_{n_2}b_{n_3} \right|^2. \tag{72}$$

We make 2 complementary estimates on (72).

Applying first Hölder's inequality *wrt* $\{b_{n_2}\}$, we get

$$N_1^{2s-2+\varepsilon} N_2^{-2s} N_3^{-2s} \sum_{n,n_2} \left| \sum_{\substack{n=n_1-n_2+n_3,\, n_2\neq n_1, n_3 \\ \langle n_2-n_1,\, n_2-n_3\rangle=\mu}} g_{n_1}(\omega)b_{n_3} \right|^2. \tag{73}$$

For fixed n, n_2, the set

$$\{(n_1, n_3) \subset \mathbb{Z}^2 \times \mathbb{Z}^2 \,|\, n_1+n_3 = n+n_2, \langle n_1-n_2, n_3-n_2\rangle = \mu, |n_1| \le N_1, |n_2| \le N_2\} \tag{74}$$

has at most N_1^ε-elements (reduces to lattice points on an N_1-circle in $\mathbb{Z}^2$).

Hence

$$(73) \ll N_1^{2s-2+\varepsilon} N_2^{-2s} N_3^{-2s} \sum_{\substack{\langle n_2-n_1,\, n_2-n_3\rangle=\mu \\ n_2\neq n_1, n_3}} |b_{n_3}|^2 \tag{75}$$

assuming $|g_n(\omega)| \ll N^\varepsilon$ for $|n| \sim N$.

To estimate (75), fix n_3 and consider

$$\{(n_1, n_2) \in \mathbb{Z}^2 \times \mathbb{Z}^2 \,|\, \langle n_2-n_1, n_2-n_3\rangle = \mu, n_2 \neq n_1, n_3 \text{ and } |n_1| \sim N_1, |n_2| \sim N_2\}. \tag{76}$$

Denote $m_1 = n_1 - n_2$, $m_2 = n_2 - n_3$, hence $|m_i| \le N_i (i = 1, 2)$ and we need to estimate the number of solutions of

$$\langle m_1, m_2\rangle = -\mu (m_i \neq 0). \tag{77}$$

Writing $\langle m_1, m_2\rangle = m_1' m_2'' + m_1' m_2''$, one gets for fixed m_2 a bound by $\frac{N_1}{N_2} \gcd(m_2', m_2'')$. Hence

$$(76) \le \frac{N_1}{N_2} \sum_{|m_2|<N_2} \gcd(m_2', m_2'') < N_1 N_2 \sum_{d<N_2} \frac{1}{d} < N_1 N_2^{1+\varepsilon}. \tag{78}$$

Hence, by (71)

$$(72), (75) \ll N_1^{2s-2+\varepsilon} N_1^{-2s} N_3^{-2s} N_1 N_2^{1+\varepsilon} < N_1^{2s-1+\varepsilon} N_2^{1-2s} N_3^{-2s}. \tag{79}$$

In this estimate, only a bound $|g_n(\omega)| \ll |n|^\varepsilon$ was involved. The next one will exploit orthogonality relations too. Estimate (72) applying Hölder's inequality *wrt* $\{b_{n_3}\}$ to get

$$N_1^{2s-2+\varepsilon} N_2^{-2s} N_3^{-2s} \sum_{n,n_3} \left| \sum_{\substack{n=n_1-n_2+n_3,\, n_2 \neq n_1, n_3 \\ \langle n_2-n_1,\, n_2-n_3 \rangle = \mu}} g_{n_1}(\omega) b_{n_2} \right|^2. \tag{80}$$

Fix n_3, $|n_3| \leq N_3$. We estimate

$$\left\{ \sum_{n \neq n_3} \left| \sum_{\substack{n_2 \neq n_3,\, \langle n_3-n,\, n_2-n_3 \rangle = \mu}} g_{n+n_2-n_3}(\omega) b_{n_2} \right|^2 \right\}^{1/2} = \|\mathfrak{S}\{b_n\}\| \tag{81}$$

where the matrix $\mathfrak{S} = \mathfrak{C}_\omega$ is defined as follows

$$\sigma_{n,n_2} = \begin{cases} g_{n+n_2-n_3}(\omega) & \text{if } n_2 \neq n_3,\, \langle n_3-n,\, n_2-n_3 \rangle = \mu \\ 0 & \text{otherwise.} \end{cases} \tag{82}$$

Estimate

$$\|\mathfrak{S}\|^2 \leq \|\mathfrak{S}\mathfrak{S}^*\| \leq \max_n \left(\sum_{n_2} |\sigma_{nn_2}|^2 \right) + \left(\sum_{n \neq n'} \left| \sum_{n_2} \sigma_{nn_2} \overline{\sigma}_{n',n_2} \right|^2 \right)^{1/2}. \tag{83}$$

The first term in (83) is bounded by $N_1^\varepsilon N_2$. Write the second down explicitly

$$\sum_{n \neq n'} \left| \sum_{\substack{n_2,\, n_2 \neq n_3 \\ \langle n_3-n,\, n_2-n_3 \rangle = \mu \\ \langle n_3-n',\, n_2-n_3 \rangle = \mu}} g_{n+n_2-n_3}(\omega) \overline{g}_{n'+n_2-n_3}(\omega) \right|^2. \tag{84}$$

In the innersum any Gaussian product (of 2 distinct Gaussians) is at most repeated twice. Hence, from orthogonality considerations, one expects for generic ω the bound

$$(84) \ll N_1^\varepsilon \cdot (\#S) \tag{85}$$

where

$$\begin{aligned} S = \{(n, n', n_2) \,|\, & n \neq n', \, n \neq n_3, \, n' \neq n_3, \, n_2 \neq n_3, \, \langle n_3 - n,\, n_2 - n_3 \rangle \\ & = \langle n_3 - n',\, n_2 - n_3 \rangle = \mu \}. \end{aligned} \tag{86}$$

For fixed n_3, since $|n| \lesssim N_1$,

$$\langle n_3 - n,\, n_2 - n_3 \rangle = \mu \tag{87}$$

has from preceding estimate on the number of solutions of (77) at most $N_1^{1+\varepsilon} N_2$ solutions in (n, n_2). Since $n_2 \neq n_3$, there are at most N_1 possible choices for n' satisfying

$$\langle n_3 - n',\, n_2 - n_3 \rangle = \mu. \tag{88}$$

Hence

$$\#S \ll N_1^{2+\varepsilon} N_2 \tag{89}$$

and therefore also

$$(84) \ll N_1^{2+\varepsilon} N_2. \tag{90}$$

Hence, the second term in (83) is bounded by $N_1^{1+\varepsilon} N_2^{1/2}$. From the preceding

$$(83) \ll N_1^{1+\varepsilon} N_2^{1/2} \tag{91}$$

$$\|\mathfrak{S}\| \ll N_1^{\frac{1}{2}+\varepsilon} N_2^{1/2} \tag{92}$$

$$(81) \ll N_1^{\frac{1}{2}+\varepsilon} N_2^{1/2}. \tag{93}$$

Thus

$$(72),(80) \ll N_1^{2s-2+\varepsilon} N_2^{-2s} N_3^{-2s} N_3^2 N_1^{1+\varepsilon} N_2^{1/2} = N_1^{2s-1+\varepsilon} N_2^{\frac{1}{2}-2s} N_3^{2-2s} \tag{94}$$

and combined with (79), we get for "generic" ω a bound $(N_1 \geq N_2 \geq N_3)$

$$(72) \ll \min(N_1^{2s-1+\varepsilon} N_2^{1-2s} N_3^{-2s}, \ N_1^{2s-1+\varepsilon} N_2^{\frac{1}{2}-2s} N_3^{2-2s})$$

$$= \min\left(N_1^\varepsilon \left(\frac{N_1}{N_2}\right)^{2s-1} N_3^{-2s}, N_1^{-\frac{1}{2}+\varepsilon} \left(\frac{N_1}{N_2}\right)^{2s-\frac{1}{2}} N_3^{2-2s}\right) \tag{95}$$

$$\ll N_1^{-s/2} \tag{96}$$

(assuming $s < \frac{1}{4}$).

The contribution (69) to (58), thus assuming (68), is at most $N_1^{-s/4}$.

The analysis of the other cases for $\{u_i\}_{i=1,2,3}$ distinguishing the 2 cases (54), (56) and size ratio's for N_1, N_2, N_3 is much in the same flavor of the preceding. The conclusion is that one may find $s > 0^{(*)}$ such that

$$(57) = \left\| \int_0^t S(t-\tau)w(\tau)d\tau \right\|_s = o(1)$$

for $u_i \in S(t)\phi_\omega + B$, for typical ω. More precisely

$$\left\| \int_0^t S(t-\tau)w(\tau)d\tau \right\|_{X^{s,\frac{1}{2}}[0,\tau]} < \tau^{c_1} \tag{97}$$

except for ω in a set of measure $< e^{-1/\tau^{c_2}} (c_1, c_2 > 0)$. Throughout the argument one encounters indeed expressions such as (59), (84) and previous estimate results from distributional properties for sums of products of independent Gaussians.

Hence also

$$\left\| \int_0^t S(t-\tau)[u|u|^2 - 2u \int |u|^2](\tau)d\tau \right\|_{X^{s,\frac{1}{2}}[0,\tau]} < \tau^{c_1} \tag{98}$$

for $u \in S(t)\phi_\omega + B$, except for ω in a set of measure $< e^{-1/\tau^{c_2}}$. Similarly, one shows that (51) is a contraction on $S(t)\phi_\omega + B$. This yields Proposition 38.

$(*)s$ has to be sufficiently small

Remarks.

(i) Denoting $S'(t)\phi$ the solution to the NLS

$$\begin{cases} iu_t + \Delta u - (u|u|^2 - 2u \int |u|^2) = 0 \\ u(0) = \phi \end{cases} \tag{99}$$

and $S(t)\phi = \sum \hat\phi(n) \, e^{i(\langle n,x\rangle + |n|^2 t)}$ the solution to the linear equation, the statement of Theorem 44 is that $u - S(t)\phi \in C_{H^s(\mathbb{T}^2)}(\mathbb{R})$ for some fixed $s > 0$ and almost all $\phi = \phi_\omega$. This property reminds of "scattering" occurring in certain dispersive models.

(ii) In the $2D$-case, the L^2-truncation argument used in $1D$ for normalization in the focusing case, does not apply, since in $2D$

$$\|\phi_\omega^N\|_2 = \left(\sum_{|n| \leq N} \frac{|g_n(\omega)^2|}{\rho + |n|^2} \right)^{1/2} \xrightarrow[N \to \infty]{} \infty \tag{100}$$

almost surely.

(iii) Consider a $2D$-nonlinear wave equation

$$u_{tt} - \Delta u + \rho u + f'(u) = 0 \qquad (u \text{ real}, \ \rho > 0) \tag{101}$$

written in Hamiltonian form as (u, v are conjugate variables)

$$\begin{cases} u_t = -v \\ v_t = -\Delta u + \rho u + f'(u) \end{cases} \tag{102}$$

with Hamiltonian

$$H(u,v) = \int \left(\frac{1}{2}|Bu|^2 + \frac{1}{2}v^2 + f(u) \right) \qquad B^2 = -\Delta + \rho. \tag{103}$$

One may prove in the defocusing case, say

$$f(u) = u^4 \tag{104}$$

a result similar to the one described in this section for the defocusing cubic $2D$ NLS

$$iu_t - \Delta u + \rho u + |u|^2 u = 0. \tag{105}$$

In the real case, one has by (8)

$$\int : \phi_N^4 := \int \phi_N^4 - 6a_N \int \phi_N^2 + 3a_N^2 \tag{106}$$

where $a_N = \sum_{|n| \leq N} \frac{1}{|n|^2 + \rho}$.

The truncated equation

$$\ddot{u}_n - \Delta u_N + \rho u_N + P_N(u_N^3) = 0 \tag{107}$$

is modified according to the Wick ordering (106) and we get

$$\ddot{u}_N - \Delta u_N + \rho u_N + P_N(u_N^3) - 3a_N u_N = 0. \tag{108}$$

Observe that in the NLW-case, we don't have the L^2-conservation as for NLS (105). For this reason, we are unable to replace (108) by truncations of the equation

$\ddot{u} - \Delta u + \rho u + (u^3 - 3u \int u^2) = 0$. One gets however an analogue of Theorem 44. The invariant Gibbs measure is given by

$$\mu(d\phi d\phi_1) = e^{-\frac{1}{4} \int :\phi^4:} \rho(d\phi) \otimes \rho_1(d\phi_1) \tag{109}$$

where ρ_1 is the image measure corresponding to random Fourier series

$$\phi_{1,\omega} = \sum_{n \in \mathbb{Z}^2} g_n(\omega)\, e^{i\langle n, x\rangle} \qquad g_{-n} = \overline{g}_n. \tag{110}$$

Theorem 111. *Consider the IVP*

$$\begin{cases} \ddot{u}_N - \Delta u_N + \rho u_N + P_N(u_N^3) - 3a_N u_N = 0 \\ u_N(0) = P_N \phi \\ \dot{u}_N(0) = P_N \phi_1. \end{cases} \tag{112}$$

Then for μ almost all (ϕ, ϕ_1), the difference

$$\{u_N(t) - P_N\big(S(t)(\phi,\phi_1)\big),\ \dot{u}_N(t) - P_N\big(\dot{S}(t)(\phi,\phi_1)\big)\} \tag{113}$$

where $w = S(t)(\phi, \phi_1)$ solves the linear problem

$$\begin{cases} \ddot{w} - \Delta w + \rho w = 0 \\ w(0) = \phi \\ \dot{w}(0) = \phi_1 \end{cases} \tag{114}$$

converges in $C_{H^s \times H^{s-1}(\mathbb{T}^2)}(\mathbb{R})$ for some fixed $s > 0$.
   Denoting

$$\big(u(t), \dot{u}(t)\big) = \big(S(t)(\phi,\phi_1), \dot{S}(t)(\phi,\phi_1)\big) + \lim_N (113) \tag{115}$$

the measure $d\mu$ defined in (109) is invariant under the map

$$(\phi, \phi_1) \mapsto \big(u(t), \dot{u}(t)\big)$$

for all time t.

   Recall that

$$\begin{aligned} S(t)(\phi,\phi_1) = \sum_n \Bigg\{ &\frac{1}{2}\big(\hat{\phi}(n) + \frac{i}{(|n|^2 + \rho)^{1/2}} \hat{\phi}_1(n)\big)\, e^{i(\langle x, n\rangle + t(|n|^2 + \rho)^{1/2})} \\ &+ \frac{1}{2}\big(\hat{\phi}(n) - \frac{i}{(|n|^2 + \rho)^{1/2}} \hat{\phi}_1(n)\big)\, e^{i(\langle x, n\rangle - t(|n|^2 + \rho)^{1/2})} \Bigg\}. \end{aligned} \tag{116}$$

The local wellposedness problem in the case of the wave equation is studied using Fourier restriction norms

$$\|u\|_{s,b} = \left(\sum_n (1 + |n|^{2s})\, (1 + ||\lambda| - |n||)^{2b}\, |\hat{u}(n, \lambda)|^2 \right)^{1/2}. \tag{117}$$

Details are easier then for the Schrödinger case because of the gain of 1-derivative for the operator $\Box^{-1}$, where $w = \Box^{-1}\phi$ solves

$$\begin{cases} \dot{w} - \Delta w + \rho w = \phi \\ w(0) = 0 \\ \dot{w}(0) = 0. \end{cases} \tag{118}$$

In fact, one has the analogue of Theorem 111 of any $f(u) = u^{2k}+$ (lower order) instead of u^4.

(iv) Related to remark (ii), one may consider truncations of the Wick ordered L^2-norm $\int : |\phi_N|^2 :$ rather than the L^2-norm (which is a.s. bounded). A. Jaffe observed that

$$e^{\int :\phi_N^3:}\chi_{[\int :\phi_N^2:\,\leq B]}\rho(d\phi) \tag{119}$$

converges to a weighted Wiener measure for any $B < \infty$.

This property just fails for $\int : |\phi_N|^4 :$ and hence does not yield a substitute for the L^2-truncation of $2D$ focusing cubic NLS (Wick ordering requires to consider for $F(|u|)$ even powers of $|u|$). The problem to consider (119) as invariant measure for a $2D$ wave equation is the lack of conservation of L^2-norm.

One may however modify (119) to

$$e^{\int :\phi_N^3:-A\left(\int :\phi_N^2:\right)^2}\rho(d\phi) \tag{120}$$

where $A > 0$ is a sufficiently large constant. Then the same property holds, i.e. (120) converges to a weighted Wiener measure (the subtraction of $A(\int : \phi_N^2 :)^2$ is optimal, up to the value of A). One may thus restate Theorem 111 for the truncated NLW equations

$$\ddot{u}_N - \Delta u_N + \rho u_N + u_N^2 - a_N - \frac{4}{3}A\left(\int u_N^2 - a_N\right)u_N = 0 \tag{121}$$

with

$$e^{\int :\phi^3:-A(\int :\phi^2:)^2}\rho(d\phi) \tag{122}$$

as invariant measure.

We include the proof of the (120)-convergence.

Since

$$: \phi_N^3 := \phi_N^3 - 3a_N\phi_N \quad \text{with} \quad a_N = \sum_{|n|\leq N}\frac{1}{|n|^2 + \rho} \tag{123}$$

we get for

$$\phi_N = \sum_n \frac{g_n(\omega)}{(|n|^2 + \rho)^{1/2}}\,e^{i\langle n,x\rangle}, \quad \bar{g}_n = g_{-n} \tag{124}$$

$$\int : \phi_N^3 := -2\frac{g_0(\omega)^3}{\rho^3} + 3\frac{g_0(\omega)}{\rho}\int : \phi_N^2 : + \sum_{\substack{n_1+n_2+n_3=0 \\ n_i\neq 0}}\frac{g_{n_1}}{|n_1|}\frac{g_{n_2}}{|n_2|}\frac{g_{n_3}}{|n_3|}. \tag{125}$$

Define for $\lambda, K > 1$

$$\Omega_{\lambda,K} = \left\{ \omega \,\middle|\, \left| \sum_{n_1,n_2,n_1+n_2\neq 0} \frac{g_{n_1} g_{n_1} g_{-n_1-n_2}}{|n_1|\,|n_2|\,|n_1+n_2|} \right| + |g_0|^3 \sim \lambda, \left| \int : \phi_N^2 : \right| \sim K \right\}.$$

$$(126)$$

We show that

$$\mathrm{mes}\,\Omega_{\lambda,K} < e^{-C_1\lambda} + e^{-c\frac{\lambda^2}{K^2}+C(\log\lambda)^2} \tag{127}$$

where C_1 may be chosen arbitrarily large. Consequently

$$\sum_{\lambda,K\ \mathrm{dyadic}} e^{\lambda - AK^2}\,\mathrm{mes}(\Omega_{\lambda,K}) < \infty. \tag{128}$$

The argument will also yield the convergence of (120) for $N \to \infty$ (stability)

Fix $N_1 = N_1(\lambda)$ and consider the set

$$\Omega' = \left\{ \omega \,\middle|\, \left| \sum_{\substack{n_1,n_2,n_1+n_2\neq 0 \\ |n_1|\vee|n_2|>N_1}} \frac{g_{n_1} g_{n_2} g_{-n_1-n_2}}{|n_1|\,|n_2|\,|n_1+n_2|} \right| > \lambda \right\}. \tag{129}$$

It follows then from hypercontractivity that

$$\mathrm{mes}(\Omega') < \exp -c\left(\lambda \frac{N_1}{\log N_1}\right)^{2/3} < e^{-c_1\lambda} \tag{130}$$

if $N_1(\lambda) > \lambda$.

Also, letting

$$\Omega'' = \left\{ \omega \,\middle|\, \left| \sum_{|n|>N_1} \frac{|g_N(\omega)|^2 - 1}{|n|^2} \right| > 1 \right\} \tag{131}$$

we have

$$\mathrm{meas}(\Omega'') < \exp(-cN_1) < e^{-c_1\lambda} \tag{132}$$

for $N_1(\lambda) \gtrsim \lambda$. If $\omega \notin \Omega''$, $\omega \in \Omega_{\lambda,K}$

$$\sum_{|n|\leq N_1} \frac{|g_n(\omega)|^2}{|n|^2+\rho} \leq \sum_{|n|\leq N_1} \frac{|g_n(\omega)|^2 - 1}{|n|^2+\rho} + \log N_1 \leq K + 1 + c\log\lambda. \tag{133}$$

It follows from the preceding that in order to establish (127), we may assume $|n_1| \vee |n_2| < N_1 = N_1(\lambda) \sim \lambda$ in (126) and

$$\sum_{|n|\leq N_1} \frac{|g_n(\omega)|^2}{|n|^2+\rho} < L \equiv K + 1 + c\log\lambda. \tag{134}$$

Hence in particular we may assume $|g_0|^3 < c(K + \log\lambda)^{3/2}$ and this term may thus be ignored in (126). We now come to the main estimate.

Define dyadic regions

$$R_j = \{n \in \mathbb{Z}^2 \,|\, 2^j \leq |n| < 2^{j+1}\}. \tag{135}$$

Consider

$$\left| \sum_{|n_i|<N} \frac{g_{n_1} g_{n_2} g_{-n_1-n_2}}{|n_1|\,|n_2|\,|n_1+n_2|} \right|. \tag{136}$$

Let $|n_1| \leq |n_2|$, $n_1 \in R_{j_1}$, $n_2 \in R_{j_2}$. If $j_1 < j_2 - 1$, then $|n_1 + n_2| \cong |n_2|$, hence $n_1 + n_2 \in R_{j_2} \cup R_{j_2-1} \cup R_{j_2+1} \equiv \tilde{R}_{j_2}$. Let $J \sim \log N_1 \sim \log \lambda$. It follows that (136) may essentially be estimated by

$$\left| \sum_{j \leq J} \int \left\{ \sum_{\substack{n_1 \in \cup R_{j'} \\ j' \leq j}} \frac{g_{n_1}}{|n_1|} e^{i\langle n_1, x\rangle} \right\} \left\{ \sum_{n_2 \in R_j} \frac{g_{n_2}}{|n_2|} e^{i\langle n_2, x\rangle} \right\} \left\{ \sum_{n \in \tilde{R}_j} \frac{g_n}{|n|} e^{i\langle n, x\rangle} \right\} dx \right|$$

$$\leq \sum_{j \leq J} \left\| \sum_{\substack{n \in \cup R_{j'} \\ j' \leq j}} \frac{g_n}{|n|} e^{i\langle n, x\rangle} \right\|_2 \left\| \sum_{n \in R_j} \frac{g_n}{|n|} e^{i\langle n, x\rangle} \right\|_2 \left\| \sum_{n \in \tilde{R}_j} \frac{g_n}{|n|} e^{i\langle n, x\rangle} \right\|_\infty \tag{137}$$

$$\leq \sum_{j \leq J} L^{1/2} L_j^{1/2} \left\| \sum_{n \in \tilde{R}_j} \frac{g_n}{|n|} e^{i\langle n, x\rangle} \right\|_\infty \tag{138}$$

where

$$L_j = \sum_{n \in R_j} \frac{|g_n(\omega)|^2}{|n|^2 + \rho} \tag{139}$$

hence

$$L = \sum L_j. \tag{140}$$

Fix $\{L_j\}_{j=1,\ldots,J}$ (the number of such systems may clearly be bounded by $N_1^{\log N_1}$).
Estimate

$$\text{meas} \left[\sum_{j \leq J} \left(\frac{L_j}{L} \right)^{1/2} \left\| \sum_{n \in R_j} \frac{g_n}{|n|} e^{i\langle n, x\rangle} \right\|_\infty > \frac{\lambda}{L} \right]. \tag{141}$$

To evaluate $\left\| \sum_{n \in R_j} \frac{g_n}{|n|} e^{i\langle n, x\rangle} \right\|_\infty$, consider a set $\mathcal{F}_j$ of $\sim 4^j$ points in $\mathbb{T}^2$ such that for all $\{a_n | n \in R_j\}$

$$\left\| \sum_{n \in R_j} a_n e^{i\langle \cdot, x\rangle} \right\|_\infty \sim \max_{x \in \mathcal{F}_j} \left| \sum_{n \in R_j} a_n e^{i\langle n, x\rangle} \right|. \tag{142}$$

Hence

$$(141) < \sum_{\{x_j\} \in \Pi_{j \leq J} \mathcal{F}_j} \sum_{\{\varepsilon_j\} \in \{1,-1\}^J} \text{meas} \left[\left| \sum_{j \leq J} \left(\frac{L_j}{L} \right)^{1/2} \varepsilon_j \left(\sum_{n \in R_j} \frac{g_n}{|n|} e^{i\langle n, x\rangle} \right) \right| > \frac{\lambda}{L} \right]. \tag{143}$$

Since by (140)

$$\sum_{j \leq J} \frac{L_j}{L} \sum_{n \in R_j} \frac{1}{|n|^2} = O(1) \tag{144}$$

$$(143) < \sum_{\{x_j\} \in \Pi \mathcal{F}_j} \sum_{\{\varepsilon_j\} \in \{1-1\}^J} \exp\left(-\frac{\lambda^2}{L^2}\right) \tag{145}$$

$$< \Pi_{j \leq J}(\#\mathcal{F}_j) \, 2^j \cdot \exp\left(-\frac{\lambda^2}{L^2}\right)$$

$$< 10^{J^2} \exp\left(-\frac{\lambda^2}{L^2}\right)$$

$$< \exp\left(C(\log \lambda)^2 - c\frac{\lambda^2}{K^2 + (\log \lambda)^2}\right). \tag{146}$$

Inequality (127) easily follows.

Comments and references related to Lecture 5

In addition to those mentioned in lecture 4, there are the following papers relating to $2D$ problems.

[B7]   J. Bourgain, *Invariant measures for the $2D$-defocusing nonlinear Schrödinger equation*, CMP 176 (1996), 421-445.

This paper deals with the invariant measure construction for the equation $iu_t + \Delta u - u|u|^2 = 0$ in $2D$ (after Wick-ordering). Some parts of the argument are reproduced here. For the general theory of normalization of the Gibbs-measure by Wick ordering in $2D$, a standard reference is

[G-J]   J. Glimm, A. Jaffe, *Quantum Physics*, Springer-Verlag (1987).

Discussion of invariant measures for $2D$ nonlinear wave equations (the defocusing and ϕ^3-case) will appear in a forthcoming paper.

[B-J-W] J. Bourgain, A. Jaffe, W. Wang, *Invariant Gibbs measures for the $2D$ nonlinear wave equation*, preprint 95.

LECTURE 6
Quasi-Periodic Solutions of Hamiltonian PDE

Introduction

The subject of the next lectures is the problem of persistency of quasi-periodic solutions after (Hamiltonian) perturbation.

Consider the equation

$$iu_t + Au + \varepsilon \, \frac{\partial H}{\partial \bar{u}} = 0 \tag{1}$$

where A is a selfadjoint operator and $H = H(u, \bar{u})$ is polynomial (or real analytic). We assume that A is diagonal in an eigenfunction basis which is "well localized" *wrt* the exponentials.

Examples.

(i) $A = -\Delta + M_\sigma \quad$ where $\quad M_\sigma \, e^{i\langle n,x\rangle} = \sigma_n e^{i\langle n,x\rangle}, \ \sigma_n \in \mathbb{R}$
 obtained by adding to the Laplacian a Fourier multiplier M_σ. In this case, the eigenfunction basis are the exponentials and the eigenvalues

$$\lambda_n = |n|^2 + \sigma_n. \tag{2}$$

We assume

$$\lim_{|n|\to\infty} \sigma_n = 0 \tag{3}$$

or

$$\lim_{|n|\to\infty} |\sigma_{n_1} - \sigma_{n_2}| = 0. \tag{4}$$

(ii) Consider the 1D Sturm Liouville operator

$$A = -\frac{d^2}{dx^2} + V(x) \tag{5}$$

where V is a real analytic 1-periodic potential
 The periodic spectrum of A is a sequence

$$\lambda_0 < \lambda_1 \leq \lambda_2 < \lambda_3 \leq \lambda_4 < \ldots < \lambda_{2n-1} \leq \lambda_{2n} < \ldots \tag{6}$$

which satisfies

$$\lambda_{2n-1}, \lambda_{2n} = \pi^2 n^2 + \int V \, dx + 0(n^{-2}) \tag{7}$$

69

$$\lambda_{2n} - \lambda_{2n-1} \to 0 \quad \text{rapidly.} \tag{8}$$

The corresponding eigenfunctions $\varphi_{2n-1}, \varphi_{2n}$ are periodic or antiperiodic according to the parity of n. Expanding φ_j in a Fourier series

$$\varphi_j(x) = \sum_m \widehat{\varphi}_j(m)\, e^{i\pi mx}. \tag{9}$$

There is the localization property *wrt* exponentials

$$|\widehat{\varphi}_j(m)| < C\, e^{-c|\,|m|-j/2\,|} \tag{10}$$

for some constants $c > 0$, $C < \infty$ depending on V.

Conversely, one has an expansion

$$e^{i\pi mx} = \sum_j \hat{e}_m(j)\, \varphi_j(x) \tag{11}$$

with again

$$|\hat{e}_m(j)| < C\, e^{-c|\,|m|-j/2\,|}. \tag{12}$$

Assume V even i.e. $V(x) = V(-x)$. Consider the spectrum of $A = -\frac{d^2}{dx^2} + V$ subject to Dirichlet boundary conditions on $[0,1]$. This gives a sequence $\{\mu_n\}$ interlacing the periodic spectrum

$$\lambda_0 < \lambda_1 \leq \mu_1 \leq \lambda_2 < \ldots < \lambda_{2n-1} \leq \mu_n \leq \lambda_{2n} < \ldots \tag{13}$$

and the corresponding eigenfunctions $\{\psi_n\}$ form a basis for the 2-periodic odd functions. Thus

$$\psi_n(x) = \sum_{m=1}^{\infty} \widehat{\psi}_n(m)\, \sin \pi mx \tag{14}$$

$$\sin \pi nx = \sum_{m=1}^{\infty} \widehat{s}_n(m)\, \psi_m(x) \tag{15}$$

where again

$$|\widehat{\psi}_n(m)|, \; |\widehat{s}_n(m)| < C\, e^{-c|n-m|}. \tag{16}$$

(iii) The analogue of the theory in 1D described above fails in higher dimension. One may however, consider the special case of a potential of the form

$$V(x) = V_1(x_1) + \cdots + V_d(x_d). \tag{17}$$

For periodic boundary conditions, one gets for eigenvalues and eigenfunctions

$$\{\lambda_{j_1}^1 + \cdots + \lambda_{j_d}^d, \; \varphi_{j_1}^1(x_1) \ldots \varphi_{j_d}^d(x_d)\}. \tag{18}$$

In the Dirichlet case, one gets

$$\{\mu_{n_1}^1 + \cdots + \mu_{n_d}^d, \; \psi_{n_1}^1(x_1) \ldots \psi_{n_d}^d(x_d)\} \tag{19}$$

where $\{\lambda_j; \varphi_j\}$ and $\{\mu_n; \psi_n\}$ are as above.

(iv)

$$A = (-\Delta + \rho)^{1/2} \tag{20}$$

with eigenvalues

$$\lambda_n = (|n|^2 + \rho)^{1/2} \tag{21}$$

and the exponentials as eigenfunctions.

Observe that the sequence $\{(k + \rho)^{1/2} \mid k \in \mathbb{Z}_+\}$ consists of rationally independent numbers for typical ρ.

Coming back to 1D, denote $\{\varphi_n, \mu_n\}$ the eigenfunctions and corresponding eigenvalues for A, i.e.

$$A\varphi_n = \mu_n \varphi_n.$$

Fix some specific indices $n_1, n_2, \ldots, n_b$ and denote $\lambda_j = \mu_{n_j}$ $(j = 1, \ldots, b)$. Then

$$u_0(x, t) = \sum_{j=1}^{b} a_j \, e^{i\lambda_j t} \, \varphi_{n_j}(x) \tag{22}$$

yields a solution of the linear equation

$$iu_t^0 + Au^0 = 0. \tag{23}$$

This solution is quasi-periodic and corresponds to a flow on a b-torus $\mathbb{T}^b = \mathcal{T}_0\{|a_j|\}$. The problem we study is which of the solutions (22) "persist" for the perturbed equation

$$iu_t + Au + \varepsilon \, \frac{\partial H}{\partial \bar{u}} = 0. \tag{24}$$

The meaning of persistency is as follows. One has a quasi-periodic solution

$$u_\varepsilon(x, t) = \sum_{n,k} \hat{u}_\varepsilon(n, k) \, e^{i\langle \lambda', k \rangle t} \, \varphi_n(x) \tag{25}$$

of (24), where k is a $\mathbb{Z}^b$-index

$$\hat{u}_\varepsilon(n_j, e_j) = a_j \qquad (e_j = j^{\text{th}} \text{ unit vector}) \tag{26}$$

$$\sum_{(n,k) \notin \mathcal{S}} \omega(n, k) \, |\hat{u}_\varepsilon(n, k)| < \sqrt{\varepsilon} \tag{27}$$

$$|\lambda' - \lambda| < C \varepsilon. \tag{28}$$

In (27),

$$\mathcal{S} = \{(n_j, e_j) \mid j = 1, \ldots, b\} \tag{29}$$

in the "resonant set".

We let ω be a weight function. For instance

$$\omega(n, k) = e^{c(|n|+|k|)} \tag{30}$$

corresponds to analytic solutions. We will mostly consider weights of the form

$$\omega(n, k) = e^{(|n|+|k|)^c} \tag{31}$$

for some $c > 0$ (because they lead to some technical simplifications later on). There is dependence of the new frequency λ' on $\{a_j\}$ and the perturbation $\varepsilon \, \frac{\partial H}{\partial \bar{u}}$. In fact, assuming $\lambda'_1, \ldots, \lambda'_b$ rationally independent, a time shift permits to assume

$a_j (1 \leq j \leq b)$ real and hence λ' depends only on $\{|a_j|\}$. The solution (25) corresponds by (27) to a perturbed torus $\mathcal{T}_\varepsilon\{|a_j|\}$ of $\mathcal{T}_0\{|a_j|\}$ in the phase space.

Consider $\lambda = (\lambda_1, \ldots, \lambda_b)$ as a parameter taken in a parameter set Λ. Persistency of quasi-periodic solutions (22) after the Hamiltonian perturbation as described above will occur for λ in a large subset of the frequency set. More precisely, the persistency will hold provided

$$\lambda \notin \Lambda'_\varepsilon(|a_j| \ (1 \leq j \leq b), \ \text{perturbation}) \tag{32}$$

where

$$\operatorname{mes} \Lambda'_\varepsilon \xrightarrow{\varepsilon \to 0} 0. \tag{33}$$

The preceding deals with perturbations of linear equations. In order to obtain families of quasi-periodic solutions with b frequencies, one needs to consider a parameter dependent linear equation with b-parameters $p = (p_1, \ldots, p_b)$, such that if $\{\lambda_n(p)\}$ are the eigenvalues of $A(p)$, then

$$\det \left(\frac{\partial \lambda_{n_j}}{\partial p_k} \right)_{1 \leq j, \, k \leq b} \neq 0. \tag{34}$$

In the previous examples, such parameters are obtained via the Fourier multiplier M_σ or by letting V range in some b-manifold of potentials. Recall here that if

$$A(V) = -\frac{d^2}{dx^2} + V(x) \tag{35}$$

then from first order variation

$$\frac{\partial \lambda}{\partial V} = |\varphi|^2 \tag{36}$$

where φ is the eigenfunction with eigenvalue λ. The use of parameter dependent equations may be avoided in various ways.
First, writing

$$\lambda' = \lambda + \varepsilon(\lambda, a) \tag{37}$$

and assuming initial nonresonance conditions (1^c Melnikov condition)

$$\sum_{j=1}^{b} k_j \lambda_j + \mu_n \neq 0 \quad (n \notin \{n, \ldots, n_b\}) \tag{38}$$

(expressing nonresonance of the normal frequencies μ_n with ($\mathbb{Z}$-combinations of) the tangential frequencies $\lambda_1, \ldots, \lambda_b$), conditions appearing later on in the process may be ensured by restricting $|a|$ to an appropriate Cantor set. This requires the dependence (37) of λ' on $|a|$ to be sufficiently nondegenerate (*amplitude-frequency modulation*).

Secondly, one may sometimes extract parameters out of the nonlinearity by putting the Hamiltonian

$$H(\phi) = \langle A\phi, \phi \rangle + \varepsilon H_1(\phi) \tag{39}$$

in an appropriate *Birkhoff normal form*.

Assume that one may write (1) in the form

$$\begin{cases} i\dot{q}_{n_j} + \lambda_j q_{n_j} + \varepsilon\, q_{n_j}\, |q_{n_j}|^2 + \varepsilon\delta\, \frac{\partial H_2}{\partial \bar{q}_{n_j}} = 0 & (j = 1, \ldots, b) \\[2mm] i\dot{q}_n + \mu_n q_n + \varepsilon\delta\, \frac{\partial H_2}{\partial \bar{q}_n} = 0 & (n \neq n_1, \ldots, n_b) \end{cases} \tag{40}$$

where

$$u(x,t) = \sum_n q_n(t)\varphi_n(x).$$

Define for $j = 1, \ldots, b$

$$I_j = |q_{n_j}|^2 \tag{41}$$

(action variables). From (40)

$$\dot{I}_j = 2\,\mathrm{Re}\,\dot{q}_{n_j}\bar{q}_{n_j}, \quad i\dot{I}_j = \varepsilon\delta\,\left(\frac{\partial H_2}{\partial q_{n_j}}\,q_{n_j} - \frac{\partial H_2}{\partial \bar{q}_{n_j}}\,\bar{q}_{n_j}\right). \tag{42}$$

Fix some values $\{I_j^0\}_{j=1}^b$ and write

$$I_j = I_j^0 + \sqrt{\delta}\, I_j'. \tag{43}$$

From (40), (42), (43) we get the system

$$\begin{cases} i\dot{q}_{n_j} + (\lambda_j + \varepsilon\, I_j^0)\, q_{n_j} + \varepsilon\sqrt{\delta}\, I_j'\, q_{n_j} + \varepsilon\delta\, \frac{\partial H_2}{\partial \bar{q}_{n_j}} = 0 & (j = 1, \ldots, b) \\[2mm] i\dot{I}_j' - \varepsilon\sqrt{\delta}\,\left(\frac{\partial H_2}{\partial q_{n_j}}\,q_{n_j} - \frac{\partial H_2}{\partial \bar{q}_{n_j}}\,\bar{q}_{n_j}\right) = 0 & (j = 1, \ldots, b) \\[2mm] i\dot{q}_n + \mu_n q_n + \varepsilon\delta\, \frac{\partial H_2}{\partial \bar{q}_n} = 0 & (n \neq n_1, \ldots, n_b) \end{cases} \tag{44}$$

which appears as $(\varepsilon\sqrt{\delta})$-perturbation of the linear system

$$\begin{cases} i\dot{q}_{n_j} + (\lambda_j + \varepsilon I_j^0)q_{n_j} = 0 & (j = 1, \ldots, b) \\[2mm] i\dot{I}_j' = 0 & (j = 1, \ldots, b) \\[2mm] i\dot{q}_n + \mu_n q_n = 0 & \end{cases} \tag{45}$$

with a b-parameter set of frequencies $\{\lambda_j + \varepsilon\, I_j^0\}_{j=1}^b$ obtained by variation of $\{I_j^0\}$. Hence the problem is reduced to a perturbation of a linear system.

A similar approach is applied in the study of perturbations of integrable systems. For instance, perturbations of the KdV-equation

$$u_t + u_{xxx} + uu_x = 0 \tag{46}$$

of the form

$$u_t + u_{xxx} + uu_x + \varepsilon\, f(u)_x = 0 \tag{47}$$

($f(u)$ polynomial or real analytic). The process of extracting the normal form here is rather involved since it is based on the Riemann surface correspondence.

Coming back to (1), we deal with the problem of persistency of finite dimensional tori in an infinite dimensional phase space. This is a generalization of the more classical KAM (Kolmogorov–Arnold–Moser) setting of persistency of n tori in $2n$-dimensional phase space. As a first method to approach (1), one may however, apply the basic KAM scheme, which consists in removing the perturbation εH in (39) by composing with canonical transformations. Given a Hamiltonian flow

$$iq_t = \varepsilon \, \frac{\partial F}{\partial \bar{q}} \qquad F = F(q, \bar{q}) \quad \text{Hamiltonian} \tag{48}$$

denote $q \to q'$ the time-1 shift, which is a symplectic transformation of phase space. Then the system

$$iq_t = \frac{\partial H}{\partial \bar{q}} \tag{49}$$

is transformed into

$$iq'_t = \frac{\partial H'}{\partial \bar{q}'} \tag{50}$$

where $H(q) = H'(q')$. Denote in complex coordinates the Poisson bracket

$$\{F_1, F_2\} = \operatorname{Im} \sum_n \frac{\partial F_1}{\partial q_n} \frac{\partial F_2}{\partial \bar{q}_n}. \tag{51}$$

Then

$$H' = H + \varepsilon \, \{H, F\} + \frac{\varepsilon^2}{2!} \, \{\{H, F\}, F\} + \cdots = H + \varepsilon \, \{H, F\} + 0(\varepsilon^2). \tag{52}$$

Hence, for H given by (39)

$$H(q, \bar{q}) = \sum \mu_n \, |q_n|^2 + \varepsilon \, H_1(q, \bar{q}) = H_0 + \varepsilon H_1 \tag{53}$$

we find

$$H'(q', \bar{q}') = \sum \mu_n \, |q_n|^2 + \varepsilon \, H_1(q, \bar{q}) + \varepsilon \, \{H_0, F\} + 0(\varepsilon^2). \tag{54}$$

The main idea is to choose F in order to reduce $H_1 + \{H_0, F\}$ and iterate the process. The drawback of this method is that it requires the spectrum of A to satisfy nonresonance conditions of the form (i.e. Melnikov conditions)

$$\sum_{j=1}^{b} k_j \lambda_j + \mu_n - \mu_{n'} \neq 0 \qquad (n \neq n' \notin \{n_1, \ldots, n_b\}) \tag{55}$$

besides (38). Condition (55) is violated in the case of multiplicities in the normal frequencies and impairs the use of KAM[*] in this situation (which in the context of (1) appears in 1D under periodic boundary conditions and in higher dimension). To illustrate the appearance of (55), assume F of the form

$$F(q, \bar{q}) = \langle Bq, q \rangle, \quad B = B^*. \tag{56}$$

Then, by (51)

$$(56) \ \{H_0, F\} = \frac{1}{2i} \sum_{n, n'} (\mu_n - \mu_{n'}) \, b_{nn'} \bar{q}_n q_{n'} \sim \langle [A, B]q, q \rangle$$

where $[A, B] = AB - BA$.

---

[*] At least in its usual form as described in [**Kuk1**]

If one rewrites (1) in the Fourier version ($wrt\ e^{i\langle\lambda',k\rangle t}\,\varphi_n(x)$), one gets

$$(-\langle\lambda',k\rangle+\mu_n)\,\hat{u}\,(n,k)+\varepsilon\,\frac{\widehat{\partial H}}{\partial\bar{u}}\,(n,k)=0. \tag{57}$$

Hence (38) is the only type of condition which appears if one tries to solve (57) by a direct perturbative approach. If one considers such a ε-series, considerable problems appear due to the small divisors

$$\mu_n-\langle\lambda',k\rangle. \tag{58}$$

They yield approximative solutions by truncation, provided we introduce only divisors such that

$$|\mu_n-\langle\lambda',k\rangle|\gg\varepsilon. \tag{59}$$

To perform their actual summation is a very delicate issue (leading to renormalization problems) and was even in finite dimensional phase space (i.e. n ranges in a finite set) only recently fully understood. The method followed here is to solve (57) by a Newton iteration scheme, converging much faster (double exponentially fast) and therefore less affected by the presence of small divisors. On the other hand, it will require to control the inverse of the linearization of (57) which is a non-diagonal operator.

We first perform a Lyapunov-Schmidt type decomposition. Consider the b equations

$$(-\lambda'_j+\lambda_j)\,a_j+\varepsilon\,\frac{\widehat{\partial H}}{\partial\bar{u}}\,(n_j,e_j)=0\quad(j=1,\dots,b) \tag{60}$$

obtained by taking $(n,k)\in\mathcal{S}$ of (26), (29).

They form the (finite) system of Q-equations. The remaining (infinite) system

$$(-\langle\lambda',k\rangle+\mu_n)\,\hat{u}(n,k)+\varepsilon\,\frac{\widehat{\partial H}}{\partial\bar{u}}\,(n,k)=0\quad(n,k)\notin\mathcal{S} \tag{61}$$

are called the P-equations. The general procedure is to determine $\hat{u}|_{(n,k)\notin\mathcal{S}}$ from (61) (depending on λ') and then substitute in (60) to obtain the new frequencies $\lambda'=(\lambda'_1,\dots,\lambda'_b)$. Now these frequencies $\lambda'_1,\dots,\lambda'_b$ need to be real. Thus (60) expresses, in fact, $2b$ conditions with b parameters. The formal solvability is a consequence of the Hamiltonian nature of (1) (result due to Poincaré). Assume $H(u,\bar{u})$ is a sum of monomials $u^j\bar{u}^k$ with real coefficients. In proving the persistency result, we may assume $a_j\in\mathbb{R}$ $(j=1,\dots,b)$, considering time shifts as observed earlier. Hence, the system $\hat{u}|_{(n,k)\notin\mathcal{S}}$ produced from solving (61) will be real and so will be $\frac{\widehat{\partial H}}{\partial\bar{u}}\,(n_j,e_j)$ in (60). Thus (60) yields a real solution in λ' and the formal solvability is clear in this case.

The system (61) cannot be treated by a standard implicit function theorem because of the appearance of small divisors. We denote

$$v=\bar{u}. \tag{62}$$

Assuming the eigenfunction basis for A in (1) given by exponentials, we have thus

$$\hat{v}(n,k)=\overline{\hat{u}(-n,-k)} \tag{63}$$

(if one would consider a real eigenfunction basis $\{\varphi_n\}$, then clearly $\hat{v}(n,k) = \overline{\hat{u}(n,-k)}$).

Duplicate the equations (61) considering the system

$$\begin{cases} \left(-\langle \lambda', k \rangle + \mu_n\right) \hat{u}(n,k) + \varepsilon\, \widehat{\frac{\partial H}{\partial v}}(n,k) = 0 & (n,k) \notin \mathcal{S} \\ \left(\langle \lambda', k \rangle + \mu_{-n}\right) \hat{v}(n,k) + \varepsilon\, \widehat{\frac{\partial H}{\partial u}}(n,k) = 0 & (n,k) \notin -\mathcal{S}. \end{cases} \tag{64}$$

We solve this system in $\left(\hat{u}|_{(n,k)\notin\mathcal{S}},\ \hat{v}|_{(n,k)\notin\mathcal{S}}\right)$ by Newton's algorithm. The relation (63) will be preserved for the approximative solutions.

Recall the formal scheme. Consider the equation

$$F(y) = 0. \tag{65}$$

Starting from y_0 (here 0), the consecutive approximations are defined by

$$\Delta_{i+1} y \equiv y_{i+1} - y_i = -[F'(y_i)]^{-1} F(y_i) \tag{66}$$

and

$$\|F(y_{i+1})\| = 0 \ (\|y_{i+1} - y_i\|^2) = 0 \ \left(\|[F'(y_i)]^{-1} F(y_i)\|^2\right). \tag{67}$$

In our case (64), the linearized operator T is given by

$$T = D + \varepsilon \mathcal{S} \tag{68}$$

where D is diagonal

$$D = \begin{pmatrix} D^+ & 0 \\ 0 & D^- \end{pmatrix} \qquad \begin{aligned} D^+_{n,k} &= -\langle \lambda', k \rangle + \mu_n \\ D^-_{n,k} &= \langle \lambda', k \rangle + \mu_{-n} \end{aligned} \tag{69}$$

and

$$\mathcal{S} = \begin{pmatrix} S_{\frac{\partial^2 H}{\partial u \partial v}} & S_{\frac{\partial^2 H}{\partial v^2}} \\ S_{\frac{\partial^2 H}{\partial u^2}} & S_{\frac{\partial^2 H}{\partial u \partial v}} \end{pmatrix}. \tag{70}$$

S_ϕ expresses ϕ-multiplication in Fourier, thus here

$$S_\phi\left((n,k),\ (n',k')\right) = \hat{\phi}(n - n',\ k - k'). \tag{71}$$

The operator S is selfadjoint and depends on the given approximative solution (u,v). Along the construction, we satisfy an estimate on their Fourier coefficients of the form

$$|\hat{u}(n,k)| < C\, e^{-(|n|+|k|)^c} \tag{72}$$

for some fixed $c > 0$. This fact clearly yields a corresponding off-diagonal decay estimate for each of the 4 matrixes S_ϕ appearing in (70)

$$S_\phi(x,x') < C\, e^{-|x-x'|^c}. \tag{73}$$

The main difficulty consists in controlling the inverse T^{-1}, due to the fact that the diagonal elements may be arbitrarily small. Fixing some $\varepsilon \ll \delta < 1$, call a site (n,k) singular if

$$|D^+_{n,k}| = |-\langle \lambda', k \rangle + \mu_n| < \delta$$

or

$$|D_{n,k}^-| = |\langle \lambda', k \rangle + \mu_{-n}| < \delta. \tag{74}$$

The geometric structure of the singular sites plays an important role in the study of T^{-1} (in particular, one exploits essentially their separation properties). Assume that one succeeds to ensure for these operators T bounds

$$(T_N)^{-1} < B(N) \tag{75}$$

on the inverse of the restriction

$$T_N = T|_{|n|, \, |k| < N} \tag{76}$$

where, for instance, $B(N)$ grows at most polynomially

$$B(N) < N^C \tag{77}$$

or at least much slower than exponential.

Assume $H(u,v)$ is a polynomial expression. At the jth step of the Newton iteration, an approximative solution y is obtained and we assume that

$$\text{supp } \hat{u}_j \text{ is contained in an } (n,k) - \text{ball of radius } M^j \tag{78}$$

for some constant M. Letting $y_j = (\hat{u}_j, \hat{v}_j)$ and $F(y)$ the left number of (64), assume

$$\|F(y_j)\| < \varepsilon_j. \tag{79}$$

Since F is expressed as a polynomial in $\hat{u}, \hat{v}$, we get

$$\text{supp } \widehat{F(y_j)} \subset CM^j. \tag{80}$$

Put $N = M^{j+1}$ and define y_{j+1} by (cf. (66))

$$y_{j+1} - y_j = -T_N^{-1}\left(F(y_j)\right). \tag{81}$$

Hence, by (77), (79)

$$\|y_{j+1} - y_j\| < N^C \varepsilon_j \tag{82}$$

and by (81)

$$\|F(y_{j+1})\| \leq \|(T - T_N)(y_{j+1} - y_j)\| + 0\left(\|y_{j+1} - y_j\|^2\right). \tag{83}$$

From (79), (82) and off-diagonal decay of T and T_N^{-1} (cf. (73)), the first term in (83) may be bounded by

$$\|(T - T_N)(y_{j+1} - y_j)\| < e^{-\frac{1}{2} N^c} \varepsilon_j. \tag{84}$$

Hence, from (83)

$$\varepsilon_{j+1} < e^{-\frac{1}{2}(M^{j+1})^c} \varepsilon_j + M^{2C(j+1)} \varepsilon_j^2. \tag{85}$$

Letting $c > 0$ be sufficiently small, one may conclude that

$$\varepsilon_j < e^{-4(M^j)^c} \tag{86}$$

$$\|y_{j+1} - y_j\| < e^{-3(M^j)^c} < e^{-2(M^{j+1})^c} \tag{87}$$

which is compatible with (72), (78).

The main difficulty is to obtain the estimate (75) and the off-diagonal estimate, say

$$|T_N^{-1}(x, x')| < e^{-\frac{1}{2}|x-x'|^c} \text{ for } |x - x'| > \frac{N}{100}. \tag{88}$$

Coming back to (38), assume (except for $(n, k) \in \mathcal{S}$)

$$|-\langle k, \lambda \rangle + \mu_n| > (1 + |k|)^{-C} \tag{89}$$

for some constant C. Assuming λ' as in (28), it follows that also

$$|-\langle k, \lambda' \rangle + \mu_n| > \frac{1}{2}(1 + |k|)^{-C} \text{ for } |k| < \varepsilon^{-1/2C}. \tag{90}$$

Thus, for

$$N < \varepsilon^{-1/2C} \tag{91}$$

one may invert $T_N = D_N + \varepsilon S_N$ by a Neumann series, i.e.

$$T_N^{-1} = D_N^{-1} + \sum_{j=1}^{\infty} (-1)^j \, \varepsilon^j \, \left(S_N D_N^{-1}\right)^j \tag{92}$$

since $\|S_N D_N^{-1}\| < \sqrt{\varepsilon}$, and (77), (88) are clearly satisfied.

In order to fulfill (77), (88) at later scales, requires to impose further conditions on (λ, λ', a); $a = (a_1, \ldots, a_b)$, considered as initial parameters in the Newton iteration process (recall that at each stage, T depends on previous approximative solution). The effect of these further conditions is to excise from the (λ, λ', a)-parameter set exceptional subsets which measure tends rapidly to zero. Since λ' will be determined by solving the Q-equations (60) as

$$\lambda' = \lambda + \varepsilon(\lambda, a) \tag{93}$$

our aim is to remain with conditions on (λ, a) as in (32), (34); in particular in a model when a is fixed, λ will be restricted to a Cantor set of positive measure. Essentially speaking, these conditions on (λ, λ', a) consist in keeping certain expressions $E(\lambda, \lambda', a)$ away from zero, where $E(\lambda, \lambda', a)$ is differentiable with more sensitivity on λ' than λ, a.

In order for $\hat{u}|_{(n,k)\notin\mathcal{S}}$ to solve the P-equations, previous conditions on $(\lambda, \lambda'a)$ need to be fulfilled. We assume however, that $\hat{u}|_{(n,k)\notin\mathcal{S}}$ is smoothly defined on the entire parameter set (λ, λ', a). Substitution in (60) allows then to solve in λ' invoking the standard implicit function theorem. One gets (93) or more precisely (assuming $a_j \in \mathbb{R}$)

$$\lambda_j' = \lambda_j + \varepsilon \, \frac{1}{a_j} \left[\frac{\partial H}{\partial \bar{u}} \left(\sum_{j=1}^{b} a_j \, e^{i\theta_j}, \, \sum_{j=1}^{b} a_j \, e^{-i\theta_j}\right)\right]^{\wedge} (n_j, e_j) + 0(\varepsilon^2) \tag{94}$$

where the second term depends on $a = (a_1, \ldots, a_b)$ only.

In order to carry out previous discussion in detail, there is a distinction between the case of time periodic solutions ($b = 1$) and quasi-periodic solutions ($b > 1$). The case $b = 1$ turns out to be significantly easier. The other aspects are the arithmetic properties of the sequence $\{\mu_n\}$, which play an essential role in the analysis. In our PDE-context, this sequence is infinite and its properties are dependent on the space dimension d. We will require in particular the property that the difference set

$$\{\mu_n - \mu_{n'} \mid n, n' \in \mathbb{Z}^d\} \tag{95}$$

has a closure of zero measure.[*] Moreover, the structure of clusters of the form

$$\{n \in \mathbb{Z}^d \mid |\mu_n - \mu| < 1\} \tag{96}$$

when $\mu \to \infty$ will play a basic role.

It turns out however, that conditions of the form (55) are unnecessary (at least in the present discussion) and appear as an artifax of the KAM approach.

Remarks.

(i) The method described above is more flexible than KAM. In fact, in the main body of the analysis, which consists in solving the P-equations, the Hamiltonian structure plays essentially no role.

There are variants. For instance, one may use a truncated perturbation series to obtain an approximative solution of the P-equation up to ε^K, for any power K, and then apply the more rapidly converging Newton method and get an actual solution, ε^K-close to the approximative one.

(The Hamiltonian counterpart of this consists in achieving a normal form with nonresonant part $0(\varepsilon^K)$).

The condition (38)

$$\sum_{j=1}^{b} k_j \, \lambda_j + \mu_n \neq 0 \quad (n \notin \{n_1, \dots , n_b\})$$

(1^e Melnikov condition) expresses nonresonance of the normal frequencies with the tangential ones. In the resonant case, one modifies the previous scheme as follows. Define now the resonant set $\widetilde{\mathcal{S}} \supset \mathcal{S}$ as

$$\widetilde{\mathcal{S}} = \{(n, k) \mid -\langle \lambda, k \rangle + \mu_n = 0\} \tag{97}$$

and consider for the Lyapunov–Schmidt decomposition

$$\begin{cases} (-\lambda'_j + \lambda_j)a_j + \varepsilon \, \widehat{\frac{\partial H}{\partial \bar{u}}} \, (n_j, e_j) = 0 \quad (j = 1, \dots , b) \\[2mm] (-\langle \lambda', k \rangle + \mu_n) \, \hat{u}(n, k) + \varepsilon \, \widehat{\frac{\partial H}{\partial \bar{u}}} \, (n, k) = 0 \text{ for } (n, k) \in \widetilde{\mathcal{S}} \backslash \mathcal{S} \end{cases} \tag{98}$$

(Q-equations)
and

$$(-\langle \lambda', k \rangle + \mu_n) \, \hat{u}(n, k) + \varepsilon \, \widehat{\frac{\partial H}{\partial \bar{u}}} \, (n, k) = 0 \text{ for } (n, k) \notin \widetilde{\mathcal{S}} \tag{99}$$

(P-equations).

Fix

$$(\lambda'_j)_{j=1,\dots ,b}, \ \{a_j\}_{j=1,\dots ,b} \text{ and } \hat{u}\big|_{(n,k)\in\widetilde{\mathcal{S}}\backslash\mathcal{S}}$$

---

[*]To treat the general quasi-periodic case $(b > 1)$.

and solve $\hat{u}|_{(n,k)\notin\tilde{\mathcal{S}}}$ from (99). One then uses (98) for the determination of λ' and $\hat{u}|_{\tilde{\mathcal{S}}\backslash\mathcal{S}}$. Examples in the context of wave equations will be discussed later on.

(ii) The Lyapunov–Schmidt approach to (1) yields a new method to solve stability problems in finite dimensional phase space as well. Observe that the Hamiltonian system

$$\dot{p}_n = \frac{\partial H}{\partial q_n}, \quad \dot{q}_n = -\frac{\partial H}{\partial p_n} \quad (n = 1,\ldots,N) \tag{100}$$

and Hamiltonian

$$H(p,q) = H(p_1,\ldots,p_N,\, q_1,\ldots,q_N) \tag{101}$$

is equivalent to

$$i\dot{u} = 2\frac{\partial H}{\partial \bar{u}} \tag{102}$$

where $u = p + iq$. Thus one gets a new proof of the KAM theorem (stability of N tori is 2N dimensional phase space) and Melnikov's theorem (more generally, stability of tori of dimension $n \leq N$) along the lines of Lyapunov's theorem (the periodic case $n = 1$). For Lyapunov's result, the nonresonance condition is

$$\mu_n \notin \lambda_1 \mathbb{Z} \tag{103}$$

corresponding to the case $b = 1$ in (38). Hence, in the finite dimensional case, we prove a Melnikov theorem for (1) under the nonresonance assumption (38), without the need of the extra assumption (55). This result is new. From the previous remark, it appears that besides multiple normal frequencies one may investigate also the case of normal frequencies resonant with the tangential ones, thus when (103), (38) are partially violated. Presently there has been no systematic study of this.

(iii) Consider a NLW

$$\partial_{tt} - \Delta y + V y + \varepsilon\, F'(y) = 0 \tag{104}$$

and denote

$$B = (-\Delta + V)^{1/2} \tag{105}$$

(assuming this makes sense).

Rewrite (104) in the following Hamiltonian form

$$\left\{ \begin{array}{l} \dot{y} = Bz \\ \dot{z} = -By - \varepsilon\, B^{-1}F'(y) \end{array} \right. \tag{106}$$

considering instead of L^2 the Hilbert space $H^{1/2}$ with scalar product

$$\langle u, v\rangle_{\frac{1}{2}} = \langle u, Bv\rangle. \tag{107}$$

Denoting $u = y + iz$, (106) yields

$$i\dot{u} - Bu + \varepsilon\, B^{-1}F'(\operatorname{Re} u) = 0 \tag{108}$$

which is of the form (1) with A replaced by $-B$. Thus the spectrum in 1D under periodic (resp. Dirichlet) boundary conditions is given by $-\sqrt{\lambda_j}$ (resp. $-\sqrt{\mu_n}$) and behaves as

$$\sqrt{\lambda_j} = \frac{\pi j}{2} + 0\left(\frac{1}{j}\right), \quad \sqrt{\mu_n} = \pi n + 0\left(\frac{1}{n}\right). \tag{109}$$

In this particular case, $V = \rho$ (constant), we get for the periodic spectrum

$$\mu_n = (|n|^2 + \rho)^{1/2} \tag{110}$$

(cf example (iv) above).

One may treat the persistency of quasi-periodic solutions of (104) by applying the preceding to (108) or it may be done directly from (104), replacing (57) by

$$\left(-\langle \lambda', k\rangle^2 + \mu_n\right)\hat{y}(n, k) + \varepsilon \,\widehat{F'(y)}\,(n, k) = 0 \tag{111}$$

where y is real. This avoids the use of a pair of equations (64) and is notationally simpler.

Next, we state some concrete results.

Theorem 112. *Consider a 1D NLS*

$$iu_t - u_{xx} + Vu + \varepsilon\,\frac{\partial H}{\partial \bar{u}} = 0 \tag{113}$$

or

$$iu_t = u_{xx} + M_\sigma u + \varepsilon\,\frac{\partial H}{\partial \bar{u}} = 0 \tag{114}$$

where $H = H(u, \bar{u})$ is polynomial or real analytic, V a real analytic periodic potential and M_σ a real Fourier multiplier (as discussed in the examples). Consider an unperturbed solution of (113), (114) with $\varepsilon = 0$

$$u_0(x, t) = \sum_{j=1}^{b} a_j\,e^{i\lambda_j t}\,\varphi_{n_j}(x) \tag{115}$$

where $\lambda_j = \mu_{n_j}$ $(j = 1, \ldots, b)$ and $\{\varphi_{n_j}\}$ are the corresponding eigenfunctions. Consider $\lambda = (\lambda_1, \ldots, \lambda_b)$ as a b-parameter (for equation (113), this may be achieved by appropriate variation of V, cf. (35)-(36)). Assume a nonresonance condition (38)

$$|\mu_n + \langle k, \lambda\rangle\,| \geq c\,(1 + |k|)^{-C} \tag{116}$$

satisfied, for $|k| < N$ sufficiently large (depending on H).

Then, for $\lambda \in \Lambda_\varepsilon(|a|)$, a subset of the parameter set of small complementary measure when $\varepsilon \to 0$, there is a perturbed solution u_ε of (113), (114) with perturbed frequency λ'

$$u_\varepsilon(x, t) = \sum_{n,k} \hat{u}_\varepsilon(n, k)\,e^{i\langle \lambda', k\rangle t}\,\varphi_n(x) \tag{117}$$

satisfying (26)-(28). Thus

$$\hat{u}_\varepsilon(n_j, e_j) = a_j \quad (j = 1, \ldots, b) \tag{118}$$

$$\sum_{(n,k)\notin S} e^{(|n|+|k|)^c}\,|\hat{u}_\varepsilon(n,k)| < \sqrt{\varepsilon} \tag{119}$$

$$|\lambda' - \lambda| < C\varepsilon \tag{120}$$

for some $c > 0$, and where $S = \{(n_j, e_j)\,(j = 1, \dots, b)\}$.

In the case of (113), with Dirichlet boundary conditions, assume V even and $H(u, \bar{u})$ even (hence $\frac{\partial H}{\partial \bar{u}}$ odd), see discussion in example (ii) above.

Theorem 121. *The analogue of Theorem 112 holds for 2D Schrödinger equations, but in (113) the potential $V(x_1, x_2)$ should be assumed of the form $V_1(x_1) + V_2(x_2)$ (see discussion in example (iii) above).*

In 1D, a result similar to Theorem 112 may be stated for NLW equations

$$u_{tt} - u_{xx} + V(x)u + \varepsilon\, f'(u) = 0 \tag{122}$$

and even derivative NLW equations

$$u_{tt} - u_{xx} + V(x)u + \varepsilon\, B f'(u) = 0 \quad B = \left(\frac{-d^2}{dx^2}\right)^{1/2}. \tag{123}$$

In the next theorem, the role of outer parameters in the equation is replaced by amplitude-frequency modulation.

Theorem 124. *Consider a 1D NLW*

$$u_{tt} - u_{xx} + \rho u + (u^3 + \text{higher order terms}) = 0 \tag{125}$$

and assume $\rho \geq 0$ is a typical number in the sense of linear independence of the sequence

$$\mu_n = (n^2 + \rho)^{1/2}. \tag{126}$$

Fix a sequence

$$0 < n_1 < n_2 < \cdots < n_b \in \mathbb{Z}_+ \tag{127}$$

and consider the solution

$$u_0(x, t) = \sum_{j=1}^{b} a_j \, \cos n_j x \cdot \cos \lambda_j t \qquad (\lambda_j = \mu_{n_j}) \tag{128}$$

of the linear equation

$$u_{tt} - u_{xx} + \rho u = 0. \tag{129}$$

There is a Cantor set $C \subset \{a = (a_j)_{j=1,\dots,b} \mid a_j > 0\}$ of positive measure, in fact, of asymptotically full measure when $|a| \to 0$, such that for $a \in C$ the solution (128) of (129) persists for (125)

$$u(x, t) = \sum_{j=1}^{b} q_j \, \cos n_j x \cdot \cos \lambda'_j t + 0(|a|^3). \tag{130}$$

The persistency problem for higher dimensional wave equations seems **more** difficult, due to the behavior of the frequencies $|n| = (n_1^2 + \dots + n_d^2)^{1/2}$ when $d \geq 2$. One may treat however, the special case of time periodic solutions in any dimension (this is also the case for NLS).

Theorem 131. *Consider the periodic wave equation in dimension d*

$$u_{tt} - \Delta u + \rho u + (u^3 + higher\ order\ terms) = 0 \tag{132}$$

where again $\rho > 0$ is a typical number. More precisely, we require ρ to satisfy a condition of the form

$$\left| \sum_{j=0}^{r} k_j\ \rho^j \right| > \left(\sum |k_j| \right)^{-C_r} \quad for\ all\ \{k_j\} \in \mathbb{Z}^{r+1}\backslash\{0\}. \tag{133}$$

Fix $n_0 \in \mathbb{Z}^d\backslash\{0\}$. There is a Cantor set C of positive measure in an interval $[0,\delta]$ and for $p_0 \in C$ a solution of (132) of the form

$$u(x,t) = p_0\ \cos\left(\langle n_0, x\rangle + \lambda t\right) + 0\left(p_0^3\right) \tag{134}$$

where

$$\lambda^2 = \lambda(p_0)^2 = |n_0|^2 + \rho + \left(\frac{3}{4} + o(1)\right) p_0^2. \tag{135}$$

The next two results are normal form reductions, bringing the problem back to perturbations of a linear problem with parameters. (cf. the discussion (39)-(45)).

Theorem 136. *Consider a 1D NLS*

$$iu_t - u_{xx} + mu + f(|u|^2)u = 0 \tag{137}$$

where f is a polynomial or real analytic and satisfying a nondegeneracy condition

$$f'(0) \neq 0. \tag{138}$$

Consider (137) with periodic boundary conditions say. Fix a sequence of positive integers

$$n_1 < n_2 < \ldots < n_b. \tag{139}$$

Then for $a = \{a_j\}_{j=1,\ldots,b}$ in a Cantor family C of positive measure, there is a quasi-periodic solution

$$u(x,t) = \sum_{j=1}^{b} q_j\ e^{i(n_j x + \lambda_j' t)} + 0(|a|^3) \tag{140}$$

with frequencies $\lambda_1', \ldots, \lambda_b'$, where

$$\lambda_j' = n_j^2 + m + 0(|a|^2) \quad (j = 1,\ldots,b). \tag{141}$$

Theorem 142. *Consider the 2D cubic NLS*

$$iu_t - \Delta u + cu|u|^2 = 0 \quad (c \neq 0) \tag{143}$$

or, more generally, an 2D NLS

$$iu_t - \Delta u + f(|u|^2)u = 0 \tag{144}$$

with f as above in Theorem 136, with periodic boundary conditions.
For the modes $n_1, \ldots, n_b \in \mathbb{Z}^2$, we fix 2 lattice points n_1, n_2 on a same circle

$$|n_j| = R \quad (j = 1,2), n_1 \neq -n_2 \tag{145}$$

(more complicated structures involving more then 2 points may be treated as well but this is the simplest case). There is a Cantor family C of positive measure such that for $a = \{a_j\}_{j=1,2} \in C$ (143) (144) has a quasi-periodic solution

$$u(x,t) = \sum_{j=1}^{2} a_j \, e^{i(\langle n_j, x\rangle + \lambda'_j t)} + 0(|a|^3) \tag{146}$$

with frequencies $\lambda' = (\lambda'_1, \lambda'_2)$

$$\lambda'_j = |n_j|^2 + 0(|a|^2) \quad (j = 1, 2). \tag{147}$$

Comments.

(i) An approach to the problems discussed in this section by a suitable reworking of the KAM scheme was developed by S. Kuksin (cf. Springer LNM 1556). He obtains for instance, Theorem 112 in 1D for (113) under Dirichlet boundary conditions and similarly for 1D wave equations. This technique, at least in the original version developed by Kuksin, seems non-compatible with the presence of multiple or nearby normal frequencies. Thus his work leaves out the case of periodic boundary conditions in 1D or higher dimensional problems.

(ii) The method described in this section, based on a Lyapunov–Schmidt decomposition is a resonant (Q-equations) and nonresonant P-equations) part and the use of Newton's method to solve the P-equation originates from the work of Craig and Wayne. Their original work deals with 1D equations

$$u_{tt} - u_{xx} + V(x)u + f'(u) = 0 \tag{148}$$

and the construction of time periodic solutions. In this PDE problem, contrary to the Lyapunov Theorem in the setting of a finite dimensional phase space, small divisor problems appear already here, since the expressions

$$|k^2 \lambda^2 - n^2| \quad k, n \in \mathbb{Z} \tag{149}$$

will for typical λ take arbitrary small values when $|n| \to \infty$. The author extended their work to the case of time quasi-periodic solutions. A first step is to treat the classical case (finite dimensional phase space) using this method.

(iii) Theorem 136 appears in a joint paper of S. Kuksin and J. Pöschel, for the case of Dirichlet boundary conditions. The problem considered in Theorem 136 may also be solved as perturbation of the 1D cubic NLS

$$iu_t - u_{xx} + c\,u|u|^2 = 0 \tag{150}$$

which is integrable, as is the KdV equation

$$u_t + u_{xxx} + uu_x = 0 \tag{151}$$

and the modified MKdV equation

$$u_t + u_{xxx} + u^2 u_x = 0. \tag{152}$$

This approach, depending heavily on Riemann surface theory underlying the integrable structure of those equations, has been carried out in earlier work by S. Kuksin. The normal form approach to Theorem 136 is much more elementary and is the only one applicable in the 2D case, since (143) is not integrable.

(iv) Results for 1D nonlinear wave equations via the KAM scheme were obtained by C. Wayne around 1990, independently of S. Kuksin.

(v) The results in $D > 1$ Theorem 121, Theorem 131, and Theorem 142 are due to the author. The specific choice of the sites $\{n_j\}_{j=1,2}$ as in Theorem 142 corresponds to a particular example analyzed in detail.

This general area is presently subject to active research which soon should lead to a more complete understanding.

Some references related to Lecture 6

[C-W1] W. Craig, C. Wayne, *Newton's method and periodic solutions of nonlinear wave equations*, Comm. Pure and Appl. Math. 46 (1993), 1409-1501.

[C-W2] W. Craig, C. Wayne, *Periodic solutions of Nonlinear Schrödinger equations and the Nash-Moser method*, (preprint).

[Kuk1] S. Kuksin, *Nearly integrable infinite-dimensional Hamiltonian systems LNM*, 1556, Springer.

[K-P] S. Kuksin, J. Pöschel, *Invariant Cantormanifolds of quasi-periodic oscillations for a nonlinear Schrödinge equation*, Annals of Math., vol. 143, N1 (1996), 149-179.

[B8] J. Bourgain, *Construction of Quasi-periodic solutions for Hamiltonian perturbations of linear equations and applications to nonlinear PDE*, IMRN (1994), N11, 475-497.

[B9] J. Bourgain, *Quasi-periodic solutions of Hamiltonian perturbations of 2D linear Schrödinger equations*, to appear in Annals of Math.

[B10] J. Bourgain, *Construction of periodic solutions of nonlinear wave equations in higher dimension*, GAFA Vol. 4 (1995), 629-639.

LECTURE 7
Time Periodic Solutions

In this lecture we treat persistency problems as discussed in previous section for time periodic solutions $(b = 1)$. We will consider the case of the wave equation

$$u_{tt} - \Delta u + \rho u + \partial_u f(u, x) = 0 \tag{1}$$

and prove Theorem 1.31 in $D = 1$ and $D > 1$.

Remark. A similar discussion may be carried out for the Schrödinger equation, but the setup for wave equation is simpler (in particular the linear equation) and more interesting, since natural NLS $iu_t - \Delta u + f(|u|^2)u = 0$ possess obvious families of time periodic solutions.

We consider as model example the equation

$$u_{tt} - \Delta u + \rho u + u^3 = 0 \tag{2}$$

(Klein-Gordon equation).

The more general case

$$u_{tt} - \Delta u + \rho u + u^3 + (\text{higher order terms}) = 0 \tag{3}$$

requires only minor changes[*] Replacing u by δu yields

$$u_{tt} - \Delta u + \rho u + \delta^2 u^3 = 0 \tag{4}$$

which is a perturbation of the linear equation

$$u_{tt} - \Delta u + \rho u = 0. \tag{5}$$

Define

$$\mu_n = (|n|^2 + \rho)^{1/2}. \tag{6}$$

Fix $n_0 \in \mathbb{Z}^d \backslash \{0\}$ and let

$$\lambda_0 = \mu_{n_0} = (|n_0|^2 + \rho)^{1/2}. \tag{7}$$

---

[*] This higher order terms may be x-dependent. In fact, if there is no x-dependence, the construction of time periodic solutions may be achieved in a much simpler way (this is in particular the case for equation (2)). The method developed here applies however to general x-dependent nonlinearity with $\rho u + u^3$ as leading part (see also comments at the end of this section).

87

Periodic solutions of (5) are given by

$$u_0(x,t) = p_0 \cos(\langle n_0, x \rangle + \lambda_0 t) \quad \text{or} \quad u_0(x,t) = p_0 \, \sin(\langle n_0, x \rangle + \lambda_0 t) \tag{8}$$

and our purpose is to prove their persistency for p_0 taken in a Cantor set of positive measure. Consider the case

$$u_0(x,t) = p_0 \cos(\langle n_0, x \rangle + \lambda_0 t). \tag{9}$$

The solution of (4) will have the form

$$u(x,t) = \sum_{n \in \mathbb{Z}^d, k \in \mathbb{Z}} \hat{u}(n,k) \, e^{i(\langle n,x \rangle + k\lambda t)} \tag{10}$$

where

$$\hat{u}(n,k) = \hat{u}(-n,-k) \in \mathbb{R} \tag{11}$$

$$\hat{u}(n_0, 1) = \frac{1}{2} p_0 \tag{12}$$

$$\sum_{(n,k) \neq (n_0,1), k \geq 0} |\hat{u}(n,k)| \, e^{(|n|+|k|)^c} < 0(\delta) \tag{13}$$

$$\lambda = \lambda_0 + c\delta^2 p_0^2 + 0(\delta^4). \tag{14}$$

Writing

$$k^2 \lambda_{n_0}^2 - \mu_n^2 = k^2(|n_0|^2 + \rho) - (|n|^2 + \rho) = (k^2 |n_0|^2 - |n|^2) + (k^2 - 1)\rho \tag{15}$$

one may clearly from assumption on ρ assume

$$|k^2 \lambda_{n_0}^2 - \mu_n^2| > (|k| + 1)^{-C} \tag{16}$$

except for the resonant set

$$\tilde{S} = \{(n,k) \in \mathbb{Z}^d \times \mathbb{Z} \mid |n| = |n_0|, k = \pm 1\} \tag{17}$$

(we are dealing here with a resonant situation).

Rewrite (10) as

$$u(x,t) = p_0 \cos(\langle n_0, x \rangle + \lambda t) + \sum_{\substack{|n|=|n_0| \\ n \neq n_0}} p_n \cos(\langle n, x \rangle + \lambda t)$$

$$+ 2 \sum_{\substack{(n,k) \notin \tilde{S} \\ k \geq 0}} \hat{u}(n,k) \cos(\langle n, x \rangle + k\lambda t). \tag{18}$$

The Q-equations are given by

$$(-\lambda^2 + |n_0|^2 + \rho)p_n + 2\delta^2 \widehat{u^3}(n,1) = 0 \quad (|n| = |n_0|) \tag{19}$$

and the P-equations

$$(-k^2\lambda^2 + \mu_n^2)\hat{u}(n,k) + \delta^2 \widehat{u^3}(n,k) = 0 \quad ((n,k) \notin \tilde{S}). \tag{20}$$

Denote

$$\sigma = \delta^{-2}(-\lambda^2 + |n_0|^2 + \rho) \tag{21}$$

and rewrite (19) as

$$\sigma p_n + 2\widehat{u^3}(n,1) = 0 \quad (|n| = |n_0|) \tag{22}$$

hence

$$\sigma p_n$$

$$+ 2 \int \left[p_0 \cos(\langle n_0, x \rangle + \theta) + \sum_{\substack{|n'|=|n_0| \\ n' \neq n_0}} p_{n'} \cos(\langle n', x \rangle + \theta) \right]^3 \cos(\langle n, x \rangle + \theta) dx d\theta$$

$$+ 0(\delta^2) = 0. \tag{23}$$

Write

$$[\cdots]^3 = p_0^3 \cos^3(\langle n_0, x \rangle + \theta)$$

$$+ 3p_0^2 \cos^2(\langle n_0, x \rangle + \theta) \left[\sum_{|n'|=|n_0|, n' \neq n_0} p_{n'} \cos(\langle n', x \rangle + \theta) \right] \tag{24}$$

$$+ 0(|p_{n'}|^2; n' \neq n_0)$$

and calculate the second term in (23).

Case $n = n_0$.

One gets from the first two terms in (24)

$$p_0^3 \int \cos^4(\langle n_0, x \rangle + \theta) dx d\theta + 3p_0^2 \int \cos^3(\langle n_0, x \rangle + \theta) \left[\sum_{\substack{|n'|=|n_0| \\ n' \neq n_0}} p_{n'} \cos(\langle n', x \rangle + \theta) \right] dx d\theta. \tag{25}$$

The first term equals $\frac{3}{8} p_0^3$. Writing $\cos^3 \alpha = \frac{3}{4} \cos \alpha + \frac{1}{4} \cos 3\alpha$, the second clearly vanishes.

Hence for $n = n_0$, (23) becomes

$$\sigma p_0 + \frac{3}{4} p_0^3 + 0(\delta^2) + 0(|p_{n'}|^2; n' \neq n_0) = 0. \tag{26}$$

Case $n \neq n_0$.

One gets for the second term in (23)

$$p_0^3 \int \cos^3(\langle n_0, x \rangle + \theta) \cos(\langle n, x \rangle + \theta) dx d\theta + \tag{27}$$

$$3p_0^2 \int \cos^2(\langle n_0, x \rangle + \theta) \cos(\langle n, x \rangle + \theta) \left[\sum_{|n'|=|n_0|, n' \neq n_0} p_{n'} \cos(\langle n', x \rangle + \theta) \right] dx d\theta. \tag{28}$$

The term (27) vanishes and (28) equals $\frac{3}{4} p_0^2 p_n$, since

$$\int \cos(2\langle n_0, x \rangle + 2\theta) \cos(\langle n, x \rangle + \theta) \cos(\langle n', x \rangle + \theta) dx d\theta = 0 \tag{29}$$

unless $2n_0 = n + n'$, thus $n_0 = n = n'$ since $|n_0| = |n| = |n'|$.

This yields for (23) the equations

$$\sigma p_n + \frac{3}{2} p_0^2 p_n + 0(\delta^2) + 0(|p_{n'}|^2; n' \neq n_0) = 0 \quad (|n| = |n_0|, n \neq n_0). \tag{30}$$

Consider the system (26), (30)

$$\begin{cases} \sigma p_0 + \frac{3}{4} p_0^3 + 0(\delta^2) + 0(|p_{n'}|^2; n' \neq n_0) = 0 \\ \sigma p_n + \frac{3}{2} p_0^2 p_n + 0(\delta^2) + 0(|p_{n'}|^2; n' \neq n_0) = 0 \end{cases} \tag{31}$$

implying

$$\sigma + \frac{3}{4} p_0^2 = 0(\delta^2) \tag{32}$$

$$p_n = 0(\delta^2) \quad \text{for} \quad n \neq n_0. \tag{33}$$

Hence, by (21), (32), the perturbed frequency is given by

$$\lambda^2 = \lambda(p_0)^2 = |n_0|^2 + \rho + \frac{3}{4} \delta^2 p_0^2 + 0(\delta^4). \tag{34}$$

Remark. Before the scaling $u \to \delta u$, one has for the solution of equation (2)

$$\lambda^2 = \lambda(p_0)^2 = |n_0|^2 + \rho + \frac{3}{4} p_0^2 + 0(p_0^4). \tag{35}$$

Next, it remains to discuss the solvability of the P-equations (20)

$$(-k^2 \lambda^2 + \mu_n^2) \hat{u}(n, k) + \delta^2 \widehat{u^3}(n, k) = 0 \tag{36}$$

in $\hat{u}|_{(n,k) \notin \tilde{S}}$, given $\{p_n | \ |n| = |n_0|\}$, based on Newton's method as discussed in previous section.

It follows from (16) and (34) that

$$|k^2 \lambda^2 - \mu_n^2| > c(1 + |k|)^{-C} \tag{37}$$

as long as $k^2 \delta^2 \ll k^{-C}$, thus

$$|k| < \delta^{-2/2+C}, \quad (n, k) \notin \tilde{S}. \tag{38}$$

We will first discuss the case $D = 1$. Here there is an elementary approach, by direct Neumann series computation (this observation is due to S. Kuksin).

For $\tilde{S}$, we have

$$\tilde{S} = \{(n, k) | n = \pm n_0, k = \pm 1\}. \tag{39}$$

If $(n, k) \notin \tilde{S}$, (36) holds as long as $|k| < \delta^{-2/2+C}$.

Linearization of (36) yields the operator

$$T = (-k^2 \lambda^2 + \mu_n^2) \mathbb{1} + 3\delta^2 S_{u^2} \tag{40}$$

where S_{u^2} is the Toeplitz operator corresponding to u^2-multiplication, where u is the approximative solution obtained at a given stage of the Newton iteration. We remove from S_{u^2} its diagonal part, thus

$$S_{u^2} = \widehat{u^2}(0, 0) \mathbb{1} + S'_{u^2}; S'_{u^2} \quad \text{with 0-diagonal} \tag{41}$$

and write

$$T = \tilde{D} + \tilde{T} \tag{42}$$

where

$$\tilde{D}_{n,k} = -k^2\lambda^2 + \mu_n^2 + 3\delta^2\, \widehat{u^2}(0,0) \quad \text{(diagonal)} \tag{43}$$

$$\tilde{T} = 3\delta^2 S'_{u^2}. \tag{44}$$

We need some estimates on $\tilde{D}_{n,k}$. First, by (37), (38), (43), it follows that

$$|\tilde{D}_{n,k}| > c(1+|k|)^{-C} \quad \text{if } |k| < \delta^{-2/2+C}. \tag{45}$$

We assume $\tilde{D}, \tilde{T}$ restricted to the complement of the resonant set $\tilde{S}$.

Write by (34)

$$\begin{aligned}
\tilde{D}_{n,k} &= -k^2\lambda^2 + \mu_n^2 + 3\delta^2\, \widehat{u^2}(0,0) \\
&= -k^2\left(n_0^2 + \rho + \frac{3}{2}\delta^2 p_0^2 + 0(\delta^4)\right) + (n^2 + \rho) + 3\delta^2\, \widehat{u^2}(0,0) \\
&= -k^2 n_0^2 + n^2 - (k^2 - 1)\rho - \frac{3}{2}k^2\delta^2 p_0^2 + 3\delta^2\, \widehat{u^2}(0,0) + 0(\delta^4).
\end{aligned} \tag{46}$$

Recall that we consider $p_0 = 0(1)$ as a parameter. The dependence of the last two terms in (46) on p_0 is $0(\delta^2)$. Hence for $|k|$ sufficiently large

$$\left|\frac{\partial(46)}{\partial p_0}\right| \sim k^2\delta^2. \tag{47}$$

It follows easily from (47) and standard measure considerations that restricting p_0 to the complement of a set of small measure permits to ensure that

$$|\tilde{D}_{n,k}| > c\delta^2(1+|k|)^{-\gamma} \tag{48}$$

where $\gamma > 0$ is an arbitrary positive number (to have (48) for all k, use also (45)). From (45), (48), it also follows that for some constant C

$$|\tilde{D}_{n,k}| > c(1+|k|)^{-C} \quad \text{for all } k \tag{49}$$

(for simplicity, we do not introduce different notations for various constants; the exponential decay constant in (73), (88) from previous section however is specified). Write for $m \neq 0$

$$\|m\lambda^2\| = \left\|m\rho + \frac{3}{2}m\delta^2 p_0^2 + 0(\delta^4)\right\|. \tag{50}$$

From assumption on ρ

$$\|m\rho\| \geq (1+|m|)^{-C} \tag{51}$$

and thus

$$\|m\lambda^2\| \geq (1+|m|)^{-C} \quad \text{if } |m| < \delta^{-2/1+C}. \tag{52}$$

On the other hand, we may again restrict p_0 to the compliment of a set of small measure such that

$$(50) > \frac{\delta^2}{|m|^{1+\gamma}}. \tag{53}$$

Hence from (52), (53) ensure that for all $m \neq 0$

$$\|m\lambda^2\| > (1+|m|)^{-C_1} \tag{54}$$

and thus

$$\|m\lambda\| > \frac{1}{|m|}\|m^2\lambda^2\| > (1 + |m|)^{-C_2}. \tag{55}$$

Finally, remark also that

$$|\tilde{D}_{n,k}| > |n^2 - k^2\lambda^2| - 1 > (|k| + |n|)(\|k\lambda\| + \|n\lambda^{-1}\|) - 1. \tag{56}$$

Our aim is to control the inverse T_N^{-1} of the restriction $T_N = T|_{|n|<N,|k|<N}$ by direct analysis of the Neumann series

$$T_N^{-1} = ((1 + \tilde{T}_N\tilde{D}_N^{-1})\tilde{D}_N)^{-1} = \tilde{D}_N^{-1}\sum_{j=0}^{\infty}(-1)^j(\tilde{T}_N\tilde{D}_N^{-1})^j$$

$$= \tilde{D}_N^{-1}(1 - \tilde{T}_N\tilde{D}_N^{-1})\sum_{j=0}^{\infty}(\tilde{T}_N\tilde{D}_N^{-1}\tilde{T}_N\tilde{D}_N^{-1})^j. \tag{57}$$

Recall that one has

$$\begin{cases} \tilde{T}(x,x') = 0 \ \text{ if } \ x = x' \\ \tilde{T}(x,x') < c\delta^2 \ e^{-|x-x'|^c} \end{cases} \quad \text{(c.f. (6.73))}. \tag{58}$$

Consider $\tilde{T}\tilde{D}^{-1}\tilde{T}\tilde{D}^{-1}$ and write explicitly

$$(\tilde{T}\tilde{D}^{-1}\tilde{T}\tilde{D}^{-1})((n,k),(n',k'))$$

$$= \sum_{(n_1,k_1)} \tilde{T}((n,k),(n_1,k_1))\tilde{D}_{n_1,k_1}^{-1}\tilde{T}((n_1,k_1),(n',k'))\tilde{D}_{n',k'}^{-1} \tag{59}$$

where, by (59), $n_1 \neq n'$ or $k_1 \neq k'$. Assume $n_1 \neq n'$. Then by (55)

$$\max(\|n_1\lambda^{-1}\|, \|n'\lambda^{-1}\|) > \frac{1}{2}\|(n_1 - n')\lambda^{-1}\| > |n_1 - n'|^{-C}. \tag{60}$$

Assume

$$\|n_1\lambda^{-1}\| > |n_1 - n'|^{-C} \tag{61}$$

then, by (56)

$$|\tilde{D}_{n_1,k_1}| > (|k_1| + |n_1|)(|k_1 - k'| + |n_1 - n'|)^{-C} \tag{62}$$

provided

$$|k_1| + |n_1| \gg (|k_1 - k'| + |n_1 - n'|)^C. \tag{63}$$

Estimate using (58), (48), (49), (62)

$$(59) \lesssim$$

$$\sum_{(n_1,k_1)(63)} \delta^2 \ e^{-(|n-n_1|^c+|k-k_1|^c)} \frac{(|k_1 - k'| + |n_1 - n'|)^C}{|k_1| + 1} \tag{64}$$

$$\delta^2 \ e^{-(|n_1-n'|^c+|k_1-k'|^c)} \ \delta^{-2}(1 + |k'|)^{\gamma}$$

$$+$$

$$\sum_{(n_1,k_1) \text{ not } (63)} \delta^2 \ e^{-(|n-n_1|^c+|k-k_1|^c)}(1 + |k_1|)^C \ \delta^2 \ e^{-(|n_1-n'|^c+|k_1-k'|^c)} \ (1 + |k'|)^C.$$

$$\tag{65}$$

Writing in (64)

$$|k'|^\gamma \le |k_1|^\gamma + |k_1 - k'|^\gamma \tag{66}$$

one obtains clearly

$$(64) < c\delta^2 \; e^{-\frac{3}{4}(|n-n'|^c + |k-k'|^c)}. \tag{67}$$

In (65), the summation ranges over (n_1, k_1) satisfying

$$|k_1| + |k'| \lesssim (|k_1 - k'| + |n_1 - n'|)^C \tag{68}$$

since (63) is violated. Consequently

$$(65) < C\delta^4 \; e^{-\frac{3}{4}(|n-n'|^c) + (|k-k'|^c)} \tag{69}$$

and hence

$$(59) < C\delta^2 \; e^{-\frac{3}{4}(|k-k'|^c + |n-n'|^c)}. \tag{70}$$

It follows now immediately from (49), (57), (70) that

$$|T_N^{-1}\big((n,k),(n',k')\big)| < N^C \; e^{-\frac{1}{2}(|k-k'|^c + |n-n'|^c)} \tag{71}$$

implying in particular (6.77), (6.88) from previous section.

Remarks.

(i) At step j of the Newton iteration, the function u appearing in (40) is the approximative solution u_{j-1} obtained at previous stage. Hence the operator $T = T^{(j)}$ is changing along the iteration and there is the issue of satisfying the conditions controlling T^{-1} at each stage. What imports here is the fact that $\{u_j\}$ is rapidly convergent

$$\|u_j - u_{j-1}\| < C \; e^{-\rho^j} \quad \text{for some} \quad \rho > 1. \tag{72}$$

Hence, also

$$\|T^{(j)} - T^{(j-1)}\| < C\delta^2 \; e^{-\rho^j}. \tag{73}$$

Conditions such as (48), (49) above relative to

$$T|_{|k| \sim K}$$

will, from (73), be fulfilled by $T^{(j)}$ for all j provided they are satisfied for $j < C(\log \log K)$. Hence, this leads to a measure excision of

$$K^{-\alpha} \cdot (C \log \log K) \tag{74}$$

(for some $\alpha > 0$) in the parameter set.

(ii) At stage j, we consider the restriction $T_N = T_N^{(j)}$ with $N = M^j$ where M is some constant (depending on the nonlinear perturbation). The control of T_N^{-1} requires to remove a subset of the parameter set, which may be covered by intervals of size N^{-C}. On the other hand, if $\Delta_j u = u_j - u_{j-1}$ is the correction of the approximative solution obtained at stage j, one has

$$\|\Delta_j u\| < e^{-N^c} \tag{75}$$

and a similar bound for the derivatives *wrt* the parameters

$$\|\partial_{p_0}^\alpha \partial_\lambda^\beta (\Delta_j u)\| < e^{-\frac{1}{2}N^c}. \tag{76}$$

In particular, there is no problem[*] in assuming $\Delta_j u$ defined on the full parameter set as a smooth function of the parameters (this point is of importance in order to solve the Q-equation).

We will next discuss the proof of Theorem 131 in higher dimension d. In order to solve the P-equations, a control of the inverse T^{-1} of the linearized operator (40)

$$T = D + 3\delta^2 S \tag{77}$$

$$D_{n,k} = -k^2\lambda^2 + \mu_n^2 \tag{78}$$

$$S = S_{u^2} \tag{79}$$

requires a more elaborate technique than used earlier. In particular, the geometry of the singular sites (n, k)

$$|D_{n,k}| < \kappa \tag{80}$$

plays an essential role.

We will use the following basic lemma.

Lemma 81. *Let $T = D + \varepsilon S$ be as above, S satisfying off-diagonal decay*

$$|S(x, x')| < e^{-|x-x'|^c}. \tag{82}$$

Let Ω be a subset of the N-ball in $\mathbb{Z}^{d+1}$. Let $0 < \varepsilon_3 < \varepsilon_2 < \varepsilon_1 < \frac{1}{100}$ be fixed. Assume $\{\Omega_\alpha\}$ a collection of subsets of Ω satisfying

$$\operatorname{diam}\Omega_\alpha < M^{\varepsilon_1} \tag{83}$$

$$\operatorname{dist}(\Omega_\alpha, \Omega_\beta) > M^{\varepsilon_2} \quad \text{for} \quad \alpha \neq \beta \tag{84}$$

$$|D(x)| > \kappa > 100\varepsilon \quad \text{if} \quad x \in \Omega \backslash \cup \Omega_\alpha. \tag{85}$$

Let $\tilde{\Omega}_\alpha$ be an M^{ε_3}-neighborhood of Ω_α in Ω such that for some c_1

$$\|(T|\tilde{\Omega}_\alpha)^{-1}\| < N^{C_1} \quad \text{for all} \quad \alpha. \tag{86}$$

Then

$$\|(T|_\Omega)^{-1}\| < \kappa^{-1} N^{C_2} \tag{87}$$

and

$$|(T|_\Omega)^{-1}(x, x')| < e^{-\frac{1}{2}|x-x'|^c} \quad \text{if} \quad |x - x'| > M^{2\varepsilon_1} \tag{88}$$

(assuming M sufficiently large).

The exponent c in (88) is the same as in (82). The exponent C_2 depends on C_1.

Remark. $\{\Omega_\alpha\}$ corresponds to a partition of the singular sites (80) in "well separated clusters".

Sketch of Proof of (81).

The argument is elementary and we only specify the main points

(i) Denote $\Omega_0 = \cup \Omega_\alpha$. If $\overline{\Omega} \subset \Omega \backslash \Omega_0$, (85) permits to control the inverse $T_{\overline{\Omega}}^{-1}$ of the restriction $T_{\overline{\Omega}}$ by a Neumann series

$$T_{\overline{\Omega}} = D_{\overline{\Omega}} + \varepsilon S_{\overline{\Omega}} = (I + \varepsilon S_{\overline{\Omega}} D_{\overline{\Omega}}^{-1}) D_{\overline{\Omega}} \tag{89}$$

$$T_{\overline{\Omega}}^{-1} = D_{\overline{\Omega}}^{-1} \sum_{j=0}^{\infty} (-1)^j \varepsilon^j (S_{\overline{\Omega}} D_{\overline{\Omega}}^{-1})^j \tag{90}$$

since $\|S_{\overline{\Omega}} D_{\overline{\Omega}}^{-1}\| \leq \|D_{\overline{\Omega}}^{-1}\| < \kappa^{-1}$.

(ii) For $\Omega = \Omega_1 + \Omega_2$ (disjoint union), one has the formal identity

$$T_{\Omega}^{-1} = (T_{\Omega_1}^{-1} + T_{\Omega_2}^{-1}) - (T_{\Omega_1}^{-1} + T_{\Omega_2}^{-1})(T_\Omega - T_{\Omega_1} - T_{\Omega_2}) T_\Omega^{-1}. \tag{91}$$

Hence, if $x \in \Omega_1$, (91) implies

$$T_{\Omega}^{-1}(x, x') = T_{\Omega_1}^{-1}(x, x') - \varepsilon \sum_{x_1 \in \Omega_1, x_2 \in \Omega_2} T_{\Omega_1}^{-1}(x, x_1) S(x_1, x_2) T_{\Omega}^{-1}(x_2, x'). \tag{92}$$

We distinguish various cases. Let $K = N^{\varepsilon_3}$.

(a) **dist$(x, \Omega_0) \geq K$**

Write $\Omega = \Omega_1 + \Omega_2$ where $\Omega_1 = \Omega \cap B(x, K)$, hence $\Omega_1 \cap \Omega_0 = \phi$ and $T_{\Omega_1}^{-1}$ is controlled as in (i).

(b) **dist$(x, \Omega_0) < K$**

Then $x \in \tilde{\Omega}_\alpha$ for some α and we apply (92) with $\Omega_1 = \tilde{\Omega}_\alpha$. Thus, by (86), $\|T_{\Omega_1}^{-1}\| < N^{C_1}$. Considering the second term of (92), there are the following 2 cases.

(b_1) **dist$(x_2, \Omega_\alpha) > 2K$**

Hence $|x_1 - x_2| > 2K - K = K$ and $|S(x_1, x_2)| < e^{-K^c}$.

(b_2) **dist$(x_2, \Omega_\alpha) \leq 2K$**

From the separation (84), it follows that $x_2 \notin \tilde{\Omega}_\beta$ for $\beta \neq \alpha$ and hence dist$(x_2, \Omega_0) > K$. This permits to estimate $T_{\Omega}^{-1}(x_2, x')$ according to (a).

The preceding yields a recursive inequality on the matrix elements of T_{Ω}^{-1}, permitting us to establish both (87), (88).

In the present situation, one has

$$\Omega_0 = \{(n, k) \in \mathbb{Z}^d \times \mathbb{Z} |\ |k| \leq N, |D_{n,k}| < \kappa\}$$

$$\subset \{(n, k) \in \mathbb{Z}^d \times \mathbb{Z} |\ |k| \leq N, |\lambda^2 k^2 - |n|^2| < 1\}. \tag{93}$$

The problems are thus the following

(i) Prove for (93) a structure of separated clusters.

(ii) Verify condition (86) on the restrictions $(T|_{\tilde{\Omega}_\alpha})^{-1}$.

Recall (37), i.e.

$$|D_{n,k}| = |\lambda^2 k^2 - \mu_n^2| > c(1 + |k|)^{-C} \quad \text{if } |k| < \delta^{-c} \tag{94}$$

(and $(n, k) \notin \tilde{S}$). Here $c > 0, C < \infty$ are some constants.

Thus in (93), we may assume $|k| > \delta^{-c'}, c' > 0$. The sets Ω_α will be contained in certain balls centered at points (n_0, k_0), $|n_0| \sim |k_0|$ and radius $0(|n_0|)$. Denoting $\Omega' = \tilde{\Omega}_\alpha$, $T|_{\Omega'}$ is a selfadjoint operator and (86) amounts to keep its eigenvalues away from zero

$$|E_{\Omega'}(\lambda, p_0)| > N^{-C_1}. \tag{95}$$

If $\lambda = \lambda(p_0)$ according to (34), one has

$$\frac{\partial E_{\Omega'}}{\partial p_0} = \frac{\partial E_{\Omega'}}{\partial \lambda} \frac{\partial \lambda}{\partial p_0} + \frac{\partial E_{\Omega'}}{\partial p_0} \tag{96}$$

$$\frac{\partial \lambda}{\partial p_0} \sim \delta^2. \tag{97}$$

From first order variation, one has from the preceding ($\varphi = E$-eigenvector)

$$\left|\frac{\partial E}{\partial \lambda}\right| = \left|\left\langle \frac{\partial T}{\partial \lambda}\varphi, \varphi \right\rangle\right| \sim |n_0|^2 \tag{98}$$

since

$$T = (-k^2\lambda^2 + \mu_n^2)\mathbb{1} + 3\delta^2 S_{u^2}. \tag{99}$$

Similarly, since the dependence on p_0 may only appear from the second term in (99)

$$\left|\frac{\partial E}{\partial p_0}\right| = 0(\delta^2). \tag{100}$$

Consequently (96), (97), (98), (100) imply

$$\left|\frac{\partial E_{\Omega'}}{\partial p_0}\right| \geq |n_0|^2\delta^2. \tag{101}$$

Appropriate restriction in the parameter set will therefore ensure for each α

$$\|(T|_{\tilde{\Omega}_\alpha})^{-1}\| < N^d\delta^{-2} < N^{C_1} \tag{102}$$

letting Ω' range over the different $\tilde{\Omega}_\alpha$. (Only the case $N > \delta^{-c}$ need to be considered, by (94)). Thus the excisions in the (λ, p_0)-parameter set to ensure (95) are compatible with (34), in the sense that the resulting conditions on p_0 will only eliminate a small measure set.

Next, consider the problem of partitioning (93) in separated clusters. Observe that this will be achieved for a fixed λ, satisfying a condition

$$\left|\sum_{j=0}^{10d} a_j \lambda^j\right| > \left(\sum |a_j|\right)^{-C} \quad \text{for all } \{a_j\} \in \mathbb{Z}^{10d+1}\backslash\{0\} \tag{103}$$

for some constant C. Observe that if in (99), $|k|, |n| < N$, one varies λ by say at most N^{-3}, the original $\{\Omega_\alpha\}$ still cover the singular sites of T_N and hence the further restrictions (95) apply with given sets $\Omega' = \tilde{\Omega}_\alpha$.

Condition (6.133) yields (103) for λ replaced by ρ and hence, by (34) again, will hold automatically provided $|a| < \delta^{-c}$, for some $c > 0$. To ensure (103) for larger values of $|a|$ will require eventually a further excision in the p_0-parameter set.

We have the following result on the separation of lattice points nearby a cone of slope λ satisfying (103).

Lemma 104. *Consider the set of lattice points*

$$\Omega_0 = \{(n, k) \in \mathbb{Z}^d \times \mathbb{Z} \mid |\lambda^2 k^2 - |n|^2| < 1\}. \tag{105}$$

Fix $K > 1$. Then there is a partition

$$\Omega_0 = \cup \Omega_\alpha \tag{106}$$

such that

$$\operatorname{diam} \Omega_\alpha < K^C \tag{107}$$

and

$$\operatorname{dist}(\Omega_\alpha, \Omega_\beta) > K \quad for \quad \alpha \neq \beta. \tag{108}$$

The constant C in (107) depends only on the constant in (103) and the dimension d.

Lemma 104 is immediate from the following.

Lemma 109. *Given $x_0 \in \Omega_0$, there is a set $\Omega' \subset \Omega_0$ such that*

$$x_0 \in \Omega' \tag{110}$$

$$\operatorname{diam} \Omega' < K^C \tag{111}$$

$$\operatorname{dist}(x, \Omega') > K \ for \ all \ \ x \in \Omega_0 \backslash \Omega' \tag{112}$$

which is derived from next statement.

Lemma 113. *Let $\{x_j\}_{j=1}^{J}$ be a sequence of distinct elements of Ω_0 such that $|x_{j+1} - x_j| < K$. Then $J < K^C$.*

Proof of (113).

   (1) Choose long interval $I_0 \subset \{1, 2, \ldots, J\}$ such that

$$\operatorname{span}[x_{j_1} - x_{j_2} | j_1, j_2 \in I] = E \subset \mathbb{R}^d \tag{114}$$

$$(E \ \text{fixed})$$

for all subinterval $I \subset I_0$, $|I| > |I_0|^\gamma = M > K$

$$(\gamma > 0 \ \text{some number}).$$

   (2) Denote $\tau_\lambda = \mathbb{R}^d \to \mathbb{R}^d$ the map

$$\tau_\lambda(x) = (x_1, \ldots, x_d, -\lambda x_{d+1}). \tag{115}$$

Hence from definition of Ω_0

$$|\langle x_j, \tau_\lambda x_j \rangle| < 1 \quad j = 1, \ldots, J \tag{116}$$

and

$$|\langle x_j - x_{j'}, \tau_\lambda x_j \rangle| < |j - j'|^2 k^2 \tag{117}$$

since $|x_j - x_{j+1}| < K$.

   (3) By construction, E has a basis $\{\xi_\ell\}$

$$\xi_\ell = x_j - x_{j'} \quad |j - j'| \leq M$$

hence
$$\|\xi_\ell\| \le MK \tag{118}$$
and for $j \in I_0$, by (117)
$$|\langle \xi_\ell, \tau_\lambda x_j \rangle| < M^2 K^2. \tag{119}$$

Since
$$|\det(\langle \xi_\ell, \xi_{\ell'} \rangle)| \ge 1 \tag{120}$$
it follows from (119), (120)
$$\|P_E(\tau_\lambda x_j)\| \lesssim (MK)^{\dim E + 1}. \tag{121}$$

Hence, for $j_1, j_2 \in I$
$$\|P_E(\tau_\lambda(x_{j_1} - x_{j_2}))\| \le (MK)^{\dim E + 1} \tag{122}$$

$$x_{j_1} - x_{j_2} \in E. \tag{123}$$
Choose $j_1, j_2 \in I_0$ such that
$$\|x_{j_1} - x_{j_2}\| > |I_0|^{\frac{1}{\dim E}}. \tag{124}$$

It follows that[*]
$$\rho(P_E \tau_\lambda P_E) \lesssim \frac{(MK)^{\dim E + 1}}{|I_0|^{\frac{1}{\dim E}}} < |I_0|^{-\frac{2}{d} + 2\gamma(d+1)}. \tag{125}$$

But also, using the basis $\{\xi_\ell\}$ for E
$$\rho(P_E \tau_\lambda P_E) \gtrsim \frac{\det(\langle \tau_\lambda \xi_\ell, \xi_{\ell'} \rangle)}{\max \|\xi_\ell\|^{2 \dim E}}. \tag{126}$$

Since $\det(\langle \tau_\lambda \xi_\ell, \xi_{\ell'} \rangle)$ is a nontrivial polynomial in $\mathbb{Z}[\lambda]$, with coefficients at most
$$C(d) \max \|\xi_\ell\|^{2 \dim E} \tag{127}$$
the hypothesis on λ yields by (118)
$$|\det(\langle \tau_\lambda \xi_\ell, \xi_{\ell'} \rangle)| > (MK)^{-C} \tag{128}$$

$$\rho(P_E \tau_\lambda P_E) > (MK)^{-C-2d}. \tag{129}$$
Therefore
$$|I_0|^{-\frac{1}{d} + 2\gamma(d+1)} > |I_0|^{-2\gamma(C+2d)} \tag{130}$$
a contradiction for γ small enough.

Since J may be chosen to be a power of K in (1), $J > K^{-1/\gamma^d}$, the result follows.

---

[*] $\rho(S)$ denoting $\|S^{-1}\|^{-1}$.

Remarks.

(i) If one considers instead of a wave equation a NLS

$$iu_t - \Delta u + Vu + \varepsilon \frac{\partial H}{\partial \overline{u}} = 0 \tag{131}$$

or

$$iu_t - \Delta u + M_\sigma u + \varepsilon \frac{\partial H}{\partial \overline{u}} = 0 \tag{132}$$

one may construct time periodic solutions following the method described above (except that the simplification for $d = 1$ does not apply here and use Lemma 81 instead). The structure of singular sites is similar. Thus one considers pairs $(n, k) \in \mathbb{Z}^d \times \mathbb{Z}$ such that

$$|k\lambda - |n|^2| < 1 \tag{133}$$

instead of

$$|k^2 \lambda^2 - |n|^2| < 1 \tag{134}$$

corresponding to lattice points close to a paraboloid. Again, one has a structure of separated clusters and (104), (109), (113) hold. In this case, we do *not* need any assumption on λ, such as (103), for this issue (this is due to the geometry of a paraboloid v/s cone). In fact, the problem is equivalent to the case $\lambda = 1$ in (133), thus the set

$$\{(n, |n|^2)|n \in \mathbb{Z}^d\} \tag{135}$$

and the result, proved along the lines of Lemma 113, is a higher dimensional generalization of Jarnick's theorem about separation of lattice points on strictly convex curves in the plane.

(ii) Lemma 31 originates in the work of J. Fröhlich, T. Spencer on the Green's function and eigenstates of lattice operators of the form (Anderson model)

$$V(n) - \varepsilon \Delta \tag{136}$$

where V is a potential on $\mathbb{Z}^d$ and Δ is the lattice Laplacian

$$\Delta(i, j) = \begin{cases} 1 & \text{if } |i - j| = 1 \\ 0 & \text{otherwise.} \end{cases} \tag{137}$$

The typical examples for V are
- random potentials $V(n) = \varepsilon_n = \pm 1$
- quasi periodic potentials, $V(n) = \cos(n\lambda + \sigma)$ for example

The general pattern is that for small ε (small disorder), (136) has pure point spectrum and eigenstates exponentially decaying (for larger ε this does not need to be the case and several problems, especially in higher dimension, remain). There is an analogy of (136) and the linearized operator (40)

$$(-k^2 \lambda^2 + \mu_n^2) + 3\delta^2 S_{u^2} \tag{138}$$

considered above. Here $V(n, k) = -k^2 \lambda^2 + \mu_n^2$ is the potential on $\mathbb{Z}^{d+1}$ and S_{u^2} has a behaviour similar to Δ (translation invariance, fast off-diagonal decay). In the next section, we consider the $1D$ quasi-periodic case (Theorem (6.124)) and the linearized operator will be

$$(-\langle k, \lambda \rangle^2 + \mu_n^2) + 3\delta^2 S_{u^2}. \tag{139}$$

There is similarity with the quasi-periodic Anderson model mentioned above. In particular, the control of its inverse is based on a multiscale analysis. Considering already the case of finite dimensional phase space, thus $\{\mu_n\}$ runs over a finite set. Fixing $\mu_n^2 = A$, the linear operator (139) becomes

$$(-\langle k, \lambda \rangle^2 + A) + 3\delta^2 S_{u^2} \tag{140}$$

on the lattice $\mathbb{Z}^b$, $\lambda = (\lambda_1, \ldots, \lambda_b)$. The structure of singular sites is clearly significantly different here from the periodic case (139) and resembles the quasi-periodic Anderson problems. The techniques developed for this purpose, in particular by [F-S], do not seem immediately applicable to our situation however. The method described in the next section in the context of (6.124) will provide the general scheme.

Comments and references related to Lecture 7

The $1D$-result for the nonlinear wave equation is due to Craig and Wayne [**C-W1**] (see lecture 6). The method of proof consists in directly applying a Newton approximation scheme to the equation in order to get consecutive approximations to a solution (see discussion in previous lecture). In the particular model of the $1D$ NLW equation, there is a more elementary approach to control the inverse operators (communicated to the author by S.Kuksin) and we exposed it here along these lines. The discussion for $D > 1$ follows [**B10**]. The same approach is used. The main ingredients of the argument are certain techniques to control the inverse of the linearized operators taking into account the geometric structure of "separated clusters" for the singular sites on the diagonal. These techniques appeared in the earlier works of J. Fröhlich and T. Spencer on localization for the Anderson model. The reader may find a streamlined exposition of the Green function estimate in

[Pos]   J. Pöschel, *On Fröhlich-Spencer estimates of Green's function*, Manuscripta Math. 70, 27-37.

In the context of the NLS, the partition of $\{(n, |n|^2) | n \in \mathbb{Z}^d\}$ in separated clusters was observed earlier by A. Granville and T. Spencer.

If $\rho = 0$ in (3), it is possible to treat in 1D the problem perturbatively (by similar techniques) starting from explicit solutions of the equation $u_{tt} - u_{xx} + u^3 = 0$. Details appear in

[B11]   J. Bourgain, *Periodic solutions of nonlinear wave equations*, preprint.

The starting point is similar to

[L-S]   B. Lidskii, E. Schulman, *Periodic solutions of the equation $\dot{u} - u_{xx} + u^3 = 0$*, Funct. Anal. Appl. 22 (1988), 332-333.

Our purpose is to describe a method to deal with the problems considered in section 6 in the case of time quasi-periodic solutions. The main difficulty is the control of the inverse operator obtained by linearizing the P-equation. We carry out the discussion in the context of Theorem (6.124) and comment on the other cases later on. Thus the equation is

$$u_{tt} - u_{xx} + \rho u + u^3 + \text{(higher order terms)} = 0 \tag{1}$$

and we consider again

$$u_{tt} - u_{xx} + \rho u + u^3 = 0 \tag{2}$$

since (1) is treated identically. The unperturbed solution considered is

$$u_0(x,t) = \sum_{j=1}^{b} a_j \cos n_j x \cdot \cos \lambda_j t \tag{3}$$

and the frequencies

$$\lambda_j = (n_j^2 + \rho)^{1/2} \tag{4}$$

where

$$0 < n_1 < n_2 < \cdots < n_b \in \mathbb{Z} \tag{5}$$

is a fixed sequence.

Observe that by time translation, (3) is as general as

$$\sum_{j=1}^{b} a_j \cos n_j x \cdot \cos(\lambda_j t + \phi_j) \tag{6}$$

or

$$\sum_{j=1}^{b} \cos n_j x (a_j' \cos \lambda_j t + a_j'' \sin \lambda_j t). \tag{7}$$

Denote $\lambda' = (\lambda_1', \ldots, \lambda_b')$ the perturbed frequency. The perturbed solution $u = u(x,t)$ appears as Fourier series

$$u(x,t) = \sum_{(n,k)\in\mathbb{Z}^{1+b}} \hat{u}(n,k)\, e^{i(nx+\langle k,\lambda'\rangle t)} \tag{8}$$

where
$$\hat{u}(n, k_1, \ldots, k_b) = \hat{u}(\pm n, \pm k_1, \ldots, \pm k_b) \tag{9}$$
for all sign combinations.

The linearized equation yields the operator (after replacement of u by δu as in previous section)
$$T = (\langle k, \lambda'\rangle^2 + \mu_n^2)\mathbb{1} + 3\delta^2 S_{u^2} \tag{10}$$
with diagonal
$$D_{n,k} = -\langle k, \lambda'\rangle^2 + \mu_n^2 \tag{11}$$

$$\mu_n^2 = n^2 + \rho. \tag{12}$$
The resonant set in the representation (8) is ($e_j = j^{\text{th}}$ basis vector in $\mathbb{Z}^b$)
$$S = \{(\pm n_j, \pm e_j)| j = 1, \ldots, b\}. \tag{13}$$

We first discuss the Q-equation. Thus we have, since by (3)
$$\hat{u}(n_j, e_j) = \hat{u}_0(n_j, e_j) = \frac{1}{4}a_j \tag{14}$$

$$(-(\lambda_j')^2 + n_j^2 + \rho)a_j + 4\delta^2 \widehat{u^3}(n_j, e_j) = 0 \text{ for } j = 1, \ldots, b. \tag{15}$$
Here

$$\widehat{u^3}(n_j, e_j) = \int \left(\sum_{j=1}^{b} a_j, \cos n_j x \cdot \cos \theta_j \right)^3 \cos n_j x \cdot \cos \theta_j \, dx d\theta + 0(\delta^2) \tag{16}$$

$$= a_j^3 \int \cos^4 n_j x \cdot \cos^4 \theta_j + 3a_j \sum_{j' \neq j} a_{j'}^2 \int \cos^2 n_j x \cdot \cos^2 \theta_j \cdot \cos^2 n_{j'} x \cdot \cos^2 \theta_{j'}$$

$$= \left(\frac{3}{8}\right)^2 a_j^3 + \frac{3}{16}a_j \sum_{j' \neq j} a_{j'}^2. \tag{17}$$

Hence (15) yields
$$(\lambda_j')^2 = n_j^2 + \rho + \frac{3}{4}\delta^2 \left(\frac{3}{4}a_j^2 + \sum_{j' \neq j} a_{j'}^2 \right) + 0(\delta^4) \quad (j = 1, \ldots, b) \tag{18}$$

(we assume $a_j \sim 0(1)$) and thus
$$\left| \det \left(\frac{\partial \lambda_j'}{\partial a_{j'}} \right)_{1 \leq j, j' \leq b} \right| \sim \delta^2 \tag{19}$$

which is the amplitude-frequency modulation condition.

We discuss next the P-equation
$$(-\langle k, \lambda'\rangle^2 + \mu_n^2)\hat{u}(n, k) + 3\delta^2 \widehat{u^3}(n, k) = 0 \quad (n, k) \notin S \tag{20}$$

which is the main novelty here. Let $\varepsilon = 3\delta^2$ and consider the operator (10)
$$T = D + \varepsilon S \tag{21}$$

$$D_{n,k} = -\langle k, \lambda'\rangle^2 + \mu_n^2. \tag{22}$$

In what follows, we discuss a method to control T^{-1}. Most of the difficulty appears already in finite dimensional phase space situation (i.e. n ranges in a finite set). We denote T_N the restriction

$$T_N|_{|k|\leq N} \tag{23}$$

where the index n ranges without restriction, $(n, k) \notin S$.

One has

$$-\langle k, \lambda \rangle^2 + \mu_n^2 = -\left(\sum_{j=1}^{b} k_j \sqrt{n_j^2 + \rho} \right)^2 - n^2 - \rho. \tag{24}$$

Since the functions of ρ

$$\{(n^2 + \rho)^{1/2} | n \in \mathbb{Z}_+\} \tag{25}$$

are linearly independent, one may ensure for typical ρ that, except in the resonant cases

$$|(24)| > (1 + |k|)^{-C} \tag{26}$$

where $C = C(b)$. Hence, since

$$|\lambda' - \lambda| = 0(\varepsilon) \tag{27}$$

we have

$$|D_{n,k}| > \varepsilon^{1/10} \quad \text{if} \quad |k| < \varepsilon^{-c_1}, (n, k) \notin S. \tag{28}$$

Consequently, for $N < \varepsilon^{-c_1}$, T_N^{-1} may be controlled by a standard Neumann series

$$T_N^{-1} = D_N^{-1} \sum_{j=0}^{\infty} (-1)^j \varepsilon^j (S_N D_N^{-1})^j. \tag{29}$$

In order to control T_N^{-1} for larger N, we will carry out a multi-scale analysis and further conditions on λ' will need to be imposed. In order to exploit translation in the k-index, we introduce an additional parameter σ and define

$$T^\sigma = D^\sigma + \varepsilon S \tag{30}$$

where

$$D_{n,k}^\sigma = -(\langle k, \lambda' \rangle + \sigma)^2 + \mu_n^2. \tag{31}$$

If $A \subset \mathbb{Z}^b$, define

$$T_A^\sigma = T^\sigma|_{k \in A}. \tag{32}$$

We let here the n index vary freely, without imposing the condition $(n, k) \notin S$. Observe that

$$T_{A+k_0}^\sigma = T_A^{\sigma + \langle k_0, \lambda' \rangle} \tag{33}$$

since in (30) the matrix $S = S_{\phi = u^2}$ is in particular invariant under k-translation. It follows from (33) that a control of $(T_N^\sigma)^{-1}$ in the (λ', σ)-parameter set will permit to control as well $(T_A^\sigma)^{-1}$, for any k-ball $A = [|k - k_0| \leq N]$. This fact will be important to carry out the inductive reasoning below for different size scales N. We will ensure a bound

$$\|(T_N^\sigma)^{-1}\| < N^{C_2} \tag{34}$$

and off-diagonal decay

$$|(T_N^\sigma)^{-1}(x, y)| < e^{-\frac{1}{10}|x-y|^c} \quad \text{for} \quad |x - y| > (\log N)^{C_3} \tag{35}$$

by imposing bounds on the reciprocal of certain polynomial expressions.

More precisely, we require

$$|p(\sigma_1)| > N^{-C_4} \tag{36}$$

where

$$\sigma_1 = \sigma + \langle k_1, \lambda' \rangle \pm \mu_{n_1} \quad (|k_1| < N) \tag{37}$$

and $p(\sigma_1)$ is a polynomial of the form

$$p(\sigma_1) = \sigma_1^d + \sum_{j=0}^{d-1} a_j(\lambda')\sigma_1^j \tag{38}$$

$$= \underbrace{\Pi_{\sum |k_j| \leq 2}(\sigma_1 + \langle k, \lambda - \lambda' \rangle)}_{\text{degree } d \leq 4b^2} + \sum_{j<d} b_j(\lambda')\sigma_1^j \tag{39}$$

where

$$|b_j|, |\partial^\alpha b_j| = 0(\varepsilon(1 + |n_1|)^{-1}). \tag{40}$$

For a given n_1, the polynomial p will range over a family of at most N^{C_5} polynomials, obtained essentially by restricting the (λ', σ)-parameters to small intervals and controlling $(T_N^\sigma)^{-1}$ on these regions by a perturbation of the determinant of a submatrix of bounded size. In the case of finite dimensional phase space, one needs to control at stage N a number of polynomial expressions bounded by a power of N. In our case, the number is infinite (since σ has unbounded range), but (40) implies specific compactness properties. More precisely, if $n_0 \to \infty$, (38) reduces in the limit to the first term in (39), which dependence on n_0 appears only through (37). From (19), the variation range of λ' is $0(\varepsilon)$. The conditions on λ' appearing below will require a measure excision of size ε^C, where C may be taken arbitrarily large. We describe the first step for $N = N_0 \sim \varepsilon^{-c_1}$ and then give the general inductive step. Denote

$$E = \{(n, k) \in \mathbb{Z} \times \mathbb{Z}^b | \ |k| \leq N \text{ and } |\sigma + \langle k, \lambda' \rangle + \nu\mu_n| < \varepsilon^{1/10} \text{ for } \nu = 1 \text{ or } \nu = -1\}. \tag{41}$$

Assume $(n_1, k_1) \in E, (n_2, k_2) \in E$, hence

$$\begin{cases} |\sigma + \langle k_1, \lambda' \rangle + \nu_1\mu_{n_1}| < \varepsilon^{1/10} \\ |\sigma + \langle k_2, \lambda' \rangle + \nu_2\mu_{n_2}| < \varepsilon^{1/10} \end{cases} \tag{42}$$

$(\nu_i = \pm 1)$ and thus by subtraction

$$|\langle k_1 - k_2, \lambda' \rangle + \nu_1\mu_{n_1} - \nu_2\mu_{n_2}| < 2\varepsilon^{1/10}. \tag{43}$$

Since $|k_1 - k_2| < 2N \sim \varepsilon^{-C_1}$, (43) implies

$$|\langle k_1 - k_2, \lambda \rangle + \nu_1\mu_{n_1} - \nu_2\mu_{n_2}| < 3\varepsilon^{1/10}. \tag{44}$$

Recall that

$$\mu_n = n + 0\left(\frac{1}{n}\right) \text{ for } n \to \infty. \tag{45}$$

Imposing on λ a condition (satisfied for typical ρ)

$$\|\langle k, \lambda \rangle\| > |k|^{-C} \text{ for all } k \neq 0 \tag{46}$$

will therefore imply (if c_1 is sufficiently small) that

$$\langle k_1 - k_2, \lambda\rangle + \nu_1\mu_{n_1} - \nu_2\mu_{n_2} = 0, \quad |k_1 - k_2| \leq 2. \tag{47}$$

Hence, if we define

$$\sigma_1 = \sigma + \langle k_1, \lambda'\rangle + \nu_1\mu_{n_1} \tag{48}$$

one has by (47)

$$\sigma + \langle k_2, \lambda'\rangle + \nu_2\mu_{n_2} = \langle k_2 - k_1, \lambda' - \lambda\rangle + \sigma_1, \quad |k_1 - k_2| \leq 2. \tag{49}$$

Write

$$\mathbb{Z} \times \{k \in \mathbb{Z}^b|\ |k| \leq N\} = E_0 \cup E \tag{50}$$

with E as above. Hence

$$|-(\sigma+\langle k, \lambda'\rangle)^2+\mu_n^2| > (1+|n|) \min_{\nu=\pm 1} |\sigma+\langle k, \lambda'\rangle+\nu\mu_n| > \varepsilon^{1/10} \ \text{if}\ (n, k) \in E_0 \tag{51}$$

and the inverse of the restriction $T_{E_0}^\sigma = T^\sigma|_{(n,k)\in E_0} = D_{E_0}^\sigma + \varepsilon S_{E_0}$ is therefore controlled by a Neumann series. In particular

$$\|(T_{E_0}^\sigma)^{-1}\| < \varepsilon^{-1/10} \tag{52}$$

and there is off-diagonal decay estimate

$$|(T_{E_0}^\sigma)^{-1}(x, y)| < \varepsilon^{-1/2}\, e^{-|x-y|^c} \ \text{for}\ x \neq y. \tag{53}$$

Writing

$$T_N^\sigma = \begin{pmatrix} T_{E_0}^\sigma & Q^* \\ Q & T_E^\sigma \end{pmatrix}$$

$(T_N^\sigma)^{-1}$ is clearly controlled by

$$(T_E^\sigma - Q(T_{E_0^\sigma})^{-1}Q^*)^{-1} \tag{54}$$

and, by (52)

$$\|(T_N^\sigma)^{-1}\| < \|(T_{E_0}^\sigma)^{-1}\|\ \|(T_E^\sigma - Q(T_{E_0}^\sigma)^{-1}Q^*)^{-1}\| < \varepsilon^{1/10}\ \|(T_E^\sigma - Q(T_{E_0}^\sigma)^{-1}Q^*)^{-1}\| \tag{55}$$

reducing the problem to controlling the inverse of a matrix of bounded size. This reduction holds as long as (52) is valid, hence restricting σ to an interval in the parameter set of size $\sim \varepsilon^{1/10}$, from definition of E (at later stages of the induction, these restrictions will apply to both λ' and σ).

Consider (54). Write from (30)

$$T_E^\sigma - Q(T_{E_0}^\sigma)^{-1}Q^* = D_E^\sigma + \varepsilon S_E - Q(T_{E_0}^\sigma)^{-1}Q^* = [-(\sigma+\langle k, \lambda'\rangle)^2+\mu_n^2]\mathbb{1}_E + 0(\varepsilon) \tag{56}$$

since $Q = \varepsilon S|_{E\times E_0}$ and (52).

(The analysis of $Q(T_{E_0}^\sigma)^{-1}Q^*$ will be more subtle at later stages of the iteration).

If $(n, k) \in E$, then for some $\nu = \pm 1$

$$-(\sigma + \langle k, \lambda'\rangle)^2 + \mu_n^2 = 0(1 + |n|)(\sigma + \langle k, \lambda'\rangle + \nu\mu_n). \tag{57}$$

Recall that from (49)

$$\sigma + \langle k, \lambda' \rangle + \nu \mu_n = \langle \Delta k, \lambda' - \lambda \rangle = \sigma_1, |\Delta k| \leq 2. \tag{58}$$

Hence, (54) is controlled by the inverse of a matrix of the form

$$(\sigma + \langle k, \lambda' \rangle + \nu_{n,k} \mu_n) \mathbb{1}_E + 0(\varepsilon (1 + |n_1|)^{-1}) \tag{59}$$

by the reciprocal of its determinant

$$\Pi_{(k,n) \in E}(\sigma + \langle k, \lambda' \rangle + \nu_{n,k} \mu_n) + 0(\varepsilon(1 + |n_1|)^{-1}) \tag{60}$$

$$= \Pi_{|k| \leq 2} (\sigma_1 + \langle k, \lambda' - \lambda \rangle) + 0(\varepsilon(1 + |n_1|)^{-1}) \tag{61}$$

(The product in (61) extends over a subset of the indices k with $|k| \lesssim 2$).

The error term in (61) is a size estimate. Similar bounds hold also for derivations in λ' and σ_1 of bounded order. Write

$$T_{E_0}^\sigma = (\sigma + \langle k, \lambda' \rangle + \nu_{n,k} \mu_n) \mathbb{1}_{E_0} \cdot \big((\sigma + \langle k, \lambda' \rangle - \nu_{n,k} \mu_n) \mathbb{1}_{E_0} + 0(\varepsilon)\big) \tag{62}$$

where

$$|\sigma + \langle k, \lambda' \rangle + \nu_{n,k} \mu_n| > 1, |\sigma + \langle k, \lambda' \rangle - \nu_{n,k} \mu_n| > \varepsilon^{1/10} \tag{63}$$

and substitute σ by σ_1 according to (48). Thus $T_{E_0}^\sigma$ appears as product of

$$(\sigma_1 + \langle k - k_1, \lambda' \rangle + \nu_{n,k} \mu_n - \nu_1 \mu_{n_1}) \mathbb{1}_{E_0} \tag{64}$$

and

$$(\sigma_1 + \langle k - k_1, \lambda' \rangle - \nu_{n,k} \mu_n - \nu_1 \mu_{n_1}) \mathbb{1}_{E_0} + 0(\varepsilon). \tag{65}$$

Using this decomposition and the derivative formula

$$\partial U^{-1} = -U^{-1} \partial U \cdot U^{-1} \tag{66}$$

we get

$$\partial (Q(T_{E_0}^\sigma)^{-1} Q^*) < 0(\varepsilon^2 N \varepsilon^{-1/5}) = 0(\varepsilon) \tag{67}$$

and similarly for higher derivatives of bounded order (decrease the power $\frac{1}{10}$ in (41) if necessary).

Hence

$$(61) = \Pi_{|k| \leq 2} (\sigma_1 + \langle k, \lambda - \lambda' \rangle) + \varphi(\lambda', \sigma_1) \tag{68}$$

where

$$|\partial^\alpha \varphi| = 0\big(\varepsilon(1 + |n_1|)^{-1}\big). \tag{69}$$

We now apply Malgrange's preparation theorem in order to replace locally (68) by a polynomial of the form (39), (40), thus

$$p(\sigma_1) = \Pi(\sigma_1 + \langle k, \lambda' - \lambda \rangle) + \sum_{j<d} b_j(\lambda') \sigma_1^j \tag{70}$$

where d is the σ_1-degree of the first term in (70) and $\{b_j\}$ satisfying (40)

$$|b_j|, |\partial^\alpha b_j| < 0(\varepsilon(1 + |n_1|)^{-1}). \tag{71}$$

Observe that we only need the analytic version of the preparation theorem, since (68) is obtained as (restriction of) a rational function in σ_1.

Thus, by (55), $(T_N^\sigma)^{-1}$ will be controlled by reciprocals of polynomials of the form (70)

$$\|(T_N^\sigma)^{-1}\| < \varepsilon^{-1/10}|p(\sigma_1)|^{-1} < N^{1/c_1}|p(\sigma_1)|^{-1} \tag{72}$$

introduced for the different intervals of size $\sim \varepsilon^{1/10}$ in our partition of the parameter set. Thus, if (36) holds, i.e.

$$|p(\sigma_1)| > N^{-C_4} \tag{73}$$

for all $\sigma_1 = \sigma + \langle k_1, \lambda' \rangle \pm \mu_{n_1}, |k_1| \le N$ and polynomials p satisfying (39), (40), then by (72)

$$\|(T_N^\sigma)^{-1}\| < N^{1/c_1+C_4} < N^{C_2}. \tag{74}$$

Recall that the polynomials p form an infinite family because of the unrestricted sequence $\{\mu_n\}$. From (39), (40) they reduce however in the limit for $n_1 \to \infty$ to the first term in (39).

Assume $|x - y| > (\log N)^{C_3}$. Then either $x \in E_0$ or $y \in E_0$. Assume $x \in E_0$, the resolvent identity

$$T_{\Lambda_1 \cup \Lambda_2}^{-1} = (T_{\Lambda_1}^{-1} + T_{\Lambda_2}^{-1}) - (T_{\Lambda_1}^{-1} + T_{\Lambda_2}^{-1})(T_{\Lambda_1 \cup \Lambda_2} - T_{\Lambda_1} - T_{\Lambda_2})T_{\Lambda_1 \cup \Lambda_2}^{-1} \tag{75}$$

applied with $\Lambda_1 = E_0, \Lambda_2 = E$ yields

$$(T_N^\sigma)^{-1}(x, y) = \varepsilon \sum_{z_1 \in E_0, z_2 \in E} (T_{E_0}^\sigma)^{-1}(x, z_1)S(z_1, z_2)(T_N^\sigma)^{-1}(z_2, y) \tag{76}$$

hence

$$|(T_N^\sigma)^{-1}(x, y)| < \varepsilon N^{C_2} \sum_{z_1 \in E_0, z_2 \in E} |(T_{E_0}^\sigma)^{-1}(x, z_1)| \, |S(z_1, z_2)| \tag{77}$$

and for an appropriate C_3, (35) follows clearly from the off-diagonal decay of S

$$|S(x, y)| < e^{-|x-y|^c} \tag{78}$$

and (52), (53).

Differentiating (76), one estimates similarly $\partial_{\lambda'}^\alpha \partial_\sigma^\beta (T_N^\sigma)^{-1}$.

Next we discuss the inductive procedure. We will occasionally drop more technical considerations for the sake of the exposition.

Starting with $N_0 \sim \varepsilon^{-c_1}$, define

$$N_{j+1} = N_j^{C_6} \tag{79}$$

where the exponent C_6 will be specified.

Assume $N_j \le N \le N_{j+1}$ and consider the restriction $T_N^\sigma = T^\sigma|_{k \in [-N,N]^b}$.

Let $A_1 = k_1 + [-N_j, N_j]^b, A_2 = k_2 + [-N_j, N_j]^b$ be 2 disjoint boxes contained in $[N, N]^b$. Thus for $i = 1, 2$

$$T_{A_i}^\sigma = T_{N_j}^{\sigma + \langle k_i, \lambda' \rangle}. \tag{80}$$

Assume for $i = 1, 2$

$$\|(T_{A_i}^\sigma)^{-1}\| > N_j^{C_2}. \tag{81}$$

It then follows that there are $k_i' \in A_i, n_i \in \mathbb{Z}$ and polynomials p_i satisfying (39), (40) (obtained at stage j) such that

$$|p_i(\sigma_i)| < N_j^{-C_4} \quad (i = 1, 2) \tag{82}$$

where

$$\sigma_i = \sigma + \langle k_i', \lambda' \rangle \pm \mu_{n_i}. \tag{83}$$

Here

$$p_i(\sigma_i) = \sigma_i^{d_i} + \sum_{j=0}^{d_i} a_{ij}(\lambda')\sigma_i^j \quad (d_i \le 4b^2, i = 1, 2). \tag{84}$$

Hence denoting

$$\Delta = \sigma_1 - \sigma_2 = \langle k_1' - k_2', \lambda' \rangle \pm \mu_{n_1} \pm \mu_{n_2} \tag{85}$$

one may find a polynomial of the form

$$p(x) = x^d + \sum_{j=0}^{d-1} a_j(\lambda')x^j \tag{86}$$

where $d \le d_1 d_2$ and the coefficients $\{a_j(\lambda')\}$ are in the ring generated by the $\{a_{ij}(\lambda')\}$ in (84), such that

$$|p(\Delta)| < C_d N_j^{-C_4}. \tag{87}$$

Observe that, since by (39), (40)

$$|\partial^\alpha a_j| < C_d \tag{88}$$

in (86), one has by (85)

$$|\nabla_{\lambda'}^{(d)} p| \sim |k_1' - k_2'|^d > 1 \tag{89}$$

provided $|k_1' - k_2'|$ is sufficiently large, $|k_1' - k_2'| > C_d$, which is clearly the case if

$$|k_1 - k_2| > 3N_j. \tag{90}$$

Our purpose is to restrict λ' so that (87) is impossible.

Fix n_1, n_2 momentarily. The number of polynomials p appearing in (87) is $0(N_j^{2C_5})$ and there are $\sim N^{2b}$ pairs (k_1', k_2'). This yields $N_j^{2C_5} N^{2b}$ conditions. From (89), it follows that in particular

$$|p(\Delta(\lambda'))| > \delta \tag{91}$$

on a λ'-set of measure at most $\delta^{1/d}$. Let in (85)

$$|n_i| < M \quad (i = 1, 2)$$

where

$$N_j^{C_4} > M > N_j \tag{92}$$

will be specified. The preceding yields then a λ'-set of measure

$$N_j^{2C_5} N^{2b} M^2 \delta^{1/d}. \tag{93}$$

Assume next (cf. (85))

$$|n_i| > \frac{1}{2}M \quad (i = 1, 2). \tag{94}$$

It follows from (39), (40), (82), (92) that for $i = 1, 2$

$$\Pi_{|k| \le 2}|\sigma_i + \langle k, \lambda' - \lambda \rangle| < N_j^{-C_4} + 0(M^{-1}) < M^{-1} \tag{95}$$

and hence

$$|\sigma_1 - \sigma_2 + \langle \bar{k}, \lambda' - \lambda \rangle| \lesssim M^{-1/d} \tag{96}$$

for some $\bar{k}, |\bar{k}| \leq 4$. Thus, by (85), we obtain

$$|\langle k_1' - k_2' + \bar{k}, \lambda' \rangle - \langle \bar{k}, \lambda \rangle \pm \mu_{n_1} \pm \mu_{n_2}| < M^{-1/d}. \tag{97}$$

Using the fact that $\mu_n = |n| + 0\left(\frac{1}{|n|}\right)$, we may assume λ' satisfies

$$(97) > |k_1' - k_2'|^{-C} \tag{98}$$

provided (90) holds (for some constant C). Hence, from (97), (98)

$$N > M^{1/dC}. \tag{99}$$

Taking thus

$$M = 0(N^{dC}) \tag{100}$$

and

$$\delta = 0(N_j^{-C_4}) \tag{101}$$

it follows from (93) that under the assumption (90) one may rule out (81) provided we perform a measure excision in the λ'-parameter set of size at most

$$N_j^{2C_5} N^{2b} N^{2dC} N_j^{-C_4/d} = N^C N_j^{2C_5 - C_4/d} < N_{j+1}^{C'} N_j^{2C_5 - C_4/d}. \tag{102}$$

Condition (92) is compatible with (100), if we let

$$C_4 > 20b^4 C\, C_6. \tag{103}$$

Assume

$$C_4 > 40\, b^4 C_5 \tag{104}$$

and

$$C_4 > 40\, b^4 C' C_b \tag{105}$$

so that in particular

$$(102) < N_{j+1}^{-1}. \tag{106}$$

Take

$$C_6 \sim C_4^{(*)} \tag{107}$$

according to (103), (105).

Let in (34)

$$C_2 = 2C_4. \tag{108}$$

It follows from the preceding that if $A = k + [-N_j, N_j]^b, |k| \leq N$, then, by (107), (108)

$$\|(T_A^\sigma)^{-1}\| < N_j^{C_2} < N_{j+1}^C \tag{109}$$

(C a constant), except possibly for k in some box of size $3N_j$. Thus there is a box

$$\Lambda_j = k_j + [-3N_j, 3N_j]^b \subset \Lambda_{j+1} = [-N, N]^b \tag{110}$$

such that for all $k \in \Lambda_{j+1} \backslash \Lambda_j$, letting $A = k + [-N_j, N_j]^b$, we have

$$\|(T_A^\sigma)^{-1}\| < N_{j+1}^C \tag{111}$$

and $(T_A^\sigma)^{-1}$ satisfying the exponential decay estimate (35).

$^{(*)}$ We consider b and d as constants

We now replace Λ_{j+1} by Λ_j, N_j by N_{j-1} and repeat the construction, yielding a box

$$\Lambda_{j-1} = k_{j-1} + [-3N_{j-1}, 3N_{j-1}]^b \subset \Lambda_j \tag{112}$$

such that for all $k \in \Lambda_j \backslash \Lambda_{j-1}$, letting $A = k + [-N_{j-1}, N_{j-1}]^b$, we have

$$\|(T_A^\sigma)^{-1}\| < N_j^C \tag{113}$$

and exponential decay estimates.

Continuing this way yields a decreasing system of boxes

$$\Lambda_{j+1} \supset \Lambda_j \supset \Lambda_{j-1} \supset \cdots \supset \Lambda_0 \tag{114}$$

where Λ_0 is an N_0-box as in the first step. The property of this construction is that if $k \in \Lambda_{r+1} \backslash \Lambda_r$ and $A = k + [-N_r, N_r]^b$, then

$$\|(T_A^\sigma)^{-1}\| < N_{r+1}^C \tag{115}$$

and exponential decay estimates. We further assume the construction done such that $\Lambda_{r+1} \supset \tilde{\Lambda}_r \cap B_N$ where $\tilde{\Lambda}_r$ is the usual doubling of Λ_r. This is clearly possible. Let

$$\mathbb{Z} \times \Lambda_0 = E_0 \cup E \tag{116}$$

be the decomposition constructed in the first step and let

$$(n_1, k_1) \in E \qquad \sigma_1 = \sigma + \langle k_1, \lambda' \rangle \pm \mu_{n_1} \tag{117}$$

be as in (48).

Consider the following decomposition *wrt* the (n, k)-index

$$B_N = E \cup F \tag{118}$$

where

$$F = \cup_{r=0}^{j+1} F_r \tag{119}$$

and

$$\begin{cases} F_0 = E_0 \\ F_r = \mathbb{Z} \times (\Lambda_r \backslash \Lambda_{r-1}) \qquad r = 1, \ldots, j+1. \end{cases} \tag{120}$$

Consider T_F^σ. Let $x_1 \in F, x_2 \in F$ and assume $x_1 \in F_r$. Hence, from construction of $\{\Lambda_r\}$,

$$\mathrm{dist}(x_1, E) > N_{r-2}. \tag{121}$$

Write

$$F = \Omega_1 + \Omega_2 \tag{122}$$

where

$$\begin{cases} \Omega_1 = x_1 + (\mathbb{Z} \times [-N_{r-1}, N_{r-1}]^b) & \text{if } r \geq 1 \\ \Omega_1 = E_0 & \text{if } r = 0. \end{cases} \tag{123}$$

Apply the resolvent identity *wrt* the decomposition (122). Thus

$$(T_F^\sigma)^{-1}(x_1, x_2) = (T_{\Omega_1}^\sigma)^{-1}(x_1, x_2)$$
$$- \varepsilon \sum_{x_3 \in \Omega_1, x_4 \in \Omega_2} (T_{\Omega_1}^\sigma)^{-1}(x_1, x_3) S(x_3, x_4)(T_F^\sigma)^{-1}(x_4, x_2) \tag{124}$$

hence

$$|(T_F^\sigma)^{-1}(x_1, x_2)| < |(T_{\Omega_1}^\sigma)^{-1}(x_1, x_2)|$$

$$+ \varepsilon \sum_{x_3 \in \Omega_1, x_4 \in \Omega_2} |(T_{\Omega_1}^\sigma)^{-1}(x_1, x_3)| \, |S(x_3, x_4)| \, |(T_F^\sigma)^{-1}(x_1, x_2)|.$$

$$(125)$$

Since $x_1 \in F_r$, it follows from (115) that

$$\|(T_{\Omega_1}^\sigma)^{-1}\| < N_r^C \tag{126}$$

and

$$|(T_{\Omega_1}^\sigma)^{-1}(x_1, x_2)| < e^{-\frac{1}{10}|x_1 - x_2|^c} \text{ if } |x_1 - x_2| > (\log N_{r-1})^{C_3} \tag{127}$$

(the case $r = 0$ yields $(T_{E_0}^\sigma)^{-1}$ considered in the first step).

Consider now the second term in (125). Observe that (if $r \geq 1$), $|x_1 - x_4| \geq N_{r-1}$, hence from (78), i.e.

$$|S(x_3, x_4)| < e^{-|x_3 - x_4|^c} \tag{128}$$

and (126), (127), one gets clearly the estimate

$$\varepsilon^{9/10} \sum_{x_4 \in \Omega_2} e^{-\frac{1}{10}|x_1 - x_4|^c} \, |(T_F^\sigma)^{-1}(x_4, x_2)| \tag{129}$$

for the second term in (125). Hence, by (121), (126), (127), (129)

$$|(T_F^\sigma)^{-1}(x_1, x_2)| <$$

$$\begin{cases} N_r^C < \varepsilon^{-\frac{1}{10}}(1 + \text{dist}(x_1, E))^{C\,C_6^2} & (130) \\[2em] e^{-\frac{1}{10}|x_1 - x_2|^c} \text{ if } |x_1 - x_2| > C_6^{C_3}(\log(2 + d(x_1, E)))^{C_3} & (131) \end{cases}$$

$$+ \sqrt{\varepsilon} \max_{x_1' \in F} e^{-\frac{1}{10}|x_1 - x_1'|^c} |(T_F^\sigma)^{-1}(x_1', x_2)|. \tag{132}$$

From (130), (132), it follows that

$$\|(T_F^\sigma)^{-1}\| < N_{j+1}^C + \sqrt{\varepsilon}\|(T_F^\sigma)^{-1}\| \tag{133}$$

hence

$$\|(T_F^\sigma)^{-1}\| < N_{j+1}^C. \tag{134}$$

We also need a more refined estimate.

In (132), x_1 is replaced by x_1' but there is a factor $e^{-\frac{1}{10}|x_1 - x_1'|^c}$. We may repeat the estimate for $(T_F^\sigma)^{-1}(x_1', x_2)$. Iteration yields that

$$|(T_F^\sigma)^{-1}(x_1, x_2)| < \varepsilon^{-\frac{1}{10}}(1 + \text{dist}(x_1, E))^{C\,C_6^2} \tag{135}$$

and

$$|(T_F^\sigma)^{-1}(x_1, x_2)| < e^{-\frac{1}{10}|x_1 - x_2|^c} \text{ if } |x_1 - x_2| > (\log N_{j+1})^{C_3}. \tag{136}$$

In the preceding, the parameters λ', σ were fixed. Clearly one may vary them in an N_{j+1}^{-C}-size neighborhood (C some constant depending on (134)), leaving previous conclusions (134), (135), (136) valid. Thus the number of such subintervals in the

(λ', σ)-parameter space is at most $N_{j+1}^{C(1+b)} < N_{j+1}^{C_5}$. We restrict in the sequel (λ', σ) to an interval. Write as in the first step

$$T_N^\sigma = \begin{pmatrix} T_F^\sigma & Q^* \\ Q & T_E^\sigma \end{pmatrix}.$$

Hence, by (134)

$$\|(T_N^\sigma)^{-1}\| < \|(T_F^\sigma)^{-1}\|^2 \ \|(T_E^\sigma - Q(T_F^\sigma)^{-1}Q^*)^{-1}\| < N_{j+1}^C \ \|(T_E^\sigma - Q(T_F^\sigma)^{-1}Q^*)^{-1}\| \tag{137}$$

and we need to analyze $Q(T_F^\sigma)^{-1}Q^*$. The estimates on $Q(T_F^\sigma)^{-1}Q^*$ are here not immediate from the smallness of ε as in the first step and we make essential use of (135).

Write for $x_1, x_2 \in E$

$$Q(T_F^\sigma)^{-1}Q^*(x_1, x_2) = \varepsilon^2 \sum_{x_3, x_4} S(x_1, x_3)(T_F^\sigma)^{-1}(x_3, x_4)S(x_4, x_2) \tag{138}$$

and estimate using (128), (135)

$$|Q(T_F^\sigma)^{-1}Q^*(x_1, x_2)| < \varepsilon \sum_{x_3, x_4} e^{-|x_1-x_3|^c - |x_4-x_2|^c}[1 + d(x_3, E)]^{C \, C_6^2} \lesssim \varepsilon \tag{139}$$

since E is bounded and thus $1 + d(x_3, E) \sim 1 + |x_3 - x_1|$.

One may similarly estimate derivatives of $(T_F^\sigma)^{-1}$ (and hence of $Q(T_F^\sigma)^{-1}Q^*$) *wrt* λ' and σ_1. One differentiates (124) and controls $\partial(T_{\Omega_1}^\sigma)^{-1}$ from an inductive hypothesis on $\partial(T_{N_{r-1}}^\sigma)^{-1}$. Observe that by (123), (48)

$$T_{\Omega_1}^\sigma = T_{N_{r-1}}^{\sigma + \langle \overline{x}_1, \lambda' \rangle} = T_{N_{r-1}}^{\sigma_1 + \langle \overline{x}_1 - k_1, \lambda' \rangle - \nu_1 \mu_{n_1}} \tag{140}$$

where $\overline{x}_1$ denotes the k-component of x_1. It is important here to have σ replaced by σ_1 in order to get, when computing $\partial_{\lambda'}$ again, estimates in terms of dist(x_1, E). Hence one obtains (139) also for λ', σ_1-derivatives. We are thus in the same situation as in the first step. i.e.

$$T_E^\sigma - Q(T_F^\sigma)^{-1}Q^* = [-(\sigma_1 + \langle k - k_1, \lambda' \rangle - \nu_1 \mu_{n_1})^2 + \mu_n^2]\mathbb{1}_E + 0(\varepsilon) \tag{141}$$

and the polynomial $p(\sigma_1)$ is gotten following the same procedure and the preparation theorem. From (137), we have

$$\|(T_N^\sigma)^{-1}\| < N_{j+1}^C |p(\sigma_1)|^{-1}. \tag{142}$$

From the preceding, the number of polynomials will for fixed n_1 be bounded by a power $N_{j+1}^{C_5}$ of N_{j+1}. Assuming

$$|p(\sigma_1)| > N_{j+1}^{-C_4} \tag{143}$$

for all $\sigma_1 = \sigma + \langle k_1, \lambda' \rangle \pm \mu_{n_1}, |k_1| \leq N$ and polynomials p, (142) implies

$$\|(T_N^\sigma)^{-1}\| < N_{j+1}^{C+C_4} < N_{j+1}^{C_2} \tag{144}$$

so that (108), i.e. $C_2 = 2C_4$ is compatible. The off-diagonal decay estimates for $(T_N^\sigma)^{-1}$ are obtained as in the first step, using the resolvent identity and the estimates (134), (136) on $(T_F^\sigma)^{-1}$.

This completes the construction of the polynomials controlling $(T_N^\sigma)^{-1}$ for various scales of N. In order to control T_N^{-1}, we let $\sigma = 0$ and remove the resonant set S from the index set $\{(n, k) | n \in \mathbb{Z}, |k| \leq N\}$.

(36) will now result in conditions on λ'. More precisely, letting

$$\sigma_1(\lambda') = \langle k_1, \lambda' \rangle \pm \mu_{n_1} \tag{145}$$

we need to ensure that

$$|p(\sigma_1(\lambda'))| > N^{-C_4} \tag{146}$$

with p as in (38), (40). Observe that we may assume $\sigma_1 = 0(1)$, hence $|n_1| \lesssim N$ in (145). If $|k_1| > C_d$, one clearly has

$$|\nabla_{\lambda'}^{(d)} p| \sim |k_1|^d > 1 \tag{147}$$

and realizing (146) may thus be achieved excising a λ'-parameter set of size N^{-C_7}, for some constant C_7, depending on C_4, C_5, d.

If $|k_1| < C_d < N_0$, the conditions are automatically fulfilled, from the nonresonance property (26),

$$|\langle k, \lambda' \rangle \pm \mu_n| > N_0^{-C} \tag{148}$$

since the set S is excluded. More precisely, one decomposes the index set as

$$(\mathbb{Z} \times B_N) \backslash S = \big((\mathbb{Z} \times B_{N_0}) \backslash S \big) \cup \cup \Omega_\alpha \tag{149}$$

where the Ω_α are intervals of the form $\mathbb{Z} \times (k_\alpha + [-N_j, N_j]^b)$ where j is increasing and applies the resolvent identity. Here $T_{\Omega_\alpha} = T_{B_{N_j}}^{\sigma = \langle k_\alpha, \lambda' \rangle}$ for which the conditions $|p(\sigma_1)| > N_j^{-C_4}$ for the corresponding polynomials may be ensured as above.

The remarks (7.72)-(7.76) from previous section apply also here in solving the P-equations. The conditions on λ', in fact on λ' and $\bar{a} = \{a_j\}_{j=1,\dots,b}$, amount to remove from the parameter set a set of measure $< \varepsilon^C$, for any fixed power C. Solving $\lambda' = \lambda'(\bar{a})$ from the Q-equations, (19) holds and the exceptional $(\bar{a}, \lambda')$-set is avoided by removing from the $\bar{a}$-parameter set a set of measure $< \varepsilon^C$. This completes the construction.

Further comments

(1) To prove persistency results for $1D$ NLS equations as considered in Theorem (6.112) for instance

$$iu_t - u_{xx} + Vu + \varepsilon \frac{\partial H}{\partial \bar{u}} = 0 \tag{150}$$

one may essentially reproduce the argument given above for the $1D$ wave equation. There are a few adjustments. Consider the case of periodic boundary conditions

(i) Denoting $v = \bar{u}$, we consider the pair of equations in u, v

$$\begin{cases} iu_t - u_{xx} + Vu + \varepsilon\frac{\partial H}{\partial v} = 0 \\ -iv_t - v_{xx} + Vv + \varepsilon\frac{\partial H}{\partial u} = 0 \end{cases} \tag{151}$$

and its linearization (leading to a (2×2)-block matrix) as discussed in section 6.

(ii) The (periodic) spectrum $\{\mu_n\} = \{\mu_n^i | n \in \mathbb{Z}_+, i = 1, 2\}$ is given by

$$\mu_n^i = \pi^2 n^2 + \int_0^1 V(x)dx + 0(n^{-2}) = \pi^2 n^2 + 0(1) \quad (i = 1, 2). \tag{152}$$

Hence

$$\mu_{n_1}^{i_1} - \mu_{n_2}^{i_2} \approx \pi^2(n_1^2 - n_2^2) \quad \text{for} \quad n_1, n_2 \to \infty \tag{153}$$

implying

$$|\mu_{n+1}^{i_1} - \mu_n^{i_2}| \sim n \quad \text{for} \quad n \to \infty \tag{154}$$

and

$$\text{dist}(\mu_{n_1}^{i_1} - \mu_{n_2}^{i_2}, \pi^2\mathbb{Z}) \to 0 \quad \text{for} \quad n_1, n_2 \to \infty. \tag{155}$$

At the first stage, one estimates from (152)

$$|\langle \lambda', k \rangle - \mu_n^i| > \|\frac{1}{\pi^2}\langle \lambda', k \rangle - \frac{1}{\pi^2}\int_0^1 V\| - 0(n^{-2}) \quad \text{for} \quad n \to \infty. \tag{156}$$

(iii) Consider the matrices $T_N^\sigma = T_{|k| \leq N}^\sigma$, where

$$T^\sigma = \begin{pmatrix} \langle \lambda', k \rangle + \sigma - \mu_n^i & 0 \\ 0 & -\langle \lambda', k \rangle - \sigma - \mu_n^i \end{pmatrix} + \varepsilon \begin{pmatrix} S_{\frac{\partial^2 H}{\partial u \partial v}} & S_{\frac{\partial^2 H}{\partial v^2}} \\ S_{\frac{\partial^2 H}{\partial u^2}} & S_{\frac{\partial^2 H}{\partial u \partial v}} \end{pmatrix} \tag{157}$$

Here S_ϕ is the matrix expressing multiplication by ϕ *wrt* the basis of periodic eigenfunctions $\{\varphi_n^i | n \in \mathbb{Z}_+, i = 1, 2\}$ of $-\frac{d^2}{dx^2} + V$, thus

$$S_\phi\big((k_1, n_1, i_1), (k_2, n_2, i_2)\big) = \int_{\Pi \times \Pi^b} \phi(x, \theta)\, e^{i\langle k_1 - k_2, \theta \rangle}\, \varphi_{n_1}^{i_1}(x)\varphi_{n_2}^{i_2}(x)dxd\theta \tag{158}$$

for $k_1, k_2 \in \mathbb{Z}^b$; $n_1, n_2 \in \mathbb{Z}_+$; $i_1, i_2 = 1, 2$. The eigenfunctions $\{\varphi_n^i\}$ are exponentially well-localized *wrt* the standard exponential system (see the discussion in section 6). Thus one has decay estimates

$$|(158)| < e^{-|k_1 - k_2|^c - |n_1 - n_2|^c} \tag{159}$$

but no translation invariance *wrt* the n-index. Remark that only the translation invariance of (158) *wrt* k-index matters in the multi-scale analysis.

(iv) Assume σ large *wrt* N. Then there is $n_0 \in \mathbb{Z}_+, n_0 \to \infty$ such that (assuming $\sigma \to 0$) the control of $(T_N^\sigma)^{-1}$ amounts to controlling the inverse of

$$(\langle \lambda', k \rangle + \sigma - \pi^2 n_0^2 - \int_0^1 V)\mathbb{1} + \varepsilon S_{\int \frac{\partial^2 H}{\partial u \partial v}dx}\big||k| \leq N + 0(\frac{1}{n_0}). \tag{160}$$

Thus in the limit for $n_0 \to \infty$ the dependence on n_0 again only appears in

$$\sigma_1 = \sigma - \pi^2 n_0^2 + \langle \lambda', k_0 \rangle. \tag{161}$$

This fact is of importance as in the discussion of the NLW case above.

(2) Also for the $2D$ NLS of the form discussed in Theorem (6.121) one may perform a similar analysis. However here there are significant technical complications. We discuss the main one. The control of $(T_N^\sigma)^{-1}$ in our method is achieved by reciprocals of perturbed "local determinants", meaning that we consider the determinant of a perturbation of a submatrix. Considering here for the diagonal elements

$$D_{n,k} = \langle \lambda', k \rangle + \sigma - |n|^2 \quad (k \in \mathbb{Z}^b, n \in \mathbb{Z}^2) \tag{162}$$

such a "minimal matrix" will at least involve the index set

$$E = \{(n,k) \mid |n| = |n_0| \text{ and } k = k_0\} \tag{163}$$

thus lattice points on the $R = |n_0|$ circle. As is well-known, this number is unbounded for $R \to \infty$ and hence one expects polynomials in σ of unbounded degree.

This makes matters more delicate, in particular when applying the preparation theorem to the perturbed polynomial

$$\Pi_{|n|=R}(\sigma + \langle k_0, \lambda' \rangle - |n|^2) + \text{ perturbation.} \tag{164}$$

One makes essential use of the following 2 basic facts on the distribution of lattice points on circles.

(i) On a circle of radius R, there are at most

$$\exp \frac{\log R}{\log \log R} \ll R^\varepsilon \tag{165}$$

lattice points (the bound comes from the divisor function, factoring R^2 in the Gaussian integers $\mathbb{Z} + i\mathbb{Z}$).

(ii) The lattice points on an R-circle may be partitioned in separated clusters $\{C_\alpha\}$ of at most 2 points, such that

$$\text{dist}(C_\alpha, C_\beta) > R^{1/10} \text{ for } \alpha \neq \beta \tag{166}$$

(this property holds also more generally for lattice points on strictly convex curves in the plane, c/o Jarnick's theorem).

The control of the inverse of $T_N^\sigma = T^\sigma|_{|k| \leq N}$ is reduced to controlling the inverse of a matrix

$$T_E^\sigma - Q^*(T_F^\sigma)^{-1}Q \tag{167}$$

with E as above (the procedure is analogous to the $1D$ case).

Write according to the preceding

$$E = \cup_\alpha (C_\alpha \times \{k_0\}) = \cup_\alpha E_\alpha \tag{168}$$

and from (166) and off-diagonal decay

$$\oplus_\alpha [T_{E_\alpha}^\sigma - (R_{E_\alpha} Q^*(T_F^\sigma)^{-1}Q R_{E_\alpha})] + 0(e^{-R^c}) \tag{169}$$

where R_{E_α} denotes the restriction to E_α.

The diagonal pieces are matrices of bounded size (size 1 or 2). Applying the preparation theorem permits then to replace

$$\det(T^\sigma_{E_\alpha} - R_{E_\alpha} Q^* (T^\sigma_F)^{-1} Q R_{E_\alpha}) \tag{170}$$

by a polynomial $p_\alpha(\sigma)$ in σ of degree 1 or 2. Hence, for (169) we get

$$\det(169) = \Pi_\alpha p_\alpha(\sigma) + 0(e^{-R^c}) \tag{171}$$

where the first term is a polynomial of degree $d \ll R^\varepsilon$. Precise quantitative versions of the preparation theorem (in the analytic case) permit thus to replace (171) itself by a σ-polynomial. It is of importance here that the error term in (171) (and sufficiently many derivatives) are exponentially small in a large power of the degree.

We seek for bounds of the form

$$\|(T^\sigma_N)^{-1}\| < \exp(\log N)^C \tag{172}$$

rather then (34). Letting $R = (\log N)^C$ in (171), this leads to polynomials p of degree $d < \exp \frac{\log \log N}{\log \log \log N} \ll (\log N)^\varepsilon$ at stage N.

(3) In dimension ≥ 3, the difficulty to carry out the preceding discussion for $2D$ is the failure of (165). In $3D$ for instance, one gets the estimate $R^{1+\varepsilon}$ for the number of lattice points on an R-sphere. The partition in separated (lower dimensional) clusters is still valid and the preparation theorem may be applied to each (170)-factor. For (171) however, the error is not sufficiently small v/s the degree and the preparation theorem does not seem applicable. It is reasonable to believe that this obstacle is of a technical nature.

(4) For the $2D$-wave equation, the frequencies $\{\mu_n\}$ are essentially

$$\mu_n = (n_1^2 + n_2^2)^{1/2} \tag{173}$$

and hence the difference set $\{\mu_{n'} - \mu_{n''} | n', n'' \in \mathbb{Z}^2\}$ is dense in $\mathbb{R}$. This invalidates the $1D$ approach, already at the first stage, in order to keep

$$\langle \lambda', k \rangle + \mu_{n'} - \mu_{n''}, |k| > 2 \tag{174}$$

away from zero. Presently there is only the persistency result for time periodic solutions, discussed in section 7. The full potential of the method discussed in sections 6-8 has not yet been completely explored however.

References and comments

This section is mainly an exposition of [**B8**], with particular emphasis to the $1D$ wave equation. Details on the construction of quasi-periodic solutions for $2D$ NLS may be found in [**B9**].

LECTURE 9
Normal Forms

In previous sections, we discussed persistency of quasi-periodic solutions of linear systems under Hamiltonian perturbations, provided the linear system satisfies certain nonresonance properties. This theory is also applicable to certain nonlinear PDE's which apriori does not appear in this form. They may sometimes be reduced to the previous case by the method of normal forms. This happens for perturbations of integrable systems but also in other more general examples. We illustrate this in the context of Theorem 6.136, Theorem 6.142 for NLS equations

$$iu_t - u_{xx} + mu + u\,f(|u|^2) = 0 \qquad (1\text{D}) \tag{1}$$

$$iu_t - \Delta u + mu + u\,f(|u|^2) = 0 \qquad (2\text{D}) \tag{2}$$

assuming f real analytic in a neighborhood of 0 and nondegenerate

$$f'(0) \neq 0 \tag{3}$$

Remarks.

    (i) Since under assumption (3), (1) may be written as

$$iu_t - u_{xx} + mu + cu|u|^2 + u\,0(|u|^4) = 0, \quad c \neq 0.$$

This example may be viewed as a perturbation of the integrable equation

$$iu_t - u_{xx} + mu + u|u|^2 = 0. \tag{4}$$

This approach does not apply to equation (2) in 2D and a more general technique is used here.

  (ii) Our discussion of (1) follows a forthcoming paper of S. Kuksin and J. Pöschel. These authors consider the case of Dirichlet boundary conditions. We treat periodic boundary conditions instead (see the discussion in section 6.)

 (iii) In what follows, the nonlinear perturbations considered will often be real analytic instead of polynomial. This situation is captured also by the methods described in the preceding sections, up to minor technicalities. In fact, they apply, most likely, in the smooth category (considering sufficiently many derivatives), although details would need to be worked out here.

117

(iv) We discuss (1), (2) in the case $f(t) = t$. The general case is completely similar. Observe also that replacement of u by $e^{im't}u$, $m' \in \mathbb{R}$, permits to consider m as a parameter in (1), (2).

Writing

$$q = \sum q_n \, e^{i\langle n, x \rangle} \tag{5}$$

equations (1), (2) have the Hamiltonian form

$$\dot{q}_n = i \, \frac{\partial H}{\partial \bar{q}_n} \tag{6}$$

where

$$H(q, \bar{q}) = \sum \lambda_n |q_n|^2 + \int g(|q|^2) dx \tag{7}$$

$$\lambda_n = |n|^2 + m \tag{8}$$

and

$$g = \int_0 f. \tag{9}$$

The general idea of the method of normal forms is to simplify the initial orders in the nonlinearity by removing the nonresonant part using symplectic transformations of the phase space.

Let $\Gamma : q \to q'$ be obtained as time-1 map under the flow of the Hamiltonian vector field X_F with Hamiltonian $F(q)$. Thus $q' = v(1)$, where v solves

$$\begin{cases} v_t = i \, \frac{\partial F}{\partial \bar{v}} \\ v(0) = q. \end{cases} \tag{10}$$

Then the equations

$$\dot{q}_n = i \, \frac{\partial H}{\partial \bar{q}_n} \tag{11}$$

are transformed into

$$\dot{q}'_n = i \, \frac{\partial H'}{\partial \bar{q}'_n} \tag{12}$$

where, by Taylor's formula

$$H' = H\Gamma^{-1} = H + \{H, F\} + \frac{1}{2!} \, \{\{H, F\}, F\} + \ldots \tag{13}$$

and

$$\{H, F\} = 2 \, \mathrm{Im} \sum_n \frac{\partial H}{\partial q_n} \frac{\partial F}{\partial \bar{q}_n} \tag{14}$$

denotes the Poisson bracket.

Consider first the 1D case, thus the equation

$$iu_t - u_{xx} + mu + \delta \, u|u|^2 = 0 \tag{15}$$

with Hamiltonian

$$\begin{aligned} H(q) &= \sum \lambda_n |q_n|^2 + \frac{\delta}{2} \int |q|^4 \\ &= \sum \lambda_n q_n \bar{q}_n + \frac{\delta}{2} \sum_{n_1 - n_2 + n_3 - n_4 = 0} q_{n_1} \bar{q}_{n_2} q_{n_3} \bar{q}_{n_4} \end{aligned} \tag{16}$$

($\delta > 0$ is a small number, which we may always introduce rescaling the equation $u \to \sqrt{\delta} u$).

In order to simplify the second term of (16), consider the resonance relation

$$\lambda_{n_1} - \lambda_{n_2} + \lambda_{n_3} - \lambda_{n_4} = n_1^2 - n_2^2 + n_3^2 - n_4^2 = 0 \tag{17}$$

with

$$n_1 - n_2 + n_3 - n_4 = 0. \tag{18}$$

The key point in 1D is the identity

$$(17) = -(n_2 - n_1)(n_2 - n_3) \tag{19}$$

assuming (18). Thus the only resonances occur in the case

$$\{n_1, n_3\} = \{n_2, n_4\} \tag{20}$$

and we will transform (16) to

$$H'(q) = \sum \lambda_n \, q_n \bar{q}_n + \frac{\delta}{2} \sum_{\{n_1, n_3\} = \{n_2, n_4\}} q_{n_1} \bar{q}_{n_2} q_{n_3} \bar{q}_{n_4} + 0(\delta^2) \tag{21}$$

$$= \sum \lambda_n \, |q_n|^2 + \frac{\delta}{2} \sum |q_n|^4 + 2\delta \sum_{n_1 < n_2} |q_{n_1}|^2 |q_{n_2}|^2 + 0(\delta^2). \tag{22}$$

This is achieved as follows. Write according to (16)

$$H = H_0 + H_1 \tag{23}$$

$$H_0 = \sum \lambda_n \, q_n \bar{q}_n \tag{24}$$

$$H_1 = \frac{1}{2} \sum_{n_1 - n_2 + n_3 - n_4 = 0} q_{n_1} \bar{q}_{n_2} q_{n_3} \, \bar{q}_{n_4}. \tag{25}$$

Replacing in (13) F by δF, we get

$$H_0 + \delta H_1 + \delta \{H_0, F\} + 0(\delta^2) \tag{26}$$

and we choose F such that

$$\{H_0, F\} = -H_1^{nr} \tag{27}$$

where

$$H_1^{nr} = \frac{1}{2} \sum_{\substack{n_1 - n_2 + n_3 - n_4 = 0 \\ \{n_1, n_3\} \neq \{n_2, n_4\}}} q_{n_1} \bar{q}_{n_2} q_{n_3} \bar{q}_{n_4} \tag{28}$$

is the nonresonant part of H_1. Thus according to (14) we let

$$F = \sum F_{n_1 n_2 n_3 n_4} \, q_{n_1} \bar{q}_{n_2} q_{n_3} \bar{q}_{n_4} \tag{29}$$

where

$$F_{n_1 n_2 n_3 n_4} = \begin{cases} \dfrac{i}{\lambda_{n_1} - \lambda_{n_2} + \lambda_{n_3} - \lambda_{n_4}} & \text{if} \quad \begin{matrix} n_1 - n_2 + n_3 - n_4 = 0 \\ \{n_1, n_3\} \neq \{n_2, n_4\} \end{matrix} \\ \qquad\qquad 0 & \text{otherwise} \end{cases} \tag{30}$$

and verify (27).

The evolution equations (6) corresponding to (22) are

$$\dot q_n = i\,\frac{\partial H'}{\partial \bar q_n} = i\left\{\lambda_n q_n + \delta\,|q_n|^2\,q_n + 2\delta\left(\sum_{n'\neq n} |q_{n'}|^2\right) q_n + \delta^2\,\frac{\partial H_2}{\partial \bar q_n}\right\}. \qquad (31)$$

Recall that the L^2-norm

$$\int |u(x)|^2 dx = \sum |q_n|^2 \qquad (32)$$

is a conserved quantity for NLS of the form (1) or (2).
Hence, denoting

$$\rho = \sum |q_n|^2 \qquad (33)$$

(31) gives

$$i\dot q_n + \left(\lambda_n + 2\delta\rho - \delta\,|q_n|^2\right) q_n + 0(\delta^2) = 0. \qquad (34)$$

Coming back to Theorem 6.136, fix integers $0 < n_1 < n_2 < \cdots < n_b$, $\mathfrak{S} = \{n_1,\dots,n_b\}$ and coefficients $\{a_j\} \subset \mathbb{R}$. Write for $n \in \mathfrak{S}$

$$|q_n|^2 = a_n^2 + \varepsilon I_n \qquad (35)$$

denoting

$$\delta = \varepsilon^2 \qquad (36)$$

and

$$\tilde\lambda_n = \lambda_n + 2\delta\rho - \delta a_n^2 = \lambda_n + 2\varepsilon^2\rho - \varepsilon^2 a_n^2 = n^2 + m + 2\varepsilon^2\rho - \varepsilon^2 a_n^2 \quad (n \in \mathfrak{S}) \ (37)$$

(the unperturbed frequencies).
  Hence, by (34)

$$i\dot I_n = \frac{2}{\varepsilon}\,i\,\mathrm{Re}\,\dot q_n \bar q_n = \frac{2}{i}\,\varepsilon^3\,\mathrm{Im}\,\frac{\partial H_2}{\partial \bar q_n}\,\bar q_n = \varepsilon^3\left(\frac{\partial H_2}{\partial q_n}\,q_n - \frac{\partial H_2}{\partial \bar q_n}\,\bar q_n\right) \qquad (38)$$

and for $n \in \mathfrak{S}$, by (35)

$$i\dot q_n + \tilde\lambda_n q_n + 0(\varepsilon^3) = 0. \qquad (39)$$

Replace q_n by εq_n if $n \notin \mathfrak{S}$, yielding the equations

$$i\dot q_n + \mu_n q_n + 0(\varepsilon^3) = 0 \qquad (40)$$

with

$$\mu_n = \lambda_n + 2\varepsilon^2\rho = n^2 + m + 2\varepsilon^2\rho. \qquad (41)$$

The term $2\varepsilon^2\rho$ in (31), (35) may be absorbed in m.
Observe that by variation of the a_n-coefficients for $n \in \mathfrak{S}$, the frequencies $\{\tilde\lambda_n | n \in \mathfrak{S}\}$ in (37) range in an ε^2-size parameter space, while in (38), (39), (40) the perturbative term is $0(\varepsilon^3)$.
  The system

$$\begin{cases} i\dot I_n + 0(\varepsilon^3) = 0 & (n \in \mathfrak{S}) \\ i\dot q_n + \tilde\lambda_n q_n + 0(\varepsilon^3) = 0 & (n \in \mathfrak{S}) \\ i\dot q_n + \mu_n q_n + 0(\varepsilon^3) = 0 & (n \notin \mathfrak{S}) \end{cases} \qquad (42)$$

may therefore be treated as perturbation of the linear system (parameter dependent)

$$\begin{cases} i\dot{I}_n = 0 & (n \in \mathfrak{S}) \\ i\dot{q}_n + \tilde{\lambda}_n q_n = 0 & (n \in \mathfrak{S}) \\ i\dot{q}_n + \mu_n q_n = 0 & (n \notin \mathfrak{S}) \end{cases} \tag{43}$$

with unperturbed solution

$$\begin{cases} I_n = 0 & (n \in \mathfrak{S}) \\ q_n = a_n \, e^{i\tilde{\lambda}_n t} & (n \in \mathfrak{S}) \\ q_n = 0 & (n \notin \mathfrak{S}). \end{cases} \tag{44}$$

The Q-equations correspond to the projection on the resonant set $\mathcal{S} = \{(n_j, e_j) \mid j = 1, \dots, b\}$, where e_j is the jth unit vector in $\mathbb{Z}^b$. Some comments about equation (38) are needed because of the $k = 0$ Fourier mode $\widehat{I}_n(0)$, where

$$I_n(t) = \sum_{k \in \mathbb{Z}^b} \widehat{I}_n(k) \, e^{i\langle k, \lambda' \rangle t}. \tag{45}$$

According to (35), one has, since

$$\widehat{q_{n_j}}(e_j) = a_j \tag{46}$$

$$\widehat{I}_n(0) = \frac{1}{\varepsilon} \sum_{(n,k) \notin \mathcal{S}} |\widehat{q}_n(k)|^2. \tag{47}$$

Starting from

$$\sum_{n \in \mathfrak{S}} a_n \, e^{i(nx + \lambda'_n t)} \tag{48}$$

with real $\{a_n\}$, the Newton iteration process will yield real Fourier coefficients $\{\widehat{q}_n(k)\}$ and $\{\widehat{I}_n(k)\}$ since the nonlinear terms in (38)–(40) are expressed in monomials $\Pi_\alpha \, q_{n_\alpha} \bar{q}_{n'_\alpha}$ with real coefficients. Hence

$$\frac{\partial H_2}{\partial q_n} q_n - \frac{\partial H_2}{\partial \bar{q}_n} \bar{q}_n \tag{49}$$

has vanishing 0-Fourier mode and equations (38) do not give additional resonances.

We next discuss the 2D-case and consider equation (2) in the form

$$iu_t - \Delta u + mu + \delta u |u|^2 = 0. \tag{50}$$

The main difference with 1D is that the resonance (17), (18), i.e.

$$n_1 - n_2 + n_3 - n_4 = 0 \tag{51}$$

$$\lambda_{n_1} - \lambda_{n_2} + \lambda_{n_3} - \lambda_{n_4} = |n_1|^2 - |n_2|^2 + |n_3|^2 - |n_4|^2 = 0 \tag{52}$$

implies

$$\langle n_2 - n_1, n_2 - n_3 \rangle = 0 \tag{53}$$

and hence $\{n_1, n_2, n_3, n_4\}$ are the consecutive vertices of a rectangle in $\mathbb{Z}^2$. Hence, previous symplectic transformation yields for the new Hamiltonian

$$H'(q) = \sum \lambda_n |q_n|^2 + \frac{\delta}{2} \sum |q_n|^4 + 2\delta \sum_{n_1 \neq n_2} |q_{n_1}|^2 |q_{n_2}|^2$$

$$+ 2\delta \sum_{\mathcal{R}} q_{n_1} \bar{q}_{n_2} q_{n_3} \bar{q}_{n_4} + \delta^2 H_2(q, \bar{q}) + 0(\delta^3) \tag{54}$$

when $\mathcal{R}$ denotes 4-tuples $\{n_1, n_2, n_3, n_4\}$ with $\{n_1, n_3\}$, $\{n_2, n_4\}$ opposite vertex pairs of a rectangle and H_2 is of degree 6 in $q_n, \bar{q}_n$.

We perform one more symplectic transformation to remove the nonresonant part of H_2, i.e. we assume H_2 contains only monomials of the form

$$q_{n_1} \bar{q}_{n_2} q_{n_3} \bar{q}_{n_4} q_{n_5} \bar{q}_{n_6} \quad \text{with} \quad \begin{cases} n_1 - n_2 + n_3 - n_4 + n_5 - n_6 = 0 \\ |n_1|^2 - |n_2|^2 + |n_3|^2 - |n_4|^2 + |n_5|^2 - |n_6|^2 = 0. \end{cases} \tag{55}$$

Next, fix two lattice points $n_1, n_2 \in \mathbb{Z}^2$ of the same norm

$$|n_j| = R \qquad (j = 1, 2), \ n_1 \neq -n_2. \tag{56}$$

The set $\mathfrak{S} = \{n_1, n_2\}$ will be our distinguished set of Fourier modes in constructing a quasi-periodic solution. Consider all rectangles in $\mathbb{Z}^2$ for which n_1, n_2 are vertices and define

$$\mathcal{V} = \bigcup_{\{n_1, n, n_2, n'\} \in \mathcal{R}} \{n, n'\} \tag{57}$$

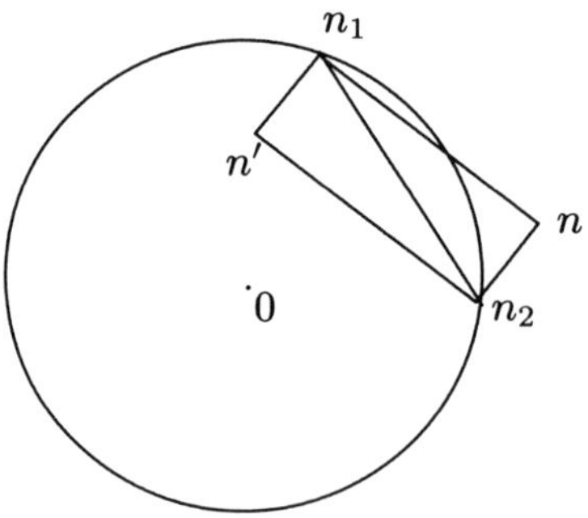

and

$$\mathcal{W} = \bigcup_{\{n_1, n_2, n, n'\} \in \mathcal{R}} \{n, n'\} \tag{58}$$

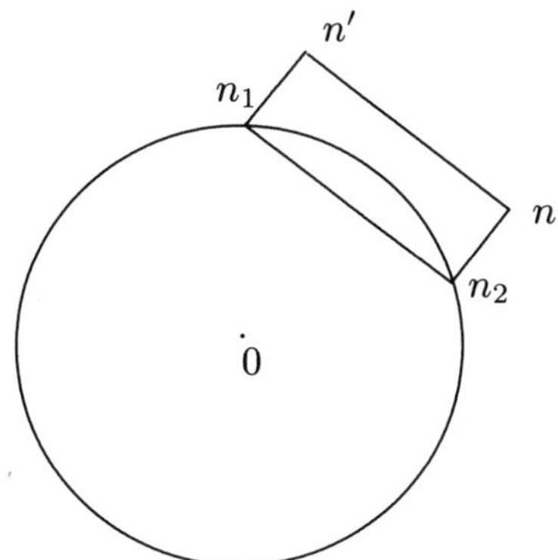

Thus $\mathcal{V}$ is finite and contains always

$$\{(n_1^1, n_2^2),\ (n_2^1, n_1^2)\} \tag{59}$$

where $n_1 = (n_1^1, n_1^2)$, $n_2 = (n_2^1, n_2^2)$.

The set $\mathcal{W}$ is infinite and contains, in particular, $\{-n_1, -n_2\}$. Clearly $\mathcal{V} \cap W = \phi$.

Let $\varepsilon^2 = \delta$. Proceeding as in the 1D case, we get by (56)

$$\begin{cases} \tilde{\lambda}_n = |n|^2 + m + 2\varepsilon^2\rho - \varepsilon^2 a_n^2 = R^2 + m - \varepsilon^2 a_n^2 & (n \in \mathfrak{S}) \\ \mu_n = |n|^2 + m + 2\varepsilon^2\rho = |n|^2 + m \end{cases} \tag{60}$$

absorbing the $2\varepsilon^2\rho$-term in m. By (54), we get the equations

$$\begin{cases} i\dot{I}_n + 0(\varepsilon^3) = 0 & (n \in \mathfrak{S}) & (61) \\ i\dot{q}_n + \tilde{\lambda}_n q_n + 0(\varepsilon^3) = 0 & (n \in \mathfrak{G}) & (62) \\ i\dot{q}_n + \mu_n q_n + \lambda\varepsilon^2 q_{n_1} q_{n_2} \bar{q}_{n'} + 0(\varepsilon^3) = 0 & (n \in \mathcal{V}) & (63) \\ i\dot{q}_n + \mu_n q_n + 2\varepsilon^2 \bar{q}_{n_1} q_{n_2} q_{n'} + 0(\varepsilon^3) = 0 & (n, n' \in \mathcal{W}) & (64) \\ i\dot{q}_{n'} + \mu_{n'} q_{n'} + 2\varepsilon^2 q_{n_1} \bar{q}_{n_2} q_n + 0(\varepsilon^3) = 0 & & \\ i\dot{q}_n + \mu_n q_n + 0(\varepsilon^3) = 0 & (n \notin \mathfrak{S} \cup \mathcal{V} \cup \mathcal{W}) & (65) \end{cases}$$

Our main concern is the presence of the $\varepsilon^2 q_{n_1} q_{n_2} \bar{q}_{n'}$ term in (63) and the $\varepsilon^2 \bar{q}_{n_1} q_{n_2} q_{n'}$, $\varepsilon^2 q_{n_1} \bar{q}_{n_2} q_n$ terms in (64) which are of the same order ε^2 as the size of the λ'-parameter set, of (60).

Next, we remove these terms. First consider (63), which we rewrite as

$$i\dot{q}_n + \mu_n q_n + 2\varepsilon^2 q_{n_1} q_{n_2} \bar{q}_{n'} + \varepsilon^3 \frac{\partial H''}{\partial \bar{q}_n} + 0(\varepsilon^5) = 0 \tag{66}$$

where H'' consists of degree 4 and degree 6 terms, which, by (55), are resonant.

Differentiate (66) and substitute (62), (66) to get

$$-\ddot{q}_n + \mu_n(i\dot{q}_n) + 2\varepsilon^2(i\dot{q}_{n_1})\,q_{n_2}\bar{q}_{n'} + 2\varepsilon^2 q_{n_1}(i\dot{q}_{n_2})\bar{q}_{n'} - 2\varepsilon^2 q_{n_1}q_{n_2}\overline{i\dot{q}_{n'}}$$

$$+ i\varepsilon^3\left(\frac{\partial H''}{\partial\bar{q}_n}\right)^{\!\bullet} + 0(\varepsilon^5) = 0$$

$$= -\ddot{q}_n + \mu_n(i\dot{q}_n) - 2\varepsilon^2(\tilde{\lambda}_{n_1} + \tilde{\lambda}_{n_2})q_{n_1}q_{n_2}\bar{q}_{n'} + 0(\varepsilon^5) + 2\varepsilon^2\mu_{n'}q_{n_1}q_{n_2}\bar{q}_{n'}$$

$$+ 4\varepsilon^4\,|q_{n_1}|^2\,|q_{n_2}|^2\,q_n + 0(\varepsilon^5) + i\varepsilon^3\left(\frac{\partial H''}{\partial\bar{q}_n}\right)^{\!\bullet}. \tag{67}$$

If Q is a (resonant) monomial from H'', it follows from (62)–(65) that

$$i\left(\frac{\partial Q}{\partial\bar{q}_n}\right)^{\!\bullet} = -(|n|^2 + m)\,\frac{\partial Q}{\partial\bar{q}_n} + 0(\varepsilon^2). \tag{68}$$

Hence, from (67), (68), we get the equation

$$-\ddot{q}_n + \varepsilon\mu_n\dot{q}_n + 4\varepsilon^4\,|q_{n_1}|^2\,|q_{n_2}|^2 q_n$$

$$-2\varepsilon^2(\tilde{\lambda}_{n_1} + \tilde{\lambda}_{n_2} - \mu_{n'})\,q_{n_1}q_{n_2}\bar{q}_{n'} - \varepsilon^3(|n|^2 + m)\,\frac{\partial H''}{\partial\bar{q}_n} + 0(\varepsilon^5) = 0 \tag{69}$$

and since by (60)

$$\tilde{\lambda}_{n_1} + \tilde{\lambda}_{n_2} - \mu_{n'} = |n|^2 + m - \varepsilon^2(a_{n_1}^2 + a_{n_2}^2) = \mu_n - \varepsilon^2(a_{n_1}^2 + a_{n_2}^2) \tag{70}$$

$$|q_n|^2 = a_n^2 + \varepsilon\,I_n \quad \text{for} \quad n \in \mathfrak{S} \tag{71}$$

we obtain from (69)

$$-\ddot{q}_n + i\mu_n\dot{q}_n + 4\varepsilon^4\,a_{n_1}^2 a_{n_2}^2 q_n - (\tilde{\lambda}_{n_1} + \tilde{\lambda}_{n_2} - \mu_{n'})\left(2\varepsilon^2 q_{n_1}q_{n_2}\bar{q}_{n'} + \varepsilon^3\,\frac{\partial H}{\partial\bar{q}_n}\right) + 0(\varepsilon^5) = 0 \tag{72}$$

Substitution of (66) in (72) yields

$$\ddot{q}_n + i\mu_n\dot{q}_n + 4\varepsilon^4\,a_{n_1}^2 a_{n_2}^2 q_n + (\tilde{\lambda}_{n_1} + \tilde{\lambda}_{n_2} - \mu_{n'})\,(i\dot{q}_n + \mu_n q_n) + 0(\varepsilon^5) = 0 \tag{73}$$

hence

$$-\ddot{q}_n + i(2\mu_n - \varepsilon^2(a_{n_1}^2 + a_{n_2}^2))\dot{q}_n + (4\varepsilon^4 a_{n_1}^2 a_{n_2}^2 + \mu_n^2 - \varepsilon^2(a_{n_1}^2 + a_{n_2}^2)\mu_n)q_n + 0(\varepsilon^5) = 0. \tag{74}$$

Replace equations (63) by (74). Writing

$$q_n = \sum_{k \in \mathbb{Z}^b} q_{nk}\,e^{i\langle k, \lambda'\rangle t} \tag{75}$$

we obtain thus for $n \in \mathcal{V}$

$$\left[\langle\lambda', k\rangle^2 - (2\mu_n - \varepsilon^2(a_{n_1}^2 + a_{n_2}^2))\langle\lambda', k\rangle + (\mu_n^2 - \varepsilon^2(a_{n_1}^2 + a_{n_2}^2)\mu_n + 4\varepsilon^4 a_{n_1}^2 a_{n_2}^2)\right]q_{n,k}$$

$$+ 0(\varepsilon^5) = 0 \tag{76}$$

hence

$$\left[\left(\langle \lambda', k\rangle - \mu_n + \frac{\varepsilon^2}{2}\,(a_{n_1}^2 + a_{n_2}^2)\right)^2 + \varepsilon^4\left(4a_{n_1}^2 a_{n_2}^2 - \frac{1}{4}(a_{n_1}^2 + a_{n_2}^2)^2\right)\right]q_{nk} \tag{77}$$
$$+\, 0(\varepsilon^5) = 0.$$

We proceed similarly for equations (64)

$$\begin{cases} i\dot{q}_n + \mu_n q_n + 2\varepsilon^2 \bar{q}_{n_1} q_{n_2} q_{n'} + \varepsilon^3\, \frac{\partial H''}{\partial \bar{q}_n} + 0(\varepsilon^5) = 0 & (78) \\[2mm] i\dot{q}_{n'} + \mu_{n'} q_{n'} + 2\varepsilon^2\, q_{n_1} \bar{q}_{n_2} q_n + \varepsilon^3\, \frac{\partial H''}{\partial \bar{q}_{n'}} + 0(\varepsilon^5) = 0 & (79) \end{cases}$$

leading to

$$\left[\left(\langle \lambda', k\rangle - \mu_n - \frac{\varepsilon^2}{2}\,(a_{n_1}^2 - a_{n_2}^2)\right)^2 - \varepsilon^4(4\,a_{n_1}^2 a_{n_2}^2) + \frac{1}{4}\,(a_{n_1}^2 - a_{n_2}^2)^2\right]q_{n,k} + 0(\varepsilon^5) = 0. \tag{80}$$

The resulting system of P-equations (61), (62), (77), (80) is thus

$$\begin{cases} -\langle \lambda', k\rangle\, I_{nk} + 0(\varepsilon^3) = 0 & (n \in \mathfrak{S},\ k \neq 0) & (81) \\[2mm] (-\langle \lambda', k\rangle + \tilde{\lambda}_n)q_{nk} + 0(\varepsilon^3) = 0 & (n \in \mathfrak{S},\ (n,k) \notin \mathfrak{S}) & (82) \\[4mm] \left[\left(\langle \lambda', k\rangle - \mu_n + \frac{\varepsilon^2}{2}(a_1^2 + a_2^2)\right)^2 + \varepsilon^4\left(4\,a_1^2 a_2^2 - \frac{1}{4}\,(a_1^2 + a_2^2)^2\right)\right]q_{n,k} + 0(\varepsilon^5) = 0 & & \\[2mm] & (n \in \mathcal{V}) & (83) \\[4mm] \left[\left(\langle \lambda', k\rangle - \mu_n \pm \frac{\varepsilon^2}{2}\,(a_1^2 - a_2^2)\right)^2 - \varepsilon^4\left(4\,a_1^2 a_2^2 + \frac{1}{4}\,(a_1^2 - a_2^2)^2\right)\right]q_{n,k} + 0(\varepsilon^5) = 0 & & \\[2mm] & (n \in \mathcal{W}) & (84) \\[4mm] (-\langle \lambda', k\rangle + \mu_n)q_{n,k} + 0(\varepsilon^3) = 0 & (n \notin \mathfrak{S} \cup \mathcal{V} \cup \mathcal{W}). & (85) \end{cases}$$

Consider for $n \in \mathcal{V}$ (by (60))

$$\langle \tilde{\lambda}, k\rangle - \mu_n = \left(\sum_{j=1}^{2} k_j\right)(R^2 + m) - |n|^2 - m + 0\,(\varepsilon^2 |k|) \tag{86}$$

and assume $|k| < \varepsilon^{-c_1} < \varepsilon^{-1/10}$. Then (86) may be kept away from 0 by variation of m, except if $\sum k_j = 1$. In that case, one uses however, the fact that

$$|n| \neq R \quad \text{for} \quad n \in \gamma \tag{87}$$

since n_1, n_2 are not symmetric. Hence one may ensure that for $|k| < \varepsilon^{-c_1}$ the q_{nk}-coefficient in (83) is at least ε say.

Consider next the coefficient in (84). We have by (60), assuming

$$k_1 + k_2 = 1, \quad |n| = R \tag{88}$$

$$\varepsilon^4\left[\left(-(k_1 a_1^2 + k_2 a_2^2) + \frac{\nu}{2}\,(a_1^2 - a_2^2)\right)^2 - 4\,a_1^2 a_2^2 - \frac{1}{4}\,(a_1^2 - a_2^2)^2\right] \tag{89}$$

where $\nu = \pm 1$. Considering (89) as a polynomial in a_1^2, a_2^2, the coefficients of $a_1^4, a_2^4, a_1^2, a_2^2$ are resp.

$$\left(-k_1 + \frac{\nu}{2}\right)^2 - \frac{1}{4} \tag{90}$$

$$\left(-k_2 - \frac{\nu}{2}\right)^2 - \frac{1}{4} \tag{91}$$

$$2\left(-k_1 + \frac{\nu}{2}\right)\left(-k_2 - \frac{\nu}{2}\right) - 4 + \frac{1}{2}. \tag{92}$$

If (90), (91) vanish, we have $-k_1 + \frac{\nu}{2} = \pm \frac{1}{2}, -k_2 - \frac{\nu}{2} = \pm \frac{1}{2}$, hence (92) $= \pm 2 \cdot \frac{1}{2} = \frac{1}{2} - 4 + \frac{1}{2} \neq 0$. Consequently, for $|k| < \varepsilon^{-c_1}$, the coefficient of (84) may be ensured at least ε^4.

Thus the nonresonance properties at the first stage of the iteration may be fulfilled. To study the linearized operator corresponding to (81)–(85) at larger scales according to the method discussed in section 8, one renormalizes dividing (81), (82), (85) by ε^2 and (83), (84) by ε^4.

Reference

[K-P]   S. Kuksin, J. Pöschel, *Invariant Cantor manifolds of quasi-periodic oscillations for a nonlinear Schrödinger equation*, Annals of Math., Vol 143, N1, 149-179.

LECTURE 10

Applications of Symplectic Capacities to Hamiltonian PDE

In Lectures 4 and 5 we discussed Gibbs measures as invariant measures for the corresponding Hamiltonian evolution equation. Other symplectic invariants are the so called symplectic capacities originating from M. Gromov's work [**Gr**] on pseudo-holomorphic curves. These symplectic capacities are introduced in finite dimensional phase space but some of the main results of this theory are dimension independent and hence expected to have an infinite dimensional generalization. For the infinite dimensional phase space, the finite dimensional normalizations induce a specific topology leading to a certain "symplectic Hilbert space" associated to a given Hamiltonian PDE. This program was initiated in a paper of S. Kuksin [**K**] and the author pursued some of these investigations. In [**K**] identifies a symplectic Hilbertspace for the various "standard" Hamiltonian evolution equations. For instance

$$\text{(NLS)} \qquad iu_t + \Delta u + \frac{\partial}{\partial \bar{u}} H_1(u, \bar{u}, t, x) = 0 \leftrightarrow L^2(\mathbb{T}^d)$$

$$\text{(NLW)} \qquad u_{tt} - \Delta u + p(u; t, x) = 0 \leftrightarrow H^{1/2}(\mathbb{T}^d) \times H^{1/2}(\mathbb{T}^d)$$

$$\text{(KdV)} \qquad u_t + \partial_x^3 u + \partial_x p(u) = 0 \leftrightarrow H^{-1/2}(\mathbb{T})$$

It turns out that this symplectic topology is essentially uniquely defined. Observe that in the NLS, NLW case, there is no dependence on the dimension. The main drawback of this theory is that this topology is not smooth. In fact some of the conclusions (on non-squeezing) are simply false in a sufficiently smooth phase space. One nevertheless obtains certain information about the longtime behavior of the dynamics which is new even in the integrable case. In what follows we briefly recall the concepts and discuss some PDE applications. For more details one is referred to the corresponding research papers.

Let $\mathbb{R}^{2n}$, $dp \wedge dq$ be equipped with the standard symplectic structure. Fix a nonempty open domain $O \subset \mathbb{R}^{2n}$. A smooth function φ will be called m-admissible ($m > 0$) provided $\varphi = m$ on a neighborhood of a boundary ∂O, $\varphi = 0$ on a

127

nonempty subdomain of O. Define the symplectic capacity

$$c_{2n}(O) = \inf\{m \mid \text{for any } m\text{-admissible } \varphi, \text{ the corresponding}$$

$$\text{Hamiltonian vectorfield } \left(-\frac{\partial \varphi}{\partial q}, \frac{\partial \varphi}{\partial p}\right) \text{ has a nontrivial periodic orbit}$$

$$\text{of period } \leq 1\} \tag{1}$$

(There are other related ones, but we restrict our attention to (1).)

Some of the basic properties are

(2) $0 < c_{2n}(O) < \infty$ if O is bounded and nonempty

(3) $c_{2n}(\tau O) = \tau^2 c_{2n}(O)$

(4) c_{2n} is a symplectic invariant

(5) c_{2n} is invariant under translation

(6) Denoting B_R an R-ball in $\mathbb{R}^{2n}$ equipped with euclidean norm

$$\|x\| = \left(\sum_{j=1}^{n} p_j^2 + q_j^2\right)^{1/2} \tag{7}$$

and $\Pi_r^{(k)}$ the cylinder of width r with the k-coordinate

$$\Pi_r^{(k)} = \left\{(p, q) \mid (p_k^2 + q_k^2)^{1/2} < r\right\} \tag{8}$$

one has

$$c_{2n}(B_R) = c_{2n}(\Pi_R^{(k)}) = \pi R^2 \tag{9}$$

and similarly for translations, by (5).

Observe that (9) is a dimension independent statement. In order to generalize to infinite dimensional phase space, one clearly proceeds by finite dimensional approximations. The first step consists in identifying a "Darboux basis" $\{\varphi_j^{\pm}\}$ for the phase space such that the Hamiltonian equation is expressed in usual Hamiltonian format

$$\begin{cases} \dot{p} = -\frac{\partial H}{\partial q} \\ \dot{q} = \frac{\partial H}{\partial p} \end{cases} \tag{10}$$

with respect to this basis. The symplectic topology is the corresponding Hilbert space Z with norm

$$\left\|\sum(p_j \varphi_j^+ + q_j \varphi_j^-)\right\| = \left(\sum p_j^2 + q_j^2\right)^{1/2} \tag{11}$$

in the case of the NLS

$$iu_t = \frac{\partial H}{\partial \bar{u}} \tag{12}$$

expand ϕ in the trigonometric system

$$\phi = \sum_{n \in \mathbb{Z}^d} a_n e^{in.x}; \quad a_n = \hat{\phi}(n) \tag{13}$$

(12) is then equivalent with

$$i\dot{a}_n = \frac{\partial H}{\partial \bar{a}_n} \tag{14}$$

as mentioned in lecture 1. Hence we let $(p_n, q_n) = (Re\, a_n, Im\, a_n)$ and the corresponding normalization is $L^2(\mathbb{T}^d)$.

Consider next the wave equation

$$u_{tt} - \Delta u + \rho u + f'(u) = 0 \tag{15}$$

which we rewrite as

$$\begin{cases} u_t = -Bv \\ v_t = Bu + B^{-1}f'(u) \end{cases} \qquad B = (-\Delta + \rho)^{1/2}. \tag{16}$$

Consider u, v as canonical variables and denote

$$H(u, v) = \tfrac{1}{2} \int \{|Bu|^2 + |Bv|^2 + f(u)\}. \tag{17}$$

Consider the Hilbertspace with inner product

$$(\phi, \psi) = (B\phi, \psi) \tag{18}$$

hence

$$\|\phi\| = \|B^{1/2}\phi\|_2 \sim \|\phi\|_{H^{1/2}(\mathbb{T}^d)} \tag{19}$$

which we therefore denote again by $H^{1/2}$. Let Z be the direct sum $H^{1/2} \oplus H^{1/2}$. Then (16) is the reformulation of (10), taking the gradient in Z. It follows that for NLW the symplectic Hilbertspace identifies with $H^{1/2}(\mathbb{T}^d) \oplus H^{1/2}(\mathbb{T}^d)$, where $H^{1/2}(\mathbb{T}^d)$ refers here to the real Sobolev space. Thus the Darboux basis is

$$\left\{ \frac{\sqrt{2}\cos n.x}{(n^2+\rho^{1/4})}, \frac{\sqrt{2}\cos n.x}{(n^2+\rho^{1/4})} \right\} \cup \left\{ \frac{\sqrt{2}\sin nx}{(n^2+\rho^{1/4})}, \frac{\sqrt{2}\sin nx}{(n^2+\rho^{1/4})} \right\}.$$

Finally, consider an equation of KdV type

$$u_t = \frac{\partial}{\partial x} \nabla H(u). \tag{20}$$

Letting $\varphi_n^+ = 2\sqrt{n}\cos nx$, $\varphi_n^- = 2\sqrt{n}\sin nx$ $(n \in \mathbb{Z}_+)$, equation (20) gets the form (10), expanding $u = \sum(p_n\varphi_n^+ + q_n\varphi_n^-)$. Hence the symplectic Hilbertspace identifies here with $H^{-1/2}(\mathbb{T})$.

In [**K**], the definition of symplectic capacity (1) in finite dimension is extended to infinite dimensional phase space and, roughly speaking, it is shown that there is conservation under Hamiltonian flow of the form

$$\text{linear operator} + \text{``compact perturbation''} \tag{21}$$

where the compactness refers to the symplectic topology as discussed above. The invariance of symplectic capacity has various consequences, in particular it implies that there is no squeezing of an R-ball in the symplectic Hilbertspace Z in a cylinder of smaller width, defined with the Darboux basis $\{\varphi_j^+, \varphi_j^-\}$. As discussed in [**K**], this fact relates to "weak turbulence", i.e. the spreading of energy from lower to higher Fourier modes as time evolves. The (non)-squeezing property implies that for any given time t the flowmap $S(t)$ cannot let part of the "energy" leave the kth Fourier mode, for any k. It also implies absence of uniform asymptotic stability in a neighborhood of a global solution, since diam $S(t)B_\varepsilon \geq \varepsilon$ hence $\varlimsup_{t\to\infty}$ diam $S(t)B_\varepsilon \neq 0$, for any ball B_ε. This statement again refers to the Z-topology. Establishing a statement such as (21) requires to study the Cauchy problem for data in the symplectic Hilbert space Z. This is the simplest in the case of the NLW, since here this topology is the strongest. Thus we consider the Cauchy problem

$$\begin{cases} u_{tt} - \Delta u + u + f(u; t, x) = 0 \\ u(0) = \phi \in H^{1/2}(\mathbb{T}^d) \\ \dot{u}(0) = \psi \in H^{-1/2}(\mathbb{T}^d) \end{cases}$$

Proposition 23. *Let $f(u; t, x)$ be a polynomial in u with smooth coefficients in x and t. Then (21) holds, local in time, in the following cases*

   (i) *dimension $d = 1$*

  (ii) *dimension $d = 2$ with f of degree ≤ 4 in u*

 (iii) *dimension $d = 3, 4$ with f of degree ≤ 2 in u.*

Prop. 23 (i) and (ii) for f of degree 2 appear in Kuksin's paper. For the other statements a bit more refined analysis is used, based mainly on Strichartz' inequality for the wave equation (see [**B**]). Recall the argument from [**B**]. We have

$$u(t) =$$

$$\sum_{\xi} \left[\frac{\widehat{u(0)}(\xi) + \widehat{iv(0)}(\xi)}{2} e^{i(\langle x, \xi\rangle + (1+|\xi|^2)^{1/2}t)} + \frac{\widehat{u(0)}(\xi) - \widehat{iv(0)}(\xi)}{2} e^{i(\langle x, \xi\rangle - (1+|\xi|^2)^{1/2}t)} \right] \quad (22)$$

$$+ \sum_{\xi} \int d\lambda \, \frac{\hat{f}(\xi, \lambda)}{\lambda^2 - 1 - |\xi|^2} e^{i\langle x, \xi\rangle}$$

$$\left[e^{i\lambda t} - \frac{1}{2}\left(1 + \frac{\lambda}{(1+|\xi|^2)^{1/2}}\right) e^{i(1+|\xi|^2)^{1/2}t} - \frac{1}{2}\left(1 - \frac{\lambda}{(1+|\xi|^2)^{1/2}}\right) e^{-i(1+|\xi|^2)^{1/2}t} \right] \quad (23)$$

where $\hat{f}$ denotes the Fourier transform of $f(u; t, x)$ on $\mathbb{T}^d \times I$. The expression for $v(t)$ is obtained as $-B^{-1}u_t$.

The assumptions on dimension d and f are the following

$$d = 2 \qquad f \text{ of the form } a_1(x, t)u + a_2(x, t)u^2 + a_3(x, t)u^3 + a_4(x, t)u^4 \qquad (24)$$

$$d = 3, 4 \qquad f \text{ of the form } a_1(x, t)u + a_2(x, t)u^2. \qquad (25)$$

Consider following norm for functions $A(x, t) = \sum_{\xi} \int d\lambda \hat{A}(\xi, \lambda) e^{i(\langle x, \xi\rangle + \lambda t)}$ on $\mathbb{T}^d \times I$, cf (5.117)

$$\|A\|_s = \left(\sum_{\xi} d\lambda (1 + |\xi|)^{2s}(1 + \||\lambda| - |\xi|\|)^{2\rho}|\hat{A}(\xi, \lambda)|^2 \right)^{1/2} \quad (26)$$

(to be understood as a restriction norm with respect to $\mathbb{T}^d \times I$).

Here ρ is chosen a bit larger than $\frac{1}{2}$. Observe that

$$\|A(t)\|_{H^s} \leq \|A\|_s \quad \text{for} \quad t \in I. \quad (27)$$

The expression (22) defines a linear operator of $u(0)$, $v(0)$; observe that

$$\|(22)\|_s + \|B^{-1}(22)_t\|_s \le c(\|u(0)\|_{H^s} + \|v(0)\|_{H^s}). \tag{28}$$

Our purpose is to show that for some $s_1 < \tfrac{1}{2} < s_2$

$$\|(23)\|_{s_2} + \|B^{-1}(23)_t\|_{s_2} \le C\sigma^c(1 + \|u\|_{s_1}^3)\|u\|_{s_1} \tag{29}$$

where $\sigma = |I|$, for some constants $0 < c$, $C < \infty$. Replacing s_2 by s_1 in the left member of (29), one gets as a first consequence, letting σ be sufficiently small,

$$\|u\|_{s_1} + \|v\|_{s_1} \le c(\|u(0)\|_{H^{s_1}} + \|v(0)\|_{H_{s_1}}). \tag{30}$$

Hence, from (27), (29), (30), it follows that for $t \in I$

$$\|(4)\|_{H^{s_2}} + \|B^{-1}(4)_t\|_{H^{s_2}} \le \|u(0)\|_{H^{s_1}} + \|v(0)\|_{H^{s_1}} \tag{31}$$

and hence the nonlinear part of the flow map acts boundedly from $H^{s_1} \times H^{s_1}$ to $H^{s_2} \times H^{s_2}$ ($s_1 < \tfrac{1}{2}$, $s_2 > \tfrac{1}{2}$) which is the required condition (21).

the $a(x, t)$-coefficients will play little role in the verification of (23) and we ignore them for simplicity's sake. In fact the relevant calculation appears for $f(u; t, x) = u^4$ in $d = 2$ and $f(u; t, x) = u^2$ in $d = 3$, $d = 4$.

Consider the expression (23) with $\|\lambda| - |\xi\| < 10$. One easily verifies because t is local (multiply the expression with a localizing function $\varphi(t)$) that the corresponding contribution to (26) is bounded by

$$\left[\sum_\xi (1 + |\xi|)^{2(s-1)} \int_{\|\lambda|-|\xi\|<10} |\hat{f}(\xi, \lambda)|^2\right]^{1/2}. \tag{32}$$

Hence $\|(23)\|_{s_2}$ may be estimated by the sum of

$$\left[\sum_\xi |\xi|^{2s_2} \int d\lambda \, \frac{|\hat{f}(\xi,\lambda)|^2}{(|\xi|+|\lambda|+1)^2(|\,|\xi|-|\lambda|\,|+1)^{2(1-\rho)}}\right]^{1/2} \tag{33}$$

$$\left[\sum_\xi |\xi|^{2(s_2-1)} \left(\int d\lambda \, \frac{|\hat{f}(\xi,\lambda)|}{|\,|\xi|-|\lambda|\,|+1}\right)^2\right]^{1/2} \tag{34}$$

and hence, from Hölder's inequality ($\rho > \tfrac{1}{2}$)

$$\left[\sum_\xi \int d\lambda \frac{|\hat{f}(\xi,\lambda)|^2}{(1+|\xi|)^{2(1-s_2)}(1+|\,|\xi|-|\lambda|\,|)^{2(1-\rho)}}\right]^{1/2} \tag{35}$$

To estimate $\|B^{-1}(4)_t\|_{s_2}$, we need to introduce an extra factor $\frac{\lambda^2}{1+|\xi|^2}$ in (23). Hence (35) is an estimate on both $\|(23)\|_{s_2}$ and $\|B^{-1}(23)_t\|_{s_2}$.

Consider the first case $d = 2$. Letting $f = u^4$, one has $\hat{f} = \hat{u} * \hat{u} * \hat{u} * \hat{u}$. According to (26), define

$$c(\xi, \lambda) = (1 + |\xi|)^{s_1}(1 + |\,|\lambda| - |\xi|\,|)^\rho |\hat{u}(\xi, \lambda)| \tag{36}$$

hence

$$\|c\|_2 = \|u\|_{s_1}. \tag{37}$$

In the sequel, we will write $|\dots|$ instead of $1 + |\dots|$.

By duality (35) may be estimated by

$$\sum_{\substack{\xi=\xi_1+\xi_2+\xi_3+\xi_4 \\ \lambda=\lambda_1+\lambda_2+\lambda_3+\lambda_4}} \int \prod_{i=1}^{4} \frac{c(\xi_i,\lambda_i)}{|\xi_i|^{s_1}||\lambda_i|-|\xi_i||^{\rho}} \frac{d(\xi,\lambda)}{|\xi|^{1-s_2}||\lambda|-|\xi||^{1-\rho}} \tag{38}$$

where $d(\xi,\lambda) \geq 0$, $\|d\|_2 \leq 1$.

Observe that in the problem of estimating (38), the discrete character of the summation plays clearly no role and we may as well replace $\sum_{\xi=\xi_1+\xi_2+\xi_3+\xi_4}$ by $\int_{\xi=\xi_1+\xi_2+\xi_3+\xi_4}$ (all denominators are taken > 1).

At this stage, we invoke Strichartz' inequality on the Fourier transform of an L^2-density carried by a cone in $\mathbb{R}^{d+1}$ (see [**S**]).

Let $q = \frac{2(d+1)}{d-1}$. Then

$$\left\| \int_{|\xi|\sim R} a(\xi)e^{i(\langle x,\xi\rangle+t|\xi|)}d\xi \right\|_{L^q(dx\,dt)} \leq CR^{1/2}\left(\int |a(\xi)|^2 d\xi \right)^{1/2}. \tag{39}$$

In our case $d = 3$, $q = 6$. It follows from (39) and Hölder's inequality that

$$\left\| \int_{|\xi|\sim R} d\xi \int d\lambda \frac{c(\xi,\lambda)}{\||\lambda|-|\xi|\|^{\frac{1}{2}+}} e^{i(\langle x,\xi\rangle+\lambda t)} \right\|_{L^6(dx\,dt)}$$
$$\leq CR^{1/2}\left(\iint |c(\xi,\lambda)|^2 d\xi\,d\lambda \right)^{1/2} \tag{40}$$

and hence, interpolating with the obvious (Parseval) L^2-inequality

$$\left\| \int d\xi \int d\lambda c(\xi,\lambda)e^{i(\langle x,\xi\rangle+\lambda t)} \right\|_{L^2(dx\,dt)} \leq \left(\iint |c(\xi,\lambda)|^2 d\xi\,d\lambda \right)^{1/2} \tag{41}$$

we get the inequality

$$\left\| \int_{|\xi|\sim R} d\xi \int d\lambda \frac{c(\xi,\lambda)}{||\lambda|-|\xi||^{\frac{9}{20}+}} e^{i(\langle x,\xi\rangle+\lambda t)} \right\|_{L^5(dx\,dt)}$$
$$\leq CR^{\frac{9}{20}}\left(\iint |c(\xi,\lambda)|^2 d\xi\,d\lambda \right)^{1/2} \tag{42}$$

which is used to bound (38). Restricting ξ_i, ξ to dyadic regions

$$\begin{cases} |\xi_i| \sim R_i & (i=1,2,3,4) \\ |\xi| \sim R \end{cases} \tag{43}$$

one gets

$$(R_1 R_2 R_3 R_4)^{-s_1} R^{-(1-s_2)} \sum_{\substack{\xi=\sum \xi_i:\lambda=\sum \lambda_i \\ |\xi_i|\sim R_i,|\xi|\sim R}} \int \prod_{i=1}^{4} \frac{c(\xi_i,\lambda_i)}{||\xi_i|-|\lambda_i||^{\rho}} \frac{d(\xi,\lambda)}{||\xi|-|\lambda||^{1-\rho}}. \tag{44}$$

Define $F_i = F_i(x,t)$, $G = G(x,t)$ letting $\hat{F}_i(\xi,\lambda) = \frac{c(\xi,\lambda)}{\||\xi|-|\lambda|\|^\rho} \chi_{[|\xi|\sim R_i]}(\xi)$, $\hat{G}(\xi,\lambda) = \frac{d(\xi,\lambda)}{\||\xi|-|\lambda|\|^{1-\rho}} \chi_{[|\xi|\sim R]}(\xi)$. Thus (44) equals

$$(R_1 R_2 R_3 R_4)^{-s_1} R^{-(1-s_2)} \int \prod_{i=1}^{4} F_i \cdot G \, dx \, dt$$

$$\leq (R_1 R_2 R_3 R_4)^{-s_1} R^{-(1-s_2)} \prod_{i=1}^{4} \|F_i\|_5 \cdot \|G\|_5. \qquad (45)$$

Assume $s_1 < \frac{1}{2} < s_2$ chosen such that $s_1, 1-s_2 > \frac{9}{20}$ and $\rho, 1-\rho > \frac{9}{20}$. It follows from (42) that $\|F_i\|_5 \leq C R_i^{\frac{9}{20}} \|c\|_2$ $(1 \leq i \leq 4)$ and $\|G\|_5 \leq C R^{\frac{9}{20}}$, so that by (37)

$$(45) \leq (R_1 R_2 R_3 R_4)^{\frac{9}{20}-s_1} R^{-\frac{11}{20}+s_2} \|u\|_{s_1}^4 \qquad (46)$$

which is summable for dyadic values of R_i, R.

Considering a small time interval I, $|I| = \sigma$, there is an extra saving of σ^c, for some $c > 0$, which is inequality (29). Consider functions u which are supported on a 2σ neighborhood of 0. It follows from the definition of the norm (26), in particular the $\||\lambda| - |\xi|\|^\rho$-multiplier, that localizing (22), (23) to I will affect the $\| \quad \|_{s_2}$-norm by a factor $\left(\frac{1}{\sigma}\right)^{\rho-\frac{1}{2}}$ $\left(\rho > \frac{1}{2}\right)$. On the other hand, repeating the previous L^5-estimate, one gets factors

$$R^{\frac{9}{20}} \left\| \frac{c(\xi,\lambda)}{\||\xi|-|\lambda|\|^{\rho-\frac{9}{20}}} \chi_{|\xi|\sim R} \right\|_{L^2_{\xi,\lambda}} = R^{\frac{9}{20}+s_1} \left\| \||\xi|-|\lambda|\|^{\frac{9}{20}} |\hat{u}(\xi,\lambda)| \chi_{|\xi|\sim R} \right\|_{L^2_{\xi,\lambda}}$$

which by interpolation are bounded by

$$R^{\frac{9}{20}+s_1} \|\hat{u}\chi_{|\xi|\sim R}\|_2^{1-\frac{9}{20\rho}} \left\| \||\xi|-|\lambda|\|^\rho |\hat{u}(\xi,\lambda)| \chi_{|\xi|\sim R} \right\|_2^{\frac{9}{20\rho}}$$

$$\leq R^{\frac{9}{20}+(1-\frac{9}{20\rho})s_1} \cdot \left(\|\hat{u}(\xi)\|_{L^2_{|\xi|\sim R} L^2_t}\right)^{1-\frac{9}{20\rho}} \cdot \|c\|_2^{\frac{9}{20\rho}}. \qquad (47)$$

Since $\operatorname{supp} u \subset \mathbb{T}^d \times I$, $\|\hat{u}(\xi)\|_{L^2_t} \leq \sigma^{1/2} \|\hat{u}(\xi)\|_{L^\infty_t} \leq \sigma^{1/2} \| \||\lambda| - |\xi|\|^\rho |\hat{u}(\xi,\lambda)| \|_{L^2_\lambda}$ by Hölder's inequality and $(47) \leq R^{\frac{9}{20}} \sigma^{\frac{1}{2}(1-\frac{9}{20}\rho)} \|c\|_2 \leq R^{\frac{9}{20}} \sigma^{\frac{1}{20}} \|u\|_{s_1}$. For $\rho > \frac{1}{2}$ close enough to $\frac{1}{2}$, this clearly implies inequality (29) with the σ^c-factor. From the earlier discussion, Proposition 23 follows in case (ii).

The proof of (iii) is completely analogous. For the argument to work, one needs the exponent $q = \frac{2(d+1)}{d-1}$ from inequality to fulfill the condition $q > k + 1$, where k is the degree of $f(u;t,x)$ in u. Thus for $d \geq 3$, $\frac{d+3}{d-1} > k \geq 2$ only permits (iii).

The problem for the NLS is more difficult because the phase space is L^2. In fact for the periodic Cauchy problem, we only have this L^2-analysis available in the 1D-case with cubic nonlinearity (see lecture 2). Statement (21) fails here. However, we shall prove an approximation result that will enable us to derive the nonsqueezing of balls in cylinders of smaller width from the finite dimensional theory. The main result here is

Proposition 48. *Consider the solutions u, v to the Cauchy problems*

$$\begin{cases} iu_t + u_{xx} + \frac{\partial G}{\partial \bar{u}}(u, \bar{u}, t, x) = 0 \\ u(x, 0) = \phi(x) \end{cases} \tag{49}$$

and

$$\begin{cases} iv_t + v_{xx} + P_N \frac{\partial G}{\partial \bar{v}}(v, \bar{v}; t, x) = 0 \\ v(x, 0) = \phi(x) \end{cases} \tag{50}$$

where $\phi = P_N \phi$ and $G(u, \bar{u}, t, x)$ of the form

$$G(u, \bar{u}, t,) = A(x, t)|u|^2 + B(x, t)|u|^4 \tag{51}$$

with A, B real sufficiently smooth functions of x, t both periodic in x.

Fix a positive integer N' and a time t. Let $\varepsilon > 0$. Then one has an approximation

$$\|P_{N'}(u(t) - v(t))\|_2 < \varepsilon \tag{52}$$

provided $N > N(N', |t|, \varepsilon, \|\phi\|_2)$.

Denote $S_N(t)$, $S_\infty(t)$ the respective flow maps corresponding to (49), (50). It follows from (48) that

$$P_k S_N(t)(P_N B_R) > N \to \infty >> P_k S_\infty(t)(P_N B_R) \quad \text{(uniformly)} \tag{53}$$

for any L^2-ball (not necessarily centered at 0). Hence for $r < R$

$$S_N(t)(P_N B_R) \not\subset \Pi_r^{(k)} \tag{54}$$

clearly implies

$$S_\infty(t)(B_R) \not\subset \Pi_r^{(k)} \tag{55}$$

which is the nonsqueezing property.

In proving Prop. 48, the significant part of the analysis appears for

$$G(u, \bar{u}, t, x) = |u|^4 \tag{56}$$

and the more general form (51) only requires minor adjustments when A, B are smooth. Observe that Prop. 48 is a refinement of lemma 41 in lecture 4. It will be derived from following facts local in time. Assume (56).

Lemma 57. *Consider the solutions u, v to the Cauchy problems*

$$\begin{cases} iu_t + u_{xx} + u|u|^2 = 0 \\ u(x, 0) = \phi(x) \end{cases} \tag{58}$$

$$\begin{cases} iv_t + v_{xx} + v|v|^2 = 0 \\ v(x, 0) = \psi(x) \end{cases} \tag{59}$$

and assume $\|\phi\|_2 = \|\psi\|_2$. Then for $|t| < \tau(\|\phi\|_2)$ one has

$$\|P_{N_0}(u(t) - v(t))\|_2 \le \|P_{N_1}(\phi - \psi)\|_2 + \varepsilon \tag{60}$$

provided $N_1 - N_0 > C_1 \, \varepsilon^{-C_1}$ (C_1 a constant).

Lemma 61. *Consider the solutions u, v to the Cauchy problems*

$$\begin{cases} iu_t + u_{xx} + u|u|^2 = 0 \\ u(x,0) = \phi(x) \end{cases} \tag{62}$$

$$\begin{cases} iv_t + v_{xx} + P_N(v|v|^2) = 0 \\ v(x,0) = \phi(x) \end{cases} \tag{63}$$

where $\phi = P_N\phi$. Then for $|t| < \tau(\|\phi\|_2)$ one has

$$\|P_{N_0}(u(t) - v(t))\|_2 \leq \varepsilon \tag{64}$$

provided $N - N_0 > C_1\varepsilon^{-C_1}$.

To deduce Proposition 48, one breaks up $[0,t]$ in time intervals $[t_i, t_{i+1}]$ of size $\tau(\|\phi\|_2)$. For fixed i, compare on $[t_i, t_{i+1}]$ the solutions to the initial value problems

$$\begin{cases} iu_t + u_{xx} + u|u|^2 = 0 \\ u(x,t_i) = u(t_i)(x) \end{cases} \tag{65}$$

$$\begin{cases} i\tilde{u}_t + \tilde{u}_{xx} + \tilde{u}|\tilde{u}|^2 = 0 \\ \tilde{u}(x,t_i) = v(t_i)(x) \end{cases} \tag{66}$$

$$\begin{cases} iv_t + v_{xx} + P_N(v|v|^2) = 0 \\ v(x,t_i) = v(t_i)(x). \end{cases} \tag{67}$$

Observe that $\|u(t_i)\|_2 = \|v(t_i)\|_2 = \|\phi\|_2$. Denoting by $\{N_i\}$ a decreasing sequence of positive integers $< N$, Lemma 57 implies that $\|P_{N_{i+1}}(u(t_{i+1}) - \tilde{u}(t_{i+1}))\|_2$ $\leq \|P_{N_i}(u(t_i) - v(t_i))\|_2 + (N_i - N_{i+1})^{-c_2}$ for some $c_2 > 0$ and Lemma 61 yields $\|P_{N_{i+1}}(\tilde{u}(t_{i+1}) - v(t_{i+1}))\|_2 \leq (N - N_{i+1})^{-c_2}$. Hence

$$\|P_{N_{i+1}}(u(t_{i+1}) - v(t_{i+1}))\|_2 \leq \|P_{N_i}(u(t_i) - v(t_i))\|_2 + (N_i - N_{i+1})^{-c_2} \tag{68}$$

and (68) implies $\|P_{N'}(u(t) - v(t))\|_2 \leq \sum(N_i - N_{i+1})^{-c_2}$. Since the number of steps is controlled by $\|\phi\|_2$, Proposition 48 follows.

Recall that (62) is equivalent to the integral equation

$$u(t) = S(t)\phi + i\int_0^t S(t - \tau)w(\tau)d\tau \quad w = u|u|^2; \; S(t) = e^{it\partial_x^2} \tag{69}$$

and the norms $X^{0,b}(I)$ defined as

$$\|u\|_{0,b} = \left(\sum_k \int d\lambda(1 + |\lambda - k^2|^2)^b|\hat{u}(k,\lambda)|^2\right)^{1/2} \tag{70}$$

for

$$u = \sum_k \int d\lambda\, \hat{u}(k,\lambda)e^{i(kx+\lambda t)} \quad \text{on} \quad \mathbb{T} \times I$$

Write the nonlinear term $w = |u|^2 u = w(u,u,u)$ as

$$w(u,u,u) = \sum_{k=k_1-k_2+k_3} \int_{\lambda=\lambda_1-\lambda_2+\lambda_3} e^{i(kx+\lambda t)}\hat{u}(k_1,\lambda_1)\overline{\hat{u}(k_2,\lambda_2)}\hat{u}(k_3,\lambda_3), \tag{71}$$

and splitting the $\sum_{k=k_1-k_2+k_3}$ summation as

$$\sum_{\substack{k=k_1-k_2+k_3 \\ k_2 \neq k_1,k_3}} - \sum_{\substack{k=k_1-k_2+k_3 \\ k_1=k_2=k_3}} + \sum_{\substack{k=k_1-k_2+k_3 \\ k_1=k_2}} + \sum_{\substack{k=k_1-k_2+k_3 \\ k_3=k_2}}$$

(71) clearly yields, since $\int |u(t)|^2 = \int |\phi|^2$,

$$\sum_{\substack{k=k_1-k_2+k_3 \\ k_2 \neq k_1,k_3}} \int_{\lambda=\lambda_1-\lambda_2+\lambda_3} e^{i(kx+\lambda t)} \hat{u}(k_1,\lambda_1)\overline{\hat{u}(k_2,\lambda_2)}\hat{u}(k_3,\lambda_3) \tag{72}$$

$$- \sum_k e^{ikx} \int_{\lambda=\lambda_1-\lambda_2+\lambda_3} e^{i\lambda t}\hat{u}(k,\lambda_1)\overline{\hat{u}(k,\lambda_2)}\hat{u}(k,\lambda_3) \tag{73}$$

$$+ 2\left(\int |\phi|^2\right) \cdot \sum_k \int d\lambda\, e^{i(kx+\lambda t)} \hat{u}(k,\lambda). \tag{74}$$

The corresponding contributions to the integral term in (69) are

$$\sum_{\substack{k=k_1-k_2+k_3 \\ k_2 \neq k_1,k_3}} \int_{\lambda=\lambda_1-\lambda_2+\lambda_3} e^{ikx} \frac{e^{i\lambda t} - e^{ik^2 t}}{\lambda - k^2} \hat{u}(k_1,\lambda_1)\overline{\hat{u}(k_2,\lambda_2)}\hat{u}(k_3,\lambda_3) \tag{75}$$

$$- \sum_k e^{ikx} \int_{\lambda=\lambda_1-\lambda_2+\lambda_3} \frac{e^{i\lambda t} - e^{ik^2 t}}{\lambda - k^2} \hat{u}(k,\lambda_1)\overline{\hat{u}(k,\lambda_2)}\hat{u}(k,\lambda_3) \tag{76}$$

$$+ 2\int |\phi|^2 \cdot \sum_k e^{ikx} \int d\lambda \frac{e^{i\lambda t} - e^{ik^2 t}}{\lambda - k^2}\hat{u}(k,\lambda). \tag{77}$$

For the subsequent argument, the terms (73), (74) appear as harmless because $\hat{u}$ appears for the same k^{th} Fourier mode. To estimate the contribution of (75), we mainly need to bound for given positive integers K, Δ the subsum

$$\sum_{\substack{k=k_1-k_2+k_3;k_2 \neq k_1,k_3 \\ |k| \leq K;\max |k_i|>K+\Delta}} \int_{\lambda=\lambda_1-\lambda_2+\lambda_3} e^{ikx} \frac{e^{i\lambda t} - e^{ik^2 t}}{\lambda - k^2}\hat{u}(k_1,\lambda_1)\overline{\hat{u}u(k_2,\lambda_2)}\hat{u}(k_3,\lambda_3). \tag{78}$$

Define

$$c(k,\lambda) = (1 + |\lambda - k^2|^{1/2})|\hat{u}(k,\lambda)| \tag{79}$$

so that

$$\|u\|_{0,b} = \|c\|_{\ell_k^2 L_\lambda^2}. \tag{80}$$

One estimates (78) by an expression (of lecture 2)

$$\sum_{\substack{k=k_1-k_2+k_3,k_2 \neq k_1,k_3 \\ |k| \leq K,\max |k_i|>K+\Delta}} \int_{\lambda=\lambda_1-\lambda_2+\lambda_3} \frac{c(k_1,\lambda_1)}{|\lambda_1 - k_1^2|^{1/2}} \frac{c(k_2,\lambda_2)}{|\lambda_2 - k_2^2|^{1/2}} \frac{c(k_3,\lambda_3)}{|\lambda_3 - k_3^2|^{1/2}} \frac{a(k,\lambda)}{|\lambda - k^2|^{1/2-}}$$

$$\tag{81}$$

where $\sum_k \int d\lambda |a(k,\lambda)|^2 \leq 1$ and $|\cdot|$ stands for $|\cdot|+1$ in the denominators. The main point is the observation that since for $k = k_1 - k_2 + k_3$

$$-k^2 + k_1^2 - k_2^2 + k_3^2 = 2(k_1 - k)(k_3 - k) \tag{82}$$

and one of the factors $k - k_1$, $k - k_3$ vanishes, one gets from the assumptions on the (81) summation

$$|(82)| > \Delta \tag{83}$$

and therefore

$$\max\left(|\lambda_1 - k_1^2|,\ |\lambda_2 - k_2^2|,\ |\lambda_3 - k_3^2|,\ |\lambda - k^2|\right) > \Delta. \tag{84}$$

On the other hand, the Strichartz bound

$$\|u\|_{L^4_{x,t}} \leq c\|u\|_{0,\frac{3}{8}} \tag{85}$$

as used in lecture 2 allows a saving on the denominators of (81) and hence, by (84), (81) may be bounded by $|I|^{c_3}\Delta^{-c_3}\|u\|^3_{0,\frac{1}{2}}$, for some $c_3 > 0$.

Writing from (69)

$$P_K u(t) = S(t)P_K \phi + i \int_0^t S(t - \tau)(P_K w)(\tau)d\tau \tag{86}$$

it follows from the preceding that

$$\|P_K u\|_{0,\frac{1}{2}} \leq \|P_K \phi\|_2 + |I|^{c_3}\left(\|\phi\|_2^2 + \|u\|^2_{0,\frac{1}{2}}\right)\|P_{K+\Delta} u\|_{0,\frac{1}{2}} + |I|^{c_3}\Delta^{-c_3}\|u\|^3_{0,\frac{1}{2}}. \tag{87}$$

Hence, choosing $|I|$ sufficiently small, depending on $\|\phi\|_2$, (87) yields

$$\|P_K u\|_{0,\frac{1}{2}} \leq \|P_K \phi\|_2 + \delta\|P_{K+\Delta} u\|_{0,\frac{1}{2}} + \Delta^{-c_3} \tag{88}$$

with $\delta > 0$ a small constant. Iteration of (83) r times gives

$$\|P u\|_{0,\frac{1}{2}} \leq \|P_{K+r\Delta}\phi\|_2 + \delta^r\|\phi\|_2 + \Delta^{-c_3} \tag{89}$$

and hence, for $N_1 > N_0$

$$\|P_{N_0} u\|_{0,\frac{1}{2}} \leq \|P_{N_1}\phi\|_2 + (N_1 - N_0)^{-c_3/2}. \tag{90}$$

One also obtains that for $t \in I$

$$\|(P_{N_0} u)(t)\|_2 \leq \|P_{N_1}\phi\|_2 + (N_1 - N_0)^{-\frac{1}{2}c_3}. \tag{91}$$

Similarly, considering the Cauchy problems with $u(0) = \phi$, $v(0) = \psi$ satisfying

$$\|\phi\|_2 = \|\psi\|_2 \tag{92}$$

the preceding yields

$$\|P_{N_0}(u - v)\| \leq \|P_{N_1}(\phi - \psi)\|_2 + (N_1 - N_0)^{-c} \tag{93}$$

and

$$\|P_{N_0}(u - v)(t)\|_2 \leq \|P_{N_1}(\phi - \psi)\|_2 + (N_1 - N_0)^{-c} \tag{94}$$

and the proofs of Lemmas 57 and 61 are easily carried out.

Remarks.

(1) Consider an equation

$$iu_t = |u|^2 u \tag{95}$$

obtained by removing the dispersive term in NLS. Then we may write for (95) explicit solutions of the form (φ smooth)

$$u(x, t) = \varphi(x)e^{-i|\varphi(x)|^2 t}. \tag{96}$$

It is clear that $u(0) = \varphi$ and for $s \in \mathbb{Z}_+$

$$\|u(t)\|_{H^s} \sim |t|^s \tag{97}$$

if $|\varphi|$ is not a constant. S. Kuksin observed [**K4**] that if in the NLS (in dimension $d \leq 3$) we consider a small dispersion

$$iu_t + \delta \Delta u + u|u|^2 = 0 \tag{98}$$

and take $\varphi \in H^s(\mathbb{T}^d)$, s sufficiently large, its solution u (which exists globally) will resemble (96) for times $|t| < \left(\frac{1}{\delta}\right)^c$ and hence satisfy (97) for such times. In fact Kuksin shows, based on this idea, a squeezing phenomenon of a ball B in H^s-space into arbitrarily narrow cylinders. Hence the nonsqueezing results discussed above may not be expected to be valid in arbitrary smooth phase space topologies. The same comment holds for wave equations (see [**K3**]). The main shortcoming of the preceding explanation for weak turbulence is that there is clearly no distinction between integrable and non-integrable equations.

(2) The symplectic Hilbert space for KdV equations (20) is $H^{-1/2}(\mathbb{T})$. Presently, there is a local wellposedness theorem in that space, which is borderline for the method discussed in lecture 3. It does not seem, however, to yield (uniform) finite dimensional approximation properties, as required for the use of the invariance of symplectic capacities.

Comments and references related to Lecture 10

[G]   M. Gromov, *Pseudo-holomorphic curves on almost complex manifolds*, Inventiones Math. 82 (1985), 307-347.

For the construction of symplectic capacities in infinite dimensional phase space, see

[K1]   S. Kuksin, *Infinite dimensional symplectic capacities and a squeezing theorem for Hamiltonian PDE*, Comm. Math. Phys. 167 (1995), 531-552.

Proposition 23 (resp. 48) for the NLW (resp. 1D NLS) appear in

[B12]   J. Bourgain, *Aspects of longtime behaviour of solutions of nonlinear Hamiltonian evolution equations*, GAFA, Vol. 5, N2 (1995), 83-112.

[B13]   J. Bourgain, *Approximation of solutions of the cubic NLS by finite-dimensional equations and non-squeezing properties*, IMRN (1994), N2, 79-90.

Squeezing phenomena for NLW with respect to higher Sobolev norms are discussed in

[K3] S. Kuksin, *On squeezing and flow of energy for nonlinear wave equations*, GAFA, Vol. 5 (1995), 668-701.

and results in a similar spirit for NLS are to appear in a forthcoming paper of the same author.

Appendix

Remarks on longtime behaviour of the flow of Hamiltonian PDE

Our aim here is to make some comments on the general and essentially open problems about the behaviour of smooth solutions of Hamiltonian PDE (with emphasis on the non-integrable situation). Consider more precisely the situation of a perturbation of a linear or integrable equation. In the context of finite dimensional phase space, one may recall Nekhoroshev's stability result for the action variables [N] which is essentially the following. Consider an unperturbed Hamiltonian

$$H_0(I) = \lambda I + c|I|^2 + \text{(higher order)} \tag{1}$$

where $I = (I_1, \ldots, I_N)$ are the action variables and either $\lambda = (\lambda_1, \ldots, \lambda_N)$ has good diophantine properties or $H_0(I)$ is strictly convex, i.e. $c \neq 0$ in (1). Consider a perturbation

$$H(I, \varphi) = H_0(I) + \varepsilon H_1(I, \varphi) \tag{2}$$

and the corresponding evolution equations

$$\begin{cases} \dot{I}_j = -\varepsilon \frac{\partial H_1}{\partial \varphi_j} \\ \dot{\varphi}_j = \frac{\partial H_0}{\partial I_j} + \varepsilon \frac{\partial H_1}{\partial I_j}. \end{cases} \tag{3}$$

Then one has exponentially long times of stability (we consider the real analytic situation) thus

$$|I(t) - I(0)| < \varepsilon^a \text{ for } |t| < \exp \frac{1}{\varepsilon^b} \tag{4}$$

where $a, b > 0$ and dependent on the dimension N of the phase space.

It seems a natural question to find analogous of this phenomenon in the PDE context. We have a result in the case of perturbations of nonresonnant linear equations, that may roughly be stated as follows.

Proposition 5. *Consider an equation written in the form of a perturbed linear Schrödinger equation*

$$iu_t + Au + \varepsilon \frac{\partial H}{\partial \overline{u}} = 0 \tag{6}$$

where we assume for simplicity $H = H(u, \overline{u})$ a polynomial. Here A is a selfadjoint operator with spectrum $\{\lambda_j\}$ and eigenfunctions $\{\varphi_j\}$, which we assume "well localized" wrt the exponential system. Let $\{\lambda_j\}$ satisfy certain nonresonance conditions, of the form

$$|k_1 \lambda_1 + \cdots + k_{j_0} \lambda_{j_0} + k_{j_1} \lambda_{j_1} + k_{j_2} \lambda_{j_2}| > \varepsilon^{c_1} \tag{7}$$

for $j_0 < \varepsilon^{-c_2}$ and $|k_1| + \cdots + |k_{j_0}| < r$, $|k_{j_1}| \leq 1$, $|k_{j_2}| \leq 1$. Here $0 < c_1 < \frac{1}{10}$, r is a fixed large number and $c_2 = c_2(r)$. Then the solution u of (6) for smooth data $u(0)$ will be ε^M-close to a quasi-periodic function of time appearing as perturbation of the linear solution[(*)] *for times $|t| < \varepsilon^{-M}$. Here $M > 0$ may be taken any fixed number, M depending on r.*

Examples are for instance the following models (in $1D$) (cf. lecture 6).

(I) A perturbation of a linear Schrödinger equation

$$iu_t - u_{xx} + V(x)u + \varepsilon\frac{\partial H}{\partial \overline{u}} = 0 \tag{8}$$

where V is a "typical" smooth potential $\big($to guarantee diophantine properties such as (7)$\big)$ and considering Dirichlet boundary conditions.

(II) A NLW equation

$$y_{tt} - y_{xx} + \rho y + F'(y) = 0 \tag{9}$$

(F polynomial or even smooth) with ρ a typical number and $F'(y) = 0(|y|^3)$ odd. Then the statement of Prop. 5 will hold for smooth, odd periodical data $y(0), y'(0)$ of size ε.

Remarks.

(1) Nekhoroshev type stability results for perturbations of nonresonnant linear systems in infinite dimensional phase space appear in the literature but for perturbations that are (essentially) finite range interactions and hence do not cover the PDE problems (cf. for instance [**B-F-G**]).

(2) We presently do not know of extensions to infinite dimensional phase space of a Nekhoroshev type result for strictly convex Hamiltonian, even endowing the phase space with the weak topology (see some examples discussed later on in this respect). It is likely however that the methods of proof of Proposition 5 permit to show stability results for certain nonlinear PDE written in a form

$$i\dot{q}_n = (\lambda_n + |q_n|^2)q_n + \varepsilon\frac{\partial H}{\partial \overline{q}_n} \tag{10}$$

considering "typical" initial data $\{q_n(0)\}$, exploiting the $|q_n|^2 q_n$-terms.

The Birkhoff-normal form discussed for the NLS $iu_t - u_{xx} + u|u|^2+$ (higher order) $= 0$ in lecture 9 would be a natural model, considering typical small initial data of size ε.

(3) The proof of Proposition 5 may be summarized as follows.
  - Construction of "approximation solutions" of (6), using for instance a truncated perturbative series.
  - Control of the linearized equation at this approximative solution.
       In particular, one needs zero-Lyapounov exponents. For this purpose the "2^e Melnikov condition" (cf. Lecture 6), excluding multiple frequencies, is necessary and restricts the applications to $1D$.
  - Control of the nonlinear difference equation.

(4) The optimal stability times resulting from Proposition 5 turn out to be of the order of $\exp(\log\frac{1}{\varepsilon})^2$ rather than $\exp\frac{1}{\varepsilon^b}$ as in (4), cf. [**B-F-G**].

Next we go over the proof of Proposition 5. See [**B14**] for more details.

[(*)]with appropriate frequency perturbation $\lambda' = \lambda + 0(\varepsilon)$

(i) Approximative solution

Fix modes $1, 2, \ldots, j_0$ and $j_1 > j_0$. Let r be a fixed positive integer.

Let $a_1, \ldots, a_{j_0}, a_{j_1} \in \mathbb{R}_+^*$ and denote $a = (a_1, \ldots, a_{j_0}, a_{j_1})$, $\mathfrak{a} = (a_1, \ldots, a_{j_0})$.

Eventually j_0 and $\mathfrak{a}$ will be fixed ($\mathfrak{a}$ with fast decay estimates) and $a_{j_1} \to 0$.

We construct an "approximative" solution u_a of (6) which is quasi periodic in time with frequencies

$$\lambda' = (\lambda'_1, \ldots, \lambda'_{j_0}, \lambda'_{j_1}). \tag{11}$$

This will essentially be achieved by a finite expansion in an ε series.

We first construct a periodic function $F = F^a(\theta_1, \ldots, \theta_{j_0}, \theta_{j_1})$ on Π^{j_0+1} and $\lambda' = \lambda'(a)$ satisfying

$$i\langle \lambda', \nabla F \rangle + AF + \varepsilon \frac{\partial H}{\partial \overline{u}}(F, \overline{F}) = \xi_a \tag{12}$$

where $\xi_a = \xi_a(\theta)$ will satisfy in particular $(r_1 \sim r)$

$$\|\xi_a\| < \varepsilon^{r_1} + 0(|a_{j_1}|^2) \tag{13}$$

$$\|\partial_{a_{j_1}}^\alpha \partial_{\mathfrak{a}}^\beta \partial_{\lambda'}^\gamma \xi_a\| < \varepsilon^{r_1} + 0(|a_{j_1}|^{2-\alpha}) \tag{14}$$

for $\alpha \leq 2$ and β, γ bounded

$$\partial_{\theta_j} \xi_a \big|_{a_j = 0} = 0. \tag{15}$$

Assume the following properties satisfied for the spectrum of A

$$|k_1 \lambda_1 + \cdots + k_{j_0} \lambda_{j_0} + k_{j_1} \lambda_{j_1} + k_{j_2} \lambda_{j_2}| > \varepsilon^{1/10}. \tag{16}$$

The new frequency $\lambda' = (\lambda'_1, \ldots, \lambda'_{j_0}, \lambda'_{j_1}) = \lambda'(a)$ will satisfy

$$|\lambda'_j - \lambda_j| \leq \varepsilon \quad \text{for} \quad j \in \{1, \ldots, j_0, j_1\}. \tag{17}$$

Hence (16) will remain valid after replacement of $\lambda_1, \ldots, \lambda_{j_0}, \lambda_{j_1}$ by λ'.

We will define

$$u_{a,\theta}(t) = F^a(\theta_1 + \lambda'_1 t, \ldots, \theta_{j_0} + \lambda'_{j_0} t, \theta_{j_1} + \lambda'_{j_1} t)$$

that will clearly satisfy

$$i\dot{u}_{a,\theta} + A u_{a,\theta} + \varepsilon \frac{\partial H}{\partial \overline{u}}(u_{a,\theta}, \overline{u}_{a,\theta}) = \xi_a(\theta_1 + \lambda'_1 t, \ldots, \theta_{j_0} + \lambda'_{j_0} t, \theta_{j_1} + \lambda'_{j_1} t). \tag{18}$$

We construct F^a as a standard perturbative series in ε truncated to order $r_1 \sim r$. Rewrite the equation

$$(-\langle \lambda', k \rangle + \lambda_j)\hat{F}_j(k) + \varepsilon \widehat{\frac{\partial H}{\partial \overline{u}_j}}(k) = 0. \tag{19}$$

Here $k = (k_1, \ldots, k_{j_0}, k_{j_1})$. Recall that H is a polynomial in $u, \overline{u}$ of degree d. We delete the set R of resonant sites $(j, k) = (j, e_j)$ for $j \in \{1, \ldots, j_0, j_1\}$ and e_j the j-unit vector in $\mathbb{Z}^{j_0+1}$ (they correspond to the Q-equations). Define for $(j, k) \in R$

$$\hat{F}_j(e_j) = a_j \qquad j \in \{1, \ldots, j_0, j_1\}. \tag{20}$$

Put also

$$\hat{F}_j(k) = 0 \quad \text{if} \quad |k| > r + 1 \quad \text{or} \quad |k_{j_1}| > 1. \tag{21}$$

Determine remaining $\hat{F}_j(k)$ inductively as series in ε

$$\hat{F}_j(k) = \sum_s F_s(j,k)\varepsilon^s \tag{22}$$

using equation (1.19) and starting from

$$F_j = a_j e^{i\theta_j} \quad \text{for} \quad j \in \{1,\ldots,j_0,j_1\} \tag{23}$$
$$= 0 \text{ otherwise.}$$

Thus

$$F_{s+1}(j,k) = \frac{1}{\langle \lambda', k\rangle - \lambda_j} \frac{1}{s!} \frac{\partial^s}{\partial \varepsilon^s} \left[\widehat{\frac{\partial H}{\partial \overline{u}_j}}(F_{(s)}, \overline{F}_{(s)}) \right]\Bigg|_{\varepsilon=0} \tag{24}$$

where

$$F_{(s)} = \sum_{j,k,s'\leq s} \varepsilon^{s'} F_{s'}(j,k)\varphi_j(x)e^{i\langle k,\theta\rangle}. \tag{25}$$

By induction, it is easily seen that in $F_s(j,k)$ the multi index k satisfies $|k| \leq s(d-1)$ where d is the degree of H. Hence, the first constraint in (21) may be ignored up to order $r_1 \sim \frac{r}{d-1}$.

By (16), the divisor in (24) remains at least $\varepsilon^{1/10}$. By induction, one gets therefore bounds

$$|F_s(j,k)| \leq \sum_{j,k} |F_s(j,k)| \leq \|F_s\| < s^{2sd}\varepsilon^{-\frac{s}{10}}. \tag{26}$$

Similar bounds hold for derivatives in a and λ' of bounded order (only order ≤ 2 will be considered). We disregard at this point the dependence of λ' on a which will be established afterwards. From (26), the s-term in (22) is controlled by $\varepsilon^{s/2}$.

By construction, equation (19) will be satisfied up to an error of order $\varepsilon^{r_1} + 0(|a_{j_1}|^2)$, except for $(j,k) \in R$. The $|a_{j_1}|^2$-term comes from the second restriction $|k_{j_1}| \leq 1$ in (21). Projecting (19) on R, one determines the new frequencies $\lambda' = (\lambda_1',\ldots,\lambda_{j_0}',\lambda_{j_1}')$. Thus

$$\lambda_j' = \lambda_j + \frac{\varepsilon}{a_j} \widehat{\frac{\partial H}{\partial \overline{u}_j}}(e_j) \quad j = 1,2,\ldots,j_0,j_1 \tag{27}$$

solved as implicit equation in λ' as function of a (λ' is real).

Observe that in particular $\widehat{\frac{\partial H}{\partial \overline{u}_j}}(e_j)\Bigg|_{a_j=0} = 0$ so that the dependence of λ' on a remains smooth up to 0. In particular, for $a_{j_1} \to 0$, we have

$$\lambda_{j_1}'(\mathfrak{a}) = \lambda_{j_1} + \varepsilon\partial_{a_{j_1}}\widehat{\frac{\partial H}{\partial \overline{u}_{j_1}}}(e_j)\Bigg|_{a_{j_1}=0}. \tag{28}$$

From the preceding, the function $u_{a,\theta}$ introduced above has a smooth dependence on a and satisfies equation (18) with right member error term

$$0(\varepsilon^{r_1} + |a_{j_1}|^2). \tag{29}$$

This approximative solution $u_{\theta,a}$ has the form

$$u_{\theta,a} = \sum_{j=1,\ldots,j_0,j_1} a_j e^{i\theta_j} e^{i\lambda_j' t} \varphi_j(x) + \varepsilon^{1/2} u_{\theta,a}' \tag{30}$$

with in particular

$$\left.\frac{\partial u'_{\theta,a}}{\partial \theta_j}\right|_{a_j=0} = 0. \tag{31}$$

One has

$$\left|\frac{\partial \lambda'_j}{\partial a_k}\right| < \varepsilon \tag{32}$$

$$\left.\frac{\partial \lambda'_j}{\partial a_{j_1}}\right|_{a_{j_1}=0} = 0. \tag{33}$$

To verify (33), distinguish the cases $j = 1, \ldots, j_0$ and $j = j_1$. Use the fact that for fixed λ', $\widehat{F^a}(k)$ is even (resp. odd) in a_{j_1} if $k_{j_1} = 0$ (resp. $k_{j_1} = \pm 1$), as verified from the inductive construction.

(ii) Estimates on the linearized equation

Differentiate (30) in θ_j and a_j. Taking (31) into account, one gets

$$a_j U_{j,a,\theta} = i a_j e^{i\theta_j} e^{i\lambda'_j t} \varphi_j(x) + 0(\varepsilon^{1/2}|a_j|) \tag{34}$$

and

$$V_{j,a,\theta} = e^{i\theta_j} e^{i\lambda'_j t} \varphi_j(x) + 0(\varepsilon^{1/2}) + 0\left(\left|\frac{\partial \lambda'}{\partial a_j}\right| |t|\right) \tag{35}$$

satisfying resp.

$$iU_t + AU + \varepsilon \frac{\partial^2 H}{\partial \overline{u}_{a,\theta} \partial u_{a,\theta}} U + \varepsilon \frac{\partial^2 H}{\partial \overline{u}^2_{a,\theta}} \overline{U} = \frac{1}{a_j} \frac{\partial}{\partial \theta_j} \xi_a(\theta_1 + \lambda'_1 t, \ldots, \theta_{j_0} + \lambda'_{j_0} t, \theta_{j_1} + \lambda'_{j_1} t) \tag{36}$$

and

$$iV_t + AV + \varepsilon \partial \overline{\partial} H.V + \varepsilon \overline{\partial}\overline{\partial} H.\overline{V} = \frac{\partial \xi_a}{\partial a_j} + \langle \nabla_{\lambda'} \xi_a, \frac{\partial \lambda'}{\partial a_j} \rangle + \langle \nabla_\theta \xi_a, \frac{\partial \lambda'}{\partial a_j} \rangle t. \tag{37}$$

Here the left side is the linearized equation at $u = u_{a,\theta}$.

We take $a_j = c_j \neq 0$ for $j = 1, \ldots, j_0$ (fixed) and let $a_{j_1} \to 0$ for $j_1 > j_0$ varying over all $j > j_0$. Denote $u_{\theta,(c,0)}$ by u_θ and denote $\lambda'_j = \lambda'_j(c_1, \ldots, c_{j_0})$ for all $j = 1, \ldots, j_0, j_0 + 1, \ldots$
(iii) Case $j = 1, \ldots, j_0$
By (32)

$$U_{j,\theta} = i e^{i\theta_j} e^{i\lambda'_j t} \varphi_j + 0(\varepsilon^{1/2}) \tag{38}$$

$$V_{j,\theta} = e^{i\theta_j} e^{i\lambda'_j t} \varphi_j + 0(\varepsilon^{1/2}) + 0(\varepsilon|t|) \tag{39}$$

satisfying an approximative linearized equation at $u = u_\theta$

$$iU_t + AU + \varepsilon \partial \overline{\partial} H.U + \varepsilon \overline{\partial}^2 H.\overline{U} = \xi_j(\theta,t) \tag{40}$$

with error term ξ_j satisfying by (14), (15)

$$\|\xi_j\| < \varepsilon^{r_1}(1 + |t|). \tag{41}$$

(iv) Case $j = j_1$

$$U_{j_1,\theta} = ie^{i\theta_{j_1}} e^{i\lambda'_{j_1} t} \varphi_{j_1} + 0(\varepsilon^{1/2}) \tag{42}$$

$$V_{j_1,\theta} = e^{i\theta_{j_1}} e^{i\lambda'_{j_1} t} \varphi_{j_1} + 0(\varepsilon^{1/2}) \tag{43}$$

by (33) and satisfying (40) with error term ξ_j satisfying

$$\|\xi_j\| < \varepsilon^{r_1}. \tag{44}$$

Thus for $j > j_0$, the error terms remain small for all time. For $j = 1, \ldots, j_0$, there is a linear growth in $|t|$.

Observe that the expression (linearization at $u = u_\theta$)

$$iU_t + AU + \varepsilon \,\partial\bar\partial H.U + \varepsilon \bar\partial^2 H.\overline{U} \tag{45}$$

is $\mathbb{R}$-linear in U.

Fixing θ, define

$$\begin{cases} U_j &= \cos\theta_j.V_{j,\theta} - \sin\theta_j\, U_{j,\theta} = e^{i\lambda'_j t}\varphi_j + 0(\varepsilon^{1/2}) + 0(\varepsilon|t|) \tag{46} \\ V_j &= \sin\theta_j.V_{j,\theta} + \cos\theta_j\, U_{j,\theta} = ie^{i\lambda'_j t}\varphi_j + 0(\varepsilon^{1/2}) + 0(\varepsilon|t|). \tag{47} \end{cases}$$

From the preceding $\{(U_j, V_j)|_{j=1,2,\ldots j_0, j_0+1,\ldots}\}$ are approximative solutions of (45)= 0, in the sense that the equation is satisfied up to error $(1 + |t|)\varepsilon^{r_1}$.

For $t = 0$, one has

$$U_j^0 = U_j(x,0) = \varphi_j(x) + 0(\varepsilon^{1/2}) \tag{48}$$

$$V_j^0 = V_j(x,0) = i\varphi_j(x) + 0(\varepsilon^{1/2}) \tag{49}$$

and hence a perturbation of the sequence $\{(\varphi_j, i\varphi_j)\}$. From a more careful analysis of the error terms in (48), (49) one verifies that $\{(U_j^0, V_j^0)\}$ is in fact a perturbed basis.

It follows indeed from the construction of $F^{a_1,\ldots,a_{j_0},a_{j_1}}(\theta_1, \ldots, \theta_{j_0}, \theta_{j_1})$ for $j_1 \gg j_0$ and $\frac{\partial\lambda'}{\partial a_{j_1}}\big|_{a_{j_1}=0} = 0$ that $\partial_{a_{j_1}} u_{a,\theta}\big|_{a_{j_1}=0}$ has a good localization *wrt* the j_1-mode in x-space. Hence, given any smooth $\varphi, \|\varphi\| \leq 1$ ($\|\ \|$ referring to a sufficiently smooth Sobolev norm), there is an expansion

$$\varphi = \varphi(x) = \sum(\alpha_j U_j^0 + \beta_j V_j^0) \qquad (\alpha_j, \beta_j \in \mathbb{R}). \tag{50}$$

The function

$$\Phi = \sum(\alpha_j U_j + \beta_j V_j) \tag{51}$$

satisfies in particular by (38), (39), (42), (43)

$$\Phi(0) = \varphi \qquad \|\Phi(t)\| \leq (1 + \varepsilon|t|)\|\varphi\| \leq 1 + |t| \tag{52}$$

and

$$i\Phi_t + A\Phi + \varepsilon \frac{\partial^2 H}{\partial\overline{u}_\theta \partial u_\theta}\Phi + \varepsilon \frac{\partial^2 H}{\partial\overline{u}_\theta^2}\overline{\Phi} = w \tag{53}$$

where by (41), (44)

$$\|w\| = 0\big(\varepsilon^{r_1}(1 + |t|)\big). \tag{54}$$

Denote $S_\theta(t)\varphi = U(x,t)$ the solution of the IVP

$$\begin{cases} iU_t + AU + \varepsilon\frac{\partial^2 H}{\partial\overline{u}_\theta \partial u_\theta}U + \varepsilon\frac{\partial^2 H}{\partial\overline{u}_\theta^2}\overline{U} = 0 \\ U(0) = \varphi. \end{cases} \tag{55}$$

Thus $S_\theta(T)^{-1}\psi = V(x,T)$, where $V(x,t)$ is obtained by solving

$$\begin{cases} -iV_t + AV + \varepsilon\frac{\partial^2 H}{\partial u_\theta \partial u_\theta}(T-t)V + \varepsilon\frac{\partial^2 H}{\partial \overline{u}_\theta^2}(T-t)\overline{V} = 0 \\ V(0) = \psi. \end{cases} \tag{56}$$

It follows from definition of u_θ that

$$u_\theta(T-t) = F^c\big(\theta_1 + \lambda_1'(T-t),\ldots,\theta_{j_0} + \lambda_{j_0}'(T-t)\big) = u_{\theta+\lambda'T}(-t) \tag{57}$$

and thus $V(t) = S_{\theta+\lambda'T}(-t)\psi$. The conclusion is that

$$S_\theta(T)^{-1} = S_{\theta+\lambda'T}(-T). \tag{58}$$

Our aim is to establish a bound on the flow map $S_\theta(t)$, for $|t| < \varepsilon^{-r_2}$, $r_2 \sim r$. Given φ, let Φ be as above satisfying (53). Define $U_1 = U - \Phi$ satisfying the IVP

$$\begin{cases} i(U_1)_t + AU_1 + \varepsilon\frac{\partial^2 H}{\partial \overline{u}_\theta \partial u_\theta}U_1 + \varepsilon\frac{\partial^2 H}{\partial \overline{u}_\theta^2}\overline{U}_1 = -w \\ U_1(0) = 0. \end{cases} \tag{59}$$

Hence

$$U_1(t) = \int_0^t S_\theta(t)S_\theta(\tau)^{-1}(iw(\tau))d\tau \tag{60}$$

and from (58)

$$S_\theta(t)\varphi = U = \Phi + \int_0^t S_\theta(t)S_{\theta+\lambda'\tau}(-\tau)(iw(\tau))d\tau. \tag{61}$$

It follows now from (52), (53) that

$$\|S_\theta(t)\varphi\| \leq (1+|t|) + |t|\left(\max_{\psi\in\Pi^{j_0},|\tau|\leq|t|}\|S_\psi(\tau)\|^2\right)(\varepsilon^{r_1}(1+|t|)). \tag{62}$$

Hence

$$\|S_\theta(t)\| \leq 1 + |t|\left\{1 + \varepsilon^{r_1}(1+|t|)\left[\max_{\psi\in\Pi^{j_0},|\tau|\leq|t|}\|S_\psi(\tau)\|^2\right]\right\} \tag{63}$$

implying that

$$\|S_\theta(t)\| < 1 + 2|t| \text{ for } |t| < \varepsilon^{-r_2} \text{ with } r_2 \sim r_1 \sim r. \tag{64}$$

(v) Estimating the solution of the difference equation
   Return to the original IVP

$$\begin{cases} iu_t + Au + \varepsilon\frac{\partial H}{\partial \overline{u}} = 0 \\ u(0) = \varphi. \end{cases} \tag{65}$$

We consider a sufficiently smooth Sobolev norm

$$\|\varphi\|_s = \sum(1+|j|)^s|\alpha_j| \tag{66}$$

for

$$\varphi = \sum \alpha_j \varphi_j. \tag{67}$$

Given a sequence $\alpha = \{\alpha_j\}_{j\leq j_0}$, $\alpha_j \neq 0$, $\|\alpha\|_s \leq 1$, write $\alpha_j = c_j e^{i\theta_j}$, $c_j > 0$. One obtains a quasiperiodic approximative solution $u_\alpha = u_{c,\theta}$, i.e.

$$i\dot{u}_\alpha + Au_\alpha + \varepsilon\frac{\partial H}{\partial \overline{u}_\alpha} = w \tag{68}$$

with
$$\|w\| < \varepsilon^{r_1} \tag{69}$$

and by (30)
$$\|u_\alpha(0) - \sum_{j \leq j_0} \alpha_j \varphi_j\|_s < \sqrt{\varepsilon}. \tag{70}$$

Observe that F^α, u_α (resp. $\lambda'(|\alpha|)$) depend smoothly on α (resp. $|\alpha_j|^2, j = 1, \dots, j_0$). Hence, the map

$$\Omega : B_{\| \ \|_s}(1) \to \mathbb{C}^{j_0}, \| \ \|_s : \alpha = (\alpha_j)_{j=1,\dots,j_0} \mapsto \alpha - ((\langle u_\alpha(0), \varphi_j \rangle)_{j=1,\dots,j_0} \tag{71}$$

satisfies $\|\Omega\| < \sqrt{\varepsilon}$ and similarly a contractive estimate.

It follows from the open map principle that there is $(c_j)_{j=1,\dots,j_0}, c_j > 0$ and $\theta \in \Pi_0$ so that
$$\langle u_{c,\theta}(0), \varphi_j \rangle = \alpha_j \qquad j = 1, \dots, j_0 \tag{72}$$

and
$$\|c\|_s \leq 2. \tag{73}$$

Denote $U = u - u_{c,\theta}$ satisfying the difference equation

$$iU_t + AU + \varepsilon \frac{\partial^2 H}{\partial \overline{u}_{c,\theta} \partial u_{c,\theta}} U + \varepsilon \frac{\partial^2 H}{\partial \overline{u}_{c,\theta}^2} \overline{U} + 0(|U|^2) = -w \tag{74}$$

$$U(0) = \varphi - u_{c,\theta}(0) = \sum_{j > j_0} (\alpha_j - \langle u_{c,\theta}(0), \varphi_j \rangle)\varphi_j \tag{75}$$

hence
$$\|U(0)\|_{s_1} < j_0^{-(s-s_1)} \quad \text{for} \quad s > s_1. \tag{76}$$

Define $U = \gamma U'$, $\gamma > 0$ to be specified, satisfying

$$iU_t' + AU' + \varepsilon \partial \overline{\partial} H.U' + \varepsilon \overline{\partial}^2 H.\overline{U'} + \gamma 0(|U'|^2) = w' \tag{77}$$

where by (76), (69)

$$\|U'(0)\|_{s_1} < \gamma^{-1} j_0^{-(s-s_1)} \quad \text{and} \quad \|w'\| < \gamma^{-1} \varepsilon^{r_1}. \tag{78}$$

One has then from the integral equation

$$U'(t) = S_\theta(t)U'(0) - \int_0^t S_\theta(t)S_\theta(\tau)^{-1}[iw'(\tau) - i\gamma 0(|U'(\tau)|^2)d\tau \tag{79}$$

by (78), (64) the following estimate

$$\|U'(t)\|_{s_1} \leq (1 + 2|t|)\gamma^{-1} j_0^{-(s-s_1)} + (1 + 2|t|)^3(\gamma^{-1}\varepsilon^{r_1} + \gamma) \tag{80}$$

provided
$$\|U'(\tau)\|_{s_1} < 1 \text{ for } |\tau| \leq |t|. \tag{81}$$

Taking $\gamma = \varepsilon^{r_1/2}$, this yields (81) and hence

$$\|U(t)\|_{s_1} < \varepsilon^{r_1/2} \tag{82}$$

provided
$$|t| < \varepsilon^{r_1/2} j_0^{s-s_1} \wedge \varepsilon^{-r_1/10}. \tag{83}$$

Fixing r, the nonresonance assumption (7), (16) may by hypothesis be fulfilled for $j_0 < \varepsilon^{-c_2(r)}$. Hence, letting s be sufficiently large, (83) will impose a restriction $|t| < \varepsilon^{-r_1/10}$, proving the proposition for $M \sim r_1 \sim r$.

Remarks.

(i) To verify condition (7) in example 8, we use the fact that for sufficiently smooth potential V, there is an asymptotic expansion for the Dirichlet spectrum

$$\lambda_j = j^2\pi^2 + c_0(V) + c_1(V)j^{-1} + \cdots + c_R(V)j^{-R} + 0(j^{-R-1}) \tag{84}$$

where for typical V the $c_s(V)$-coefficients may be treated as independent parameters

$$(c_0(V) = \int_0^{'} V(x)dx \text{ and in general certain multilinear expressions in } V).$$

It is important in the preceding argument that the smoothness assumptions on V do not interfere with (7), hence the dependence of $\{\lambda_j\}$ wrt $\hat{V}$ may not be used here.

In the case of example (9), write the equation in the form (6) as

$$\begin{cases} y_t = -Bz \\ z_t = By + B^{-1}F'(y) \end{cases} \tag{85}$$

where $B^2 = -\frac{d^2}{dx^2} + \rho$. Hence $\lambda_j = \sqrt{\pi^2 j^2 + \rho}$ and we use the fact that $\left\{\sqrt{\pi^2 j^2 + \rho}\right\}_{j\geq 1}$ are linearly independent functions of ρ to fulfill (7).

(ii) It is easily verified that the argument proving Proposition 5 yields in the case of finite dimensional phase space stability times of the order $\exp . \frac{1}{\varepsilon^c}, c = c(\lambda, d)$, as in Nekhoroshev's theorem. The first step in proving Proposition 5 may be performed also by the standard method of consecutive symplectic transformations of phase space and reduction of the nonlinearity.

Our next purpose is to sketch the construction of some examples (of NLW equations) illustrating an energy transition from low to high Fourier modes and a growth of higher derivatives in time. We consider an equation of the form

$$y_{tt} + B^2 y + P_N f'(y) = 0 \tag{86}$$

where for instance

$$f(y) = y^4 \tag{87}$$

and B will be a certain perturbation of $\sqrt{-\Delta}$, B diagonal wrt the exponential basis with eigenvalues $\mu_n = \mu_{-n}$ satisfying

$$\mu_n = \rho_n + \sigma_n \tag{88}$$

$$\rho_n = |n| + \rho \text{ for } |n| \neq 2, 3 \tag{89}$$

$$\rho, \rho_2, \rho_3 \text{ are independent} \tag{90}$$

$$\sigma_n = 0(\gamma^2) \tag{91}$$

where $\gamma = \gamma_N$ is a parameter, say

$$\gamma = N^{-10}. \tag{92}$$

Thus (86) may be written as

$$\begin{cases} \dot{y} = Bz \\ \dot{z} = -By - B^{-1}P_N f'(y) \end{cases} \tag{93}$$

or, letting $q = y + iz$, as

$$i\dot{q} = B^{-1}\frac{\partial H}{\partial \overline{q}} \tag{94}$$

with $(q_n^{-1} \equiv \overline{q}_n)$

$$H = \int [|Bq|^2 + f(\mathrm{Re}q)] = \sum \mu_n^2 |q_n|^2 + \sum_{\sum \varepsilon_j n_j = 0} q_{n_1}^{\varepsilon_1} q_{n_2}^{\varepsilon_2} q_{n_3}^{\varepsilon_3} q_{n_4}^{\varepsilon_4} \quad (\varepsilon_j = \pm 1) \tag{95}$$

the Hamiltonian. Here (94) is the gradient in the symplectic Hilbert space with inner product $\langle \xi, B\eta \rangle$ identified with $H^{1/2}$. In this symplectic Hilbert space, the Poisson bracket is given by

$$\{H_1, H_2\} = \mathrm{Im}\sum_n \frac{1}{\mu_n}\frac{\partial H_1}{\partial q_n}\frac{\partial H_2}{\partial \overline{q}_n}. \tag{96}$$

The monomial $\Pi q_{n_j}^{\varepsilon_j}$ with $\sum \varepsilon_j n_j = 0$ is called resonant provided

$$\sum \varepsilon_j \rho_{n_j} = 0. \tag{97}$$

Let

$$\delta = \gamma^3 = N^{-30} \tag{98}$$

and rescale

$$q_n \to \gamma q_n \text{ for } n = 0, |n| = 1 \tag{99}$$
$$q_n \to \gamma\delta q_n \text{ for } |n| \geq 2. \tag{100}$$

Eliminate the nonresonnant part in the Hamiltonian (95). This yields after the rescaling (99)-(100) a Hamiltonian of the form

$$\begin{aligned}
H_1(q, \overline{q}) = {}& \gamma^2 \sum_{|n|=0,1} \mu_n^2 |q_n|^2 + \gamma^2\delta^2 \sum_{|n|\geq 2} \mu_n^2 |q_n|^2 \\
& + \gamma^4 \Phi_1(|q_j|^2 | j = 0, 1, -1) + \gamma^4\delta^2 \Phi_2(|q_j|^2 | j = 0, 1, -1) . \sum_{|n|\geq 2} |q_n|^2 \\
& + \gamma^4\delta^2 \mathrm{Re}\{q_0\overline{q}_1 \big(\sum_{n\geq 4} \overline{q}_n q_{n+1}\big) + q_0\overline{q}_{-1}\big(\sum_{n\leq -4} \overline{q}_n q_{n-1}\big)\} \\
& + 0(\delta^3\gamma^2 + \delta^2\gamma^6 + \delta\gamma^8)
\end{aligned} \tag{101}$$

using (88)-(91).

The corresponding equations are

$$\begin{cases} i\dot{q}_n = \dfrac{1}{\mu_n\gamma^2}\dfrac{\partial H_1}{\partial \overline{q}_n} & (n = 0, 1, -1) \\[2mm] i\dot{q}_n = \dfrac{1}{\mu_n\gamma^2\delta^2}\dfrac{\partial H_1}{\partial \overline{q}_n} & (|n| \geq 2) \end{cases} \tag{102}$$

and thus, by (101)

$$
\begin{cases}
i\dot{q}_n = \left(\mu_n + \frac{\gamma^2}{\mu_n}\frac{\partial\phi_1}{\partial\xi_n}(|q_j|^2)\right)q_n + 0(\delta^2 + \delta\gamma^6) \qquad (|n| = 0, 1) & (103)\\[2mm]
i\dot{q}_n = \left(\mu_n + \frac{\gamma^2}{\mu_n}\phi_2\right)q_n + 0(\delta + \gamma^4 + \frac{\gamma^6}{\delta}) \qquad (|n| = 2, 3) & (104)\\[2mm]
i\dot{q}_4 = \left(\mu_4 + \frac{\gamma^2}{\mu_4}\phi_2\right)q_4 + \frac{\gamma^2}{\mu_4}q_0\bar{q}_1 q_5 + 0(\delta + \gamma^4 + \frac{\gamma^6}{\delta}) & (105)\\[2mm]
i\dot{q}_{-4} = \left(\mu_4 + \frac{\gamma^2}{\mu_4}\phi_2\right)q_{-4} + \frac{\gamma^2}{\mu_4}q_0\bar{q}_{-1}q_{-5} + 0(\delta + \gamma^4 + \frac{\gamma^6}{\delta}) & (106)\\[2mm]
i\dot{q}_n = \left(\mu_n + \frac{\gamma^2}{\mu_n}\phi_2\right)q_n + \frac{\gamma^2}{\mu_n}\left(q_0\bar{q}_1 q_{n+1} + \bar{q}_0 q_1 q_{n-1}\right) + \frac{1}{|n|}0(\delta + \gamma^4 + \frac{\gamma^6}{\delta}) & \\[1mm]
\qquad\qquad (n \geq 5) & (107)\\[2mm]
i\dot{q}_n = \left(\mu_n + \frac{\gamma^2}{\mu_n}\phi_2\right)q_n + \frac{\gamma^2}{\mu_n}\left(q_0\bar{q}_{-1}q_{n-1} + \bar{q}_0 q_{-1}q_{n+1}\right) + \frac{1}{|n|}0(\delta + \gamma^4 + \frac{\gamma^6}{\delta}) & \\[1mm]
\qquad\qquad (n \leq -5). & (108)
\end{cases}
$$

We assume

$$|q_n| \leq 2. \tag{109}$$

It follows from (103) that for $j = 0, 1, -1$

$$\widehat{|q_j|^2} = 0(\delta^2 + \delta\gamma^6). \tag{110}$$

Hence may assume

$$|q_j|^2 = 1 + 0(\delta^2 + \delta\gamma^6)|t|. \tag{111}$$

From (88), (111), eqs (103) for $n = 0, 1, -1$ yield

$$i\dot{q}_n = \left(\rho_n + \sigma_n + 0(\gamma^2)\right)q_n + 0(\delta^2 + \delta\gamma^6)(1 + \gamma^2|t|) \tag{112}$$

ans by appropriate choice of $\sigma_n = 0(\gamma^2)$

$$i\dot{q}_n = \rho_n q_n + 0(\delta^2 + \delta\gamma^6)(1 + \gamma^2|t|) \qquad (n = 0, 1, -1). \tag{113}$$

Similarly, one gets from (104)

$$i\dot{q}_n = \rho_n q_n + 0\left(\delta + \gamma^4 + \frac{\gamma^6}{\delta} + \gamma^2(\delta^2 + \delta\gamma^6)|t|\right) \qquad (|n| = 2, 3) \tag{114}$$

and from (105), (106), using (98), (91)

$$
\begin{aligned}
i\dot{q}_4 &= (\rho_4 + \sigma_4')q_4 + \gamma^2\rho_4^{-1}q_0\bar{q}_1 q_5 + 0(\cdots)\\
i\dot{q}_{-4} &= (\rho_4 + \sigma_4')q_{-4} + \gamma^2\rho_4^{-1}q_0\bar{q}_{-1}q_{-5} + 0(\cdots)
\end{aligned}
\tag{115}
$$

and from (107), (108)

$$i\dot{q}_n = (\rho_n + \sigma_n')q_n + \gamma^2\rho_n^{-1}(\bar{q}_0 q_1 q_{n-1} + q_0\bar{q}_1 q_{n+1}) + \frac{1}{|n|}0(\cdots) \quad (n \geq 5)$$

$$i\dot{q}_n = (\rho_n + \sigma_n')q_n + \gamma^2\rho_n^{-1}(q_0\bar{q}_{-1}q_{n-1} + \bar{q}_0 q_{-1}q_{n+1}) + \frac{1}{|n|}0(\cdots) \quad (n \leq -5)$$

$$\tag{116}$$

where $\{\sigma_n'|4 \leq n \leq N\}$ are the new parameters.

Replacement of q_n by $e^{-i\rho_n t}q_n$ eliminates the $\rho_n q_n$ terms in (113)-(116).

From (113), take

$$q_n = 1 + 0(\delta^2 + \delta\gamma^6)(1 + \gamma^2|t|)|t|. \tag{117}$$

From (114), take

$$q_n = 0\left(\left(\delta + \gamma^4 + \frac{\gamma^6}{\delta}\right)|t| + \gamma^2(\delta^2 + \delta\gamma^6)|t|^2\right) \quad (|n| = 2, 3). \tag{118}$$

From (115), (117)

$$\begin{cases} i\dot{q}_4 = \sigma'_4 q_4 + \gamma^2 \rho_4^{-1} q_5 + 0(\delta^2 + \delta\gamma^6)(1 + \gamma^2|t|)^2 + 0(\delta + \gamma^4 + \frac{\gamma^6}{\delta}) \\ i\dot{q}_{-4} = \sigma'_4 q_{-4} + \gamma^2 \rho_4^{-1} q_{-5} + 0(\cdots). \end{cases} \tag{119}$$

From (116), (117)

$$\begin{cases} (n \geq 5) \quad i\dot{q}_n = \sigma'_n q_n + \gamma^2 \rho_n^{-1}(q_{n-1} + q_{n+1}) + \frac{1}{|n|}0(\delta^2 + \delta\gamma^6)(1 + \gamma^2|t|)^2 + \\ \frac{1}{|n|}0(\delta + \gamma^4 + \frac{\gamma^6}{\delta}) \\ (n \leq -5) \quad i\dot{q}_n = \sigma'_n q_n + \gamma^2 \rho_n^{-1}(q_{n-1} + q_{n+1}) + \frac{1}{|n|}0(\cdots). \end{cases} \tag{120}$$

Define $Q_n = \rho_n^{1/2} q_n$. Since $\delta = \gamma^3$, we get from (119) - (120)

$$\begin{cases} i\dot{Q}_4 = \sigma_4 Q_4 + \gamma^2(\rho_4 \rho_5)^{1/2} Q_5 + 0\left(\gamma^3 + \gamma^6(1 + \gamma^2|t|)^2\right) \\ i\dot{Q}_{-4} = \sigma'_{-4} Q_{-4} + \gamma^2(\rho_{-4}\rho_{-5})^{-1/2} Q_{-5} + 0(\cdots) \\ (|n| \geq 5) \quad i\dot{Q}_n = \sigma'_n Q_n + \gamma^2\left((\rho_n \rho_{n-1})^{-1/2} Q_{n-1} + (\rho_n \rho_{n+1})^{-1/2} Q_{n+1}\right) + 0(\cdots) \end{cases} \tag{121}$$

Write

$$Q_+ = Q_n|_{n \in [4, N]} \qquad Q_- = Q_n|_{n \in [-N, -4]} \tag{122}$$

and define the selfadjoint matrix

$$A_+ = A = D + \gamma^2 B \tag{123}$$

with

$$D_n = \sigma'_n \text{ for } n \in [4, N] \tag{124}$$

$$\begin{cases} B_{n-1,n} = B_{nn-1} = (\rho_{n-1}\rho_n)^{1/2} \text{ for } 5 \leq n \leq N \\ B_{nn'} = 0 \text{ otherwise} \end{cases} \tag{125}$$

and similarly A_-.

Rewrite (121) as

$$i\dot{Q}_\pm = A_\pm Q_\pm + \Omega_\pm(t) \tag{126}$$

where

$$\Omega_\pm = 0\left(\gamma^3 + \gamma^6(1 + \gamma^2|t|)^2\right). \tag{127}$$

Specify an eigenvector $\varphi = \varphi^+$ of A_+ with eigenvalue $\lambda = 0$, such that

$$\begin{cases} \varphi_4 = 1 \\ \varphi_n = 0(\tau) \quad (5 \leq n \leq N) \end{cases} \tag{128}$$

with $\tau = \tau_N$ to be specified. Thus by (123) - (125), this yields the equations

$$\begin{cases} \sigma'_4 \varphi_4 + \gamma^2(\rho_4 \rho_5)^{-1/2}\varphi_5 = 0 \\ \sigma'_n \varphi_n + \gamma^2\left((\rho_n \rho_{n-1})^{-1/2}\varphi_{n-1} + (\rho_n \rho_{n+1})^{-1/2}\varphi_{n+1}\right) = 0 \quad (5 \leq n < N) \\ \sigma'_N \varphi_N + \gamma^2(\rho_{N-1}\rho_N)^{-1/2}\varphi_{N-1} = 0. \end{cases} \tag{129}$$

Define parameters

$$\begin{cases} \xi_4 = \frac{\varphi_4}{\varphi_5} \\ \xi_n = \frac{\varphi_n}{\varphi_{n+1}} \qquad 5 \le n \le N-1 \end{cases} \tag{130}$$

hence by (129)

$$\begin{cases} \sigma_4' = -\gamma^2(\rho_4\rho_5)^{-1/2}\xi_4^{-1} \\ \sigma_n' = -\gamma^2\big((\rho_n\rho_{n-1})^{-1/2}\xi_{n-1} + (\rho_n\rho_{n+1})^{-1/2}\frac{1}{\xi_n}\big) \qquad (5 \le n < N) \\ \sigma_N' = -\gamma^2(\rho_N\rho_{N-1})^{-1/2}\xi_{N-1}. \end{cases} \tag{131}$$

Denote $(\varphi,0)$ and (ψ^λ,λ) the normalized eigenfunctions and eigenvalues of A_+, considered as $(\xi_4,\dots,\xi_{N-1})$-dependent. From first eigenvalue variation, one gets

$$\frac{\partial\lambda}{\partial\xi_j} = \langle\frac{\partial A}{\partial\xi_j}\psi^\lambda,\psi^\lambda\rangle = \sum_{n=4}^{N}\frac{\partial\sigma_n'}{\partial\xi_j}\psi_n^2. \tag{132}$$

Observe that by (128)

$$0 = \langle\varphi,\psi^\lambda\rangle = \varphi_4\psi_4^\lambda + 0(\tau\sqrt{N}) = \psi_4^\lambda + 0(\tau\sqrt{N})$$

hence

$$\psi_4^\lambda = 0(\tau\sqrt{N}). \tag{133}$$

By (131), (132)

$$\begin{cases} \frac{\partial\lambda}{\partial\xi_4} = \gamma^2(\rho_4\rho_5)^{-1/2}\left(\frac{\psi_4^2}{\xi_4^2} - \psi_5^2\right) \\ \frac{\partial\lambda}{\partial\xi_n} = \gamma^2(\rho_n\rho_{n+1})^{-1/2}\left(\frac{\psi_n^2}{\xi_n^2} - \psi_{n+1}^2\right) \qquad (5 \le n < N). \end{cases} \tag{134}$$

Differentiate at $\xi_4 = \frac{1}{\tau}, \xi_5 = \cdots = \xi_{N-1} = 1$. It follows from (133), (134) that

$$(\rho_{N-1}\rho_N)^{1/2}\frac{\partial\lambda}{\partial\xi_{N-1}} + 2(\rho_{N-2}\rho_{N-1})^{1/2}\frac{\partial\lambda}{\partial\xi_{N-2}} + \cdots +$$

$$(N-4)(\rho_4\rho_5)^{1/2}\frac{\partial\lambda}{\partial\xi_4} \tag{135}$$

$$= \gamma^2[(\psi_{N-1}^2 - \psi_N^2) + 2(\psi_{N-2}^2 - \psi_{N-1}^2) + \cdots + (N-5)(\psi_5^2 - \psi_6^2)+$$

$$(N-4)(\tau^2\psi_4^2 - \psi_5^2)] \tag{136}$$

$$= -\gamma^2(\psi_N^2 + \psi_{N-1}^2 + \cdots + \psi_5^2 - \tau^2(N-4)\psi_4^2)$$

$$= -\gamma^2\big(1 - (1 + \tau^2(N-4))\psi_4^2\big)$$

$$= -\gamma^2\left(1 - 0(\frac{1}{N})\right) \tag{137}$$

letting

$$\tau = \tau_N = \frac{1}{N}. \tag{138}$$

It easily follows from (137) that we may choose $(\xi_4,\dots,\xi_{N-1})$ such that

$$|\lambda| > \frac{\gamma^2}{N^5} \text{ for } \lambda \ne 0 \tag{139}$$

for all eigenvalues $\ne 0$ of A_+.

Consider an initial data $Q(0)$ defined by

$$\begin{cases} Q(0)_4 = 1 \\ Q(0)_n = 0 \quad \text{for} \quad N \geq n > 4, -4 \geq n \geq -N \end{cases} \tag{140}$$

and write

$$Q(0) = c\varphi + \sum_{\lambda \neq 0} c_\lambda \psi^\lambda$$

expanding in the eigenfunctions of A_+. Thus by (128), (138)

$$\varphi_4 = \langle Q(0), \varphi \rangle = c = 1 - 0(N\tau^2) = 1 - 0\left(\frac{1}{N}\right). \tag{141}$$

The corresponding solution $Q = Q(t) = (Q_4, \dots, Q_N)$ of the linear equation

$$i\dot{Q} = AQ \tag{142}$$

is

$$Q(t) = c\varphi + \sum_{\lambda \neq 0} c_\lambda \psi^\lambda e^{-i\lambda t}. \tag{143}$$

It follows from (139) that

$$\frac{1}{T} \int_0^T Q(t) = c\varphi + \sum_{\lambda \neq 0} |c_\lambda| 0\left(\frac{1}{|\lambda T|}\right) = c\varphi + 0(N^{1/2} N^5 \gamma^{-2} T^{-1}). \tag{144}$$

Taking

$$T = N^7 \gamma^{-2} \tag{145}$$

(144), (138), (128) imply

$$\max_{|t| < T} |Q_N(t)| > \frac{1}{2} \varphi_N = \frac{1}{N}. \tag{146}$$

The difference with the solution of the equation (126) is bounded by

$$\left| \int_0^t e^{-i(t-\tau)A} \Omega(\tau) d\tau \right| < T \|\Omega\|$$
$$< 0(\gamma^3 T + (1 + \gamma^2 T)^2 \gamma^6 T)$$
$$< \gamma N^7 + \gamma^4 N^{21} \tag{147}$$

from (127), (145). Take

$$\gamma = N^{-10} \tag{148}$$

so that (146) remains valid for the solution of (126) with initial data (140) for $t = 0$.

It follows from (146), (145), (148) that the system (103) - (108) restricted to $|n| \leq N$ with initial data for $t = 0$

$$\begin{cases} q_n(0) = 1 & (n = 0, 1, -1) \\ q_n(0) = 0 & (|n| = 2, 3) \\ q_4(0) = \rho_4^{-1/2} \\ q_n(0) = 0 \quad \text{otherwise} \end{cases} \tag{149}$$

yields by (145), (148)

$$\max_{|t| \leq N^{27}} |q_N(t)| > \rho_N^{-1/2} \frac{1}{N} \sim N^{-3/2} r. \tag{150}$$

Notice that by (143)

$$|q_n(t)| = \rho_n^{-1/2}|Q_n(t)| < |c| + \sum_{\lambda \neq 0} |c_\lambda|\,|\psi_n^\lambda| < 1 + \left(\sum c_\lambda^2\right)^{1/2} < 2 \qquad (151)$$

for $t < T$, so that (109) holds.

Since $\gamma = N^{-10}, \delta = \gamma^3 = N^{-30}$, it follows that for the original equation $(*)$ (94) with $f(t) = t^4, B_n = \mu_n \approx |n|$, the data at $t = 0$

$$\begin{cases} q_n(0) = \gamma = N^{-10} & (n = 0, -1, -1) \\ q_n(0) = 0 & (|n| = 2, 3) \\ q_4(0) = \rho_4^{-1/2}\gamma\delta \sim N^{-40} \\ q_n(0) = 0 & \text{otherwise} \end{cases} \qquad (152)$$

evolve to

$$\sup_t |q_N(t)| \geq \max_{|t| \leq N^{27}} |q_N(t)| > N^{-40}N^{-3/2} > N^{-42}. \qquad (153)$$

A slight elaboration of the preceding construction permits to define a perturbation B os $\sqrt{-\Delta}$ as above such that for the equation

$$y_{tt} + B^2 y + y^4 = 0$$

one gets

$$\inf_{s,\langle \infty,\gamma\rangle > 0} \overline{\lim}_{|t| \to \infty} |t|^{-as}\left[\sup_{\|y(0)\|_{H^{s_1}}} + \|\dot{y}(0)\|_{H^{s_1}} \leq \gamma \,\|y(t)\|_{H^s}\right] = \infty \qquad (154)$$

for some constant $a > 0$, provided $s > s_0$.

The preceding example shows in particular the importance of the diophantine properties of the spectrum $\{\mu_n\}$ of the operator A in (6) for Proposition 5 to hold. Observe that moreover the nonlinearity y^4 is natural and nondegenerate (thus one should not expect Nekhoroshev stability in a finer topology than controlled by the twist term of the Hamiltonian, at least for general data (cf. remark 2, p. 134)).

It does not seem clear how to obtain an NLS analogue of this construction, because of the "better" arithmetic properties of $\{n^2\}$.

Remarks.

(i) Conversely, there is an analogue of the longtime estimates of higher Sobolev norms discussed for NLS at the end of section 2 in the case of the wave equation. Consider a NLW

$$y_{tt} + B^2 y + f'(y) = 0 \qquad (155)$$

say in $D1$ or $D2$, where B has eigenvalues $\{\mu_n\}$, $\mu_n = \mu_{-n}$, $\mu_n \sim |n|$ and assuming $f(y)$ a polynomial. Assume further the Hamiltonian

$$\int [|By|^2 + |Bz|^2 + f(y)] \qquad (\dot{y} = Bz) \qquad (156)$$

yields an apriori control of $\|y\|_{H^1}$. Thus one gets for smooth data $y(0)$, $\dot{y}(0)$ bounds of the form (compare with (2.122) for NLS)

$$\|y(t)\|_{H^s} + \|\dot{y}(t)\|_{H^{s-1}} \lesssim T^{s-1+} \qquad (157)$$

in higher Sobolev norms, in particular polynomial in time (compare with (154)).

Denoting

$$\|q\|_s = (\|y\|_{H^s}^2 + \|z\|_{H^s}^2)^{1/2} \tag{158}$$

we derive (157) from the following inequality relative to the flow map $t \to t + 1$

$$\|q(t+1)\|_s \le \|q(t)\|_s(1 + C\|q(t)\|_s^{-\frac{1}{s-1}+}) \tag{159}$$

assuming $\|q\|_1 \le 1$. The proof will be an easy consequence of the order - 1 smoothing property of $\Box^{-1}$, where

$$\Box^{-1}w = y \quad \text{solves} \quad \begin{cases} \Box y = y_{tt} + B^2 y = w \\ y(0) = \dot{y}(0) = 0. \end{cases} \tag{160}$$

The flow in phase space $(y, B^{-1}\dot{y}) = (y, z)$ is given by

$$q(t) = (y(t), z(t)) = \left(S(t)(a,b), B^{-1}\dot{S}(t)(a,b)\right) - \left(\Box^{-1}f'(y), B^{-1}\frac{d}{dt}\Box^{-1}f'(y)\right) \tag{161}$$

where $a = y(0)$, $b = z(0)$ and

$$S(t)(a,b) = \frac{1}{2}\sum_n [(a_n - ib_n)\, e^{i(n,x+\mu_n t)} + (a_n + ib_n)\, e^{i(nx-\mu_n t)}]. \tag{162}$$

Let $|t| \le 1$. Observe that the first term in (161) has unitary behaviour

$$(\|S(t)(a,b)\|_{H^s}^2 + \|B^{-1}\dot{S}(t)(a,b)\|_{H^s}^2)^{1/2} = \left(\sum |n|^{2s}(|a_n|^2 + |b_n|^2)\right)^{1/2} = \|q(0)\|_s. \tag{163}$$

Consider next the nonlinear term (161). Both factors are estimated in the norm

$$\|u\|_{s,\rho} = \left(\sum_n \int d\lambda |n|^{2s}\,||\lambda| - |n||^{2\rho}\,|\hat{u}(n,\lambda)|^2\right)^{1/2} \tag{164}$$

letting ρ be slightly $> \frac{1}{2}$. Write $f'(y) = \sum_{K \text{ dyadic}} P_k f'(y)$ with $P_K \phi = \sum_{|n|\sim K} \hat{\phi}(n)e^{inx}$. There is an apriori bound on $\|y\|_{1,\rho}$, hence $\|y\|_{L^q(x,t)}$ for all $q < \infty$ since we are in $D \le 2$. One gets for fixed K

$$\|\Box^{-1}P_K f'(y)\|_{s,\rho} \lesssim K^{s-2+} \wedge K^{-1+}\|y\|_{s,\rho} \lesssim K^{s-2+} \wedge K^{-1+}\|q(0)\|_s \tag{165}$$

and hence

$$\|\Box^{-1}f'(y)(t)\|_{H^s} \le \|\Box^{-1}f'(y)\|_{s,\rho} \ll \|q(0)\|_s^{1-\frac{1}{s-1}+}. \tag{166}$$

A similar estimate holds for $B^{-1}\frac{d}{dt}\Box^{-1}f'(y)$. This yields inequality (159).

(ii) More recent investigations did confirm the comments made in Remark 2, p.144, regarding equation (10). The stability property was verified in [**Ba**] (for typical data) with respect to the H^1-topology and for exponentially long times. Using a different point of view, the author also checked stability in smooth topology (again for typical data) for times $|t| < T$, $\log T \gg \log \frac{1}{\varepsilon}$.

References to Appendix

[N]   N.W. Nekhoroshev, *An exponential estimate of the time of stability of nearly integrable Hamiltonian systems*, Uspekhi Math. Nauk. 32; 1 (1977), 5-66.

[B-F-G]   G. Benettin, J. Fröhlich, A. Giorgilli, *A Nekhoroshev-type theorem for Hamiltonian systems with infinitely many degrees of freedom*, CMP 119 (1989), 95-108.

[B14]   J. Bourgain, *Construction of approximative and quasi-periodic solutions of perturbed linear Schrödinger and wave equations*, Gafa, Vol. 6, N2 (1996).

[Ba]   D. Bambusi, *On the long time behavior of the majority of small amplitude solutions in some nonlinear Schrödinger equations*, preprint.

[B-N]   D. Bambusi, N.N. Nekhoroshev, *A property of exponential stability in nonlinear wave equations near the fundamental linear mode*, preprint 1996.

Harmonic Analysis, Wavelets and Applications

Ingrid C. Daubechies
Anna C. Gilbert

IAS/Park City Mathematics Series
Volume 5, 1999

Harmonic Analysis, Wavelets and Applications

Ingrid C. Daubechies
Anna C. Gilbert

LECTURE 1
Introduction

Wavelets constitute a tool to decompose, analyze, and synthesize functions, with emphasis on time-frequency localization. We decompose a function f of time t by writing

$$f(t) = \sum_{j,k \in \mathbb{Z}} c_{j,k}\, \psi_{j,k}(t)\,.$$

Here, the variable t is one-dimensional, but we can also deal with higher dimensions (we will discuss this later). In the above sum, the functions $\psi_{j,k}$ are the wavelets; they are generated by scaled and translated versions of a "parent" function ψ, in the following way:

$$(1.1) \qquad \psi_{j,k}(t) = 2^{j/2}\psi(2^j t - k) \quad \text{for} \quad j, k \in \mathbb{Z}\,.$$

We shall restrict our attention to square integrable wavelets ψ with certain properties. The basic properties are that ψ is oscillating in some sense (i.e., $\psi \in L^1$ and $\int \psi = 0$) and that ψ is "well-localized" (to be made more precise later). Figure 1.1 shows some diagrams of a possible generating wavelet ψ and two scaled and translated versions of it. Notice that the dilation affects the wavelet by shrinking or stretching its support, and also by increasing or decreasing its amplitude (so as to preserve the L^2-norm). Also, $\psi(2^j t - k) = \psi(2^j(t - 2^{-j}k))$, so that at each "scale" j, we have adapted our translation step to the scale. In other words, the wavelets in formula (1.1) are in fact translated by steps of size 2^{-j} and not by integer-sized steps. (There exists wavelet families where the dilation is not systematically by

[1]Department of Mathematics and Program in Applied and Computational Mathematics, Princeton University, Princeton, New Jersey 08544
E-mail address: `ingrid@math.princeton.edu`
[2]Department of Mathematics, Princeton University, Princeton, New Jersey 08544
E-mail address: `agilbert@math.princeton.edu`

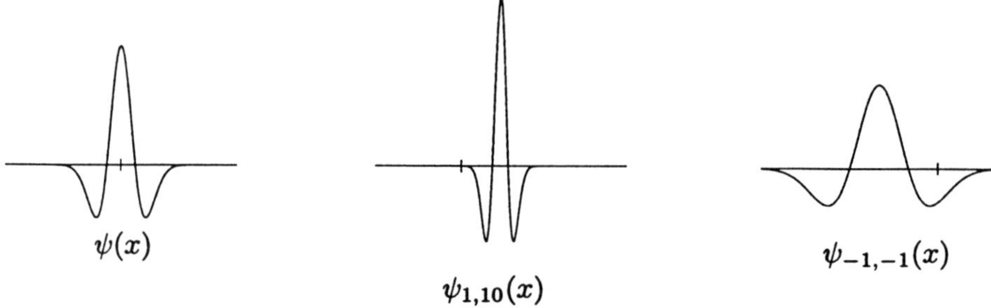

Figure 1.1. Translates and dilates of the "Mexican hat" wavelet $\psi(x) = (1 - x^2)e^{-x^2/2}$. (In each case, the tick corresponds to $x = 0$.) For this ψ the $\{\psi_{j,k}\}$ do not constitue an orthonormal basis.

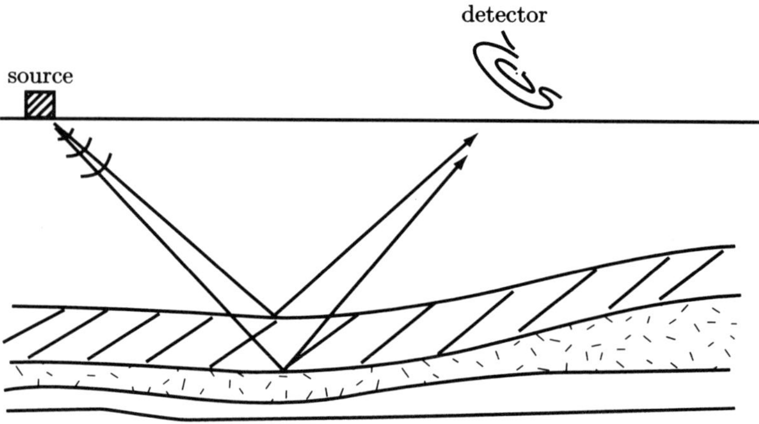

Figure 1.2. Seismography

factors 2, and many other generalizations. See, e.g. [**DTL**]. But (1.1) is widely used, and we shall stick to it here.)

Why are these wavelet decompositions useful? Why would time-frequency localizations be of interest? The answer is simple: time-frequency localizations turn up in many settings.

One setting where we encounter time-frequency decomposition is geophysics or seismography. If one places a source of sound waves (generated by a vibrator or by a detonation) on the earth's surface and also places a receiver at another spot, one will hear or detect the signal after it has traveled through the earth (see Figure 1.2). The problem with trying to analyze this signal, and trying to determine from this signal what the earth looks like beneath the surface, is that this signal is very complicated. Different frequencies will arrive at the receiver at different times, waves will get reflected many times, etc. Time-frequency analysis helps to disentangle all of this information.

A second setting where we find time-frequency decomposition is in music. A musical score tells a musician what note (i.e., what frequency) to play when and for

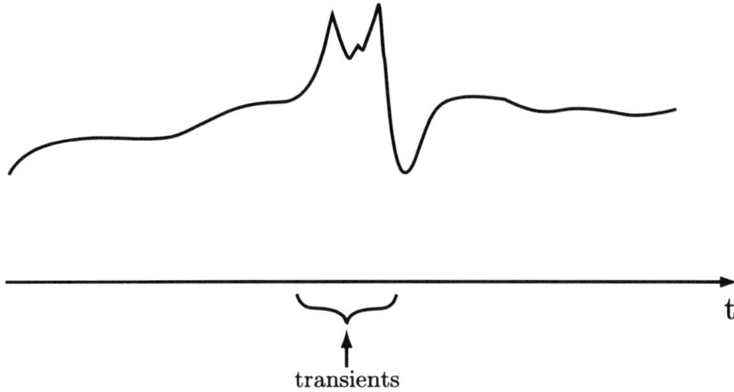

Figure 1.3. Signal with a sharp transient

how long. The cellist should play an A for a quarter note followed by one measure of rest, for example. This type of time-frequency decomposition has been around for centuries!

These were just two examples. Many more frameworks use time-frequency type decompositions, including physics, engineering, and (yes!) mathematics. (The principal symbol of a pseudo-differential operator is a time-frequency representation of the operator, and micro-local analysis is essentially time-frequency localization.) Ideas from these different directions all played a role in the development of wavelet theory.

How does a wavelet transform differ from a Fourier transform? What does a wavelet transform do that a Fourier transform cannot do? Suppose we have a signal that has a sudden transient as in Figure 1.3. We would like to know that this sharp transient happens at time t_0; if we were to examine the Fourier transform of this signal, we would, from the decay rate of the Fourier transform, be able to determine only that a singularity happened, not at what time the sharp transient appeared. We could try to fix this by taking a smooth window function and looking at the Fourier transform of the window function times the signal (i.e., a piece of our signal). Then we could repeat this after moving the window to look at the next piece of signal, each time computing "windowed Fourier coefficients". This is frequently done in applications. But even though this will give a better time localization than a straightforward Fourier transform, it still cannot pin down times of events very finely, since the precision is limited by the width of the window.

What if we also wanted to find out how smooth our signal is? We can't characterize exactly the smoothness (or the singularity) with the Fourier transform. We do know the following, however:

Theorem. *If f is a function on $[0,1]$ which is k times continuously differentiable, then its Fourier coefficients $\widehat{f}(n)$ satisfy*

$$(1.2) \qquad\qquad |\widehat{f}(n)| \le C|n|^{-k}.$$

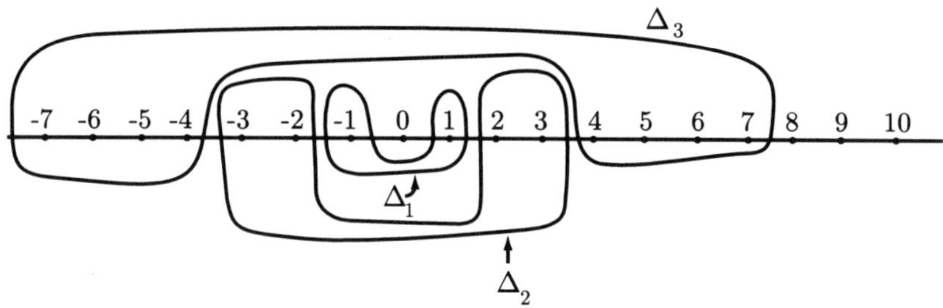

Figure 1.4. Littlewood-Paley dyadic blocks

To prove (1.2), integrate $\widehat{f}(n) = \int_0^1 f(t)e^{-2\pi itn}dt$ by parts to obtain $\widehat{f}(n) \sim \left(\frac{1}{n}\right)^k \widehat{f^{(k)}}(n)$. Because the $\widehat{f^k}(n)$ are bounded, (1.2) follows. (In fact, we even know that $\widehat{f^k}(n)$ tends to zero as $n \to \infty$, so that $n^k\widehat{f}(n)$ does as well.)

In other words, smoothness of our function f carries a signature in the Fourier coefficients. The following theorem in the other direction is also easy to prove:

Theorem. *If $|\widehat{f}(n)| \le C|n|^{-1-\epsilon}$ then $f \in C([0,1])$ and if $|\widehat{f}(n)| \le C|n|^{-1-k-\epsilon}$ then $f \in C^k([0,1])$.*

Notice that we have lost a one in the exponent here when we try to characterize smoothness from the Fourier coefficients.

However, we can do better. We can use the Littlewood-Paley theory (which was developed in the 1920's by two pure mathematicians) to characterize local smoothness exactly. Assume that

$$f(x) = \lim_{N\to\infty} \sum_{n=-N}^{N} \widehat{f}(n)\,e^{2\pi inx}\,;$$

then write

$$f(x) = \widehat{f}(0) + \lim_{J\to\infty} \sum_{j=1}^{J}(\Delta_j f)(x)\,,$$

where

$$(\Delta_j f)(x) = \sum_{2^{j-1}\le|n|<2^j} \widehat{f}(n)\,e^{2\pi inx}\,.$$

In other words, bunch together parts of the sum in dyadic blocks. Notice that the width in n of these blocks grows exponentially (see Figure 1.4). We can now characterize exactly local smoothness in terms of these $\Delta_j f$. If for some $\epsilon > 0$, we have $\sup_x |(\Delta_j f)(x)| \le C2^{-j\epsilon}$, then f is continuous. Here is a sketch of the proof (fill in the details yourself):

Proof. (sketch) We know that the sum $\sum_{j=1}^{\infty}(\Delta_j f)(x)$ automatically converges because of the bound on $\Delta_j f$. Furthermore, the inequality

$$\sum_{j=J+1}^{\infty} |(\Delta_j f)(x)| \le C2^{-J\epsilon}$$

holds uniformly in x. Then,

$$|f(x+h) - f(x)| \leq \sum_{j=1}^{J} |(\Delta_j f)(x+h) - (\Delta_j f)(x)| + C2^{-J\epsilon}.$$

Also,

$$|(\Delta_j f)(x+h) - (\Delta_j f)(x)| \leq \sum_{2^{j-1} \leq |n| < 2^j} |\widehat{f}(n)e^{2\pi i n x}(e^{2\pi i n h} - 1)|$$

$$\leq C \sum_{2^{j-1} \leq |n| < 2^j} |nh|$$

$$\leq C2^j|h|.$$

(Note: $f \in L^1$, so $\widehat{f}(n)$ is bounded.) Putting these estimates together, we have

$$|f(x+h) - f(x)| \leq C \sum_{j=1}^{J} 2^j|h| + C2^{-J\epsilon} \leq C(2^J|h| + 2^{-J\epsilon}),$$

and this can be made arbitrarily small by first choosing J sufficiently large, and then picking small h. $\qquad\square$

There is also a converse. (Note that one can do in fact far more detailed analysis than this with Littlewood-Paley type estimates.)

Why does the Littlewood-Paley technique work when the Fourier coefficient method fails? What is the "miracle"? Notice that when we sum $\sum_{n=N}^{\infty} \frac{1}{n^k}$, its value is approximately $C\frac{1}{N^{k-1}}$ but when we sum $\sum_{n=N}^{\infty} 2^{-n\epsilon}$, its value is approximately $C2^{-N\epsilon}$. So, the exponential character of the Littlewood-Paley blocks buys us something that straightforward use of the Fourier coefficients, one by one, can't.

Wavelets have this Littlewood-Paley character as well. Because of this, the wavelet transform can also characterize smoothness exactly. The dilation and translation of the basic wavelet corresponds to dilation and multiplication by $e^{ik2^{-j}\xi}$ on the Fourier transform side:

$$2^{j/2}\psi(2^j x - k) \longleftrightarrow 2^{-j/2}\widehat{\psi}(2^{-j}\xi)e^{ik2^{-j}\xi}$$

A serviceable wavelet ψ has integral 0 and so $\widehat{\psi}(0) = 0$. Its Fourier transform looks like Figure 1.5. When we express a function g in a wavelet series $g = \sum_{j,k\in\mathbb{Z}} c_{j,k}\psi_{j,k}$ (where $\{\psi_{j,k}\}_{j,k\in\mathbb{Z}}$ is an orthonormal basis for $L^2(\mathbb{R})$), the coefficients $c_{j,k}$ are the inner products of g with each wavelet $\psi_{j,k}$:

$$c_{j,k} = \langle g, \psi_{j,k} \rangle = 2^{-j/2} \int \widehat{g}(\xi)\widehat{\psi}(2^{-j}\xi)e^{i2^{-j}\xi k}d\xi,$$

and $\widehat{\psi}(2^{-j}\xi)e^{i2^{-j}\xi k}$ is concentrated in the region $2^j\pi \leq |\xi| < 2^{j+1}\pi$. In other words, the $c_{j,k}$ are "cutting out" the corresponding region from the support of $\widehat{g}$, just like the Littlewood-Paley theory used dyadic frequency blocks to define $\Delta_j f$.

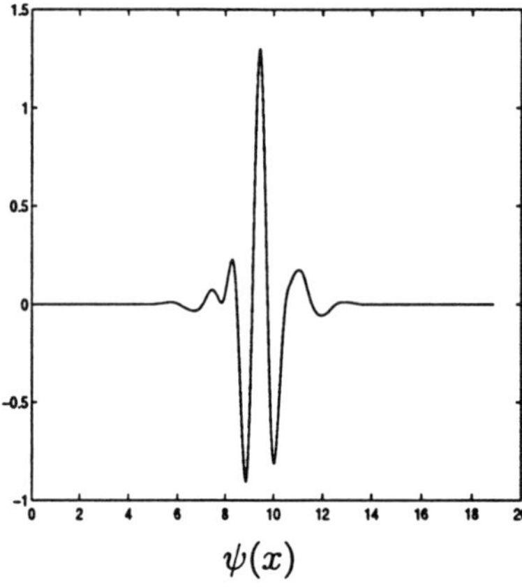
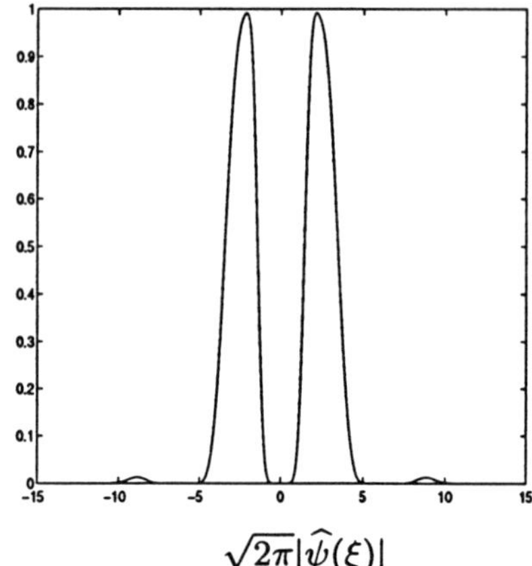

$$\psi(x) \qquad\qquad \sqrt{2\pi}|\widehat{\psi}(\xi)|$$

Figure 1.5. A wavelet ψ for which the $\psi_{j,k}$ constitute an orthonormal basis, and the absolute value $|\widehat{\psi}|$ of its Fourier transform.

What's the difference between Littlewood-Paley theory and wavelets? When we write

$$(\Delta_j f)(x) = \sum_{2^{j-1}\le|n|<2^j} \widehat{f}(n)e^{2\pi i n x}$$

for some $f \in L^2([0,1])$, we have different layers corresponding to different octaves of frequencies but each $\Delta_j f$ is a function that lives on the whole interval. With wavelets, we also have different layers in our expansion of f,

$$\sum_{j\in\mathbb{Z}}\left(\sum_{k\in\mathbb{Z}} c_{j,k}\psi_{j,k}\right) ;$$

again each j-layer lives on the whole interval $[0,1]$ but within each layer we have broken up things even finer, because each $\psi_{j,k}$ is a function localized in an interval of width proportional to 2^{-j}.

To conclude this introductory lecture, let's review one very useful theorem known to electrical engineers which mathematicians should also know. It is the Shannon sampling or interpolation theorem and it tells us the minimum number of samples per wavelength necessary to reconstruct a band-limited function. Band-limited functions are L^2-functions with compactly supported Fourier transform. If support $\widehat{f} \subset [-W, W]$, then we say that f is band-limited with bandwidth $\frac{W}{2\pi}$.

Theorem (Shannon Sampling). *Assume that f is a band-limited function; i.e., that the support of its Fourier transform is contained in $[-\pi, \pi]$. Then f is determined by its samples at the integers; i.e.,*

$$f(t) = \sum_{n\in\mathbb{Z}} f(n)\frac{\sin(t-n)\pi}{(t-n)\pi}.$$

Proof. Since $\widehat{f}$ is compactly supported, $\widehat{f}$ has a Fourier series. That is, we can write $\widehat{f}(\xi) = \sum c_n e^{-in\xi}$ where $c_n = \frac{1}{2\pi} \int_{-\pi}^{\pi} \widehat{f}(\xi) e^{in\xi} d\xi$. But $c_n = \frac{1}{2\pi} \int_{-\infty}^{\infty} \widehat{f}(\xi) e^{in\xi} d\xi = \frac{1}{\sqrt{2\pi}} f(n)$ by the Fourier Inversion theorem. This tells us that the Fourier transform $\widehat{f}$ is

$$\widehat{f}(\xi) = \left(\sum f(n) e^{-in\xi} \right) \left(\frac{1}{\sqrt{2\pi}} \right).$$

Again by the Fourier Inversion formula, the value of f at any point t is

$$f(t) = \frac{1}{\sqrt{2\pi}} \int_{-\pi}^{\pi} \widehat{f}(\xi) e^{it\xi} d\xi = \frac{1}{2\pi} \sum f(n) \int_{-\pi}^{\pi} e^{i\xi(t-n)} d\xi = \sum_n f(n) \frac{\sin(t-n)\pi}{(t-n)\pi}.$$

$\square$

Exercise. We have left out the justifications of some arguments—as an exercise, prove this rigorously. As another exercise, generalize the theorem to $\widehat{f}$ limited to the band $[-W, W]$.

Exercise. We could take as our first example of a wavelet the function

$$\widehat{\Psi}_{Sh}(\xi) = (\chi_{\pi \le |\xi| < 2\pi})\widehat{};$$

i.e., $\widehat{\Psi}_{Sh}$ Fourier transform of the characteristic function of the ring $\pi \le |\xi| < 2\pi$. In electrical engineering terms, this is a perfect filter (it brutally chops off the function $\widehat{f}(\xi)$), but it is not used in practice because it has bad localization in x. That is, as a function of x, it decays slowly. Each piece $\widehat{f}(\xi)\widehat{\Psi}_{Sh}(2^j\xi)$ is then a band-limited function, and one can apply Shannon's theorem to it. This results in an expansion for f with respect to the orthonormal family of functions

$$\frac{\sin\left((2^{-j}t - n)\pi\right)\left(\cos\left((2^{-j}t - n)\pi\right) - 1\right)}{(2^{-j}t - n)\pi}.$$

This is a wavelet expansion.

LECTURE 2
Constructing Orthonormal Wavelet Bases:
Multiresolution Analysis

First a few remarks to complement the first lecture before we begin with new material.

The first remark is a more precise statement of Littlewood-Paley theory. For $0 < \alpha < 1$, one has that $|(\Delta_j f)(x)| \le C2^{-j\alpha}$ implies $f \in C^\alpha(\mathbb{R})$; the converse also holds. By the space $C^\alpha(\mathbb{R})$, we mean those functions g such that $|g(x+t) - g(x)| \le C|t|^\alpha$ (this is Hölder continuity). This statement is also true for $\alpha > 1$, provided one defines the $\Delta_j f$ by less brutal cuts in frequency, and provided the definition of the C^α-spaces is also adapted. (For $n < \alpha < n+1$, we'll say that $f \in L^\infty$ is in C^α if f is n times continuously differentiable, and $f^{(n)} \in C^{\alpha-n}$.) A reference in which to find more discussion of Littlewood-Paley theory is the monograph by Frazier, Jawerth, and Weiss [**FJW**].

The second remark concerns the importance for the smoothness of a function f of the phases of its Fourier coefficients $\widehat{f}(n)$. When we looked at the Fourier series of a function f

$$f(x) = \sum \widehat{f}(n)e^{-2\pi i n x},$$

it was hard to determine, from the magnitude of the Fourier coefficients $|\widehat{f}(n)|$, the smoothness of f. In fact, the phases of $\widehat{f}(n)$ play a key role here—if we change the phases, we can drastically affect the smoothness. The following example from Zygmund's "Trigonometric Series" [**Zyg**] shows this phenomenon:

$$\text{if} \quad f(x) = \sum_{n=2}^{\infty} n^{-\frac{1}{4}} e^{2\pi i n x}, \quad \text{then as} \quad |x| \to 0, f(x) \sim |x|^{-\frac{3}{4}},$$

but if we change the phases, we get different behavior. Set

$$\widetilde{f}(x) = \sum_{n=2}^{\infty} n^{-\frac{1}{4}} e^{i\sqrt{n}} e^{2\pi i n x}, \quad \text{then as} \quad |x| \to 0, \widetilde{f}(x) \sim |x|^{-2}.$$

So we have altered the nature of the singularity at $x = 0$ simply by adjusting the phases of $\widehat{f}(n)$. The interplay of the phases of the different $\widehat{f}(n)$ can cause subtle cancellations or interferences. In fact, as made clear by Littlewood-Paley theory, we need not consider the phases of all the $\widehat{f}(n)$ simultaneously. Considering the

$\widehat{f}(n)$, with their phases, within each dyadic block, is sufficient; if we change the overall phase of every individual block by multiplying all the coefficients within the block by the same phase factor (but a different one for each block), then this does not affect the overall smoothness class of the function, nor can it cause the nature of a singularity to change drastically as in the example above.

A third remark elaborates on the Shannon wavelet exercise at the end of the first lecture. We used Littlewood-Paley theory to motivate dyadic scaling of the frequency; translation enters via the sampling theorem. We have $\psi_{j,k}(x) = 2^{j/2}\psi(2^j x - k)$ where the factor 2^j corresponds to scaling or to frequency and $2^{-j}k$ to translation. The functions $\Delta_j f$ are obtained from the Fourier transform of f by cutting out dyadic blocks by multiplying it with the characteristic function $\chi_{\pi \le |\xi| < 2\pi}(2^{-j}\xi)$, i.e.,

$$(\Delta_j f)(x) = \left[\widehat{f}(\xi)\chi_{\pi \le |\xi| < 2\pi}(2^{-j}\xi) \right]^{\vee}$$

where $^\vee$ denotes the inverse Fourier transform. We can write this as a convolution in the space-time variable: if we set

$$\Psi_{Sh}(x) = \left(\chi_{\pi \le |\xi| < 2\pi}\right)^{\vee}(x) = (2\pi)^{1/2}\frac{\sin \pi x(\cos \pi x - 1)}{\pi x},$$

then

$$(\Delta_j f)(x) = \int f(y)\Psi(2^j(x-y))\,dy = \int f(y)\Psi(2^j(y-x))\,dy$$

since Ψ is symmetric. This tells us that $(\Delta_j f)(x)$ is given by an inner product with wavelets. However, we have a continuous variable x in our expression instead of samples at the points $2^{-j}k$, as we would expect from wavelets. The Shannon sampling theorem tells us that we can replace $(\Delta_j f)(x)$ by the samples of this function at the points $2^{-j}k$, so that $(\Delta_j f)(x)$ is completely characterized by $(\Delta_j f)(2^{-j}k)$ or

$$(\Delta_j f)(2^{-j}k) = \int f(y)\Psi(2^j y - 2^j \cdot 2^{-j}k)\,dy = \int f(y)\Psi(2^j y - k)\,dy.$$

We noted in the last lecture that Ψ_{Sh} was a bad candidate for a wavelet basis (in practice) because it chops off frequencies brutally. If we let $|\widehat{\psi}|$ go to zero more gracefully (see Figure 1.5b), then neighboring blocks will start to overlap, and be wider, so that we cannot simply apply Shannon's theorem with the same bandwidth as before. So we need other tools to construct better wavelet bases. We want an orthonormal basis $\{\psi_{j,k}\}_{j,k\in\mathbb{Z}}$, where $\psi_{j,k}(x) = 2^{j/2}\psi(2^j x - k)$, so that for every $f \in L^2(\mathbb{R})$, $f = \sum_{j,k\in\mathbb{Z}}\langle f, \psi_{j,k}\rangle \psi_{j,k}$. This decomposition is true for many other function spaces (even for tempered distributions) but we will stick to the ordinary space $L^2(\mathbb{R})$. We will construct other bases within a Multiresolution Analysis (MRA) framework [**Mall**] [**Mey**].

If we write

$$f = \sum_{j\in\mathbb{Z}}\left(\sum_{k=-\infty}^{\infty} \langle f, \psi_{j,k}\rangle \psi_{j,k}\right),$$

then we can view this as a decomposition of f into scales j; at each scale j, we define a projection operator Q_j associated with that scale by

$$Q_j f = \sum_{k=-\infty}^{\infty} \langle f, \psi_{j,k} \rangle \, \psi_{j,k} \, .$$

We shall interpret these projections Q_j as giving the additional information needed to go from one approximation (in a whole hierarchy) to the next one, which has some more detail. That is, we imagine having a ladder of closed linear subspaces V_j of $L^2(\mathbb{R})$

$$(2.1) \qquad \qquad \cdots \subset V_{-2} \subset V_{-1} \subset V_0 \subset V_1 \subset V_2 \subset \cdots$$

such that

$$(2.2) \qquad \qquad \overline{\bigcup_{j \in \mathbb{Z}} V_j} = L^2(\mathbb{R}) \ , \ \bigcap_{j \in \mathbb{Z}} V_j = \{0\}$$

If we write $P_j f$ to denote the orthogonal projection of f onto the space V_j, then we can write the hoped-for connection with our Q_j as

$$P_{j+1} f = P_j f + Q_j f \, .$$

A ladder of spaces that satisfies the two requirements (2.1) and (2.2) above is called a multiresolution analysis (MRA) if it satisfies in addition the following two requirements:

1. Every space is a scaled version of the central space:

$$(2.3) \qquad \qquad f \in V_j \iff f(2^{-j} \cdot) \in V_0$$

   (This requirement tells us that successive V_j correspond to resolutions increasing by a factor 2 each time.)

2. Also, there is a function $\varphi \in V_0$ (called a scaling function) such that

$$(2.4) \qquad \qquad \varphi_{0,n}(x) = \varphi(x - n), n \in \mathbb{Z} \, ,$$

   is an orthonormal basis for V_0. (In particular, V_0 is invariant under translations by integer steps; each V_j is invariant under translations by integer multiples of 2^{-j}.) We shall require that both φ and $\widehat{\varphi}$ have some decay.

Here are two important examples:

Examples.
- Let V_0 be the space of square-integrable piecewise constant functions with breaks at the integers. (See Figure 2.1.) The space V_1 is obtained by compressing the elements of V_0 by a factor of 2 so it is the space of piecewise constant functions with breaks at the half integers. Clearly, $V_0 \subset V_1$.
- Let V_0 be the space of functions f so that $\mathrm{supp}\widehat{f} \subset [-\pi, \pi]$, then V_1 is the space of functions so that $\mathrm{supp}\widehat{f} \subset [-2\pi, 2\pi]$. Again, $V_0 \subset V_1$.

Exercises.
1. Check that the function ψ with $\widehat{\psi}(\xi) = \chi_{\pi \le |\xi| < 2\pi}(\xi)$ generates an orthonormal basis for $L^2(\mathbb{R})$. (To say that ψ generates an orthonormal basis for

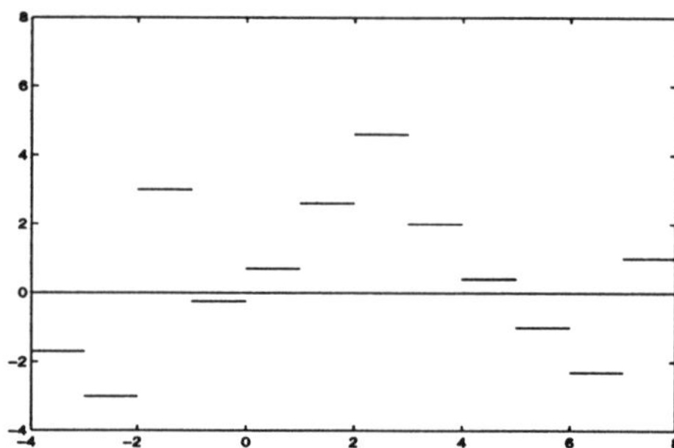

Figure 2.1. Piecewise constant function with breaks at the integers

$L^2(\mathbb{R})$, means that the dilates and translates $\{\psi_{j,k}\}_{j,k\in\mathbb{Z}}$ constitute an orthonormal basis.)

2. If $\varphi(x) = \frac{1}{\sqrt{2\pi}}\left(\chi_{[-\pi,\pi]}\right)^\wedge(x)$, check that $\{\varphi_{0,n}\}_{n\in\mathbb{Z}}$ is an orthonormal basis for V_0 in example (2).

3. If $\varphi(x) = \chi_{[0,1]}(x)$, check that $\{\varphi_{0,n}\}_{n\in\mathbb{Z}}$ is an orthonormal basis for V_0 in example (1).

4. Check that the Haar basis with

$$\psi = \begin{cases} 1, & x \in [0,1/2) \\ -1, & x \in [1/2,1) \\ 0, & \text{otherwise} \end{cases}$$

also generates an orthonormal basis for $L^2(\mathbb{R})$. (Hint: show that, for fixed j, the $\psi_{j,k}$ constitute an orthonormal basis for the orthogonal complement of V_j in V_{j+1}, with V_j as defined in example (1)).

We will see in what follows that any MRA, that is, any ladder of approximation spaces V_j that satisfies the requirements (2.1–2.4) listed above (nested ladder of spaces, that together span all of L^2, and with intersection 0, related to each other by scaling by a factor 2, and generated by a reasonably nice scaling function) gives rise to a wavelet basis, that there are many examples, and that we can construct fast algorithms for applications.

One could ask if, conversely, every orthonormal basis of wavelets gives rise to an MRA which is linked to that basis. The answer is yes, under certain conditions. If we have an orthonormal basis $\{\psi_{j,k}\}$ and if ψ satisfies the three conditions on ψ

$$|\psi(x)| \le C(1+|x|)^{-1-\epsilon}, \quad \widehat{\psi}(0) = 0, \quad \text{and} \quad |\widehat{\psi}(\xi)| \le C|\xi|^\alpha(1+|\xi|)^{-1-\alpha-\epsilon},$$

then the wavelet basis is associated to an MRA, where the wavelets for scale j correspond to the difference between projections on V_j and V_{j+1}, as described above [**LMRA**].

We have introduced the scaling function φ but we may have left the false impression that it can be any function. We will now explain what conditions φ must satisfy (as a result of the definition of an MRA) and we will show how the scaling function and the wavelet are connected.

Since the MRA spaces V_j are rescaled copies of the central space V_0, we know that $\{2^{j/2}\varphi(2^j x - k)\}_{k \in \mathbb{Z}}$ is an orthonormal basis for each V_j. Therefore, since $V_0 \in V_1$ and since the $\varphi_{1,k}$ are an orthonormal basis in V_1, $\varphi \in V_0$ may be expressed as

$$\varphi = \sum_n h_n \, \varphi_{1,n}$$

where $h_n = \langle \varphi, \varphi_{1,n} \rangle$ and $\{h_n\}_{n \in \mathbb{Z}}$ is a square summable sequence. In other words, $\varphi(x) = \sqrt{2} \sum_n h_n \varphi(2x - n)$; i.e. φ must be a linear combination of its translates and dilates $\varphi(2x - n)$. Another way to view this expression is to look at the Fourier transform of φ:

$$\widehat{\varphi}(\xi) = \frac{1}{\sqrt{2}} \left(\sum_n h_n e^{-\frac{in\xi}{2}} \right) \widehat{\varphi}\left(\frac{\xi}{2}\right).$$

We let m_0 be $\frac{1}{\sqrt{2}} \sum_n h_n e^{-in\xi} = m_0(\xi)$ and observe that m_0 is a 2π-periodic function. This tells us that there is a 2π-periodic function such that $\widehat{\varphi}(\xi) = m_0\left(\frac{\xi}{2}\right) \widehat{\varphi}\left(\frac{\xi}{2}\right)$; in other words, $\frac{\widehat{\varphi}(2\xi)}{\widehat{\varphi}(\xi)}$ is a 2π-periodic function. It should be clear now that we can't choose an arbitrary function to be our scaling function!

Another property φ must possess is that the functions $\{\varphi_{0,n}\}$ are all orthonormal. This imposes restrictions on m_0 or leads to special properties of h_n and m_0. We have the following property of h_n:

$$\begin{aligned}
\delta_{k,0} &= \int \varphi(x) \, \overline{\varphi(x - n)} \, dx \\
&= 2 \sum_{n,m} h_n \, \overline{h_m} \int \varphi(2x - n) \, \overline{\varphi(2x - 2k - m)} \, dx \\
&= \sum_m h_{m+2k} \, \overline{h_m} \quad \text{(by a change of variables)}.
\end{aligned}$$

(2.5)

On the Fourier transform side, we have

$$\begin{aligned}
\delta_{k,0} &= \int_{-\infty}^{\infty} \widehat{\varphi}(\xi) \, \overline{\widehat{\varphi}(\xi)} \, e^{i\xi k} \, d\xi \\
&= \int_{-\infty}^{\infty} |\widehat{\varphi}(\xi)|^2 e^{i\xi k} \, d\xi \\
&= \int_0^{2\pi} e^{ik\xi} \sum_{\ell \in \mathbb{Z}} |\widehat{\varphi}(\xi + 2\pi\ell)|^2 \, d\xi, \quad \text{(using the 2π-periodicity)}
\end{aligned}$$

so that

$$\sum_{\ell \in \mathbb{Z}} |\widehat{\varphi}(\xi + 2\pi\ell)|^2 = \frac{1}{2\pi}.$$

Splitting the sum into even and odd ℓ we find that

$$\frac{1}{2\pi} = \sum_{\ell \in \mathbb{Z}} |\widehat{\varphi}(\xi + 2\pi\ell)|^2 = \sum_{k \in \mathbb{Z}} \left(|\widehat{\varphi}(\xi + 4\pi k)|^2 + |\widehat{\varphi}(\xi + 4\pi k + 2\pi)|^2 \right)$$

$$= \left| m_0 \left(\frac{\xi}{2} \right) \right|^2 \sum_k \left| \widehat{\varphi} \left(\frac{\xi}{2} + 2\pi k \right) \right|^2 + \left| m_0 \left(\frac{\xi}{2} + \pi \right) \right|^2 \sum_k \left| \widehat{\varphi} \left(\frac{\xi}{2} + 2\pi k + \pi \right) \right|^2$$

$$= \left| m_0 \left(\frac{\xi}{2} \right) \right|^2 \frac{1}{2\pi} + \left| m_0 \left(\frac{\xi}{2} + \pi \right) \right|^2 \frac{1}{2\pi} \,.$$

Hence the function m_0 must satisfy

$$(2.6) \qquad\qquad |m_0(\xi)|^2 + |m_0(\xi + \pi)|^2 = 1 \,.$$

Exercises.

5. Given that the coefficients h_n are the Fourier coefficients of m_0 (up to a factor $\sqrt{2}$) , verify that the two conditions (2.5) and (2.6) are equivalent.

6. Check that these conditions are verified for $\varphi = \chi_{[0,1]}$. Here $h_0 = \frac{1}{\sqrt{2}}$ and $h_1 = \frac{1}{\sqrt{2}}$ while the other $h_n = 0$. So $m_0(\xi) = \frac{1}{2} + \frac{1}{2} e^{-i\xi}$.

We have not connected the scaling function φ to the wavelet ψ yet; that will come in the next lecture. Note that the scaling function is much more than a technical trick to help us prove theorems. We do need it for two reasons—in practice, φ is the means by which we define the spaces V_j (it makes the V_j concrete) and φ corresponds to a sequence $\{h_n\}$ which is crucial for our fast algorithm. We will use the scaling function to generate our wavelet; in particular we shall see how we can achieve the following three properties for ψ:

- we want ψ to be smooth,
- we want ψ to decay (so as to localize phenomena), and
- we want a fast algorithm.

As we shall see from several examples of types of wavelets, we have to have some trade-offs. The Meyer wavelets are C^∞ and have rapid decay. Spline wavelets are C^k functions (piecewise polynomials) with exponential decay. Also, compactly supported wavelets are C^k functions but they are not piecewise polynomial functions. In fact, there is no analytic expression for such wavelets.

LECTURE 3
Wavelet Bases: Construction and Algorithms

We will restate the definition of an MRA and the conditions we derived from this definition: We have a ladder of closed subspaces V_j of $L^2(\mathbb{R})$ such that

- $\cdots \subset V_{-1} \subset V_0 \subset V_1 \subset \cdots$,
- $\overline{\bigcup_{j \in \mathbb{Z}} V_j} = L^2(\mathbb{R})$,
- $\bigcap_{j \in \mathbb{Z}} V_j = \{0\}$,
- $f \in V_j \iff f(2^{-j}\cdot) \in V_0$ (the scaling property), and
- there is a function $\varphi \in V_0$ such that $\{\varphi_{0,n}\}_{n \in \mathbb{Z}}$ is an orthonormal basis for V_0.

From these restrictions, we derived the following properties:

- $\varphi(x) = \sqrt{2} \sum_n h_n \varphi(2x - n)$ where $\sum_n h_n \overline{h_{n+2k}} = \delta_{k,0}$,
- $\widehat{\varphi}(\xi) = m_0\left(\frac{\xi}{2}\right) \widehat{\varphi}\left(\frac{\xi}{2}\right)$ for $m_0(\xi) = \sum_n h_n e^{-in\xi}$, and
- $|m_0(\xi)|^2 + |m_0(\xi + \pi)|^2 = 1$.

We can also prove that $\int \varphi = 1$, while $\int \psi = 0$. This simple fact should highlight how different the functions φ (the scaling function) and ψ (the wavelet) are.

Now we will examine a class of examples—the spline wavelets. One of our first examples for the space V_0 was the space of piecewise constant functions (square-integrable, with breaks at the integers). We saw that the associated scaling function φ for this space was the characteristic function of the unit interval $\varphi(x) = \chi_{[0,1]}(x)$. We can extend this idea or example to higher order spline wavelets.

Let us first discuss what splines are. Suppose we are given a collection of data (not necessarily equi-spaced) which we want to "spline" together smoothly, then we would build a piecewise polynomial function connecting these points which is as smooth as possible. This piecewise polynomial will break at the knots but the links between pieces should be as smooth as possible. We can match the function, its derivative, and second derivative at the knots leading to $C^2(\mathbb{R})$ functions for cubic splines, for example. See Figure 3.1 for an example of a (cubic) spline.

A B-spline is a spline of order L (where L is the degree of the polynomial plus one) such that its support is as small as possible, allowing knots only at the integers. (See Figure 3.2.) We denote a B-spline of order L by B_L.

175

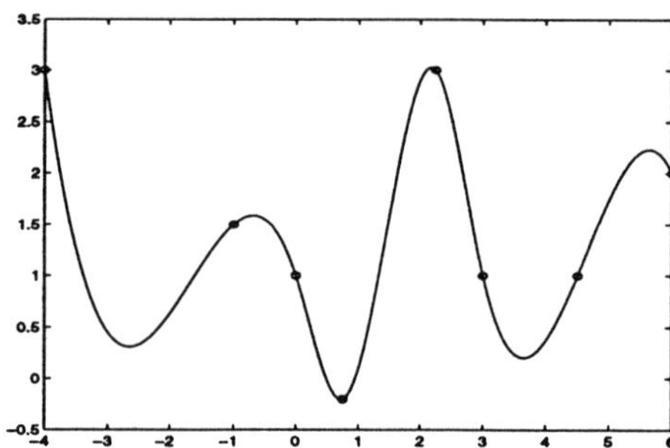

Figure 3.1. Unequally-spaced data "splined" together

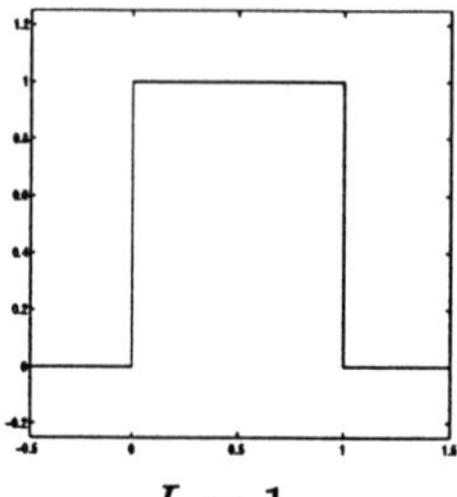

$$L = 1$$

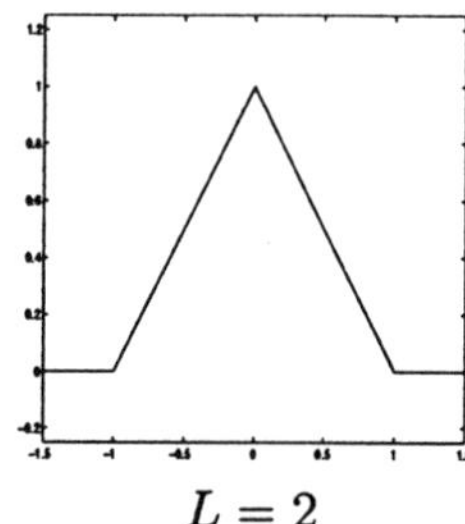

$$L = 2$$

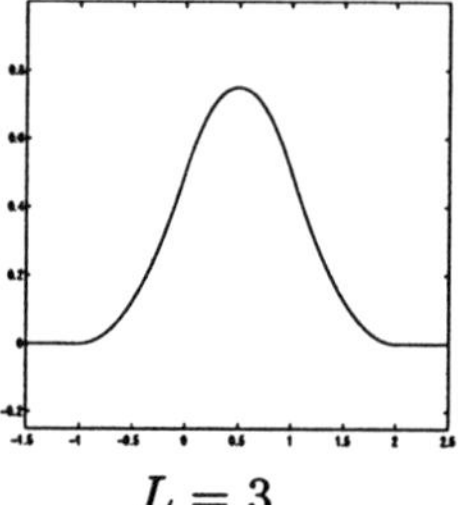

$$L = 3$$

Figure 3.2. B-splines of order L

Note that $B_2 = B_1 * B_1$ ($*$ denotes convolution), $B_3 = B_2 * B_1$, or more generally, $B_L = \underbrace{B_1 * \cdots * B_1}_{L \text{ times}}$. In Fourier space, we have

$$\widehat{B}_1(\xi) = \frac{1}{\sqrt{2\pi}} \frac{1 - e^{-i\xi}}{i\xi} \quad \text{and} \quad \widehat{B}_L(\xi) = \frac{1}{\sqrt{2\pi}} \left(\frac{1 - e^{-i\xi}}{i\xi} \right)^L .$$

Because $\widehat{B}_1(2\xi)/\widehat{B}_1(\xi) = (1 + e^{-i\xi})/2$ is a 2π-periodic function, the same is true for the B_L. It then follows immediately that the spaces V_j spanned by the $B_L(2^j x - n)$, $n \in \mathbb{Z}$, satisfy the desired ladder property. For our purposes, the B-splines for $L > 1$ have a problem which we must fix: they aren't orthogonal to their integer translates. Recall that the integer translates of the scaling function $\{\varphi(x - n)\}_{n \in \mathbb{Z}}$ are orthogonal so $\sum_\ell |\widehat{\varphi}(\xi + 2\pi\ell)|^2 = \frac{1}{2\pi}$ but we have $\sum_\ell |\widehat{B}_L(\xi + 2\pi\ell)|^2$ equal to a non-constant 2π-periodic function. Since this is a 2π-periodic function we know that we can express this sum as a Fourier series

$$\sum_{\ell} |\widehat{B}_L(\xi + 2\pi\ell)|^2 = \sum_{n} \beta_n e^{-in\xi} \quad \text{where}$$

$$\beta_n = \frac{1}{2\pi} \int_0^{2\pi} \left(\sum_{\ell} |\widehat{B}_L(\xi + 2\pi\ell)|^2 \right) e^{in\xi}\, d\xi$$

$$= \int_{-\infty}^{\infty} B_L(x)\, \overline{B_L(x - n)}\, dx\,.$$

Observe now that the last integral is non-zero for only finitely many n so we have an explicit trigonometric polynomial for $\sum_{\ell} |\widehat{B}_L(\xi + 2\pi\ell)|^2$. On the other hand, $\sum_{\ell} |\widehat{B}_L(\xi + 2\pi\ell)|^2$ doesn't vanish for any ξ, because the only candidate zeros are $\xi = 2k\pi$; since $|\widehat{B}_L(0)| > 0$, we have no zero there either. Hence, the following inequality holds

$$0 < A \le \sum_{\ell} |\widehat{B}_L(\xi + 2\pi\ell)|^2 \le B < \infty\,.$$

The important bound here is the lower bound—we have bounded this function away from zero so we can define a scaling function φ as follows:

$$\widehat{\varphi}(\xi) = \frac{\widehat{B}_L(\xi)}{\sqrt{2\pi}\left(\sum_{\ell} |\widehat{B}_L(\xi + 2\pi\ell)|^2 \right)^{1/2}} = \widehat{B}_L(\xi) b_L(\xi)\,;$$

the integer translates of this φ form an orthogonal family. Notice that we have written $\widehat{\varphi}(\xi)$ as a product $\widehat{\varphi}(\xi) = \widehat{B}_L(\xi)\, b_L(\xi)$ or $\varphi(x) = \sum \alpha_n B_L(x - n)$ where α_n is the nth Fourier coefficient of b_L. It turns out (see e.g. [**DTL**]) that φ has exponential decay; it is no longer compactly supported like the B_L themselves.

We have shown that our modified spline functions are orthogonal, that they are orthonormal bases for the appropriate space V_0, and that these spaces form a ladder, but we haven't shown how to compute the coefficients h_n. Recall that the quotient

$$\frac{\widehat{\varphi}(2\xi)}{\widehat{\varphi}(\xi)} = \frac{1}{\sqrt{2}} \sum_{n} h_n e^{-in\xi} = m_0(\xi) \quad \text{is a } 2\pi\text{-periodic function.}$$

Then we may express $m_0(\xi)$ as

$$m_0(\xi) = \frac{\widehat{B}_L(2\xi)\, b_L(2\xi)}{\widehat{B}_L(\xi)\, b_L(\xi)} = \left(\frac{1 - e^{-2i\xi}}{2(1 - e^{-i\xi})} \right)^L \frac{b_L(2\xi)}{b_L(\xi)} = \left(\frac{1 + e^{-i\xi}}{2} \right)^L \frac{b_L(2\xi)}{b_L(\xi)}\,.$$

The functions $b_L(\xi)$ have explicit formulas for general L and can be found in [**Chui**].

Why would we want to use these smoother spline wavelets? The answer is that they allow us to characterize smooth functions nicely. If we take a spline wavelet of order L and $\alpha < L$, then

$$\|f - P_j f\|_{L^\infty} \le C 2^{-\alpha j} \iff f \in C^\alpha\,, \quad \text{and} \quad \|f - P_j f\|_{L^2} \le C 2^{-\alpha j} \iff f \in H^\alpha\,.$$

Let us now return to the general construction of wavelet bases. We want an MRA structure to give us an associated wavelet basis in the sense that this basis connects the two projections $P_{j+1} f$ and $P_j f$; i.e., $P_{j+1} f = P_j f + \sum_k \langle f, \psi_{j,k} \rangle\, \psi_{j,k}$. We have projected f onto a fine scale $j + 1$ and written that projection as the projection onto a coarser scale j plus the difference in information. We would like this difference in information to be represented by an orthonormal wavelet basis. In particular, we want

$$P_1 f = P_0 f + \sum_k \langle f, \psi_{0,k} \rangle \, \psi_{0,k}$$

$$\text{and} \quad \text{if} \quad f \in W_0 = V_1 \cap (V_0)^\perp, \quad \text{then} \quad P_1 f = \sum_k \langle f, \psi_{0,k} \rangle \, \psi_{0,k} \, .$$

Can we characterize W_0? Can we get a natural orthonormal basis for W_0? Observe that W_0 is invariant under translation by integers. Also, V_1 is generated by $\varphi_{1,n} = \sqrt{2}\varphi(2x - n)_{n \in \mathbb{Z}}$, so it is generated by the integer translates of the functions $\sqrt{2}\varphi(2(x - k))$ and $\sqrt{2}\varphi(2(x - k)) - 1$. In this sense V_1 is generated by integer translations of two functions while V_0 is generated by integer translates of only one function. This gives us a "hand-wavy" argument for why W_0 should be generated by integer translates of one function. Let us make this more precise now.

Take a function $F \in W_0$. Since $W_0 \subset V_1$, we have $F \in V_1$ and F may be written as $F = \sqrt{2}\sum_n F_n \varphi_{1,n}$ or

$$\widehat{F}(\xi) = \frac{1}{\sqrt{2}} \sum_n F_n \, e^{-\frac{in\xi}{2}} \, \widehat{\varphi}\left(\frac{\xi}{2}\right) = M\left(\frac{\xi}{2}\right) \widehat{\varphi}\left(\frac{\xi}{2}\right) \, .$$

On the other hand, F is orthogonal to V_0, so $\int F(x)\varphi(x - n)dx = 0$ for all $n \in \mathbb{Z}$. Looking at the Fourier transform, we have

$$\text{for all } n \in \mathbb{Z}, \quad 0 = \int \widehat{F}(\xi) \, e^{in\xi} \widehat{\varphi}(\xi) \, d\xi$$

$$= \sum_\ell \int_{2\pi\ell}^{2(\ell+1)\pi} \widehat{F}(\xi)\widehat{\varphi}(\xi)e^{in\xi} \, d\xi$$

$$= \int_0^{2\pi} \sum_\ell \widehat{F}(\xi + 2\pi\ell)\widehat{\varphi}(\xi + 2\pi\ell)e^{in\xi} \, d\xi \, .$$

From the last statement, we determine that all the Fourier coefficients of the function $\sum_\ell \widehat{F}(\xi + 2\pi\ell)\widehat{\varphi}(\xi + 2\pi\ell)$ are zero, so the function

$$\sum_\ell \widehat{F}(\xi + 2\pi\ell)\widehat{\varphi}(\xi + 2\pi\ell) = 0 \quad \text{almost everywhere.}$$

Using the same arguments as we did for m_0 and φ (i.e., splitting into odd and even sums), we arrive at the following equality:

$$0 = \sum_\ell M\left(\frac{\xi}{2} + \pi\ell\right) \widehat{\varphi}\left(\frac{\xi}{2} + \ell\pi\right) \overline{m_0\left(\frac{\xi}{2} + \ell\pi\right)} \overline{\widehat{\varphi}\left(\frac{\xi}{2} + \ell\pi\right)}$$

$$= M\left(\frac{\xi}{2}\right) \overline{m_0\left(\frac{\xi}{2}\right)} \frac{1}{2\pi} + M\left(\frac{\xi}{2} + \pi\right) \overline{m_0\left(\frac{\xi}{2} + \pi\right)} \frac{1}{2\pi} \, .$$

We want to determine M, so let us guess that M is of the form $M(\xi) = e^{-i\xi}\alpha(\xi)$, then the above equality yields

$$\alpha(\xi)\,\overline{m_0(\xi)} = \alpha(\xi + \pi)\,\overline{m_0(\xi + \pi)}\,.$$

We know that the functions $\overline{m_0(\xi)}$ and $\overline{m_0(\xi + \pi)}$ can't both vanish together on a set of non-zero measure since $|m_0(\xi)|^2 + |m_0(\xi + \pi)|^2 = 1$ so there is a π-periodic function λ such that $\alpha(\xi) = \overline{m_0(\xi + \pi)}\,\lambda(\xi)$. We can recast λ as $\lambda(\xi) = \nu(2\xi)$ where ν is a 2π-periodic function. These calculations determine M now:

$$M(\xi) = e^{-i\xi}\,\overline{m_0(\xi + \pi)}\,\nu(2\xi)\,.$$

Thus, we can characterize the space $W_0 = V_1 \cap (V_0)^\perp$ as

$$W_0 = \left\{ F \mid \widehat{F}(\xi) = e^{-\frac{i\xi}{2}}\,\overline{m_0\left(\frac{\xi}{2} + \pi\right)}\,\widehat{\varphi}\left(\frac{\xi}{2}\right)\nu(\xi) \right\}$$

$$= \left\{ F \mid F = \sum_n \nu_n \psi(x - n) \right\}$$

where we define $\widehat{\psi}(\xi) = e^{-\frac{i\xi}{2}}\,\overline{m_0\left(\frac{\xi}{2} + \pi\right)}\,\widehat{\varphi}\left(\frac{\xi}{2}\right)$ and the coefficients $\{\nu_n\}_{n\in\mathbb{Z}}$ are the Fourier coefficients of ν.

Exercise. As an exercise, check that if $F \in L^2(\mathbb{R})$, then $\{\nu_n\} \in \ell^2$ or $\nu \in L^2(\mathbb{R})$. The functions $\{\psi(x - n)\}_{n\in\mathbb{Z}}$ span W_0; check that $\psi(x - n)$, $n \in \mathbb{Z}$ are orthonormal.

By the exercise, we know that W_0 has an orthonormal basis, so we get orthonormal bases for W_j by dilation. Recall that $W_j = V_{j+1} \cap (V_j)^\perp$ and that the spaces V_{j+1} and V_j are simply rescaled versions of V_0. This interlocking structure of spaces gives us the following decompositions:

$$\begin{aligned}
V_J &= W_{J-1} \oplus V_{J-1} \\
&= W_{J-1} \oplus W_{J-2} \oplus V_{J-2} \\
&= W_{J-1} \oplus W_{J-2} \oplus W_{J-3} \oplus V_{J-3} \\
&\quad \cdots \\
&= W_{J-1} \oplus \cdots \oplus W_{-K} \oplus V_{-K}\,.
\end{aligned}$$

Or, in terms of projections of functions,

$$P_J f = \sum_{j=-K}^{J-1} \sum_k \langle f, \psi_{j,k}\rangle\,\psi_{j,k} + P_{-K} f\,.$$

If we let $J \to \infty$, then $P_J f \to f$ (in L^2) and $P_{-K} f \to 0$ as $K \to \infty$ (again, in L^2) so the functions $\{\psi_{j,k}\}_{j,k\in\mathbb{Z}}$ are indeed an orthonormal basis for $L^2(\mathbb{R})$.

Exercise. In fact, you need some (weak) conditions on φ to ensure that $P_J f \to f$, $P_{-K} f \to 0$. Try to prove these two convergences; what conditions on φ do you need? (See also [**DTL**].)

We will now change gears and discuss how to compute with our wavelet functions. We will show why the coefficients $\{h_n\}_{n\in\mathbb{Z}}$ are so crucial. In practice, we are given samples of a function f (say, $f(2^{-J}n)$ for some scale J). We have to use these to determine the inner products $\langle f, \varphi_{J,k} \rangle$; i.e., we must determine $P_J f$. Some books or papers on practical applications of wavelets do not distinguish between the samples $f(2^{-J}n)$ and the coefficients $\langle f, \varphi_{J,k} \rangle$ (i.e., one begins computing directly with the samples, assuming $P_J f = f$, or that the error in this assumption is negligible). Before one does this, one should check that the results don't differ to much from those obtained by the following, more exact procedure. We'll work here with the scale $J = 0$ but we can easily rescale.

If we only know the sampled values $f(n)$ of f, then we must determine $\langle f, \varphi_{0,k} \rangle$ by a convolution operation. We begin with the decomposition $f = \sum_k \langle f, \varphi_{0,k} \rangle \varphi_{0,k}$ from which we can evaluate this function at each integer n:

$$f(n) = \sum_k \langle f, \varphi_{0,k} \rangle \varphi_{0,k}(n - k).$$

Therefore,

$$\sum_n f(n)\, e^{-in\xi} = \left(\sum_k \langle f, \varphi_{0,k} \rangle\, e^{-ik\xi} \right) \left(\sum_m \varphi(m)\, e^{-im\xi} \right).$$

That is, the inner products $\langle f, \varphi_{0,k} \rangle$ are the Fourier coefficients of the function $\left(\sum_n f(n)e^{-in\xi} \right) \left(\sum_m \varphi(m)e^{-im\xi} \right)^{-1}$. It follows that $\langle f, \varphi_{0,k} \rangle = \sum_n a_{k-n} f(n)$ where

$$a_m = \frac{1}{2\pi} \int_0^{2\pi} \left(\sum_\ell \varphi(\ell)e^{-i\ell\xi} \right)^{-1} e^{im\xi}\, d\xi.$$

Now that we have a way of determining the coefficients $\langle f, \varphi_{J,k} \rangle$, how do we determine the inner products $\langle f, \psi_{J-1,k} \rangle$? Recall that $\hat\psi$ is $\hat\psi(\xi) = m_1 \left(\frac{\xi}{2} \right) \hat\varphi \left(\frac{\xi}{2} \right)$ so we may write $\psi(x) = \sqrt{2} \sum_n g_n \varphi(2x - n)$, or $\psi_{j,k} = \sum_n g_{n-2k}\, \varphi_{j+1,n}$. Here $m_1(\xi) = \frac{1}{\sqrt{2}} \sum_n g_n e^{-in\xi}$ is given by $m_1 \left(\frac{\xi}{2} \right) = e^{-\frac{i\xi}{2}} m_0 \left(\frac{\xi}{2} + \pi \right)$.

Exercise. Check the last implication!

It follows that our inner products are

$$\langle f, \psi_{J-1,k} \rangle = \sum_n g_{n-2k} \langle f, \varphi_{J,n} \rangle$$

$$\langle f, \varphi_{J-1,k} \rangle = \sum_n h_{n-2k} \langle f, \varphi_{J,n} \rangle.$$

These formulas illustrate how crucial the sequences (or filters) $\{h_n\}_{n\in\mathbb{Z}}$ and $\{g_n\}_{n\in\mathbb{Z}}$ are. They are the tools we use to pass between different levels of resolution.

For the Haar basis example, we have

$$\varphi(x) = \chi_{[0,1]}(x) \quad \text{and}$$

$$\psi(x) = \begin{cases} 1, & x \in \left[0, \dfrac{1}{2}\right) \\[2ex] -1, & x \in \left[\dfrac{1}{2}, 1\right) \\[2ex] 0, & \text{elsewhere.} \end{cases}$$

In this case, the filters $\{h_n\}_{n \in \mathbb{Z}}$ and $\{g_n\}_{n \in \mathbb{Z}}$ are given by

$$h_0 = h_1 = \frac{1}{\sqrt{2}}, \quad \text{and} \quad h_n = 0 \quad \text{for all other} \quad n; \quad \text{and}$$

$$g_0 = \frac{1}{\sqrt{2}}, g_1 = \frac{-1}{\sqrt{2}}, \quad \text{and} \quad g_n = 0 \quad \text{for all other} \quad n.$$

LECTURE 4
More Wavelet Bases

We saw in Lecture 3 how the definition of an MRA led to special properties of the scaling function φ, the wavelet ψ, and their Fourier transforms. Briefly:

- $\varphi(x) = \sqrt{2} \sum_n h_n \varphi(2x - n)$ where $h_n = \sqrt{2} \int \varphi(x)\varphi(2x-n)\,dx$,
- $\widehat{\varphi}(\xi) = m_0\left(\frac{\xi}{2}\right)\widehat{\varphi}\left(\frac{\xi}{2}\right)$ where $m_0(\xi) = \frac{1}{\sqrt{2}}\sum_n h_n e^{-in\xi}$,
- $\widehat{\psi}(\xi) = e^{-i\xi/2}\,\overline{m_0\left(\frac{\xi}{2} + \pi\right)}\widehat{\varphi}(\xi)$ with $m_1\left(\frac{\xi}{2}\right)\,e^{-i\xi/2}\,\overline{m_0\left(\frac{\xi}{2}+\pi\right)}$, and
- $\psi(x) = \sqrt{2}\sum_n g_n \varphi(2x-n)$ where $g_n = (-1)^n\,\overline{h_{-n+1}}$.

Note that the function we have chosen for m_1 is not unique—we can multiply this particular choice by any 2π-periodic function with absolute value equal to one.

Exercise. If we take $m_1(\xi) = e^{-i\xi}\,\overline{m_0(\xi + \pi)}$, prove that $g_n = (-1)^n\,\overline{h_{-n+1}}$.

We also saw one example (or one family of examples) in the previous lecture, the spline wavelets. Recall that we defined the scaling function via its Fourier transform and the Fourier transform of the Lth B-spline:

$$\widehat{\varphi}(\xi) = \frac{\widehat{B}_L(\xi)}{\sqrt{2\pi}\left(\sum_\ell \left|\widehat{B}_L(\xi + \pi\ell)\right|^2\right)^{1/2}}.$$

The associated wavelet ψ is a function in $C^k(\mathbb{R})$ for $k = L - 1$ but the scaling function, the wavelet and the associated sequences $\{h_n\}_{n\in\mathbb{Z}}$ and $\{g_n\}_{n\in\mathbb{Z}}$ are infinitely supported, with exponential decay. In order to use these wavelets in a fast algorithm, one must cut off these sequences (or filters) and keep only a finite number of terms. (As a historical note, Battle and Lemarié discovered these independently in the spring of 1986 [**Bat**] [**Lem**]. It later turned out that Stromberg had constructed in 1982 a different basis of spline wavelets, [**Str**], which can be viewed as stemming for a different choice of m.)

We will now discuss two more examples—the Meyer wavelets (which are C^∞ functions and have rapid decay) and the wavelets of compact support (C^k functions). In the construction of Meyer's wavelets, we begin with the Fourier transform of φ. We should point out that this construction gives a smooth version of the Littlewood-Paley construction with the same kind of concentration in Fourier

183

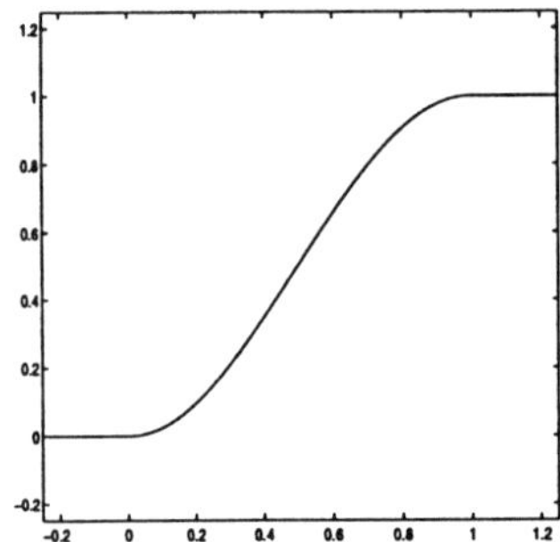

Figure 4.1. The function $\nu(x)$—notice it is symmetric about $x = \frac{1}{2}$

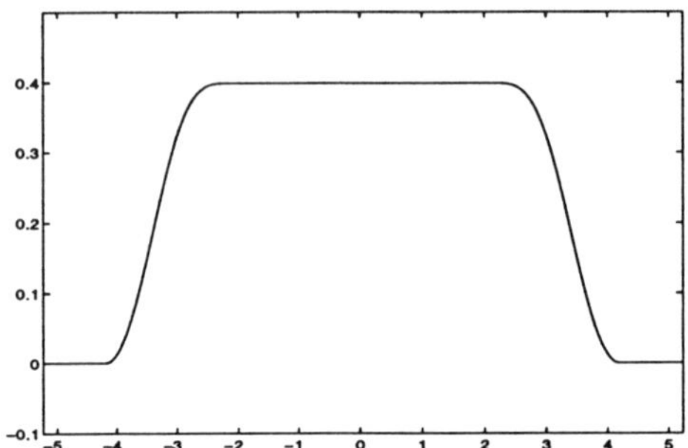

Figure 4.2. The function $\widehat{\varphi}(\xi)$

space. Define φ by

$$\widehat{\varphi}(\xi) = \begin{cases} \dfrac{1}{\sqrt{2\pi}}, |\xi| \leq \dfrac{2\pi}{3} \\[2ex] \dfrac{1}{\sqrt{2\pi}} \cos\left(\dfrac{\pi}{2}\nu\left(\dfrac{3}{2\pi}|\xi| - 1\right)\right), \dfrac{2\pi}{3} \leq |\xi| \leq \dfrac{4\pi}{3} \\[2ex] 0, \quad \text{otherwise} \end{cases}$$

where ν is a smooth function such that $\nu(x) = 0$ if $x \leq 0$ and $\nu(x) = 1$ if $x \geq 1$ and $\nu(x) + \nu(1 - x) = 1$ (see Figure 4.1). Figure 4.2 is a graph of $\widehat{\varphi}(\xi)$.

It is easy to verify that the symmetry properties of ν lead to $\sum_{\ell} |\widehat{\varphi}(\xi + 2\pi\ell)|^2 = \frac{1}{2\pi}$, which is equivalent to orthonormality of the integer translates of φ. Define V_0 to be the closed subspace spanned by the functions $\{\varphi(x - n)\}_{n \in \mathbb{Z}}$. Also, define V_j to be the closed space spanned by the function $\{\varphi_{j,k}\}_{k \in \mathbb{Z}}$. To verify that we have a ladder of spaces $\cdots \subset V_{-1} \subset V_0 \subset V_1 \subset \cdots$ we must show that $\varphi \in V_1$. Alternatively, we must show that there is a 2π-periodic function m_0 such that $\widehat{\varphi}(2\xi) = m_0(\xi)\widehat{\varphi}(\xi)$. We can construct m_0 from $\widehat{\varphi}$ itself: set $m_0(\xi) = \sqrt{2\pi} \sum_{\ell} \widehat{\varphi}(2(\xi + 2\pi\ell))$. We can

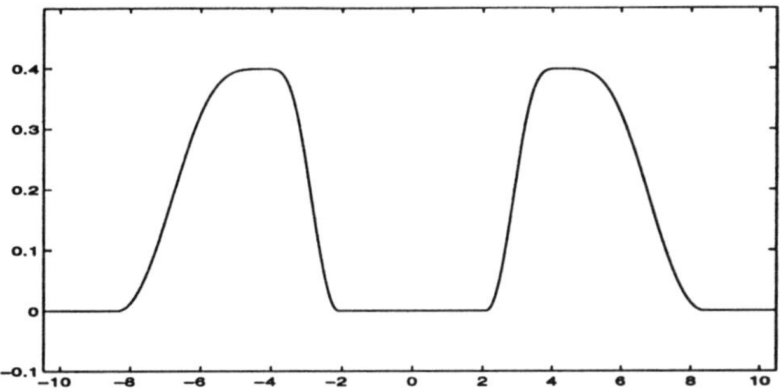

Figure 4.3. The function $|\widehat{\psi}(\xi)|$

easily check that m_0 is 2π-periodic and that

$$m_0(\xi)\widehat{\varphi}(\xi) = \sqrt{2\pi}\sum_{\ell}\widehat{\varphi}(2(\xi+2\pi\ell))\widehat{\varphi}(\xi)$$

$$= \sqrt{2\pi}\widehat{\varphi}(2\xi)\widehat{\varphi}(\xi)$$

$$= \widehat{\varphi}(2\xi).$$

Exercise. Check the other properties of an MRA for this example.

We now define the associated wavelet ψ:

$$\widehat{\psi}(\xi) = e^{i\xi/2}\,\overline{m_0\left(\frac{\xi}{2}+\pi\right)}\,\widehat{\varphi}\left(\frac{\xi}{2}\right)$$

$$= \sqrt{2\pi}e^{i\xi/2}\sum_{\ell}\widehat{\varphi}(\xi+2\pi(2\ell+1))\widehat{\varphi}\left(\frac{\xi}{2}\right)$$

$$= \sqrt{2\pi}e^{i\xi/2}\left[\widehat{\varphi}(\xi+2\pi)+\widehat{\varphi}(\xi-2\pi)\right]\widehat{\varphi}\left(\frac{\xi}{2}\right).$$

See Figure 4.3 for a graph of $|\widehat{\psi}(\xi)|$.

Because $\widehat{\varphi}$ and $\widehat{\psi}$ are C^∞ functions with compact support, we know that φ and ψ are also C^∞ functions but with rapid decay. This means that for any N, we have the following bound:

$$|F(x)| \le C_N(1+|x|)^{-N} \quad \text{where} \quad F = \varphi \quad \text{or} \quad \psi.$$

While these wavelets are often useful for theoretical purposes, the fact that they have only rapid decay is not so good for some practical purposes—the bound C_N does grow with N!

One might wonder if one could construct a wavelet which has both exponential decay and infinite smoothness. The answer is that if the wavelets $\psi_{j,k}$ are orthonormal, then it is impossible for ψ to have both properties. Here is a sketch of that argument—if the wavelets $\psi_{j,k}$ are C^k functions and orthonormal, then the wavelet

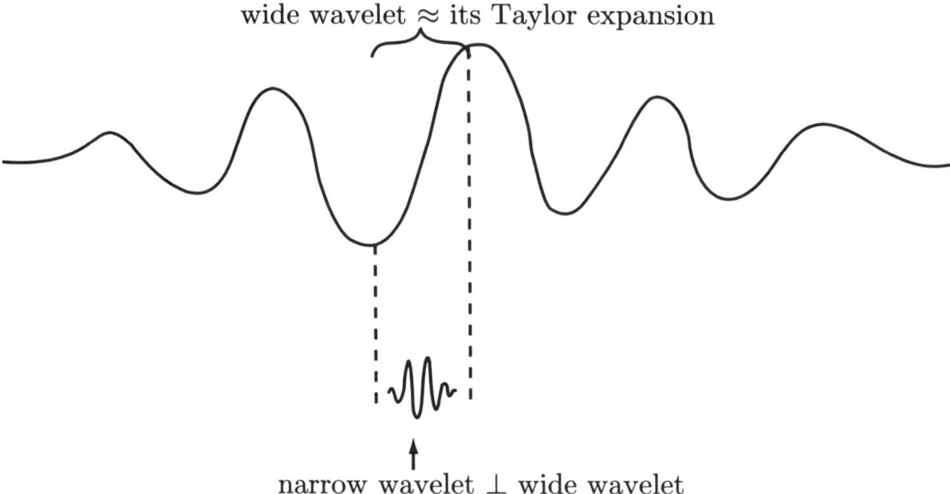

Figure 4.4. A large scale wavelet integrated against a small scale wavelet

ψ must be orthogonal to polynomials of orders $\ell = 0, \dots, k$; i.e.,

$$(4.1) \qquad \int \psi(x) x^\ell dx = 0 \quad \text{for} \quad \ell = 0, \dots, k.$$

To see this, integrate a very large scale wavelet against one with a tiny scale (see Figure 4.4): on the small scale, the large scale wavelet is accurately described by its local Taylor series expansion, to which the small scale wavelet must be orthogonal.

On the Fourier transform side, (4.1) implies that we must have $\widehat{\psi}^{(\ell)}(0) = 0$ for $\ell = 0, \dots, k$ (i.e., $\widehat{\psi}(0)$ must vanish up to order k). For the wavelet ψ to be a C^∞ function, $\widehat{\psi}(0)$ must vanish up to infinite order. Furthermore, requiring ψ to have exponential decay forces $\widehat{\psi}$ to be analytic in a strip $|\mathrm{Im}\xi| < \lambda$. So we have an analytic function with a zero of infinite order. Hence, $\widehat{\psi} \equiv 0$ on such a strip, which implies $\psi(x) = 0$ for all x, in contradiction with our requirement that the $\psi_{j,k}$ constitute an orthonormal basis.

Notice, however, that the above argument gives us some additional constraints for wavelets which are in $C^k(\mathbb{R})$—namely, vanishing moments. Here we begin our second family of examples. For this family, we shall require that ψ, φ are compactly supported C^k functions, which means that we want only finitely many $h_n \neq 0$. Since $\widehat{\psi}(\xi) = m_0 \left(\frac{\xi}{2} + \pi \right) \widehat{\varphi} \left(\frac{\xi}{2} \right) e^{-i\xi/2}$ and $\widehat{\psi}$ should have a zero of high order at zero, m_0, now a trigonometric polynomial, must have the form $m_0(\xi) = \left(\frac{1+e^{i\xi}}{2} \right)^k q(\xi)$. This fact has a few interesting consequences. For instance, φ satisfies:

$$\widehat{\varphi}(2\pi(2k+1)) = m_0(\pi(2k+1))\widehat{\varphi}(\pi(2k+1)) = 0$$

$$\widehat{\varphi}(2\pi 2^\ell(2k+1)) = m_0(\pi 2^{\ell-1}(2k+1))\cdots m_0(\pi(2k+1))\widehat{\varphi}(\pi(2k+1)) = 0$$

$$\widehat{\varphi}(2\pi n) = \delta_{n,0}\frac{1}{\sqrt{2\pi}}\,.$$

We can also derive a similar property for the derivatives of $\widehat{\varphi}$; for $n \neq 0$ we have:

$$\widehat{\varphi}'(2\pi n) = \widehat{\varphi}'(2\pi 2^m(2k+1))$$

$$= \frac{1}{2}m_0'(\pi 2^m(2k+1))\widehat{\varphi}(\pi 2^m(2k+1))$$

$$+ \frac{1}{2}m_0(\pi 2^m(2k+1))\widehat{\varphi}'(\pi 2^m(2k+1))$$

$$= \frac{1}{2}\widehat{\varphi}'(\pi 2^m(2k+1)) = \ldots = \frac{1}{2^m}m_0(\pi(2k+1))\widehat{\varphi}'(\pi(2k+1)) = 0\,;$$

similarly

$$\widehat{\varphi}^{(\ell)}(2\pi n) = \delta_{n,0}\widehat{\varphi}^{(\ell)}(0) \quad \text{for} \quad \ell = 0, \ldots, k\,.$$

A third property is that the integer translates of φ sum to 1, $\sum_k \varphi(x-k) = 1$ which we will derive by the Poisson summation formula. Observe that the function $\sum_k \varphi(x-k)$ is a 1-periodic function, so it has a Fourier series expansion $\sum_k \varphi(x-k) = \sum_n C_n e^{-2\pi i n x}$ where $C_n = \int_0^1 \sum_k \varphi(x-k)e^{2\pi i n x}dx$. Thus, the Fourier coefficients C_n are

$$C_n = \int_{-\infty}^{+\infty} \varphi(x-k)e^{2\pi i n x}dx = \sqrt{2\pi}\widehat{\varphi}(2\pi n) = \delta_{n,0}\,.$$

Similarly, the function $\sum_k (x-k)\varphi(x-k) = \widehat{\varphi}'(0)\sqrt{2\pi}$. In general, we have

$$\sum_k (x-k)^\ell \varphi(x-k) = \int y^\ell \varphi(y)dy \quad \text{for} \quad \ell = 0, \ldots, k\,.$$

We will now discuss how the factorization of m_0 determines the smoothness properties of our wavelets and we will use a corresponding factorization of $\widehat{\varphi}$ to derive these. Recall that we have

$$\widehat{\varphi}(\xi) = m_0\left(\frac{\xi}{2}\right)\widehat{\varphi}\left(\frac{\xi}{2}\right), \quad \text{and} \quad \widehat{\varphi}(\xi) = m_0\left(\frac{\xi}{2}\right)m_0\left(\frac{\xi}{4}\right)\widehat{\varphi}\left(\frac{\xi}{4}\right), \ldots$$

Therefore, we can write $\widehat{\varphi}$ as the infinite product $\widehat{\varphi}(\xi) = \prod_{j=1}^{\infty} m_0(2^{-j}\xi) \cdot \frac{1}{\sqrt{2\pi}}$. Note that this product converges pointwise in ξ, because, for fixed ξ, the factor $m_0(2^{-j}\xi)$ approaches 1 exponentially fast as $j \to \infty$ (since $\sum_n h_n = \sqrt{2}$, or $m_0(0) = 1$, and we have only finitely many $h_n \neq 0$). The factorization also tells us something about the form of $\widehat{\varphi}$. Notice that

$$\frac{1+e^{-i\xi}}{2} = \frac{1-e^{-2i\xi}}{2(1-e^{-i\xi})}\,,$$

so a product of factors of the above form gives

$$\prod_{j=1}^{N}\left(\frac{1+e^{-i\xi 2^{-j}}}{2}\right) = \frac{1-e^{-i\xi}}{2^N(1-e^{-i2^{-N}\xi})} \longrightarrow \frac{1-e^{-i\xi}}{i\xi} \quad \text{as} \quad N \to \infty.$$

Thus, the product $\prod_{j=1}^{\infty} m_0(2^{-j}\xi)$ can be rewritten as $\left(\frac{1-e^{-i\xi}}{i\xi}\right)^k \prod_{j=1}^{\infty} q(2^{-j}\xi)$. Since $|1 - e^{-i\xi}|^k$ is bounded, we find therefore that the factorization of m_0 buys us decay for the infinite product defining $\widehat{\varphi}$, provided that we can control $\prod_{j=1}^{\infty} q(2^{-j}\xi)$. We will have to check whether we can find appropriate $q(\xi)$, and how to bound their infinite product. We have thus the following plan for the construction of our second family of examples: to construct C^k wavelets of compact support, we will start with the sequence $\{h_n\}_{n\in\mathbb{Z}}$ (and not with φ, as in the spline example). Our goal is to have a finite number of non-zero coefficients h_n. Beginning with the coefficients h_n is equivalent to constructing m_0. We have two constraints on m_0:

$$|m_0(\xi)|^2 + |m_0(\xi + \pi)|^2 = 1 \quad \text{and} \quad m_0(\xi) = \left(\frac{1+e^{-i\xi}}{i\xi}\right)^k q(\xi).$$

These constraints give us the magnitude squared of m_0, $|m_0(\xi)|^2 = \left(\cos^2 \frac{\xi}{2}\right)^k Q(\xi)$, where $Q(\xi)$ is a trigonometric polynomial. Can we find these polynomials? The answer is yes. For each $k \geq 1$, there exists a unique symmetric trigonometric polynomial $Q(\xi)$ of degree $k - 1$ so that both conditions are satisfied. One can write $Q(\xi)$ as a polynomial of degree $(k - 1)$, in $\sin^2(\frac{\xi}{2})$; its coefficients are exactly the first k coefficients in the Taylor expansion of $(1 - y)^{-k}$ near $y = 0$. That is,

$$(4.2) \qquad\qquad Q(\xi) = \sum_{\ell=0}^{k-1} \binom{k-1+\ell}{\ell} \left(\sin^2 \frac{\xi}{2}\right)^{\ell}.$$

(For details concerning this derivation, see §6.1 in [**DTL**].) For $k = 1$, this gives $Q \equiv 1$, so that $m_0(\xi) = \frac{1+e^{-1\xi}}{2}$, corresponding to the Haar example. For $k = 2$, we have $Q(\xi) = 1 + 2\sin^2 \frac{\xi}{2}$. For each k, there also exist other solutions for Q, if we allow Q to be of a higher degree.

Once we have Q, we still need to "extract its square root" to find $q(\xi)$ so that $|q(\xi)|^2 = Q(\xi)$. For this to be possible, we need of course that Q is a positive trigonometric polynomial. Fortunately, (4.2) is positive everywhere. But even then, we cannot simply define $q(\xi) = Q(\xi)^{1/2}$, because we want $q(\xi)$ itself to be a trigonometric polynomial as well. For instance, if $k = 2$, then $\left(1 + 2sin^2(\xi/2)\right)^{1/2}$ is not a good choice for $q(\xi)$, but $\left(\frac{1+\sqrt{3}}{2} + \frac{1-\sqrt{3}}{2} e^{-i\xi}\right)$ is fine.

Exercise. Check that $\left|\frac{1+\sqrt{3}}{2} + \frac{1-\sqrt{3}}{2} e^{-i\xi}\right|^2 = 1 + 2\sin^2 \frac{\xi}{2}$.

In fact, given a symmetric, positive trigonometric polynomial $P(\xi)$ of degree D, one can always find a polynomial $p(e^{-i\xi})$ in $e^{-i\xi}$ with real coefficients, so that $|p(e^{-i\xi})|^2 = P(\xi)$. This is called spectral factorization in the electrical engineering literature, and it is used in many different applications, both in mathematics and engineering. For details, see e.g. [**DTL**].

Finally, we need to bound infinite products of these $q(\xi)$, to see how smooth the corresponding φ can be. One can prove (see [**DTL**]) that the $Q(\xi)$ given by (4.2) satisfy

$$\prod_{j=1}^{\infty} Q(2^{-j}\xi) \leq C(1 + |\xi|)^{2k\lambda},$$

where $\lambda = 2 - \frac{3}{4}\frac{\log 3}{\log 2} \approx .8114$. It follows that

$$|\widehat{\varphi}(\xi)| \leq C(1 + |\xi|)^{-k\mu},$$

with $\mu = 1 - \lambda \approx .1887$. By choosing k sufficiently large, one can therefore achieve arbitrary inverse polynomial decay for $\widehat{\varphi}$, or arbitrary smoothness for φ. There is much more to the story of the smoothness of these functions, but that is outside the scope of this lecture; other approaches can be found in [**DTL**] or [**CDM**].

LECTURE 5
Wavelets in Other Functional Spaces

In this lecture, we will cover several convergence theorems and discuss wavelets on the interval. Two results from the previous lecture will be useful for us: recall that if the wavelet $\psi \in C^\ell(\mathbb{R})$, then $\int x^k \psi(x)dx = 0$ for $k = 0, \dots, \ell$; that is, the wavelet has ℓ vanishing moments. Equivalently, the sum $\sum_k (x-k)^n \varphi(x-k)$ is identically constant for $n = 0, \dots, \ell$.

Assume that f is a continuous function and in $L^2(\mathbb{R})$. We already know that the sum $\sum_{j,k} \langle f, \psi_{j,k} \rangle \, \psi_{j,k}$ converges in $L^2(\mathbb{R})$ to f but we would like to know what happens to this sum pointwise. We will restrict our attention to wavelets of compact support but one can modify the arguments below for other wavelets if they have good decay. Since $\psi_{j,k}(x) = 2^{j/2}\psi(2^j x - k)$ and ψ is compactly supported, there is a finite number of integers k for which $2^{j/2}\psi(2^j x - k)$ is non-zero. In fact, we only need the integers k for which $|2^j x - k| < R$, if support $\psi \subset [-R, R]$. The number of such k is bounded uniformly in j by $2R$.

Theorem. *Assume that ψ is compactly supported and that $f \in C(\mathbb{R}) \cap L^2(\mathbb{R})$. Then for all x,*

$$f(x) = \lim_{J \to \infty} \sum_{j \leq J} \sum_k \langle f, \psi_{j,k} \rangle \, \psi_{j,k}(x) \, .$$

Proof. Let us examine the difference

$$f(x) - \sum_{j \leq J}(Q_j f)(x) = f(x) - (P_{J+1}f)(x) = f(x) - \sum_k \langle f, \varphi_{J+1,k} \rangle \, \varphi_{J+1,k}(x)$$

$$= f(x) - \int f(y) 2^{J+1} \sum_k \overline{\varphi(2^{J+1}y - k)} \, \varphi(2^{J+1}x - k) \, dy \, .$$

Define $K(x,y) = \sum_k \overline{\varphi(y-k)} \, \varphi(x-k)$, which will serve as an integral kernel. Note that

$$\int K(x,y) \, dy = \int \sum_k \overline{\varphi(y-k)} \, \varphi(x-k) \, dy = \sum_k \varphi(x-k) \int \overline{\varphi(y-k)} \, dy = 1 \, .$$

Therefore, the difference is

$$f(x) - \int f(y) 2^{J+1} K\left(2^{J+1}x, 2^{J+1}y\right) dy$$

$$= \int \left(f(x) - f(y)\right) 2^{J+1} K\left(2^{J+1}x, 2^{J+1}y\right) dy.$$

Let $y = x + 2^{-J-1}t$ and change variables. Our integral is now

$$\int \left(f(x) - f\left(x + 2^{-J-1}t\right)\right) K\left(2^{J+1}x, 2^{J+1}x + t\right) dt.$$

We want to determine the behavior of that integral as $J \to \infty$ (in particular, we'd like the limit to be zero!) so we will bound $K(z, z + t)$ independently of t and apply the Dominated Convergence Theorem. Observe that

$$|K(z, z + t)| \leq \sum_k |\varphi(z - k)||\varphi(z + t - k)|$$

and that we only need integers k such that $|z - k| \leq R$ and $|z + t - k| \leq R$. Therefore $|t|$ is restricted to $|t| \leq 2R$, and $|K(z, z + t)| \leq C \chi_{|t| \leq 2R}$. The absolute value of the integral is:

$$\int \left|f(x) - f(x + 2^{-J-1}t)\right| \left|K(2^{J+1}x, 2^{J+1}x + t)\, dt\right|$$

$$\leq C \int |f(x) - f(x + 2^{-J-1}t)| \chi_{|t| \leq 2R}\, dt.$$

If we take the limit as $J \to \infty$ and observe that f is continuous, we see that the limit is indeed zero. $\qquad\square$

The following theorem characterizes global regularity of a function based on the size of its wavelet coefficients. Recall that a function f is said to be in $C^\alpha(\mathbb{R})$ if $|f(x + t) - f(x)| \leq C|t|^\alpha$, for all $x \in \mathbb{R}$.

Theorem. *Assume φ and ψ are C^s functions with compact support. In addition, assume f is bounded. Then, for $\alpha \in (0,1)$, we have $f \in C^\alpha(\mathbb{R})$ if and only if $|\langle f, \psi_{j,k}\rangle| \leq 2^{-j(\alpha+1/2)}$ for $j > 0$ and $\alpha < s$.*

Proof. The inner products of f with $\psi_{j,k}$ are

$$\langle f, \psi_{j,k}\rangle = \int f(x) 2^{j/2}\, \overline{\psi(2^j x - k)}\, dx.$$

Because the wavelet $\psi_{j,k}$ is centered about $2^{-j}k$ and has integral zero, we have a bound on the magnitude of the inner product

$$|\langle f, \psi_{j,k}\rangle| = \left| \int \left(f(x) - f(2^{-j}k) \right) \overline{\psi(2^j x - k)}\, 2^{j/2}\, dx \right|$$

$$\leq C \int |x - 2^{-j}k|^\alpha\, 2^{j/2}\, |\psi(2^j x - k)|\, dx$$

$$\leq C 2^{-j/2} 2^{-j\alpha} \int |t|^\alpha |\psi(t)|\, dt$$

$$\text{(by letting} \quad t = 2^j x - k \quad \text{and changing variables)}$$

$$\leq C 2^{-j(\alpha + \frac{1}{2})}\,.$$

This proves one direction of the implication.

To prove the other direction of the implication, we will write

$$(F_J f)(x) = \sum_k \langle f, \varphi_{0,k}\rangle\, \varphi_{0,k}(x) + \sum_{j=0}^{J} \sum_k \langle f, \psi_{j,k}\rangle\, \psi_{j,k}(x)\,,$$

then take the limit as $J \to \infty$, and show that the limiting function is in $C^\alpha(\mathbb{R})$. The difference between $F_j f(x + t)$ and $F_j f(x)$ is

$$|(F_J f)(x + t) - (F_J f)(x)| \leq \sum_k C |\varphi(x - k) - \varphi(x + t - k)|$$

$$+ \sum_{j=0}^{J} \sum_k C 2^{-j(\alpha + \frac{1}{2})} |\psi_{j,k}(x + t) - \psi_{j,k}(x)|\,.$$

Since both terms are clearly bounded uniformly in J, we need not prove anything for $|t| > 1$; it's enough to check $|t| \leq 1$. We can bound the first term on the right hand side by $C|t|^s$ since $\varphi \in C^s(\mathbb{R})$ for $s > \alpha$. For the second term, we consider two regimes: small scales (scales with resolution smaller than $|t|$), and larger scales. Find j_0 so that $2^{-j_0 - 1} \leq |t| \leq 2^{-j_0}$ and break the sum into these two regimes:

$$\sum_{j=0}^{J} \sum_k C 2^{-j(\alpha + 1/2)} |\psi_{j,k}(x + t) - \psi_{j,k}(x)|$$

(5.1)
$$= \sum_{j=0}^{j_0} \sum_k C 2^{-j(\alpha + 1/2)} 2^{j/2} |\psi(2^j x + 2^j t - k) - \psi(2^j x - k)|$$

$$+ \sum_{j=j_0 + 1}^{J} \sum_k C 2^{-j(\alpha + 1/2)} 2^{j/2} |\psi(2^j x + 2^j t - k) - \psi(2^j x - k)|\,.$$

The first simplification we will make is to get rid of the sums over k—there is only a finite number of k which are necessary (their number is bounded independently of scale). Replace these sums over k by constants. Therefore the right-hand-side of equation (5.1) is bounded by

$$(5.2) \qquad \sum_{j=0}^{j_0} C2^{-j(\alpha+1/2)}2^{j/2}|2^j t|^s + \sum_{j=j_0+1}^{J} C2^{-j(\alpha+1/2)}2^{j/2},$$

since ψ is Hölder continuous. It follows that the difference $F_J f(x+t) - F_J f(x)$ is bounded by

$$\begin{aligned}
|F_J f(x+t) - F_J f(x)| &\leq C|t|^s + \sum_{j=0}^{j_0} C2^{j(s-\alpha)}|t|^s + \sum_{j=j_0+1}^{J} C2^{-j\alpha} \\
&\leq C|t|^s + C2^{(1+j_0)(s-\alpha)}|t|^s + C2^{-j_0\alpha} \\
&\leq C|t|^s + C2^{-j_0\alpha} \\
&\leq C|t|^s + C|t|^\alpha .
\end{aligned}$$

$\square$

We should remark at this point that we can similarly characterize higher order C^α spaces. For instance, for $\alpha \in (1,2)$ we define $f \in C^\alpha$ by the uniform requirement $|f(x+t) - f(x) - f'(x)t| \leq C|t|^\alpha$; we will then also use higher order vanishing moments of ψ, i.e., $\int t\psi(t)\, dt = 0$. Also, note that in our proof we used the vanishing moments of ψ in one direction and the smoothness of ψ in the other.

We can use the previous result to motivate the following result for spline wavelets of order L: if $\alpha < L$, then

$$\|f - P_j f\|_{L^\infty} \leq C2^{-j\alpha} \iff f \in C^\alpha(\mathbb{R}).$$

This follows easily from the previous result:

$$\begin{aligned}
|f(x) - P_J f(x)| &= |\sum_{j\geq J}\sum_k \langle f, \psi_{j,k}\rangle\, \psi_{j,k}^L(x)| \\
&\leq \sum_{j\geq J}\sum_k |\langle f, \psi_{j,k}^L\rangle|\, |\psi_{j,k}^L(x)| \\
&\leq \sum_{j\geq J} C2^{-j(\alpha+1/2)}C2^{j/2} \\
&\leq C2^{-J\alpha} .
\end{aligned}$$

Note, however, that we are not really allowed to use the theorem for spline wavelets – it assumed that the wavelet $\psi_{j,k}^L$ had compact support (allowing us to assume a finite number of k at each scale, for instance) but that doesn't hold for orthonormal spline wavelets.

Exercise. Fill in the mathematical details in the sketch of proof that we just gave, by using the exponential decay of the orthonormal spline wavelets.

For the other direction of implication, observe that the inner product $\langle f, \psi_{j,k}^L \rangle$ is bounded by

$$\left| \langle f, \psi_{j,k}^L \rangle \right| = \left| \langle f - P_j f, \psi_{j,k}^L \rangle \right| \leq C \int 2^{-j\alpha} 2^{j/2} \left| \psi^L(2^j x - k) \right| \, dx$$

$$\leq C 2^{-j(\alpha + 1/2)} .$$

Exercise. Prove the following theorem.

Theorem. *A function f lies in $H^s(\mathbb{R}) = \{ f \mid \int |\hat{f}|^2(\xi)(1 + |\xi|^2)^s \, d\xi < \infty \}$ if and only if $\sum_{j,k} |\langle f, \psi_{j,k} \rangle|^2 (1 + 2^{2j})^s < \infty$.*

Yet another remark is to note that we can also use the wavelet coefficients of a function to characterize its local regularity properties:

Theorem. *If f is Hölder continuous with exponent α ($\alpha \in (0,1)$) at x_0; i.e., $|f(x) - f(x_0)| \leq C|x - x_0|^\alpha$, then*

$$|\langle f, \psi_{j,k} \rangle| \leq C 2^{-j(\alpha + 1/2)} \left(1 + \text{dist } (x_0 - 2^j k)^\alpha \right) \quad \text{for all} \quad j .$$

Conversely, if the above holds and if f is known to be in C^ϵ for some $\epsilon > 0$, then

$$|f(x) - f(x_0)| \leq C|x - x_0|^\alpha \log \frac{2}{|x - x_0|} .$$

We won't prove this theorem due to S. Jaffard [**Jaf**], but we will remark that one should be careful in trying to determine numerically from the decay of the wavelet coefficients the local regularity of a function. It may be that very large values of j are needed to determine α reliably. (See Figure 9.2 in [**DTL**].)

A very important remark is that we have characterized the spaces $L^2(\mathbb{R})$, $C^\alpha(\mathbb{R})$, $H^s(\mathbb{R})$ by using only the **absolute value** of the wavelet coefficients. This means that the wavelets $\{\psi_{j,k}\}_{j,k \in \mathbb{Z}}$ are an *unconditional basis* for all of these spaces. Recall that the Fourier basis is an unconditional basis for $L^2(\mathbb{R})$ but not for $C^\alpha(\mathbb{R})$ or $L^p(\mathbb{R})$, $p \neq 2$. (The example given at the start of Lecture 2 shows that one can change the range of p for which a 2π-periodic function lies in L^p by tweaking only the phases of its Fourier coefficients: $f(x) = \sum_{n=2}^\infty n^{-\frac{1}{4}} e^{2\pi i n x}$ lies in L^p for $p < 4/3$, whereas $\widetilde{f}(x) = \sum_{n=2}^\infty n^{-\frac{1}{4}} e^{i\sqrt{n}} e^{2\pi i n x}$ doesn't). Similar things can happen if one deletes some Fourier coefficients and not others. With an unconditional basis this kind of thing cannot happen: you can change the phases of the different coefficients, and, although this will change the function, the new function will still lie in the same Banach space and control the norm of the new function. You can also delete some coefficients and/or shrink others without leaving the Banach space. The fact that wavelets give us an unconditional basis for the C^α, H^s spaces as well as $L^p(\mathbb{R})$, $1 < p < \infty$, and many other spaces, has both theoretical and practical applications. One practical application is that we can apply thresholding schemes to the wavelet coefficients of a function (i.e., set to zero all those coefficients smaller in absolute value that a cut-off value λ) and know that our "thresholded" function is in the same space as our original function. (This is not true for the Fourier basis.)

We now come to wavelets on the interval which we will discuss briefly. It is clear that if we place our wavelets for the whole real line on an interval, some will spill out as in Figure 5.1. (Many will also have no overlap with the interval, but

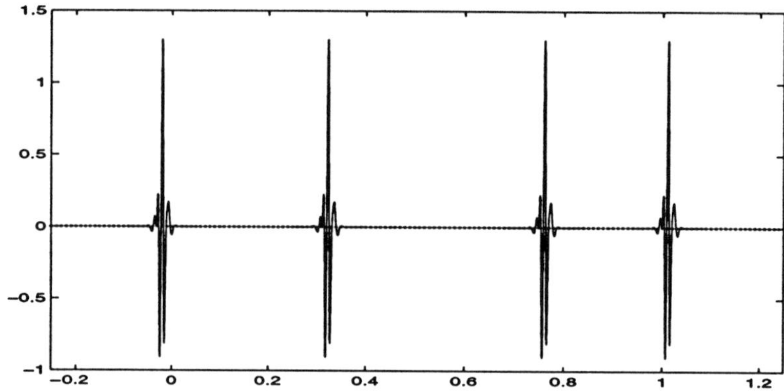

Figure 5.1. Notice how the wavelets at the edges of the interval $[0, 1]$ "spill" outside the interval

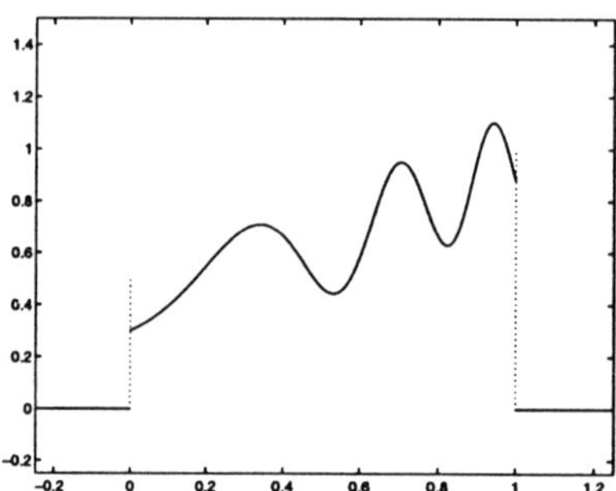

Figure 5.2. Extending the function outside the interval $[0, 1]$ by setting it equal to zero

these don't matter - only the ones that straddle endpoints can potentially cause a problem for us.) There are several possible fixes.

1. We could truncate our wavelets at the edges of our interval. This is equivalent to extending our function outside the interval by zero. However, this is not such a good solution because we will introduce local discontinuities in our function (see Figure 5.2) for which we will pay a price in the decay of the wavelet coefficients around those discontinuities (remember the local characterization theorem).

2. We could periodize our interval if our problem has periodic boundary conditions. For example, let $[a, b] = [0, 1]$ and define $\varphi_{j,k}^{\mathrm{per}}(x) = \sum_\ell \varphi_{j,k}(x - l)$. We will then have 2^j wavelets at each scale j. However, if the functions in which we are interested are not periodic (as in Figure 5.2), then we have again to pay a price for the discontinuity that we have introduced by periodizing.

3. If we don't have periodic boundary conditions, we must do something else. We would like wavelets on the interval to have all the properties of our wavelets on $\mathbb{R}$—smoothness, vanishing moments, etc. It is clear that the wavelets which are supported away from the edges of interval (within the interval) don't need to be modified since they don't "feel" the edges. We will modify the scaling functions at the edges of the interval so that we preserve all of those properties enjoyed by wavelets and scaling functions on $\mathbb{R}$. In particular, as we saw in the characterization of $C^s(\mathbb{R})$ via wavelets, it is important to preserve both the smoothness of the wavelet and its vanishing movements. This corresponds to introducing modified filters (sequences $\{h_n\}$ and $\{g_n\}$) at the edges of the interval, but otherwise our algorithm essentially remains the same. For details, see [**CDV**].

LECTURE 6
Pointwise Convergence for Wavelet Expansions

In the previous lecture, we discussed how an orthonormal basis of $C^s(\mathbb{R})$ and compactly supported wavelets could be used to characterize both local and global regularity of a function. We also proved that the wavelet expansion of a continuous function (in $L^2(\mathbb{R})$) converged pointwise to that function. In this lecture, we will show that for any function in $L^2(\mathbb{R})$ we have pointwise almost everywhere convergence of the wavelet expansion. This theorem falls under the category of linear approximation theory. We will also prove a result about the pointwise almost everywhere convergence of a "non-linear" wavelet expansion.

The pointwise almost everywhere convergence of the Fourier series of an arbitrary $L^2([0,1])$ function is a **hard** theorem but the analogous theorem for wavelets is quite straightforward. The proof mimics Lebesgue's theorem that

$$\lim_{r \to 0} \frac{1}{2r} \int_{|x-y|<r} f(y)\, dy = f(x)$$

for almost all x. Readers not familiar with the concepts introduced in the following proof can find much more in [**Ste**].

Theorem. *If $f \in L^2(\mathbb{R})$, then for almost all $x \in \mathbb{R}$, we have*

$$\lim_{J \to \infty} \sum_{j<J} \sum_{k} \langle f, \psi_{j,k} \rangle \psi_{j,k}(x) = f(x).$$

Proof. We will first introduce the Hardy-Littlewood Maximal function:

$$(Mf)(x) = \sup_{r>0} \frac{1}{2r} \int_{|y-x|<r} |f(y)|\, dy.$$

The maximal function has two important properties:

$$\|Mf\|_{L^\infty} \le \|f\|_{L^\infty} \quad \text{and} \quad \|Mf\|_{L^1_{\text{weak}}} \le A\|f\|_{L^1}.$$

The first inequality is simple to check and the second inequality is proved with the Vitali covering lemma. By splitting f into two parts, and applying the L^1-result to the "large" part (which is always in L^1), one can also derive a result for intermediate p:

$$\|Mf\|_{L^p} \le C\|f\|_{L^p} \quad \text{for} \quad 1 < p < \infty.$$

In particular, $\|Mf\|_{L^2} \le C\|f\|_{L^2}$ will be an important ingredient of our proof.

Let us define what it means for a function g to be in the space $L^1_{\text{weak}}(\mathbb{R})$ and discuss the "norm" on this space. For g to be an $L^1_{\text{weak}}(\mathbb{R})$ function we must have

$$m\left(\left\{\, x \in \mathbb{R} \mid |g(x)| > \alpha \,\right\}\right) \le \frac{c}{\alpha} \quad \text{for all} \quad \alpha.$$

The "norm" on this space is the smallest c such that the above inequality holds. This is not a norm in that it does not satisfy the triangle inequality (hence the quotes). Observe that the function $g(x) = \frac{1}{x}$ is not in $L^1(\mathbb{R})$ but it is in $L^1_{\text{weak}}(\mathbb{R})$. It's clear that $L^1(\mathbb{R}) \subset L^1_{\text{weak}}(\mathbb{R})$. Note that $Mf \notin L^1(\mathbb{R})$—one simple motivational argument is to calculate the maximal function of the characteristic function of an interval (try this yourself!).

We have

$$(P_J f)(x) = \sum_{j<J} \sum_{k} \langle f, \psi_{j,k} \rangle\, \psi_{j,k}(x) = \sum_{k} \langle f, \varphi_{J,k} \rangle\, \varphi_{J,k}(x)$$

and we will define a new maximal function

$$(\widetilde{M}f)(x) = \sup_{j} |P_j f(x)| \,.$$

We'll show that $\widetilde{M}f$ is pointwise bounded by Mf using the integral kernel we defined in the previous lecture:

$$(P_j f)(x) = \sum_{k} \int f(y)\, 2^j\, \overline{\varphi(2^j y - k)}\, \varphi(2^j x - k)\, dy$$

$$= \int f(y)\, 2^j K(2^j x, 2^j y)\, dy \,.$$

Now,

$$|K(x,y)| = \left| \sum_{k} \overline{\varphi(y-k)}\, \varphi(x-k) \right|$$

$$\le \sum_{k} |\varphi(x-k)|\, |\varphi(y-k)|$$

$$\le C\, \chi_{|y-x|\le 2R}$$

since φ and ψ have compact support. Thus,

$$|P_J f(x)| \le 2^J C \int |f(y)|\, \chi_{|y-x|\le 2^{1-J}R}\, dy$$

$$\le \frac{C}{2r} \int_{|x-y|<r} |f(y)|\, dy \quad \text{where} \quad r = 2^{-J} 2R$$

$$\le C(Mf)(x)\,, \quad \text{independent of} \quad J\,.$$

Thus, $(\widetilde{M}f)(x) \le C(Mf)(x)$, so that $\|\widetilde{M}f\|_{L^2} \le C\|f\|_{L^2}$.

Finally, assume that f is real-valued and square integrable. For any $\delta > 0$, we can write $f = g + h$ where $g \in C_0^\infty(\mathbb{R})$ and h is smaller than δ in L^2-norm, i.e.,

$\|h\|_2 < \delta$. If we let

$$(\Omega f)(x) = \limsup_{j \to \infty} P_j f(x) - \liminf_{j \to \infty} P_j f(x),$$

then we know that

$$(\Omega f)(x) \leq (\Omega g)(x) + (\Omega h)(x).$$

From our convergence results from the previous lecture, we know that $(\Omega g)(x) = 0$. We want to show that $m\left(\{\, x \mid \Omega f(x) > 0 \,\}\right) = 0$. We know that

$$\left\{\, x \mid (\Omega f)(x) \geq \frac{1}{n} \,\right\} \subset \left\{\, x \mid (\Omega h)(x) \geq \frac{1}{n} \,\right\}, \quad \text{so}$$

$$m\left(\left\{\, x \mid (\Omega f)(x) \geq \frac{1}{n} \,\right\}\right) \leq m\left(\left\{\, x \mid (\Omega h)(x) \geq \frac{1}{n} \,\right\}\right).$$

If we note that $\Omega h(x) \leq 2\widetilde{M}h(x)$, we have

$$m\left(\left\{\, x \mid (\Omega h)(x) \geq \frac{1}{n} \,\right\}\right) \leq m\left(\left\{\, x \mid (\widetilde{M}h)(x) \geq \frac{1}{2n} \,\right\}\right)$$

$$\leq C(2n)^2 \|\widetilde{M}h\|_{L^2}^2$$

$$\leq C(2n)^2 \|h\|_{L^2}^2$$

$$< C\delta(2n)^2.$$

It follows that for arbitrary $\delta > 0$, $m(\{x|\Omega f(x) \geq 1/n\}) \leq C\delta(2n)^2$, so that $m(\{x|\Omega f(x) \geq 1/n\}) = 0$. Since $\{x|\Omega f(x) > 0\} = \cup_n \{x; \Omega f(x) \geq 1/n\}$ we have therefore $m(\{x|\Omega f(x) > 0\}) = 0$. $\qquad\square$

In the above string of inequalities, we have used Chebyschev's inequality: for $h \in L^2$,

$$\|h\|_{L^2}^2 = \int |h|^2 \geq \int_{\{\, x \,\big|\, |h(x)| > \lambda \,\}} |h|^2 \geq \lambda^2 m\left(\{\, x \mid |h(x)| > \lambda \,\}\right) \quad \text{hence}$$

$$m\left(\{\, x \mid |h(x)| > \lambda \,\}\right) \leq \frac{1}{\lambda^2} \|h\|_{L^2}^2.$$

The theorem that we have just proved is called a *linear* approximation result because we have ordered our basis elements and we have projected onto a linear space of these basis elements. Take, for example, wavelets on the unit interval. At scale j_0, there are K_0 scaling functions and L_0 wavelets. At scale j_1, there are K_1 scaling functions and L_1 wavelets, etc. When we sum in the form

$$\sum \langle f, \varphi_{j_0,k} \rangle \, \varphi_{j_0,k}(x) + \sum_{j_0 < j < J} \sum_k \langle f, \psi_{j,k} \rangle \, \psi_{j,k}(x),$$

we have ordered our basis elements into shells (or scales) and we sum in order by scales. In general terms, we have a basis of functions $\{F_1, F_2, \dots\}$ and we are looking at $P_N f = \text{Proj}_{\text{span}\{F_1,\dots,F_N\}}$ and determining the convergence of $P_N f(x)$. Note that span $\{F_1, \dots, F_N\}$ is a linear space and we are projecting our function f onto this linear space.

In *non-linear* approximation theory we also take N terms in our approximation of f but these N terms don't correspond to projection onto a linear space. For example, we could define

$$(S_N f)(x) = \sum_{n \in I_N f} \langle f, F_n \rangle \, F_n(x)$$

where the index set

$$I_N f = \big\{ \, n \mid |\langle f, F_n \rangle| \ \text{ is among the } N \text{ largest coefficients } \ |\langle f, F_k \rangle| \, \big\} \, .$$

We have another way of creating a non-linear approximation to a function—define

$$(T_\lambda f)(x) = \sum_{\substack{h \\ |\langle f, F_n \rangle| > \lambda}} \langle f, F_n \rangle \, F_n(x) \, .$$

This thresholding of the coefficients is common in practical applications. If we truncate the coefficients which fall below a certain threshold, then we have reduced the number of coefficients needed to compute with, making a faster algorithm. We would like to know if the adaptive algorithms we create and use which threshold the wavelet coefficients of a solution will converge to the true solution as we let the value of the threshold λ go to zero (assuming perfect precision of a computer). We do, indeed, have an answer to this question.

Theorem (T. Tao – 1995). *Let $f \in L^p(\mathbb{R}), 1 < p < \infty$, and assume that we have an orthonormal basis of wavelet functions $\{\psi_{j,k}\}_{j,k \in \mathbb{Z}}$ which have compact support. Then*

$$\lim_{\lambda \to 0} (T_\lambda f)(x) = \lim_{\lambda \to 0} \sum_{\substack{j,k \\ |\langle f, \psi_{j,k} \rangle| > \lambda}} \langle f, \psi_{j,k} \rangle \, \psi_{j,k}(x) = f(x)$$

for almost all $x \in \mathbb{R}$.

Proof. We introduce a new maximal function $(M^{\#} f)(x) = \sup_{\lambda > 0}(T_\lambda f)(x)$. To prove this theorem, it is sufficient to prove the following lemma since the previous result will complete the proof for us. $\qquad\square$

Lemma. $|(M^{\#} f)(x)| \le C |(Mf)(x)|$

Proof. Let us try to bound $|(T_\lambda f)(x) - (P_J f)(x)|$ for J chosen appropriately. Choose J such that $\lambda \simeq 2^{J/2}(Mf)(x)$. Then

$$|(T_\lambda f)(x) - (P_J f)(x)| \le \sum_{\substack{j > J \\ |\langle f, \psi_{j,k} \rangle| > \lambda}} \sum_{k} |\langle f, \psi_{j,k} \rangle| \, |\psi_{j,k}(x)|$$

$$+ \sum_{\substack{j \le J \\ |\langle f, \psi_{j,k} \rangle| \le \lambda}} \sum_{k} |\langle f, \psi_{j,k} \rangle| \, |\psi_{j,k}(x)| \, .$$

Let us first estimate the second term in the right hand side of the above inequality. As before, we can replace the sum over k with a constant factor since we have assumed that our wavelets have compact support. Then

$$\sum_{\substack{j \leq J \\ |\langle f, \psi_{j,k}\rangle| \leq \lambda}} \sum_{k} |\langle f, \psi_{j,k}\rangle| \, |\psi_{j,k}(x)| \leq C\lambda \sum_{j=0}^{J} 2^{j/2} \leq C\lambda 2^{J/2} \leq C(Mf)(x) \, .$$

For the first term, we know that $|\psi_{j,k}(x)| \leq 2^{j/2}\chi_{|2^j x - k| \leq R}$ and

$$|\langle f, \psi_{j,k}\rangle| \leq 2^{j/2} \int_{|2^j y - k| \leq R} |f(y)| \, |\psi(2^j y - k)| \, dy$$

$$\leq 2^{j/2} \int_{2^j |y - x| \leq 2R} |f(y)| \, |\psi(2^j y - k)| \, dy$$

on the region $|2^j x - k| \leq R$. Therefore,

$$|\langle f, \psi_{j,k}\rangle| \, |\psi_{j,k}(x)| \leq C(Mf)(x) \, \chi_{|2^j x - k| \leq R} \, .$$

Thus,

$$|(T_\lambda f)(x) - (P_J f)(x)| \leq \sum_{\substack{j > J \\ |\langle f, \psi_{j,k}\rangle| > \lambda}} \sum_{k} |\langle f, \psi_{j,k}\rangle| \, \psi_{j,k}(x) + C(Mf)(x)$$

$$= C \sum_{J < j < J + L} (Mf)(x) \leq C(Mf)(x)$$

since we are looking at a range of scales j where

$$C2^{-j/2}(Mf)(x) > \lambda \simeq 2^{J/2}(Mf)(x)$$

which is only possible if $2^{(j-J)/2} < C$; i.e., $j < J + L$. $\qquad \square$

There are several remarks here:
- The proof of this result can be modified for wavelets which aren't compactly supported.
- There are other methods of non-linear approximation (or more precisely, methods of summation) which do not converge pointwise, although these methods are a bit contrived.
- This result is false for Fourier series!

In the next lecture, we will discuss how to represent operators of the form

$$(Tf)(x) = \int K(x, y) f(y) dy$$

in wavelet bases. Here, the kernel $K(x, y)$ has special properties (which we will discuss) and we will give two different representations for these kinds of operators and their associated kernels.

LECTURE 7
Two-Dimensional Wavelets and Operators

In the previous lectures, we have restricted our attention to one-dimensional wavelets but we will need two-dimensional wavelets to discuss the representation of operators in wavelet bases. We begin with $\{\psi_{j,k}\}_{j,k\in\mathbb{Z}}$, our orthonormal wavelet basis for $L^2(\mathbb{R})$ and we shall build a basis for $L^2(\mathbb{R}^2)$ from this. There are two approaches we can take:

- Let the basis for $L^2(\mathbb{R}^2)$ be a tensor product of one-dimensional bases. Then our basis functions are

$$\psi_{j_1,k_1}(x_1)\,\psi_{j_2,k_2}(x_2) \quad \text{for} \quad j_1, j_2, k_1, k_2 \in \mathbb{Z}.$$

  Here we dilate and translate independently in the coordinates $(x_1, x_2) \in \mathbb{R}^2$.
- Build a Multiresolution Analysis (MRA) for $L^2(\mathbb{R}^2)$ by taking a tensor product of the MRA for $L^2(\mathbb{R})$. Define $\mathbf{V}_j = V_j \otimes V_j$ where $\{V_j\}_{j\in\mathbb{Z}}$ is the MRA for $L^2(\mathbb{R})$. One can check that we have the ladder of subspaces $\cdots \subset \mathbf{V}_{-1} \subset \mathbf{V}_0 \subset \mathbf{V}_1 \subset \cdots$, and, as before, we can define the "difference" space $\mathbf{W}_j$:

$$\begin{aligned}
\mathbf{W}_j &= \mathbf{V}_{j+1} \cap (\mathbf{V}_j)^\perp \\
&= [V_{j+1} \otimes V_{j+1}] \cap [V_j \otimes V_j]^\perp \\
&= [(V_j \otimes W_j) \otimes (V_j \otimes W_j)] \cap (V_j \otimes V_j)^\perp \\
&= (V_j \otimes W_j) \oplus (W_j \otimes V_j) \oplus (W_j \otimes W_j).
\end{aligned}$$

Therefore, the family $\{\varphi_{j,k_1}(x_1)\varphi_{j,k_2}(x_2)\}_{k_1,k_2\in\mathbb{Z}}$ is a basis for $\mathbf{V}_j$, and likewise the family (really three families put together)

$$\{\varphi_{j,k_1}(x_1)\psi_{j,k_2}(x_2), \psi_{j,k_1}(x_1)\varphi_{j,k_2}(x_2), \psi_{j,k_1}(x_1)\psi_{j,k_2}(x_2)\}_{k_1,k_2\in\mathbb{Z}}$$

is a basis for $\mathbf{W}_j$.

Notice that in the first basis we have scaling and translation independently in each coordinate so that our wavelets have support in rectangles of widely varying aspect ratios, whereas in the second case, we dilate in both coordinates x_1 and x_2 simultaneously, so our wavelets are supported on squares. (See Figure 7.1.)

In the second construction of two-dimensional wavelets, we have created three functions whose integer translates form a basis for $\mathbf{W}_j$ ($\varphi_{j,k}\psi_{j,k'}$, $\psi_{j,k}\varphi_{j,k'}$, and $\psi_{j,k}\psi_{j,k'}$). Figure 7.2 shows where these different functions are concentrated in

205

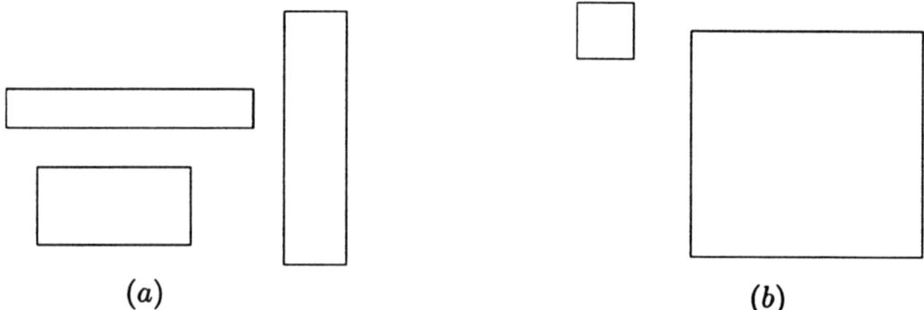

Figure 7.1. (a) Some possible supports of wavelets in a tensor wavelet basis; (b) Some possible supports (all squares) of wavelets from a tensor product MRA.

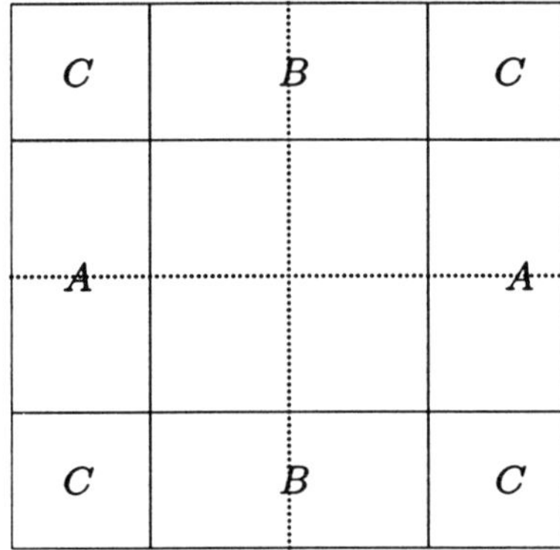

Figure 7.2. The supports of the functions $\widehat{\psi}_{j,k}(\xi_1)\widehat{\varphi}_{j,k'}(\xi_2)$ (the two regions marked A), $\widehat{\varphi}_{j,k}(\xi_1)\widehat{\psi}_{j,k'}(\xi_2)$ (the two regions marked B), and $\widehat{\psi}_{j,k}(\xi_1)\widehat{\psi}_{j,k'}(\xi_2)$ (the four regions marked C), for one fixed value of j.

Fourier space; note that this is only a sketch of where they are mostly concentrated since typically these Fourier transforms are not compactly supported.

If we have the inner products $\int f(x,y)2^{j/2}\varphi(2^j x - k_1)2^{j/2}\varphi(2^j y - k_2)\,dx dy$, then to compute the other coefficients (or other inner products), we must apply our filters $\{h_n\}_{n\in\mathbb{Z}}$ and $\{g_n\}_{n\in\mathbb{Z}}$ to the k_1 and k_2 indices independently. Recall that for one dimension, we convolved our initial sequence $\{\langle f, \varphi_{j_1 k}\rangle\}_{k\in\mathbb{Z}}$ with the two filters $\{h_n\}$ and $\{g_n\}$ to determine two new sequences $\{\langle f, \psi_{j-1,k}\rangle\}$ and $\{\langle f, \varphi_{j-1,k}\rangle\}$. We must do the same in two dimensions, only with an additional variable. This gives us four new sequences of inner products, instead of two; three of the four are wavelet coefficients, while the fourth gives the coarser approximation from which we can compute the next level.

Both constructions of bases for $L^2(\mathbb{R}^2)$ yield useful results. We will see how to use both of them in the representation of operators. Assume that $f \in L^2(\mathbb{R})$ and that T is a bounded operator on L^2. Then, we can write

$$Tf = \sum_{\substack{j,k \\ j',k'}} \langle f, \psi_{j,k} \rangle \langle T\psi_{j,k}, \psi_{j',k'} \rangle \psi_{j',k'}\,,$$

where $\langle T\psi_{j,k}, \psi_{j',k'} \rangle$ is the "matrix" (possibly infinite) corresponding to the operator T. If T is given by an integral operator

$$(Tf)(x) = \int f(y) K(x,y)\,dy\,,$$

$$\text{then} \quad \langle T\psi_{j,k}, \psi_{j',k'} \rangle = \iint K(x,y)\psi_{j,k}(y)\psi_{j',k'}(x)\,dx\,dy\,.$$

In other words, the matrix corresponding to the operator T is simply the expansion of the integral kernel in a tensor product wavelet basis for $L^2(\mathbb{R}^2)$. We will call this the standard representation for T. Notice that this representation gives us the correlation (or interaction) between scales j and j'. For a large class of operators (pseudo-differential operators, for example), the interaction between different scales j, j' decreases as the distance $|j - j'|$ between the scales increases.

Another representation for T is the non-standard form. This representation corresponds to expanding the integral kernel K into the second wavelet basis for $L^2(\mathbb{R}^2)$. We begin any sort of numerical algorithm with a discretization, corresponding to some fine (but finite) resolution; this corresponds to the restriction of T to a fine scale V_J, so that we have $P_J T P_J = T_J$. We will also consider similar restrictions of T to other (coarser) scales, and define these likewise by $T_j = P_j T P_j$. Now, at any scale j we can write T_j as a sum of different pieces (representing the interaction of scales): recall that $P_j = P_{j-1} + Q_{j-1}$, hence

$$T_j = Q_{j-1} T Q_{j-1} + Q_{j-1} T P_{j-1} + P_{j-1} T Q_{j-1} + T_{j-1}\,.$$

Denote

$$\begin{cases} A_{j-1} = Q_{j-1} T Q_{j-1} & (A_{j-1} : W_{j-1} \to W_{j-1}) \\ B_{j-1} = Q_{j-1} T P_{j-1} & (B_{j-1} : V_{j-1} \to W_{j-1}) \\ \Gamma_{j-1} = P_{j-1} T Q_{j-1} & (\Gamma_{j-1} : W_{j-1} \to V_{j-1}) \end{cases}$$

For each j, the operators A_j, B_j, and Γ_j are all matrices with elements given by

$$A_j = \{\alpha^j_{k,\ell}\} = \{\langle T\psi_{j\ell}, \psi_{jk} \rangle\}$$
$$B_j = \{\beta^j_{k,\ell}\} = \{\langle T\varphi_{j\ell}, \psi_{jk} \rangle\}$$
$$\Gamma_j = \{\delta^j_{k,\ell}\} = \{\langle T\psi_{j\ell}, \varphi_{jk} \rangle\}\,,$$

where we have implicitly assumed (as we have elsewhere in these lectures) that φ, ψ are real valued. In particular, if $(Tf)(x) = \int f(y) K(x,y)\,dy$, then we must compute

$$\alpha^j_{k,\ell} = \iint K(x,y)\psi_{j,k}(x)\psi_{j,\ell}(y)\,dx\,dy\,,$$

$$\beta^j_{k,\ell} = \iint K(x,y)\psi_{j,k}(x)\varphi_{j,\ell}(y)\,dx\,dy\,, \quad \text{and}$$

$$\delta^j_{k,\ell} = \iint K(x,y)\varphi_{j,k}(x)\psi_{j,\ell}(y)\,dx\,dy\,.$$

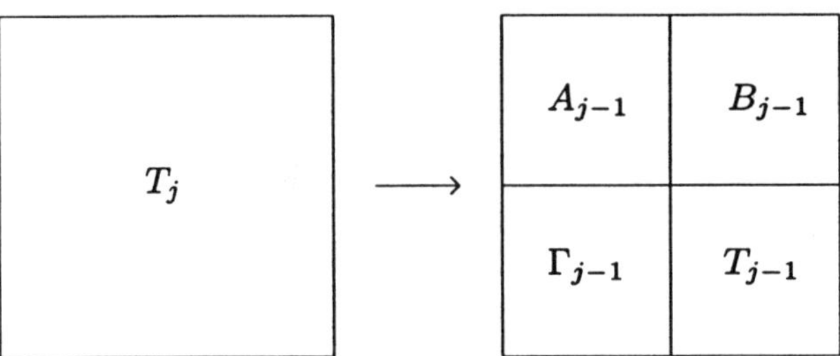

Figure 7.3. How we "organize" the different pieces of the non-standard form of the operator T_j after one decomposition step.

Since we have an MRA, we can iterate the above expansion to obtain an expression for T_j:

$$T_J = T_0 + \sum_{j=0}^{J-1} (A_j + B_j + \Gamma_j).$$

How do we "organize" or assemble both the standard and non-standard forms to apply them to functions? For the non-standard form, we begin with the operator T_j which is a large matrix. (See Figure 7.3.)

We calculate the submatrices A_{j-1}, B_{j-1}, Γ_{j-1}, and T_{j-1} by applying a filtering operation and organize the matrix as above. Note that the submatrices A_{j-1}, B_{j-1}, and Γ_{j-1} are of size $\frac{n}{2} \times \frac{n}{2}$ for T_j of size $n \times n$. As we said earlier, we can iterate this process, now decomposing T_{j-1}. We organize the matrix now as in Figure 7.4. We apply this matrix to the column vector in Figure 7.5 (which corresponds to the expansion of a function f).

This column vector representing our function f is a redundant representation of our function (we don't need the projection $P_{j-1}f$ to reconstruct f if we have $Q_{j-2}f$ and $P_{j-2}f$; another way to express this is to observe that the entries in this representation are not independent, but have to satisfy some linear relations.) Applying the matrix in Figure 7.4 to the column vector in Figure 7.5 corresponds exactly to our decomposition of T above: we have three different types of pieces,

$$\sum_l \alpha_{k,\ell}^j \langle f, \psi_{j,\ell} \rangle = Q_j T Q_j f$$

$$\sum_l \beta_{k,\ell}^j \langle f, \varphi_{j,\ell} \rangle = Q_j T P_j f$$

$$\sum_l \gamma_{k,\ell}^j \langle f, \psi_{j,\ell} \rangle = P_j T P_j f$$

and, at the coarsest level, a remaining T_0 piece. The big column vector we obtain afterwards is however no longer of the same type as in Figure 7.5: it consists of coefficients of different pieces in the V_j and W_j spaces, which, when added, yield the correct $P_J T P_J f$; to obtain the correct expansion for this vector, we have to first add all the pieces, and then expand that again.

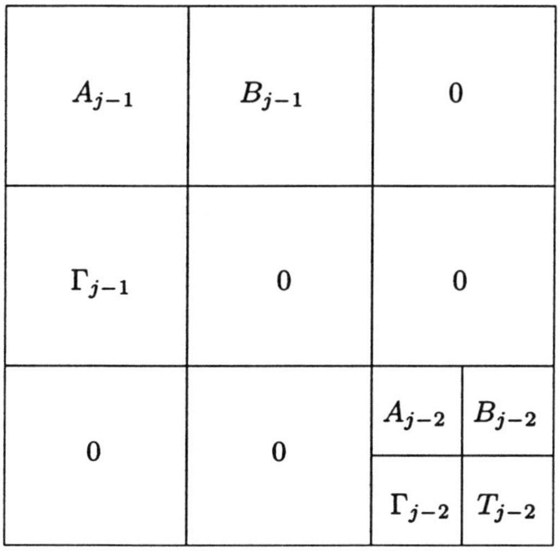

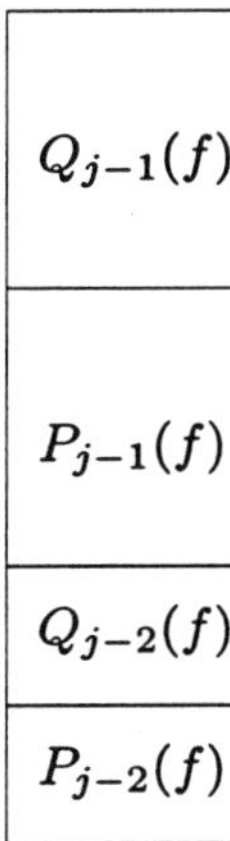

Figure 7.4. How we assemble the non-standard form of the operator T_j after two successive decomposition steps.

Figure 7.5. The redundant expansion of the function f, to be used with the nonstandard form of the operator as given in Figure 7.4

We do not have such extra work in the expansions of f and Tf if we use the standard form of our operator T—instead, the "work" is incorporated in the organization of the matrix for the standard form. We begin with the matrix A_{j-1}. Now, we must work with the operators $B_{j-1}^{j'} : W_{j'} \to W_{j-1}$ and $\Gamma_{j-1}^{j'} : W_{j-1} \to W_{j'}$ which tell us how the operator T correlates the scales $j-1$ and j' (see Figure 7.6). Here $j' = j - 1, \cdots, 0$ where V_0 is our coarsest scale. This matrix is applied to the

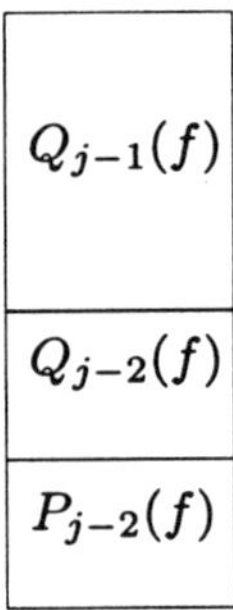

A_{j-1}	B_{j-1}^{j-2}	$\tilde{B}_{j-1}^{j-2}$
Γ_{j-1}^{j-2}	A_{j-2}	B_{j-2}
$\tilde{\Gamma}_{j-1}^{j-2}$	Γ_{j-2}	T_{j-3}

Figure 7.6. How we organize the standard form of the operator T

$Q_{j-1}(f)$
$Q_{j-2}(f)$
$P_{j-2}(f)$

Figure 7.7. The non-redundant expansion of the function f, to be used with the standard from of the operator as given in Figure 7.6

column vector for f (shown in Figure 7.7) which is not a redundant expansion for f.

For this case, the resultant vector is indeed the wavelet expansion for the new function Tf so we do not have any additional work to be done. On the other hand, to calculate the sub-matrices $B_{j-1}^{j'}$, $\Gamma_{j-1}^{j'}$, etc., we take the submatrices B_{j-1}, Γ_{j-1}, etc. which we calculated for the non-standard form, and apply our filters $\{h_n\}$ and $\{g_n\}$ to each row of the operators B_j and to each column of the operators Γ_j—that is, we must calculate the one-dimensional wavelet transform of each row of B_j and each column of Γ_j. For details, see [**Beyl**] [**BCR**].

To see why these forms are useful (in particular, why we have a fast algorithm for the application of an operator to a function), let us examine the example $T = \frac{d}{dx}$. We will assume that we are using compactly supported wavelets here. To compute the matrix elements $\alpha_{k,\ell}^j$, $\beta_{k,\ell}^j$, and $\delta_{k,\ell}^j$, it's sufficient to compute the inner products

$\langle \varphi', \varphi_{0,\ell} \rangle = r_\ell$ since

$$\alpha^j_{k,\ell} = 2^j \int \psi(2^j x - k)\psi'(2^j x - \ell)2^j \, dx = 2^j \alpha_{k-\ell},$$

$$\beta^j_{k,\ell} = 2^j \int \psi(2^j x - k)\varphi'(2^j x - \ell)2^j \, dx = 2^j \beta_{k-\ell},$$

$$\gamma^j_{k,\ell} = 2^j \int \varphi(2^j x - k)\psi'(2^j x - \ell)2^j \, dx = 2^j \gamma_{k-\ell},$$

$$r^j_{k,\ell} = 2^j \int \varphi(2^j x - k)\varphi'(2^j x - \ell)2^j \, dx = 2^j r_{k-\ell}, \quad \text{where}$$

$$\alpha_\ell = 2 \sum_k \sum_m g_k g_m r_{2\ell+k-m},$$

$$\beta_\ell = 2 \sum_k \sum_m g_k h_m r_{2\ell+k-m}, \quad \text{and}$$

$$\gamma_\ell = 2 \sum_k \sum_m h_k g_m r_{2\ell+k-m}.$$

We have repeatedly used the fact that $\psi_{j,k}$ and $\varphi_{j,k}$ can be written as a linear combination of $\varphi_{j+1,k}$ (i.e., the scaling function at a finer scale). Now,

$$r_\ell = \langle \varphi', \varphi_{0,\ell} \rangle = 4 \sum_n \sum_m h_n h_m \int \varphi'(2x - n)\varphi(2x - 2\ell - m) \, dx$$

$$= 2 \sum_n \sum_m h_n h_m \int \varphi'(x)\varphi(x - 2\ell - m + n) \, dx$$

$$\text{(by a change of variables)}$$

$$= 2 \sum_{n,m} h_n h_m r_{2\ell+m-n}.$$

Because φ and ψ are functions of compact support, only a finite number of r_ℓ are non-zero, and our system above is thus a finite linear system that we can solve with arbitrarily high accuracy. We should note also that since φ and ψ have compact support, a large number of the matrix entries $\alpha^j_{k,\ell}$, $\beta^j_{k,\ell}$, $\delta^j_{k,\ell}$, $r^j_{k,\ell}$ will be zero so that our matrices (both standard and non-standard forms) will be banded, sparse structures. These matrix entries depend on $|k - \ell|$, and if $|k - \ell|$ is large, we have a zero entry.

If we denote non-zero matrix entries with black shading and zero entries by white, the two forms of our derivative operator look like Figure 7.8.

If we turn to more general operators (integral operators with integral kernels possibly singular along the diagonal but smooth away from the singularities), we still get matrices which are sparse, leading to fast algorithms. Assume that T is a bounded operator from $L^2(\mathbb{R})$ to $L^2(\mathbb{R})$ with integral kernel $K(x,y)$ such that:

$$|K(x,y)| \leq \frac{c}{|x-y|} \quad \text{and} \quad \left| \partial_x^m \partial_y^n K(x,y) \right| \leq \frac{x}{|x-y|^{m+n+1}} \quad \text{for} \quad 0 \leq m+n \leq L.$$

If we assume our wavelets have L vanishing moments then we can determine for $|k - \ell| \geq 2R$ that $\beta^j_{k,\ell}$ is small (where R is the width of support of our wavelets).

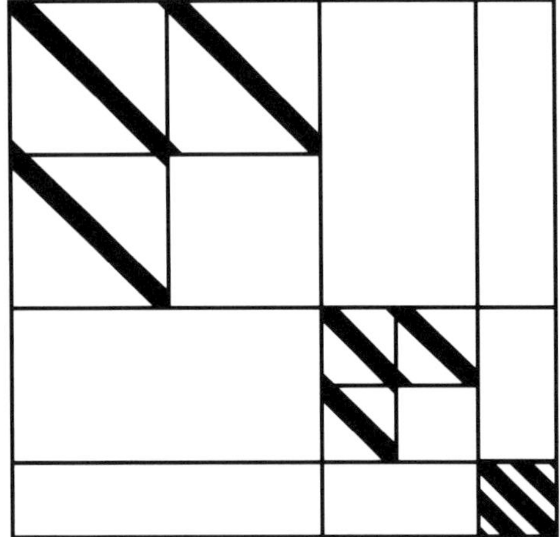

Figure 7.8. The two forms of the derivative operator: the non-standard form on the left, and the standard form on the right

We will briefly motivate this decay in the matrix entries off the diagonal:

$$\left|\beta_{k,\ell}^{j}\right| = \left|\iint K(x,y)2^{j}\varphi(2^{j}x - k)\psi(2^{j}y - \ell)\,dx\,dy\right| ;$$

expand the kernel K in a Taylor series centered about the point $2^{-j}\ell$ and use the vanishing moments in ψ to determine that

$$\left|\beta_{k,\ell}^{j}\right| \leq C \sup_{\substack{|2^{j}x-k|\leq R \\ |2^{j}y-\ell|\leq R}} \left|\partial_{y}^{L}K(x,y)\right| 2^{j} \int \left|y - 2^{-j}\ell\right|^{L} \left|\psi(2^{j}y - \ell)\right| \left|\varphi(2^{j}x - k\right| \,dx\,dy$$

$$\leq \frac{C}{|k - \ell|^{L+1}}\,.$$

One can check this for the other matrix entries as well. We can use these decay estimates to "threshold" the matrix entries in our operator—if we choose some ϵ and discard (or set to zero) all entries smaller that ϵ, then we will have a very sparse operator (in fact, we will have about n entries which are non-zero). One might wonder if we can throw away n^{2} entries below our threshold ϵ and still have an accurate representation of our operator. The answer is yes! An easy estimate (try it as an exercise) leads to

$$\|Tf - T^{\epsilon}f\|_{L^{2}} \leq CJR^{-L-1}$$

where $T^{\epsilon}f$ is the approximate operator corresponding to threshold ϵ, J is the total number of scales in our matrix, R is the width of support of our wavelets, and L is the number of vanishing moments. A smarter (but harder) estimate allows us to get rid of the J term in the error bound (provided we are a little bit more careful about the thresholding, so that certain averages are preserved); it turns out that the argument that permits this is in fact the proof celebrated $T(1)$ theorem of David and Journé [**DJ**]; details about this connection can be found in [**BCR**].

LECTURE 8
Wavelets and Differential Equations

In this final lecture, we will put together all of the pieces we have discussed to show how wavelets have been used to solve numerically Burger's equation (which is a parabolic equation). As a side remark, the algorithms developed last time for representing operators can be used to solve elliptic problems numerically. We will also discuss why, for real wave phenomena, wavelets are not so useful. We will construct several other wavelet-like bases (which are not as fully developed) that work better for such phenomena.

The form of Burger's equation with which we will work is

$$\partial_t u = \nu \partial_{xx} u + u \partial_x u$$

on the interval $[0, 1]$, with periodic boundary conditions and with the initial condition $u(x, 0) = \sin(2\pi x)$. Note that since we have periodic boundary conditions, we can use periodized wavelets on the interval. We know that the solution $u(x, t)$ is initially quite smooth and over time develops a sharp transition. From previous lectures, we know that for the initially smooth solution, we only need coarse wavelets to represent the solution because the wavelet coefficients corresponding to fine scales decay rapidly for smooth functions. However, over time we need finer and finer scales of wavelets around the transition point as it gets sharper. (See Figure 8.1). Here is a sketch of the algorithm:

For illustrative purposes, we will use a simple forward Euler scheme to calculate our solution $u(x, t_n) = u^n(x)$ where $t_n = n\Delta t$. (In practice, one would use a more sophisticated time scheme.) Thus,

$$u^{n+1} = u^n + \Delta t \left[\nu \left(u^n \right)'' + \frac{1}{2} \left((u^n)^2 \right)' \right].$$

If we define $d_{j,k}^n = \langle u^n, \psi_{j,k} \rangle$, then we have

$$d_{j,k}^{n+1} = d_{j,k}^n + \nu \Delta t \left(\langle u^n, \psi_{j,k}'' \rangle \right) - \frac{1}{2} \Delta t \left(\langle (u^n)^2, \psi_{j,k}' \rangle \right).$$

In other words, we have an equation for the wavelet coefficients of the solution at time t_{n+1} in terms of the coefficients corresponding to time t_n. At first, only a few coefficients $d_{j,k}^o$ are needed to describe the initial condition u^0 well. By allowing the solution some "room" to enter new wavelet localization space near the

215

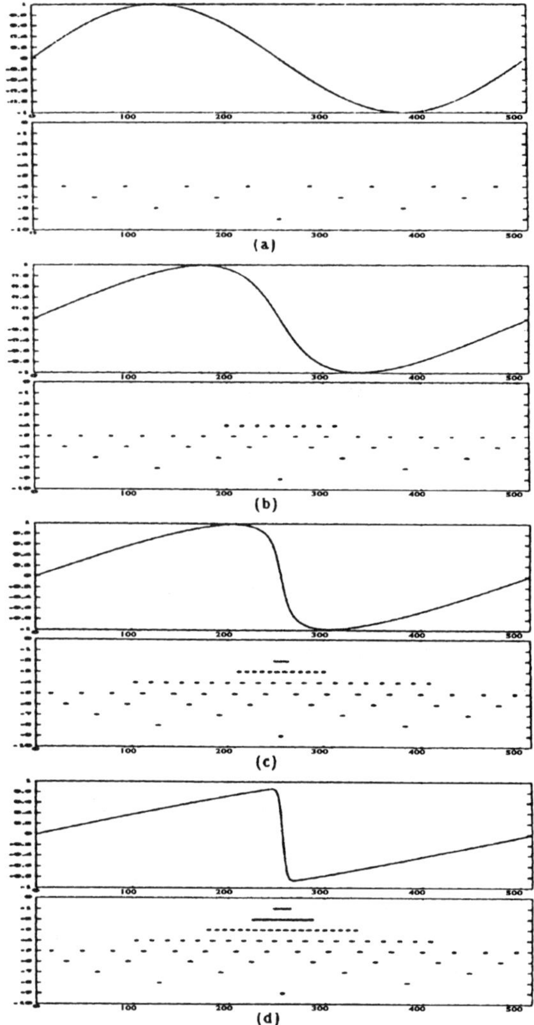

Figure 8.1. Four snapshots, at four consecutive times, of the solution to Burger's equation, and a schematic representation (underneath each frame) of the wavelet coefficients needed for an accurate description of the solution.

"occupied" spots at a previous level, we adaptively select to use wavelet coefficients that "follow" the shock formation. (See Figure 8.1.) There is one problem: we know from the previous lecture how to compute the term involving the second derivative since that is simply

$$\sum_{j',k'} d_{j',k'}^{n} \langle \psi_{j',k'}, \psi_{j,k}'' \rangle$$

and we know the matrix entries for this operator, but the non-linear term is more complicated. We must compute $\langle (u^n)^2, \psi_{j,k} \rangle$ at each time step n, or more generally,

$\langle f^2, \psi_{j,k} \rangle$. There are several ways to do this:

1. Consider integrals of the form

$$\int \psi_{j_1,k_1}(x)\psi_{j_2,k_2}(x)\psi_{j,k}(x)\,dx$$

We can reduce the number of integrals to compute by reducing these integrals to integrals of scaling functions at some finest scale. We then have a finite linear system which we can solve, since our functions have compact support. (This is analogous to how we determined the matrix for the derivative in the preceding lecture.)

Let's look in frequency space at what happens to the areas of concentration of $\widehat{\psi}_{j_1,k_1}$ and $\widehat{\varphi}_{j_2,k_2}$ when we multiply them. If $\widehat{\psi}_{j,k}(\xi)$ is concentrated on the two intervals, then the product $\psi_{j_1,k_1}(x)\psi_{j_2,k_2}(x)$ corresponds to the convolution $\widehat{\psi}_{j_1,k_1}(\xi) * \widehat{\psi}_{j_2,k_2}(\xi)$. The convolution in ξ-space $\widehat{\psi} * \widehat{\psi}$ will now be concentrated on intervals wider than the original intervals. If $j_1 = j_2$, then the positive and negative frequency halves of this concentration regions reach all the way to 0, and $\widehat{\psi}_{j,k_1} * \widehat{\psi}_{j,k_2}$ covers a wide range of frequencies, from 0 up. Now, we take the inner product of $\psi_{j_1,k_1'}(x) \cdot \psi_{j_2',k_2}(x)$ with $\psi_{j,k}(x)$ (since we had triple integrals):

$$\int \left(\widehat{\psi}_{j_1,k_1} * \widehat{\psi}_{j_2,k_2'} \right)(\xi)\; \widehat{\psi}_{j,k}(\xi)\,d\xi$$

so we have to have a detailed picture of the Fourier transform of our wavelets to know which matrix entries are large and which are not (i.e., those which must be computed and those which can be discarded). It should be clear that we will get a mixing and spreading of scales. For more details on this material, see [**DM**].

2. A second method is to try to decouple the scales in a problem by using a "telescoping" series. We have a fine scale approximation for f given by $P_J f$. We write

$$(P_J f)^2 = \sum_{j=1}^{J} \left[(P_j f)^2 - (P_{j-1} f)^2 \right] + (P_0 f)^2$$

$$= \sum_{j=1}^{J} (P_j f - P_{j-1} f)(P_j f + P_{j-1} f) + (P_0 f)^2$$

$$= \sum_{j=1}^{J} (Q_{j-1} f)(2P_{j-1} f + Q_{j-1} f) + (P_0 f)^2 .$$

Notice that all of the products we have in this sum, namely $(Q_{j-1} f)(P_{j-1} f)$ and $(Q_{j-1} f)(Q_{j-1} f)$, are products of terms at the same scale. Using the same kind of argument as in the first method, the width of the support of

$(Q_j f)^\wedge * (Q_j f)^\wedge$ will be twice the width of the support of $Q_j f$. Thus,

$$(Q_j f)(Q_j f) \in V_{j+1}$$
$$(Q_j f)(P_j f) \in V_{j+1} \,.$$

These are theoretical estimates for the spreading of scales—in practice, there is a bit more spreading than this.

For the Haar basis, products turn out to be especially simple, and we don't see as much of this spreading. But even there, a product of wavelets can have a very different frequency concentration than the wavelet itself:

$$\begin{aligned}
[\psi_{j,k}(x)]^2 &= 2^{j/2}\psi(2^j x - k) \cdot 2^{j/2}\psi(2^j x - k) \\
&= 2^j \varphi(2^j x - k) \\
&= 2^{j/2}\varphi_{j,k}(x) \,, \\
[\varphi_{j,k}(x)]^2 &= 2^{j/2}\varphi(2^j x - k)2^{j/2}\varphi(2^j x - k) \\
&= 2^{j/2}\varphi_{j,k}(x) \,, \quad \text{and} \\
[\psi_{j,k}(x)\varphi_{j,k}(x)] &= 2^{j/2}\psi(2^j x - k)2^{j/2}\varphi(2^j x - k) \\
&= 2^{j/2}\psi_{j,k}(x) \,.
\end{aligned}$$

The products $(Q_j f)(Q_j f)$ and $(Q_j f)(P_j f)$ in the summation above still need to be expanded into wavelet coefficients. To do this, we consider, for each scale j, the bilinear mappings

$$\begin{aligned}
M_{vw}^j &: V_j \times W_j \to L^2(\mathbb{R}) \\
M_{vv}^n &: V_n \times V_n \to L^2(\mathbb{R}) \\
M_{ww}^j &: W_j \times W_j \to L^2(\mathbb{R})
\end{aligned}$$

which map a product of two functions into its expansion into the wavelet basis. These mappings are tabulated by computing coefficients of the form:

$$M_{vww}^{j,j'}(k, k', \ell) = \int \varphi_{j,k}(x)\psi_{j,k'}(x)\psi_{j',\ell}(x) \, dx$$

$$M_{www}^{j,j'}(k, k', \ell) = \int \psi_{j,k}(x)\psi_{j,k'}(x)\psi_{j',\ell}(x) \, dx$$

$\cdots$ etc.

We can reduce the number of integrals to be computed by rescaling so that we have integrals of ψ, φ, etc. at the finest scale—for example,

$$M_{www}^{j,j'}(k, k', \ell) = 2^{-j'/2} \int \psi_{j-j',0}(x)\psi_{j-j',k-k'}(x)\psi_{0,2^{j-j'}k-\ell}(x) \, dx$$

so that

$$M_{www}^{j,j'}(k, k', \ell) = 2^{-j'/2} M_{www}^{j,j'}(k - k', 2^{j-j'}k - \ell) \,.$$

As in the discussion of the representation of $T = \frac{d}{dx}$, we get a finite number of non-zero coefficients (assuming compactly supported wavelets) and a linear system to solve for these coefficients. However, this process is kind of messy!

A more detailed discussion of this is in G. Beylkin's INRIA Notes (available by anonymous ftp from University of Colorado, Boulder).

3. A naive method is a pointwise multiplication scheme which, although naive, can be quite simple and straightforward. We have the values $\{\langle f, \psi_{j,k}\rangle\}$ and $\{\langle f, \varphi_{0,k}\rangle\}$ so we can apply our fast algorithm to reconstruct f approximated at a fine scale; i.e., $\{\langle f, \varphi_{J,k}\rangle\}$. From these inner products, we can compute quickly the sample values $f(2^{-J}k)$. Now just square these sample values $\left[f\left(2^{-J}k\right)\right]^2$ and apply all the algorithms in reverse to determine:

$$\{f^2(2^{-J}k)\} \longrightarrow \{\langle f^2, \varphi_{J,k}\rangle\} \longrightarrow \{\langle f^2, \psi_{j,k}\rangle\}$$
$$\{\langle f^2, \varphi_{0,k}\rangle\} \ .$$

There are several comments and/or warnings for this method: First, we hope that the approximation of f^2 can be computed exactly at the scale J where we approximated f. Our function f might be such that while $P_J f$ is a good approximation, $P_J f^2$ might not be such a good approximation. We must use some a priori knowledge about f to decide that our error in $P_J f^2$ is an acceptable error. Second, to compute $\{\langle f, \varphi_{J,k}\rangle\}$ from the sample values $\{f(2^{-J}k)\}$ (and vice versa), we must use smooth wavelets otherwise we will generate unacceptable errors in this procedure. That is, we need $\widehat{\varphi}$ to have reasonable good decay.

4. The final method is an adaptive pointwise multiplication scheme. In the above pointwise scheme, we calculate all the inner products $\{\langle f, \varphi_{J,k}\rangle\}$ at the finest scale but if regions of our function f are quite smooth (as in the solution of Burger's equation at some time t), then f^2 in those regions will also be quite smooth and we will only need coarse scale wavelets to represent f^2—calculating the coefficients for f^2 in those regions at a very fine scale would be overkill. The idea here is to calculate the inner products $\{\langle f, \varphi_{j,k}\rangle\}$ and corresponding sample values $\{f(2^{-j}k)\}$ at very fine scales only where necessary. To do this, we need scaling functions with good interpolating properties (e.g., spline wavelets). Several research groups have worked out schemes to do this.

Returning to our discussion of Burger's equation, we know how to compute the solution at each time step now. We can do this computation non-adaptively and adaptively. A non-adaptive (and time consuming!) method would be to compute every wavelet coefficient in the expansion of the solution at every scale and every time step. An adaptive method (and faster) is to begin with a wavelet decomposition of the initial data $u(x,0)$, keeping only those coefficients above a threshold value ϵ. We put in a virtual layer of labels (or coefficients) (j,k) around the ensemble of labels (or coefficients) which correspond to non-zero coefficients. This virtual layer allows our algorithm to enlarge this ensemble of large coefficients at finer scales if any coefficients should grow in time—if a sharp transition develops in our solution over time, the virtual layer will adjust accordingly and we will have refined our mesh accordingly. Note that the non-linear approximation theorem of T. Tao tells us that as our threshold value $\epsilon \to 0$, our solution converges pointwise almost everywhere to the true solution. A list of references covering material similar to this is:

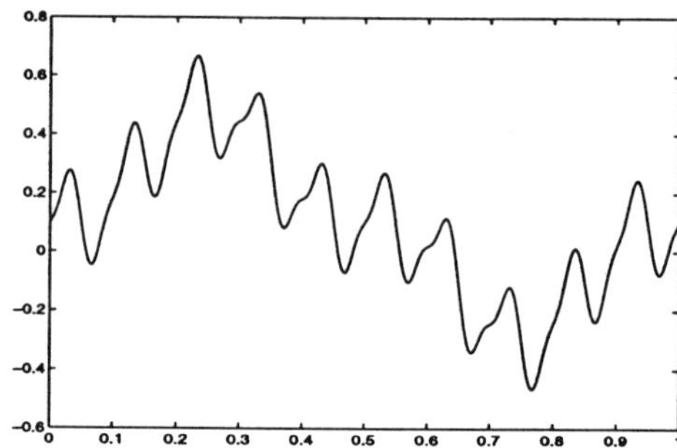

Figure 8.2. A function that can be represented much more efficiently by a Fourier decomposition than by a wavelet decomposition

1. C. Basdevant, M. Holschneider, et V. Perrier, C.R. Acad. Sci. Paris, Série I, 310, 647 (1990).
2. J. Liandrat, V. Perrier, et P. Tchamitchian, Comptes Rendus de la Conférence (CBS-NSF) Wavelets and Their Applications (Lowell, 1990), p. 14.
3. Y. Maday, V. Perrier, J.C. Ravel, C.R. Acad. Sci. Paris, Série I, 312, 405(1991).
4. E. Bacry, S.G. Mallat, and G. Papanicolaou, Mathematical Modelling and Numerical Analysis 26, 793 (1992).
5. S. Jaffard and P. Laurençot, Wavelets–A Tutorial in Theory and Applications, edited by C. Chui, (Academic Press—Boston, 1992), p. 543.
6. J. Keiser, PhD Thesis (Boulder 1995)–available by anonymous ftp, as well as papers by G. Beylkin and J. Keiser.
7. P. Ponenti, PhD Thesis (Marseille 1994), as well as papers by Liandrat, Ponenti and Tchamitchian.

Although wavelets have many practical applications, they are not well-suited for analyzing waves which have very high frequencies with long correlation length compared to their wavelength. To represent the function in Figure 8.2, we would have to take many fine scale wavelets but a Fourier transform could do much better (it would capture the same information with fewer coefficients—in this case it would need just eight coefficients).

There are several kinds of orthonormal bases which have been developed that are better adapted to oscillatory problems. The first kind is the wavelet packets of Coifman, Meyer, and Wickerhauser. These can be related to "standard" wavelets in the following way. Our algorithm for tabulating the wavelet coefficients can be schematically represented as in Figure 8.3. The arrows labeled by h, g denote the application of the filters $\{h_n\}$ and $\{g_n\}$. The decomposition of f or its fine-scale approximation $P_J f$ consists in applying successive filters, after which we retain $Q_{J-1}f, Q_{J-2}f, \cdots$, as well as some scale $P_{J-k}f$.

If our function f has some wave-like structure, then we don't really want a Littlewood-Paley type decomposition with its exponentially wide frequency block—we would want to divide some Littlewood-Paley "blocks" of frequencies to resolve

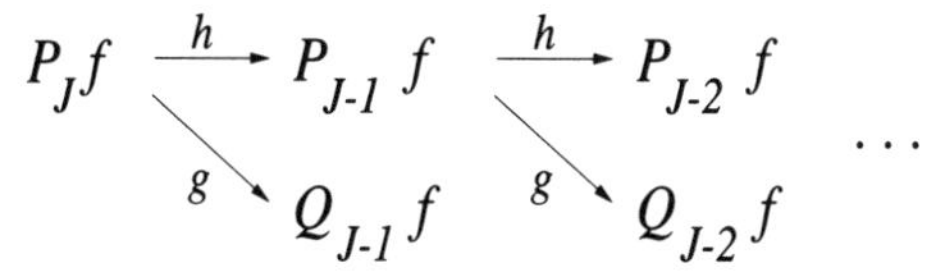

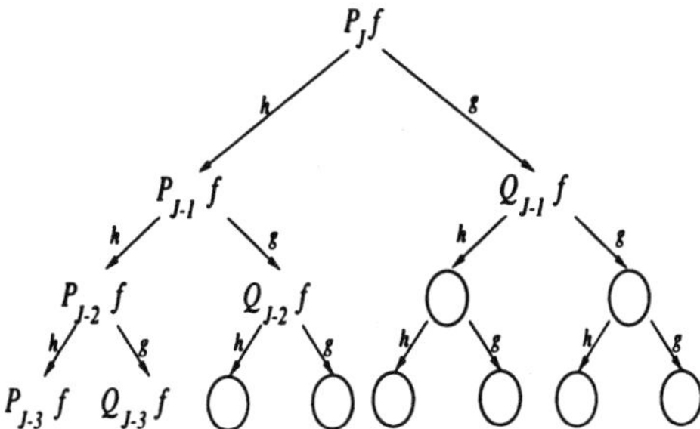

$$P_J f \xrightarrow{\ h\ } P_{J-1} f \xrightarrow{\ h\ } P_{J-2} f \quad \cdots$$

Figure 8.3. Schematic representation of the wavelet decomposition algorithm

Figure 8.4. The wavelet packet tree: every branch gets filtered and down-sampled

the frequency components more clearly, even if this means giving up spatial resolution. This corresponds to filtering the projection $Q_{J-1}f$ as well as the projection $P_{J-1}f$. If each branch gets filtered, then the algorithm resembles the tree in Figure 8.4.

Figure 8.4 shows the "full" tree, where every branch gets split again, until we reach a certain depth in the tree and we stop. This represents a certain decomposition of the full space, with projection operators defined by the filters. We can also look at a "pruned" tree, where we remove pairs of branches below the node from which they originate. For instance, the wavelet scheme in Figure 8.3 can be viewed as such a pruned tree. Figure 8.5 shows a few other pruned trees. Each of these trees corresponds to a different decomposition of the original space. At every one of the branchings in the tree in Figure 8.4, it may or may not be more advantageous to do the extra filtering round, depending on the function that we are decomposing: we can pick which nodes (i.e., which projections into different spaces) give the best decomposition of f using the smallest number of basis elements. This gives us the "best basis" concept. If we start with n coefficients at the original level, then we can go at most $\log n$ levels deep in our trees; at each level we compute $O(n)$ coefficients for the full tree. Therefore our total algorithm decomposes f into **many** possible representations with only $O(n \log n)$ computations. The best basis concept is very good for data compression because the adaptive pruning leads us to the basis elements that most accurately reconstruct our data set (using as few as possible basis functions). We should point out that all these bases are still orthonormal and that we have a fast algorithm for decomposing a function. (As an aside, one implementation of the best basis algorithm uses a notion of entropy from

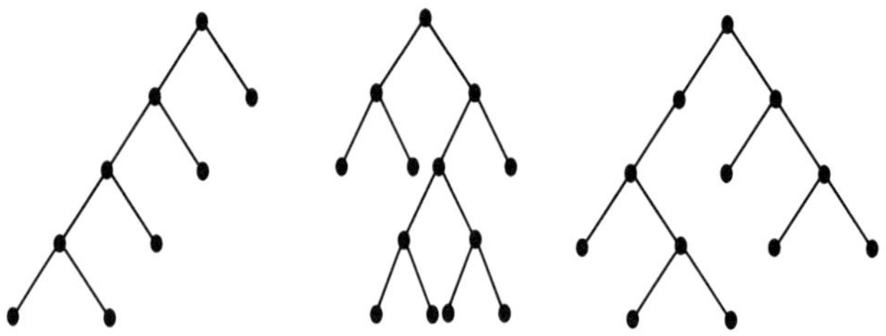

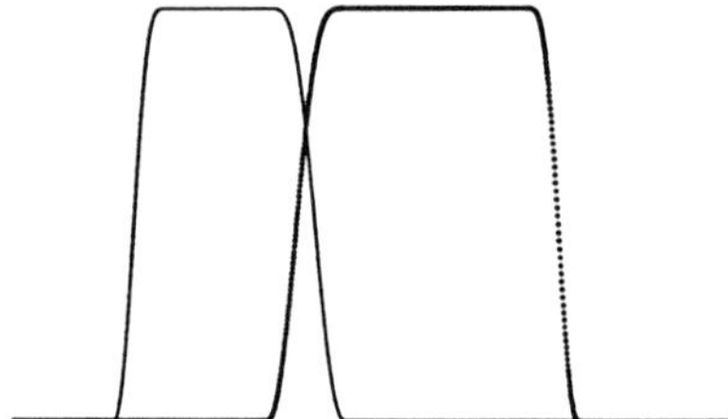

Figure 8.5. Left: A pruned tree that corresponds to a wavelet decomposition; Middle and Right: Two other pruned trees

Figure 8.6. An example of two adjacent window functions W_{n-1} and W_n.

the setting of information theory—another contribution from Shannon.) There are no theorems here about the convergence of non-linear approximations of this form and no one has yet used these ideas for solving PDEs numerically, so there are some open problems here!

A second basis that was developed is the local trigonometric basis of Coifman and Meyer. This construction is equivalent to that of Malvar (in the electrical engineering community). We begin with an arbitrary partition of the line and introduce windowing functions on each partition. Each window function $W_n(x)$ is a smooth cut-off function whose support is slightly larger than each interval; these W_n are constructed as follows.

Around each point a_n in our partition we have a transition region, a symmetric interval $(a_n - \epsilon_n, a_n + \epsilon_n)$. These transition regions shouldn't overlap, should be symmetric, and do not necessarily have the same widths. In between the transition regions, we want our window functions to be identically one, outside the transition regions zero. In the transition region $[a_n - \epsilon_n, a_n + \epsilon_n]$, W_n is given by $\sin\left[\frac{\pi}{2}\nu\left(\frac{x-a_n+\epsilon_n}{2\epsilon_n}\right)\right]$; here ν is a function similar to the one used in the construction of the Meyer wavelet basis (see Figure 4.1 in Lecture 4).

At the other edge of the interval, i.e. in $(a_{n+1} - \epsilon_{n+1}, a_{n+1} + \epsilon_{n+1})$, W_n is given by $W_n(x) = \cos\left[\frac{\pi}{2}\nu\left(\frac{x-a_{n+1}+\epsilon_{n+1}}{2\epsilon_{n+1}}\right)\right]$.

Now, we define

$$F_{n,k}(x) = \sqrt{\frac{2}{\ell_n}}W_n(x)\sin\left[\pi\left(k+\frac{1}{2}\right)\frac{(x-a_n)}{\ell_n}\right]$$

for $n, k \in \mathbb{Z}$, with $\ell_n = a_{n+1} - a_n$. We leave it to the reader to verify that this is an orthonormal basis; more details can be found in [**Wic**]. This basis has a re-grouping or merging property: if we remove a point in our partition, we only affect the basis on the two intervals which share this point, and we can replace the two bases on the intervals by the orthonormal basis on the larger interval formed by erasing the point, using the correspondingly wider window function bridging the two intervals. This gives us an adaptive method (similar to the best basis concept) for decomposing a function.

References

[Bat]   G. Battle, *A block spin construction of ondelettes, Part I: Lemarié functions*, Comm. Math. Phys. **110** (1978), 601–615.

[Beyl]   G. Beylkin, *On the representation of operators in bases of compactly supported wavelets*, SIAM J. Num. Anal. **29** (1992), 1716–1740.

[BCR]   G. Beylkin, R. Coifman and V. Rokhlin, *Fast Wavelet Transforms and Numerical Algorithms I*, Comm. Pure and Appl. Math. **44** (1991), 141–183.

[CDM]   Cavaretta, W. Dahmen and C. Micchelli, *Stationary subdivision*, Memoirs Am. Math. Soc **93** (1991), 1–186.

[Chui]   C.K. Chui, *An Introduction to wavelets*, Academic Press, Boston, 1992.

[CDV]   A. Cohen, I. Daubechies and P. Vial, *Wavelets on the interval and fast wavelet transforms*, Applied and Computational Harmonic Analysis **1** (1939), 54–81.

[DM]   W. Dahmen and C. Micchelli, *Using the refinement equation for evaluating integrals of wavelets*, SIAM Jour. Num. Anal. **30** (1993), 507.

[DTL]   I. Daubechies, *Ten lectures on wavelets*, SIAM, Philadelphia, PA, 1992.

[DJ]   G. David and J. L. Journé, *A boundedness criterion for generalized Calderòn-Zygmund operators*, Ann. Math. **120** (1984), 371–397.

[FJW]   M. Frazier, B. Jawerth and G. Weiss, *Littlewood-Paley theory and the study of function spaces*, – Regional Conference Series in Mathematics, vol 179, Amer. Math. Soc., Providence, RI, 1991.

[Jaf]   S. Jaffard, *Exposants de Hölder en des points donnés et coéfficients d'ondelettes*, C.R. Acad. Sci. Paris **308, Série 1** (1989), 79–81.

[Lem]   P.G. Lemarié, *Une nouvelle base d'ondelettes de $L^2(\mathbb{R}^n)$*, J. de Math. Pures et Appl. **67** (1988), 227–236.

[LMRA]   P.G. Lemarié-Rieusset, *Sur l'existence des analyses multirésolution en théorie des ondelettes*, Revista Mat. Iberoamericana **8** (1993), 457–474.

[Mall]   S. Mallat, *Multiresolution approximation and wavelet orthonormal bases of $L^2(\mathbb{R})$*, Trans. Amer. Math. Soc. **315** (1989), 69–88.

[Mey]   Y. Meyer, *Ondelettes et Opérateurs, I*, Hermann, Paris, France, 1990.

[Ste]   E. Stein, *Singular Integrals and Differentiability Properties of Functions*, Princeton University Press, Princeton, NJ, 1970.

[Str]   J.O. Stromberg, *A modified Franklin system and higher order spline systems on $\mathbb{R}^n$ as unconditional bases for Hardy spaces*, Conference in Harmonic Analysis in Honor of A. Zygmund, Wadsworth Math. Series II, Belmont, California, 1982, pp. 475–493.

[Wic]   M.V. Wickerhauser, *Adapted wavelet analysis from theory to software*, A.K. Peters, Wellesley, Massachusetts, 1994.

[Zyg]   A. Zygmund, *Trigonometric series (2nd edition)*, Cambridge University Press, Cambridge, United Kingdom, 1968.

Lectures on Stability and Instability of an Ideal Fluid

Susan Friedlander

IAS/Park City Mathematics Series
Volume 5, 1999

Lectures on Stability and Instability of an Ideal Fluid

Susan Friedlander

Introduction

These lectures will focus on the stability properties of *ideal* fluids, which permit the simplest fundamental description. The assumptions pertaining to an ideal fluid are largely met in a variety of real world phenomena, such as in the description of water or air flow under normal conditions and away from boundaries. An expansive mathematical structure has been built up to describe ideal fluid motion, drawing from the fields of partial differential equations, functional analysis, dynamical systems, statistical mechanics, and differential geometry. (See, for example, [88], [24], [77], [11], [23], and [26]). We aim to expose some of the beautiful mathematical structure that underlies fluid mechanics, while we hasten to note that much remains to be satisfactorily understood, particularly in the notorious problem of turbulence.

The issue of stability of fluid flows presents a wonderful example of a physical question which may be addressed through sophisticated, yet elegant, mathematical techniques. The answers have a direct physical interpretation: stable flows are robust under the inevitable disturbances in a laboratory or the environment, while unstable flows break up rapidly and will not be observed under generic conditions. Coherent structures which emerge from a turbulent flow, for example, must be stable under the types of perturbations which the rapidly varying flow induces. These lectures will start from the basics of ideal fluid motion, present some classical stability results, then turn to the modern mathematics which has been applied to the outstanding stability problems. In our attempts to understand the partial differential equations that describe the motion of an ideal fluid, we will appeal to contributions from functional analysis, spectral theory, differential geometry, and dynamical systems. Particular use will be made of asymptotic techniques and the method of geometrical optics.

The notes are organized as follows. First we discuss the two basic models for the mathematical study of fluid motion, namely the Euler equations and the Navier-Stokes equations. The emphasis in these lectures is placed on the Euler

---

[1]Department of Mathematics (M/C 249), University of Illinois-Chicago, 851 S. Morgan Street, Chicago, Illinois 60607.

E-mail address: `susan@math.nwu.edu`.

equations which describe the motion of an inviscid fluid. The effects of viscosity are incorporated in the Navier-Stokes equations. Since all fluids are at least very weakly viscous, it could be argued that the Navier-Stokes equations and not the Euler equations are physically relevant. However this is not the case. The limit of vanishing viscosity is well known to be a subtle and singular limit (see, for example, [25],[76], and [23]). The mathematical challenges raised by the Euler equations are different and sometimes more difficult than those connected with the Navier-Stokes equations. A brief discussion is given concerning the state of our understanding to date of existence and uniqueness theorems for the Euler equations. The fundamental differences between the complexity and difficulty of the problem in two and three dimensions are stressed. It is observed that the "type" of the Euler equations is nonstandard: they are degenerate, non-elliptic, and well-known existence and uniqueness theorems do not apply.

Next we introduce the important concept of the vorticity of a fluid motion, again emphasizing the fundamental differences between two and three dimensions and the critical phenomenon of vortex tube stretching that may generate vorticity in three dimensions. Classical vortex theorems are discussed together with their implications for the role of vorticity in the PDE theory of the Euler equations. The nature of steady flows, or equilibrium states, is considered in the context of two extreme cases of steady flows, namely integrable flows and chaotic flows.

We then turn to the main topic of these notes, the stability/instability of an equilibrium state. The concepts of linear stability and nonlinear (Lyapunov) stability are defined. Using various examples, we illustrate the central theme that stability/instability is crucially dependent on the metric in which the growth of a perturbation is measured and the spatial dimension of the fluid motion. The "Arnold" criterion based on the Hamiltonian nature of the Euler equations may be used to prove nonlinear stability but only for certain two-dimensional flows, and never for three-dimensional flows. We then consider in detail one of the fundamental "simplest" flows, namely plane parallel shear flow. We recall known results concerning the stability/instability of such flows and point out that there are open problems connected with the spectral theory even of this "simple" flow. For more general flows, very little indeed is known in detail about the structure of the spectrum of the linearized Euler operator. It is therefore particularly useful to have a sufficient condition for (linear) instability that circumvents this lack of knowledge. We discuss a recent theorem that provides a lower bound on the growth rate of the linearized Euler operator in terms of a geometric quantity that can be considered a Lyapunov type exponent for fluid flow. The positivity of this exponent is a sufficient condition for linear instability. It is an effective criterion that can be computed from a system of ODE's to demonstrate the instability of large classes of flows. We illustrate the power of the criterion with a number of examples ranging from "chaotic" ABC flows to generalized baroclinic instabilities in a geophysical model. Finally we describe a result which proves, under certain conditions, that linear instability of the Euler equations implies nonlinear instability.

These notes are based on lectures given at the Institute for Advanced Study, Princeton in May 1995 as a preparatory course for the 1995 IAS/Park City Summer School on Nonlinear Waves. Our discussion is intended to be introductory, and thus accessible to graduate students as well as researchers in pure and applied mathematics. The fluid dynamical discussion is self-contained, but some knowledge of differential equations and differential geometry is assumed. References to appropriate background texts are provided as needed. Some of the main techniques used in our lectures, namely high frequency asymptotics and geometric optics, are presented in more detail in several of the lectures of the summer school and may be found in other articles in this volume.

Acknowledgements

The author's research in the area of fluid stability is joint work with Misha Vishik. She much appreciates what she has found to be a wonderful collaboration and she very gratefully acknowledges his crucial contribution to research presented in these notes. She is also most grateful to Barbara Keyfitz, Peter Kramer, and Victor Yudovich for very useful and illuminating discussions concerning aspects of these lecture notes.

Support of the author's research by NSF grants DMS 91-23946, 95-00466 is gratefully acknowledged.

LECTURE 1
Equations of Motion

Let us proceed now to spell out the characteristics of and assumptions behind an ideal fluid. In general, a fluid is supposed to be a material which fills the volume of its container, and which cannot sustain a shear stress. A precise definition is hard to come by, since some materials (such as paint or glass) behave in some ways like fluids, and in other ways like solids. See the first section of Batchelor's book [**14**] for further discussion. It is safe, for our purposes, to think of a fluid as any normal gas or liquid one would encounter.

1.1. Ideal fluid model

The restriction to an *ideal* fluid amounts to the following mathematical model:

- The state of an ideal fluid is completely characterized by knowledge of its domain $D \in \mathbb{R}^n$ of motion and the value of its velocity vector $\mathbf{q}(\mathbf{x}, t) \in \mathbb{R}^n$ at every spatial location $\mathbf{x} \in D$ and time t. When considered as a global function, $\mathbf{q}(\mathbf{x}, t)$ is often referred to as the *velocity field*. In these lectures, we consider only the dimensions $n = 2$ and $n = 3$. We include the case of two-dimensional (2-D) flow both because it is interesting mathematically, and because under certain external constraints or forces, real-world fluids behave approximately two-dimensionally.
- The fluid is *incompressible,* meaning that the motion of the fluid is volume-preserving. To describe this more visually, let us define a *fluid element* or *fluid particle* to denote a parcel of fluid which retains its identity as it moves around. So we are not thinking of the fluid occupying a certain region of space as time evolves, but rather of a collection of fluid molecules which are transported by the flow. Incompressibility says that the volume of any fluid element is invariant in time.
- The density of the fluid is initially homogenous. Since the mass of a fluid element is by definition conserved in time, and incompressility implies that its volume is also conserved, we discover that the density of a fluid particle is equal to its initial density. Consequently, the density of the fluid is constant in both space and time. By an appropriate choice of mass units, we may assume that the density has a numerical value of unity, thereby removing it from the mathematical formulation as irrelevant.

233

- The fluid does not exert any internal shear stresses. Consequently, the force exerted by one fluid element (meaning some very small region of the fluid) on another is directed perpendicularly to their surface of contact. We assume the fluid is locally isotropic and homogenous, meaning that we are not in the exotic situation of, say, liquid crystals, where the molecules of the fluid tend to align with each other. Isotropy of an ideal fluid means that the force acting across any surface element passing through some point is parallel to the normal vector, and that the magnitude of this internal force is independent of the normal and tangential directions of this surface element. Let S be the bounding surface of a fluid element, $\mathbf{x}$ be a point on S, $\mathbf{n}$ be the inward normal at $\mathbf{x}$ to S. Then the internal force $\mathbf{f}$ exerted on the fluid element across an infinitesimal area δS of the surface S at $\mathbf{x}$ can be described in terms of a single *scalar* function $P(\mathbf{x}, t)$, the *pressure*, which is defined at all points $\mathbf{x}$ in the domain $\mathcal{D}$ and times t. The relation between $\mathbf{f}$ and $P(\mathbf{x}, t)$ is given by:

$$\mathbf{f} = P(\mathbf{x}, t)\mathbf{n}\delta S. \tag{1.1}$$

Note that the internal forces are local, and that we leave open the possibility of forces applied by agents external to the fluid such as gravity.

1.2. Euler equations

From these premises and basic conservation laws, Euler [**33**] derived in 1755 the equations of motion for an ideal fluid (see also [**2**]):

$$\frac{D\mathbf{q}(\mathbf{x}, t)}{Dt} = -\nabla P(\mathbf{x}, t) + \mathbf{F}(\mathbf{x}, t) \tag{1.2a}$$

$$\nabla \cdot \mathbf{q}(\mathbf{x}, t) = 0. \tag{1.2b}$$

The evolution of the velocity field $\mathbf{q}(\mathbf{x}, t)$ is determined by its current state, as well as by the internal pressure forces $-\nabla P(\mathbf{x}, t)$ and external forces $F(\mathbf{x}, t)$ acting at the same moment of time. We have introduced a notation for a commonly appearing combination of derivatives:

$$\frac{D}{Dt} = \frac{\partial}{\partial t} + (\mathbf{q}(\mathbf{x}, t) \cdot \nabla), \tag{1.3}$$

and this symbol is variously referred to as "total derivative," "material derivative," or "derivative following the fluid." Conceptually, it measures the rate of change of a fluid quantity (such as velocity) evaluated at a particular particle of fluid as it moves along. To see how the extra term arises, simply apply the chain rule to compute the time derivative of a fluid quantity $\phi(\mathbf{x}, t)$ observed at location $\mathbf{x} = \mathbf{X}(t)$ which is moving with fluid:

$$\begin{aligned}
\frac{d\phi(\mathbf{X}(t), t)}{dt} &= \frac{\partial \phi(\mathbf{X}(t), t)}{\partial t} + \frac{d\mathbf{X}(t)}{dt} \cdot \nabla \phi(\mathbf{X}(t), t) \\
&= \frac{\partial \phi(\mathbf{X}(t), t)}{\partial t} + \mathbf{q}(\mathbf{X}(t), t) \cdot \nabla \phi(\mathbf{X}(t), t) \tag{1.4} \\
&= \frac{D\phi}{Dt}\bigg|_{\mathbf{X}(t), t}
\end{aligned}$$

The first Euler equation (1.2a) is simply Newton's second law of motion written for a fluid element. The change in the velocity of the fluid element is the resultant of the forces acting upon it, which consists of the gradient of the internal pressure field $P(\mathbf{x}, t)$ and any externally applied forces $\mathbf{F}(\mathbf{x}, t)$. Since the acceleration in Newton's law refers to the change in velocity of a physical entity, the time derivative appearing must be the material derivative. The second Euler equation (1.2b) is precisely the statement of incompressibility, which may be deduced by considering the condition a fluid must satisfy so that infinitesimal fluid elements conserve their volume. We see that it is equivalent to the velocity field being divergence-free; such vector fields are often called *solenoidal.*

The observant reader might note that the Euler equations (1.2) do not at first glance appear to be closed. For the pressure $P(\mathbf{x}, t)$ is needed to determine the evolution of the velocity field, but we have not specified how the pressure field evolves. But in fact, the incompressibility condition (along with boundary conditions to be discussed later) implicitly determines the pressure (up to an additive irrelevant constant) in terms of the present configuration of the fluid velocity $\mathbf{q}(\mathbf{x}, t)$. More precisely, the mathematical role of the pressure is to keep the evolution of $\mathbf{q}(\mathbf{x}, t)$ within the space of divergence-free vectors satisfying the specified boundary conditions. See [**24**, pp.37, 38] for a discussion. The Euler equations are thus really a closed system.

1.3. Lagrangian trajectories and streamlines

We introduce in this paragraph a handier way to visualize a *steady* velocity field $\mathbf{q}(\mathbf{x}, t) = \mathbf{q}(\mathbf{x})$ than as a vector associated to every point of the fluid. (By the way, [**89**] is a wonderful book which contains many vivid pictures of fluid flow). We can condense the information contained in the velocity vector field by defining streamlines as the integral curves of the velocity field. The locus $\mathbf{X}(t)$ of a streamline thus satisfies the autonomous ordinary differential equation system:

$$\frac{d\mathbf{X}(t)}{dt} = \mathbf{q}(\mathbf{X}(t)).$$

Knowledge of all the streamlines of the fluid at a given point of time indicates at every spatial location the direction of fluid flow, since the velocity field is everywhere tangent to the streamlines. The magnitude of the velocity, however, cannot be deduced from the streamlines alone. Nevertheless, streamlines contain the important qualitative features of the velocity field, and are easily graphed.

For time-dependent flows, one needs to define streamlines as the integral curves of the *instantaneous* velocity field; see [**2**]. The solutions to the differential equation system:

$$\frac{d\mathbf{X}(t)}{dt} = \mathbf{q}(\mathbf{X}(t), t)$$

are called *Lagrangian trajectories,* and describe the location of an infinitesimal fluid element as it is carried along by the flow. For steady flows, these are identical to the streamlines, but this is not the case for non-steady flows.

1.4. Vorticity

A very important quantity associated with fluid motion is the *vorticity*, defined as the curl of the velocity field:

$$\omega(\mathbf{x}, t) = \nabla \times \mathbf{q}(\mathbf{x}, t).$$

Some of the roles which vorticity plays in fluid mechanics will be illuminated in
Section 4.

LECTURE 2
Initial-Boundary Value Problem

Let us now complete the mathematical framework for the Euler equations (1.2) by supplying the appropriate initial and boundary conditions. Since the Euler equations are first order in time, we specify our initial data as:

$$\mathbf{q}(\mathbf{x}, 0) = \mathbf{q}_0(\mathbf{x}), \tag{2.1}$$

and the initial pressure field $P(\mathbf{x}, t = 0)$ is then also determined, as previously discussed.

We still need to supply spatial boundary conditions, and we do this in different ways according to the type of our domain $D \in \mathbb{R}^n$:

1. Suppose the fluid is confined to a bounded region D with rigid walls. How should the velocity be specified at the boundary? Since the fluid may not flow outside of the region D, its velocity on the boundary must be tangent to the boundary. That is, it may have no component along the normal $\mathbf{n}$ direction to the boundary:

$$\mathbf{q} \cdot \mathbf{n}|_{\partial D} = 0. \tag{2.2a}$$

   In fact, this is the only boundary condition needed.

2. Sometimes it is useful to consider periodic flows, so that, for example, we can utilize Fourier series. Consider a flow in $\mathbb{R}^2$ which is 2π periodic in each of the two coordinate directions $\hat{\mathbf{e}}_1, \hat{\mathbf{e}}_2$,

$$\mathbf{q}(\mathbf{x}, t) = \mathbf{q}(\mathbf{x} + 2\pi \hat{\mathbf{e}}_j, t), \ j = 1, 2. \tag{2.2b}$$

   The fluid thus looks the same within any two domains which are translates of each other by a vector in the lattice $2\pi \mathbf{Z}^2$. One may thus equivalently picture the flow as occurring within a square with sides 2π long, but with opposite edges glued together. Such a domain looks like a donut, and is just a 2-D torus ($D = \mathbf{T}^2$). One may define periodic flow in three dimensions as flow on a torus $D = \mathbf{T}^3$ in the same way, even though the resulting object can't be visualized. No extra boundary conditions are necessary to specify the evolution of a periodic flow (in part because a torus doesn't have boundaries!).

3. Finally, we might wish to consider flow in free space with no walls or obstacles, $D = \mathbb{R}^n$. There are no boundaries, but it is necessary to restrict the

237

behavior of the fluid at infinity. An appropriate condition is that the fluid velocity decay rapidly enough to remain square-integrable:

$$\int_{\mathbb{R}^n} |\mathbf{q}|^2 d^n \mathbf{x} < \infty. \tag{2.2c}$$

2.1. Existence and uniqueness theorems

It is not obvious whether the initial-boundary value problem we have set up for the Euler equations is well-posed. The initial condition seems reasonable enough, but why, for example, should there be a single boundary condition on the velocity field, instead of, say, one for every component of the velocity field? The mathematics behind the well-posedness of the Euler equations is deep, and still not completely resolved. We shall very briefly report the history and current situation of the existence-uniqueness theory.

In two series of papers published in the 1920's, Lichtenstein [69],[70] and Gunther [47],[48],[49] obtained *local* existence and uniqueness results for *classical* (strong) solutions of the Euler equations valid in both two and three dimensions. A more recent discussion is given in [88]. By a classical solution, we mean that the velocity field has one continuous time derivative and Holder continuity in space:

$$\mathbf{q}(\mathbf{x}, t) \in C^{\lambda,1}(\mathbb{R}^n),$$

with $0 < \lambda < 1$. These results only proved existence and uniqueness *locally* in time, leaving open the possibility that a singularity might develop, and the solution fail to exist, after some finite time. All that is guaranteed is that a unique solution to the Euler equations (with initial and boundary conditions given) exists for some finite amount of time.

In 1933, *global* well-posedness for classical solutions of the 2-D Euler equations was proven by Wolibner [93]; he showed that the solutions never break down in finite time. Kato [59] presented this result in a modern form and extended it to include external forces.

These 2-D global existence results were extended by Yudovich [94] to generalized solutions which have their initial vorticity $\nabla \times \mathbf{u}_0 \in L^p$ for some $p > 1$. He could, however, only prove uniqueness for flows with bounded initial vorticity. Recently, Yudovich [96] extended the uniqueness result to some cases where the vorticity is unbounded but slowly growing. Detailed references are given in an excellent review by Gerard [45].

There is an important gap in the theory, which is an answer to the following question: Does an initially smooth velocity field (which is rapidly decaying at infinity if the flow is unbounded) continue to be smooth for all time as it evolves under the Euler equations (without external forcing)? In two dimensions an affirmative answer can be given. But in three dimensions, the question remains highly controversial. Some mathematicians and physicists believe that singularities will arise from an initially smooth velocity field within a finite amount of time. But no unambiguous examples to support this contention exist, and no theorems have been developed to rule out the possibility of such a singularity. This question is of great importance in the understanding of how fluids become turbulent.

2.2. Navier-Stokes equations

We briefly introduce here an important modification of the Euler equations necessary for many real fluids. Recall that in the definition of an ideal fluid, we prescribed that the only force acting between two neighboring fluid elements is a pressure force acting normally to their surface of contact. For any real fluid, there is an additional frictional force acting between the two fluid elements when they move at different velocities. Said another way, the fluid elements can transfer momentum to each other through molecular diffusion processes. If the incompressibility and constant density assumption is retained, one can show (see for example [**2**]) that the resulting equations of motion read:

$$\frac{D\mathbf{q}(\mathbf{x},t)}{Dt} = -\nabla P(\mathbf{x},t) + \nu \nabla^2 \mathbf{q}(\mathbf{x},t) + \mathbf{F}(\mathbf{x},t) \tag{2.3a}$$

$$\nabla \cdot \mathbf{q}(\mathbf{x},t) = 0. \tag{2.3b}$$

These equations are called the *Navier-Stokes equations,* and differ from the Euler equations by the addition of the term $\nu \nabla^2 \mathbf{q}(\mathbf{x},t)$ which incorporates the internal frictional force. In accordance with how a frictional force should behave, this term is large when the velocity has sharp gradients. The coefficient ν of this term is called the *viscosity*, and is purely determined by the fluid material and thermodynamic conditions. For fluids under many conditions, the viscosity ν may be considered "small" in some unambiguous way, argued by non-dimensionalization of the equations (see [**24**, p.35]). It is then natural to wonder whether the frictional term may be, to a good approximation, dropped from the equations, thereby reducing the dynamics to the Euler equations.

This is a very tricky issue. Most importantly, the viscosity is multiplying the most highly differentiated term in the equation, so the $\nu \to 0$ (*inviscid*) limit is singular. One manifestation of this singularity is the change in boundary conditions; for the Navier-Stokes equations the velocity must exactly match the velocity $\mathbf{V}(\mathbf{x},t)$ of any wall it contacts:

$$\mathbf{q}|_{\partial D} = \mathbf{V}(\mathbf{x},t),$$

while the Euler equations need only satisfy the weaker condition (2.2a). (For rigid, unmoving walls, $\mathbf{V}(\mathbf{x},t) \equiv 0$.) So there is, in some sense, a radical difference between a small viscosity and zero viscosity, particularly in a fluid with boundaries. On the other hand, it is often useful and appropriate to analyze fluids with small viscosity via the Euler equations, for example, far away from boundaries. But one must always carry the caveat about the $\nu \to 0$ limit being singular, particularly for mathematical studies.

A recent article of Constantin [**25**] illustrates some of the subtlety of this inviscid limit. In the context of a finite viscosity fluid arranging itself into a turbulent configuration, he argues that the physically important problem to study is the question of whether the *Euler* equations exhibit a singularity in finite time.

LECTURE 3
The Type of the Euler Equations

We observed above that the Euler equations do not seem to be amenable to classical existence and uniqueness theorems which can be routinely proven for many hyperbolic or elliptic systems. Let us investigate the "type" of the Euler equations to see that they do not fall into a nice category. The type is determined by the behavior of the principal symbol, and we shall forthwith compute it according to the prescription in [**57**].

3.1. Linearized Euler equations

It will in fact be more convenient for us to compute with the Euler equations *linearized* about some steady state $\mathbf{u}(\mathbf{x})$:

$$\mathbf{u}(\mathbf{x})\cdot\nabla\mathbf{u}(\mathbf{x}) \;=\; -\nabla P_0(\mathbf{x}), \tag{3.1a}$$

$$\nabla\cdot\mathbf{u}(\mathbf{x}) \;=\; 0, \tag{3.1b}$$

where P_0 is some corresponding pressure, uniquely determined by $\mathbf{u}$ up to additive constant. The linearized Euler equations result from assuming that the velocity and pressure field can be written as the sum of this steady state and a perturbation:

$$\begin{aligned}
\mathbf{q}(\mathbf{x},t) &= \mathbf{u}(\mathbf{x}) + \varepsilon\mathbf{v}(\mathbf{x},t),\\
P(\mathbf{x},t) &= P_0(\mathbf{x}) + \varepsilon p(\mathbf{x},t),
\end{aligned} \tag{3.2}$$

where the ε is interpreted as a small parameter. Substitute this expression for the velocity field into the Euler equations (1.2). The leading order terms on each side are those independent of ε, and these are just (3.1). Thus, they may be cancelled out of the expansion, and the leading order nontrivial behavior is given by the terms multiplying a single power of ε. These give the linearized Euler equations:

$$\frac{\partial\mathbf{v}(\mathbf{x},t)}{\partial t} + \mathbf{u}(\mathbf{x})\cdot\nabla\mathbf{v}(\mathbf{x},t) + \mathbf{v}(\mathbf{x},t)\cdot\nabla\mathbf{u}(\mathbf{x},t) \;=\; -\nabla p(\mathbf{x},t), \tag{3.3a}$$

$$\nabla\cdot\mathbf{v}(\mathbf{x},t) \;=\; 0. \tag{3.3b}$$

The initial and boundary conditions are the same as for the full Euler equations.

3.2. Computation of principal symbol

So we turn now to determine the principal symbol for (3.3) The reader may check that the principal symbol for this linearized problem is very close to that of the

241

full nonlinear Euler equations; the discussion subsequent to the computation of the principal symbol and characteristic form is, however, much cleaner for the linearized equations. It will clarify the presentation to write out the Euler equations in component form. The vector components will be denoted:

$$\mathbf{u} = \begin{pmatrix} u_1 \\ u_2 \\ u_3 \end{pmatrix},$$

$$\mathbf{v} = \begin{pmatrix} v_1 \\ v_2 \\ v_3 \end{pmatrix},$$

$$\mathbf{x} = \begin{pmatrix} x_1 \\ x_2 \\ x_3 \end{pmatrix}.$$

Then:

$$\frac{\partial v_1}{\partial t} + \left(u_1 \frac{\partial}{\partial x_1} + u_2 \frac{\partial}{\partial x_2} + u_3 \frac{\partial}{\partial x_3} \right) v_1 + \left(v_1 \frac{\partial}{\partial x_1} + v_2 \frac{\partial}{\partial x_2} + v_3 \frac{\partial}{\partial x_3} \right) u_1 = -\frac{\partial p}{\partial x_1},$$

$$\frac{\partial v_2}{\partial t} + \left(u_1 \frac{\partial}{\partial x_1} + u_2 \frac{\partial}{\partial x_2} + u_3 \frac{\partial}{\partial x_3} \right) v_2 + \left(v_1 \frac{\partial}{\partial x_1} + v_2 \frac{\partial}{\partial x_2} + v_3 \frac{\partial}{\partial x_3} \right) u_2 = -\frac{\partial p}{\partial x_2},$$

$$\frac{\partial v_3}{\partial t} + \left(u_1 \frac{\partial}{\partial x_1} + u_2 \frac{\partial}{\partial x_2} + u_3 \frac{\partial}{\partial x_3} \right) v_3 + \left(v_1 \frac{\partial}{\partial x_1} + v_2 \frac{\partial}{\partial x_2} + v_3 \frac{\partial}{\partial x_3} \right) u_3 = -\frac{\partial p}{\partial x_3},$$

$$\frac{\partial v_1}{\partial x_1} + \frac{\partial v_2}{\partial x_2} + \frac{\partial v_3}{\partial x_3} = 0.$$

After defining the column vector:

$$\Psi = \begin{pmatrix} v_1 \\ v_2 \\ v_3 \\ p \end{pmatrix},$$

we can write this compactly as a quasi-linear system:

$$\left(A_0 \frac{\partial}{\partial t} + \sum_{j=1}^{3} A_j(\mathbf{x}, t) \frac{\partial}{\partial x_j} \right) \Psi(\mathbf{x}, t) + B(\mathbf{x}, t) \Psi(\mathbf{x}, t) = 0, \qquad (3.4)$$

with the matrices:

$$A_0 = \begin{pmatrix} 1 & 0 & 0 & 0 \\ 0 & 1 & 0 & 0 \\ 0 & 0 & 1 & 0 \\ 0 & 0 & 0 & 0 \end{pmatrix},$$

$$A_1 = \begin{pmatrix} u_1 & 0 & 0 & 1 \\ 0 & u_1 & 0 & 0 \\ 0 & 0 & u_1 & 0 \\ 1 & 0 & 0 & 0 \end{pmatrix},$$

$$A_2 = \begin{pmatrix} u_2 & 0 & 0 & 0 \\ 0 & u_2 & 0 & 1 \\ 0 & 0 & u_2 & 0 \\ 0 & 1 & 0 & 0 \end{pmatrix},$$

$$A_3 = \begin{pmatrix} u_3 & 0 & 0 & 0 \\ 0 & u_3 & 0 & 0 \\ 0 & 0 & u_3 & 1 \\ 0 & 0 & 1 & 0 \end{pmatrix},$$

$$B = \begin{pmatrix} \frac{\partial u_1}{\partial x_1} & \frac{\partial u_1}{\partial x_2} & \frac{\partial u_1}{\partial x_3} & 0 \\ \frac{\partial u_2}{\partial x_1} & \frac{\partial u_2}{\partial x_2} & \frac{\partial u_2}{\partial x_3} & 0 \\ \frac{\partial u_3}{\partial x_1} & \frac{\partial u_3}{\partial x_2} & \frac{\partial u_3}{\partial x_3} & 0 \\ 0 & 0 & 0 & 0 \end{pmatrix}.$$

Recall that the principal symbol of a system of PDE's is simply the matrix of polynomials which results from replacing each partial derivative of the differential operator by an independent variable, and keeping only the most highly differentiated terms. Thus, all the terms except for $B(\mathbf{x}, t)$ from (3.4) will enter the principal symbol. Making the substitution:

$$\begin{pmatrix} \frac{\partial}{\partial x_1} \\ \frac{\partial}{\partial x_2} \\ \frac{\partial}{\partial x_3} \\ \frac{\partial}{\partial t} \end{pmatrix} \longrightarrow \begin{pmatrix} \xi_1 \\ \xi_2 \\ \xi_3 \\ \tau \end{pmatrix},$$

in the differential operator:

$$L_{pr} = A_0 \frac{\partial}{\partial t} + \sum_{j=1}^{3} A_j \frac{\partial}{\partial x_j}$$

results in the principal symbol matrix:

$$\mathcal{L}_{pr} = \begin{pmatrix} \bar{\tau} & 0 & 0 & \xi_a \\ 0 & \bar{\tau} & 0 & \xi_b \\ 0 & 0 & \bar{\tau} & \xi_c \\ \xi_a & \xi_b & \xi_c & 0 \end{pmatrix},$$

where we have defined:

$$\bar{\tau} = \tau + \mathbf{u}\cdot\xi$$

The determinant is easily computed:

$$Q(\xi, \tau) = -\bar{\tau}^2 |\xi|^2;$$

this is called the characteristic form. The type of the PDE system is indicated by the locus of zeroes of $Q(\xi, \tau)$. We clearly see that the characteristic form for our system vanishes precisely (in (ξ, τ) space) on the hyperplane

$$\bar\tau = \tau + \mathbf{u}(\mathbf{x}, t)\cdot\xi = 0 \tag{3.5a}$$

and on the line:

$$\xi = \mathbf{0}. \tag{3.5b}$$

These are the characteristic hypersurfaces; the line is a degenerate characteristic hypersurface. Since there are nontrivial zeroes, the system is certainly not elliptic. Neither is it hyperbolic, since there exist no unit four-dimensional vectors $\mathbf{N}$, for which the quartic equation

$$Q(\lambda\mathbf{N} + \sigma) = 0,$$

has a full set of 4 real roots for all unit four-dimensional vectors σ not in the span of $\mathbf{N}$. We leave the details to the reader; one need only convince oneself that in four-dimensional Euclidean space, there is no line such that all its parallel displacements intersect both the characteristic line (3.5b) and the characteristic hypersurface (3.5a). (Hint: Think about the three-dimensional (3-D) analogue.) So the linearized Euler equations do not fall into a standard category. One easily checks that the characteristic form for the nonlinear Euler equations is:

$$(\tau + \mathbf{q}(\mathbf{x}, t)\cdot\xi)^2|\xi|^2 = 0, \tag{3.6}$$

so clearly we may make the same conclusions about the full Euler equations.

What's worse, the initial hyperplane $t = 0$ is a characteristic surface, since its normal

$$(\xi, \tau) = (\mathbf{0}, 1)$$

is a zero of the characteristic form. This is a manifestation of the previously mentioned fact that the pressure is in fact uniquely specified by the velocity field at that instant of time. Hence, the velocity and pressure data on the initial hyperplane cannot be specified independently, and that is why we find the initial hyperplane to be characteristic. This does not say the Euler equations are ill-posed; certainly the initial pressure and velocity field must be related in a definite way for any chance of well-posedness. But when such a relation is satisfied, we cannot say anything for or against existence and uniqueness without further study. The natural thing to do is to eliminate the pressure from the Euler equations by replacing it with its explicit expression in terms of the velocity field. Leray [**64**],[**65**], [**66**] did this; the resulting equation however is no longer a PDE; it contains a singular integral operator. We will not digress to study the Leray formulation here.

3.3. Wave motion supported by linearized Euler equations

The characteristic hypersurfaces of a linear PDE are typically associated with certain waves. We can get a bit more insight into the Euler equations by examining the modes of propagation corresponding to the characteristic hypersurfaces of its linearization. Derivations and a fuller discussion can be found in [**63**] and [**74**]; we only report the facts here. The characteristic hyperplane

$$\bar\tau + \mathbf{u}(\mathbf{x})\cdot\xi = 0, \tag{3.7}$$

is actually a double root of the characteristic form and corresponds to two *trans-verse* waves: shear waves and vorticity waves. These modes have the velocity vector directed perpendicularly to the direction of propagation. Shear waves are really 2-dimensional, they propagate in one direction and slosh back and forth in a perpendicular direction. Vorticity waves on the other hand have their velocity vector steadily rotate about the plane perpendicular to the direction of propagation. The reason that we have two transverse waves is that there are two ($= 3 - 1$) inde-pendent transverse directions in three dimensions; a 2-D fluid would only support one ($= 2 - 1$) transverse mode. The characteristic line $\xi = 0$ is not associated with a propagating wave for the Euler equations. It is the residual of the (nondegen-erate) characteristic hypersurface related to the acoustic (sound) waves belonging to the *compressible* Euler equations, which permit density fluctuations. When the incompressible limit is taken, this mode of propagation vanishes, leaving behind a degenerate characteristic hypersurface. One way to see why there are only two waves (shear and vorticity) in the incompressible Euler equations, instead of a full set of four wave modes, is to note that the incompressibility condition requires that all waves be transverse. For, if the perturbation velocity field has the wave form:

$$\mathbf{q}(\mathbf{x}, t) = \mathbf{f}(\xi \cdot \mathbf{x} - \tau t)$$

for some vector function $\mathbf{f}(s)$, then incompressibility amounts to

$$\xi \cdot \mathbf{f} = 0.$$

Thus, the velocity perturbation is always perpendicular to the direction ξ in which the wave is traveling. Therefore the two acoustic modes, which are longitudinal (velocity directed parallel to ξ), cannot appear in the incompressible Euler equa-tions.

So how shall we describe the incompressible Euler equations? Properly, they are classified as *degenerate, non-elliptic*; the degeneracy refers to the collapse of two coincident characteristic hypersurfaces into a line. On the other hand, the incompressible Euler equations can be viewed as a singular limit of the compressible Euler equations, which are hyperbolic. After the incompressible limit is taken the acoustic modes of propagation disappear, but shear and vorticity wave motion are still supported. So in some sense, we might consider the Euler equations as "morally hyperbolic."

Later in these lectures, we will return to study the linearized Euler equations for specific steady states $\mathbf{u}(\mathbf{x})$.

LECTURE 4

Vorticity

We shall now develop some classical mathematical themes pertaining to the vorticity. First, we can rewrite the momentum balance equation (1.2a) to incorporate the vorticity. We assume the external body force $\mathbf{F}(\mathbf{x}, t)$ is conservative:

$$\mathbf{F}(\mathbf{x}, t) = -\nabla\Phi(\mathbf{x}, t)$$

for some scalar potential $\Phi(\mathbf{x}, t)$. Then the vector identity:

$$\mathbf{q}\cdot\nabla\mathbf{q} = \frac{1}{2}\nabla|\mathbf{q}|^2 + (\nabla\times\mathbf{q}) \times \mathbf{q} \tag{4.1}$$

allows the momentum equation to be recast as:

$$\frac{\partial\mathbf{q}(\mathbf{x}, t)}{\partial t} + \omega(\mathbf{x}, t) \times \mathbf{q}(\mathbf{x}, t) = -\nabla H(\mathbf{x}, t) \tag{4.2}$$

where the Bernoulli (Head) function is defined:

$$H(\mathbf{x}, t) = P(\mathbf{x}, t) + \frac{1}{2}|\mathbf{q}|^2 + \Phi(\mathbf{x}, t).$$

We can obtain an equation of motion for the vorticity by taking the curl of (4.2):

$$\frac{\partial\omega(\mathbf{x}, t)}{\partial t} + \nabla\times(\omega(\mathbf{x}, t) \times \mathbf{q}(\mathbf{x}, t)) = 0$$

$$\Longrightarrow \frac{\partial\omega(\mathbf{x}, t)}{\partial t} + (\mathbf{q}(\mathbf{x}, t)\cdot\nabla)\omega(\mathbf{x}, t) - (\omega(\mathbf{x}, t)\cdot\nabla)\mathbf{q}(\mathbf{x}, t) = 0. \tag{4.3}$$

The second equation follows from the vector identity:

$$\nabla\times(\mathbf{A} \times \mathbf{B}) = \mathbf{A}(\nabla \cdot \mathbf{B}) - \mathbf{B}(\nabla \cdot \mathbf{A}) + (\mathbf{B}\cdot\nabla)\mathbf{A} - (\mathbf{A}\cdot\nabla)\mathbf{B}, \tag{4.4}$$

and the fact that the vorticity and velocity fields are both divergence-free. Note that the bilinear differential operator

$$(\omega(\mathbf{x}, t)\cdot\nabla)\mathbf{q}(\mathbf{x}, t) - (\mathbf{q}(\mathbf{x}, t)\cdot\nabla)\omega(\mathbf{x}, t) \equiv \{\mathbf{q}, \omega\} \tag{4.5}$$

is a Poisson bracket, a special case of a Lie bracket (see [11, Secs. 39 and 40]). The connection of this Lie bracket operation with the Euler equations was exploited by Arnold (see for example [11, App. 2]) in the 1960's when he developed a new geometric approach for ideal fluid dynamics.

247

4.1. Vorticity and stream function in two dimensions

In two dimensions, the introduction of the vorticity concept provides a very nice alternative formulation of the Euler equations. Following convention, we will *visualize* the flow as occurring within a subset of the (x_1, x_2) plane lying in three dimensions although abstractly the relevant domain is a subset of $\mathbb{R}^2$. The advantage to this imbedding in three dimensions is twofold:

1. It will be convenient to have at our disposal the unit vector $\hat{\mathbf{k}}$, which is the direction normal to the plane of the flow.
2. We can formally view the 2-D plane as a domain in three dimensions, and thus 2-D flow is just a special case of 3-D flow. Consequently, we can specialize results about the 3-D Euler equations to apply in two dimensions as well.

Let us now proceed with the development of the role of vorticity in two dimensions. First, note that the incompressibility condition (1.2b)

$$\nabla \cdot \mathbf{q}(\mathbf{x}, t) = 0$$

implies that $\mathbf{q}(\mathbf{x}, t)$ may be locally expressed in terms of the gradient of a scalar field $\psi(\mathbf{x}, t)$:

$$\mathbf{q}(\mathbf{x}, t) = \hat{\mathbf{k}} \times \nabla \psi(\mathbf{x}, t). \tag{4.6}$$

This representation can be shown either by:

- vector calculus: the system of two differential equations implied by (4.6) is locally solvable due to the incompressibility condition, or
- by the theory of differential forms: $q_2(\mathbf{x}, t)dx_1 - q_1(\mathbf{x}, t)dx_2$ is a closed form by incompressibility, and is therefore exact on any simply connected region [85], [11, Sec. 36].

If the domain D in which the fluid flows is simply connected, one can show that $\psi(\mathbf{x}, t)$ may be defined globally over all of D. For multiply-connected domains, such as a vortex ring, we can still define a $\psi(\mathbf{x}, t)$ so that (4.6) holds for all of $\mathbf{x} \in D$, as long as we permit $\psi(\mathbf{x}, t)$ to be a multi-valued function (like the inverse trigonometric functions).

The scalar field $\psi(\mathbf{x}, t)$ is called the stream function. To see why, note that (4.6) implies that the velocity field is always directed perpendicularly to the gradient of $\psi(\mathbf{x}, t)$. Consequently the velocity field is everywhere tangent to the contours, or level sets, of $\psi(\mathbf{x}, t)$. But this means that the contours are integral curves of the velocity field, which are precisely the streamlines.

The expression of vorticity in terms of the stream function may be obtained simply by taking the curl of (4.6), and using the vector identity (4.4):

$$\omega(\mathbf{x}, t) = \nabla^2 \psi(\mathbf{x}, t)\hat{\mathbf{k}}. \tag{4.7}$$

Note that the vorticity in two-dimensions is always directed along the axis $\hat{\mathbf{k}}$ orthogonal to the plane. So to specify the vorticity in two-dimensions, one need only specify one component. Indeed, vorticity behaves as a pseudoscalar in two-dimensions, the pseudo prefix reminding us that vorticity changes sign under parity inversion in the 2-D plane.

Now consider the equation for vorticity (4.3). In two dimensions, the third term vanishes, since $\omega(\mathbf{x}, t)$ is directed along $\hat{\mathbf{k}}$, but the velocity field can only have gradients directed along the plane orthogonal to $\hat{\mathbf{k}}$. So, using the notation for material derivative (1.3), the vorticity equation in two dimensions reads:

$$\frac{D\omega(\mathbf{x},t)}{Dt} = 0. \tag{4.8}$$

That means that the vorticity of a fluid particle is conserved as it moves around. Now this vorticity equation may be expressed solely in terms of the stream function via (4.7) and (4.6):

$$\begin{aligned}
\frac{D(\nabla^2\psi\,\hat{\mathbf{k}})}{Dt} &= \hat{\mathbf{k}}\frac{\partial\nabla^2\psi(\mathbf{x},t)}{\partial t} + \hat{\mathbf{k}}\left(\left(\hat{\mathbf{k}}\times\nabla\psi(\mathbf{x},t)\right)\cdot\nabla\right)\nabla^2\psi(\mathbf{x},t) \\
&= \hat{\mathbf{k}}\left(\frac{\partial\nabla^2\psi(\mathbf{x},t)}{\partial t} + \left(\nabla\psi(\mathbf{x},t)\times\nabla\nabla^2\psi(\mathbf{x},t)\right)\cdot\hat{\mathbf{k}}\right) \\
&= 0.
\end{aligned}$$

Pulling off the vector component, we have a formally closed PDE for the evolution of the stream function alone:

$$\frac{\partial\nabla^2\psi(\mathbf{x},t)}{\partial t} + \left(\nabla\psi(\mathbf{x},t)\times\nabla\nabla^2\psi(\mathbf{x},t)\right)\cdot\hat{\mathbf{k}} = 0. \tag{4.9}$$

The boundary condition (2.2a) induces the condition that $\psi(\mathbf{x},t)$ should be constant on each component of the boundary; it may however assume different constants on different boundaries. The initial data for the stream function is chosen to be compatible with the initial (incompressible) velocity field:

$$\hat{\mathbf{k}}\times\nabla\psi(\mathbf{x},0) = \mathbf{q}_0(\mathbf{x}). \tag{4.10}$$

Thus, we have derived a closed PDE for the stream function which is equivalent to the Euler equations.

Note that the pressure does not need to appear in these equations because, as we have said, its only role is to enforce the incompressibility of the velocity field. The stream function handles this constraint automatically, because the derived velocity field

$$\mathbf{q}(\mathbf{x},t) = \hat{\mathbf{k}}\times\nabla\psi(\mathbf{x},t)$$

is necessarily divergence-free.

Remark. Suppose the vorticity were initially zero:

$$\nabla^2\psi(\mathbf{x},0) = 0.$$

Then the conservation of vorticity along fluid particle trajectories clearly implies that the vorticity vanishes at all later times:

$$\nabla^2\psi(\mathbf{x},t) = 0.$$

So the initial-boundary value problem has been reduced to the problem of finding a harmonic function $\psi(\mathbf{x},t)$ which satisfies the appropriate boundary conditions. $\psi(\mathbf{x},t)$ is certainly constant along any walls, but in the case of an unbounded domain, there may be additional conditions imposed at infinity which describe the far-field flow configuration impressed upon the fluid.

4.2. Extra complications in three dimensions

Can this formulation of the Euler equations in terms of the stream function be cast in three-dimensions? The solenoidality of the velocity field $\mathbf{q}(\mathbf{x}, t)$ implies that it can be locally expressed as the curl of a *vector* potential:

$$\mathbf{q}(\mathbf{x}, t) = \nabla \times \mathbf{A}(\mathbf{x}, t).$$

One could go on to write down PDE's for $\mathbf{A}(\mathbf{x}, t)$ but there is no real simplification; the potential is a vector just like the velocity field is. In two dimensions, the simplification rested on the relation (4.6) between a vector $\mathbf{q}(\mathbf{x}, t)$ and a *scalar* field $\psi(\mathbf{x}, t)$.

The vorticity equation in three dimensions also suffers the presence of the third term in (4.3) which disappeared in two dimensions. Written in terms of the material derivative, the 3-D vorticity equation reads:

$$\frac{D\omega(\mathbf{x}, t)}{Dt} = (\omega(\mathbf{x}, t) \cdot \nabla)\mathbf{q}(\mathbf{x}, t). \tag{4.11}$$

So vorticity is not conserved along particle trajectories in three dimensions. Instead, the term on the right hand side is extremely important in both generating and amplifying vorticity; (see [**75**] for some physical insight into the action of this "vortex stretching" term.) We shall discuss 3-D vortex dynamics further in Subsection 4.4.

4.3. Vorticity theorems

Vorticity enjoys a number of attractive properties, which we shall now list in the form of a sequence of theorems.

Recall that vorticity was conserved along particle trajectories in two dimensions, but not in three dimensions. Nonetheless, there is a quantity intimately connected to the vorticity which is conserved in both two and three dimensions [**61**]:

Theorem 4.1 (Kelvin's Circulation Theorem). *Let $C(t)$ be a closed curve which always consists of the same fluid particles; that is the curve $C(t)$ is moved by the fluid. The circulation about $C(t)$ is defined as the line integral:*

$$\Gamma(t) = \oint_{C(t)} \mathbf{q}(\mathbf{x}, t) \cdot d\mathbf{s_x}, \tag{4.12}$$

where $d\mathbf{s_x}$ is the local differential tangent vector to the curve $C(t)$ at $\mathbf{x}$. Then

$$\frac{d\Gamma(t)}{dt} = 0; \tag{4.13}$$

the circulation is exactly conserved by the fluid flow.

Proof. Let a parametrization of the contour $C(t)$ be given as:

$$\{\mathbf{x} = \mathbf{X}(s, t) | 0 \leq s \leq 1; \mathbf{X}(0, t) = \mathbf{X}(1, t)\}.$$

We demand in this parametrization that $\mathbf{X}(s, t)$ corresponds to the same fluid particle when s is fixed and t varies; thus $\mathbf{X}(s, t)$ may be interpreted as the location at time t of the particle labeled s. Then using the change-of-variables theorem for integrals, we can write the circulation as an integral over a fixed interval:

$$\Gamma(t) = \int_0^1 ds\, \mathbf{q}(\mathbf{X}(s, t), t) \cdot \frac{\partial \mathbf{X}(s, t)}{\partial s}.$$

This permits us to compute the time derivative of $\Gamma(t)$ by simply differentiating under the integral; this is not possible when the domain is itself time-dependent (as in the original definition (4.12)).

$$\frac{d\Gamma(t)}{dt} = \int_0^1 ds \, \frac{d\mathbf{q}(\mathbf{X}(s,t),t)}{dt} \cdot \frac{\partial \mathbf{X}(s,t)}{\partial s} + \mathbf{q}(\mathbf{X}(s,t),t) \cdot \frac{\partial^2 \mathbf{X}(s,t)}{\partial s \partial t}.$$

Now, since for fixed s, $\mathbf{X}(s,t)$ is the location of a specific fluid particle, we have by (1.4):

$$\frac{d\mathbf{q}(\mathbf{X}(s,t),t)}{dt} = \left.\frac{D\mathbf{q}}{Dt}\right|_{\mathbf{X}(s,t)} = -\nabla P(\mathbf{X}(s,t)).$$

The last equality follows from the Euler equation (1.2a). Another consequence of the fact that $\mathbf{X}(s,t)$ describes the trajectory of a fluid particle is that its time derivative must be the local velocity of the fluid:

$$\frac{\partial \mathbf{X}(s,t)}{\partial t} = \mathbf{q}(\mathbf{X}(s,t),t).$$

Putting these last two identities into the integral, and using the chain rule:

$$\frac{d\Gamma(t)}{dt} = \int_0^1 ds \, \left(-\nabla P(\mathbf{X}(s,t),t)\right) \cdot \frac{\partial \mathbf{X}(s,t)}{\partial s} + \mathbf{q}(\mathbf{X}(s,t),t) \cdot \frac{\partial \mathbf{q}(\mathbf{X}(s,t),t)}{\partial s}$$

$$= \int_0^1 ds \, \frac{d}{ds} \left(-P(\mathbf{X}(s,t),t) + \frac{1}{2}|\mathbf{q}(\mathbf{X}(s,t),t)|^2\right)$$

$$= -P(\mathbf{X}(1,t),t) + \frac{1}{2}|\mathbf{q}(\mathbf{X}(1,t),t)|^2 - \left(-P(\mathbf{X}(0,t),t) + \frac{1}{2}|\mathbf{q}(\mathbf{X}(0,t),t)|^2\right).$$

Since $\mathbf{X}(1,t) = \mathbf{X}(0,t)$, the right hand side vanishes, verifying the theorem.

Note that this proof did not require the fluid region to be simply connected. It is also straightforward to extend this proof for the compressible Euler equations, so long as the flow is isentropic (see [**2**]).

The relation between the circulation and vorticity can be seen by applying Stokes' theorem to the definition (4.12):

$$\Gamma = \int_{C(t)} \mathbf{q}(\mathbf{x},t) \cdot d\mathbf{s_x} = \int_{S(t)} \nabla \times \mathbf{q}(\mathbf{x},t) \cdot \mathbf{n_x} \, dA = \int_{S(t)} \omega(\mathbf{x},t) \cdot \mathbf{n_x} \, dA \qquad (4.14)$$

where $S(t)$ is any surface which is bounded by the loop $C(t)$, and $\mathbf{n_x}$ is the local normal. This connection is used to prove the following theorem about the persistence of irrotational (zero vorticity) flow [**2**]:

Theorem 4.2 (Cauchy-Lagrange). *If a region of the fluid has zero vorticity at some time (say at $t = 0$), then the region occupied by the same fluid particles will continue to be irrotational at all subsequent times.*

Remark. Note that the spatial region originally occupied by the irrotational fluid may pick up vorticity at a later time; the theorem says that the *fluid particles* which were originally vorticity-free will continue to be so. Thus, the irrotational region will in general move around with the flow.

Proof. From the formula (4.14), it is clear that the circulation will be zero about any loop drawn within the originally ($t = 0$) irrotational fluid. By Kelvin's theorem, the circulation about all loops within this same fluid at later times must continue to be zero. (Any loop in the later fluid can be mapped backwards in time to a loop in the $t = 0$ fluid.) If there were any vorticity in this fluid region at later times, one could draw a very small loop bounding a surface $S(t)$ such that the vorticity along the surface all formed an acute angle with the local unit normal. Consequently, the circulation around the loop bounding this surface would be positive, violating Kelvin's theorem. We conclude that there can be no vorticity in this fluid region at any later time.

4.4. Helmholtz Vortex Theorems

We now develop another geometric consequence of Kelvin's circulation theorem. First we define some terminology which helps us visualize vorticity. Let a *vortex line* $\mathbf{W}(s,t)$ be defined as an integral curve of the vorticity field:

$$\frac{\partial \mathbf{W}(s,t)}{\partial s} = \omega(\mathbf{W}(s,t),t).$$

Here the time variable t plays a passive role, simply giving the time at which the vorticity configuration is viewed. Vortex lines have the same relation to the vorticity field as streamlines have to the velocity field. Now take any curve (open or closed) $C(t)$ in the fluid which is nowhere tangent to the vorticity field. Then the surface formed by the vortex lines passing through each point of $C(t)$ is called a *vortex surface*. Note that the vorticity field is tangent to any vortex surface at all points of contact. If the initial curve $C(t)$ is closed, then the resulting vortex surface will be cylindrical. The fluid enclosed by such a vortex surface is called a *vortex tube*. These vortex structures have the following remarkable properties (see [51],[2]):

Theorem 4.3 (Helmholtz). *The fluid particles which form a vortex line at some time continue to form a vortex line at all later times. Colloquially, the vortex lines are "frozen" into the fluid. It also follows that vortex surfaces evolve into vortex surfaces at later time.*

Theorem 4.4 (Helmholtz). *Let a vortex tube be given at $t = 0$; by the previous theorem the fluid particles forming this vortex tube continue to form a vortex tube $T(t)$ at all later times. The quantity:*

$$\Gamma(t, S(t)) = \int_{S(t)} \omega(\mathbf{x},t) \cdot \mathbf{n}\, dA(\mathbf{x})$$

is the same for any two-dimensional cross section $S(t)$ of the vortex tube. Also, Γ is independent of time.

Consequently, Γ is an intrinsic invariant of each vortex tube; it is called the *strength* of the vortex tube.

Proof of First Helmholtz Theorem. We will present a proof based on differential equations as suggested in [24, p.26]. For a geometric argument, see for example [2]. The main point is that the differential equation for the advection of vorticity:

$$\frac{D\omega(\mathbf{x},t)}{Dt} = (\omega(\mathbf{x},t) \cdot \nabla)\mathbf{q}(\mathbf{x},t) \tag{4.15}$$

is the same differential equation as for the mapping of a *tangent vector* $\mathbf{t}(\mathbf{x},t)$ by the fluid flow $\mathbf{q}(\mathbf{x},t)$. To see this, consider a curve drawn in the fluid at time $t = 0$; let it be parametrized as $\mathbf{x} = \gamma(s,0)$. Now let $\mathbf{X}(\mathbf{x}_0,t)$ denote the position at time

t of a fluid particle originally at position $\mathbf{x}_0$ at time 0. Then the fluid particles comprising the initial curve will be located along the curve:

$$\gamma(s,t) = \mathbf{X}(\gamma(s,0),t)$$

at time t. The tangent vector to this curve is (see [**11**, Sec. 18C]):

$$\mathbf{t}(s,t) = \frac{\partial \gamma(s,t)}{\partial s} = \frac{d\gamma(s,0)}{ds}\cdot\nabla_{\mathbf{x}_0}\mathbf{X}(\mathbf{x}_0,t)|_{\mathbf{x}_0=\gamma(s,0)} = \mathbf{t}(s,0)\cdot\nabla_{\mathbf{x}_0}\mathbf{X}(\mathbf{x}_0,t)|_{\mathbf{x}_0=\gamma(s,0)},$$

which is just the well-known law for the transformation of a vector under a mapping. Differentiating both sides with respect to time, and remembering that:

$$\frac{\partial \mathbf{X}(\mathbf{x}_0,t)}{\partial t} = \mathbf{q}(\mathbf{X}(\mathbf{x}_0,t),t),$$

we find:

$$\begin{aligned}
\frac{d\mathbf{t}(s,t)}{dt} &= \mathbf{t}(s,0)\cdot\nabla_{\mathbf{x}_0}\mathbf{q}\left(\mathbf{X}(\mathbf{x}_0,t),t\right)|_{\mathbf{x}_0=\gamma(s,0)} \\
&= \mathbf{t}(s,0)\cdot\nabla_{\mathbf{x}_0}\mathbf{X}(\mathbf{x}_0,t)|_{\mathbf{x}_0=\gamma(s,0)}\cdot\nabla\mathbf{q}(\mathbf{x},t)|_{\mathbf{x}=\mathbf{X}(\gamma(s,0))} \\
&= \mathbf{t}(s,t)\cdot\nabla\mathbf{q}(\gamma(s,t),t).
\end{aligned}$$

Since $\gamma(s,t)$ describes the location of a given fluid particle when s is fixed, the vorticity equation (4.15) implies:

$$\frac{d\omega(\gamma(s,t),t)}{dt} = \omega(\gamma(s,t),t)\cdot\nabla\mathbf{q}(\gamma(s,t),t).$$

We see therefore that the vorticity vector and tangent vector attached to a fluid particle obey the same linear first-order evolution equation. It is thus clear that if they are initially parallel, they will remain parallel for all times. So if the vorticity vector is tangent to a given curve at $t = 0$, the vorticity along that curve will always remain tangent to the curve described by the same fluid particles at a later time. This is just what Helmholtz's first theorem says.

Proof of Second Helmholtz Vortex Theorem. (see [**2**]) Let $S_1(t)$ and $S_2(t)$ be two cross sections of the vortex tube $T(t)$. Then there is some cylindrical region $R(t)$ bounded by $S_1(t)$, $S_2(t)$, and a subset $U(t)$ of the surface of the tube (Figure 1). Using the divergence theorem and the fact that the vorticity field is solenoidal:

$$\begin{aligned}
0 &= \int_{R(t)} dV\,\nabla\cdot\omega(\mathbf{x},t) \\
&= \int_{S_1(t)} \omega(\mathbf{x},t)\cdot\mathbf{n}\,dA - \int_{S_2(t)} \omega(\mathbf{x},t)\cdot\mathbf{n}\,dA + \int_{U(t)} \omega(\mathbf{x},t)\cdot\mathbf{n}\,dA \\
&= \Gamma(S_1(t),t) - \Gamma(S_2(t),t) + \int_{U(t)} \omega(\mathbf{x},t)\cdot\mathbf{n}\,dA.
\end{aligned}$$

Since $U(t)$ is part of a vortex surface, $\omega(\mathbf{x},t)\cdot\mathbf{n} = 0$ on $U(t)$, so we immediately see that $\Gamma(S(t),t)$ is independent of the cross section $S(t)$ chosen. To show that $\Gamma(S(t),t)$ is independent of time, let the cross section $S(t)$ be one which is carried with the fluid. As Γ is independent of the cross section, there is no loss of generality in choosing such a cross section. Then the intersection of the cross section $S(t)$ with the vortex surface bounding the tube, $C(t)$, is a closed curve and:

$$\Gamma(S(t),t) = \int_{C(t)} \mathbf{q}(\mathbf{x},t)\cdot d\mathbf{s}_{\mathbf{x}}$$

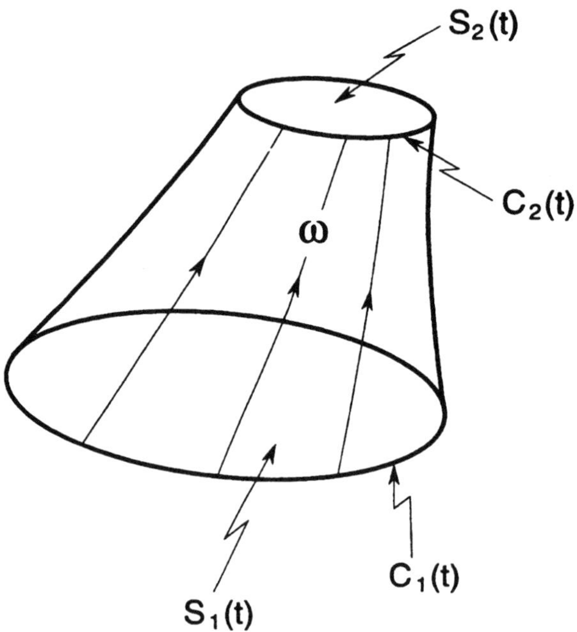

Figure 1. Vortex tube and cross sections used in proof of second Helmholtz vortex theorem

is the circulation around this curve. By Kelvin's circulation theorem, this quantity is independent of time, and the proof is complete.

Suppose now that a vortex tube is somehow stretched by the flow. That is, imagine that two given material cross sections of the vortex tube become more distant. Since the surface of the vortex tube is also given by a fixed set of fluid particles, incompressibility implies that the volume of fluid within the vortex tube and between the two given cross sections must remain constant in time. Therefore, stretching of the tube implies that its cross section must decrease in area. But the strength of a vortex tube must remain constant, and this is the area integral of the vorticity piercing a cross section. Since the area of this cross section is diminishing, the magnitude of the vorticity field must increase to keep the strength of the vortex tube constant. Hence, the stretching of a vortex amplifies its magnitude. This phenomenon is often found in nature. For example, when water flows down a bathtub drain, the vortex tubes are stretched by gravitational effects, leading to the strong vortices which can be seen. More dramatically, vortex tube stretching caused by updrafts of hot air in the atmosphere can create a whirlwind or tornado. The mathematical source for this vortex stretching is located in the term on the right hand side of (4.11). This term is absent in two dimensions, where there can be no stretching of vortex lines or tubes. This striking difference between 2-D and 3-D fluid flow manifests itself in the fundamental theory of the Euler equations, as we shall now discuss.

4.5. Role of vorticity in PDE theory of Euler equations

As Yudovich ([**94**]-[**97**]) observes, the questions of existence and uniqueness for the Euler equations are closely connected with constraints on the growth of the vorticity. An estimate of the vorticity in a uniform (L^∞) norm may be sufficient to convert local existence theorems to global ones. In fact, a theorem of Beale, Kato, & Majda [**15**] shows that a solution to the Euler equations exists in the Sobolev space $H^s, s > \frac{n}{2} + 1$ for all times t such that

$$\int_0^t d\tau \, \|\omega(\cdot, \tau)\|_{L^\infty}$$

is finite. In other words, no singularities can occur before this time integral becomes infinite. In two dimensions, the vorticity of a fluid particle does not change in time, so clearly the supremum of the vorticity magnitude at all later times is the same as that for the initial flow configuration. Thus if the initial vorticity is bounded, it will stay bounded and no singularities could occur. Consequently, we can successfully generate global solvability theorems in two dimensions. In three dimensions, however, the only constraints on the vorticity really come from the Helmholtz theorems. The issue of finite time blowup of the vorticity is equivalent to the question of whether an infinitesimal area of fluid surface which is initially positive can shrink to zero area within finite time. This is also equivalent to the possibility of infinite stretching of a vortex line. These difficult questions are so far unanswered, although the last 50 years have seen a number of attempts to find a collapse of vorticity in three dimensions. (See, for example, [**87**], [**78**], [**80**]).

LECTURE 5

Steady Flows

Flows for which the velocity field is independent of time $(\mathbf{q}(\mathbf{x},t) = \mathbf{q}(\mathbf{x}))$ are referred to as *steady states* or *equilibrium configurations* for the Euler equations. Let $\mathbf{u}(\mathbf{x})$ be a steady velocity field; then by (4.2), the PDE's which $\mathbf{u}(\mathbf{x})$ must satisfy read:

$$
\begin{aligned}
(\nabla \times \mathbf{u}(\mathbf{x})) \times \mathbf{u}(\mathbf{x}) &= -\nabla H(\mathbf{x}), & (5.1\text{a}) \\
\nabla \cdot \mathbf{u}(\mathbf{x}) &= 0, & (5.1\text{b})
\end{aligned}
$$

with the same boundary conditions as one would prescribe for a time-dependent flow. Here $H(\mathbf{x})$ is the head function, which is not specified ahead of time. The condition for $\mathbf{u}(\mathbf{x})$ to be a steady flow is simply that there exists *some* $H(\mathbf{x})$ such that the two equations (5.1) are satisfied.

5.1. Two dimensional case

Consider now the case of steady flow in two dimensions. Then the solenoidality of $\mathbf{u}(\mathbf{x})$ implies the existence of a stream function $\psi(\mathbf{x})$ such that:

$$\mathbf{u}(\mathbf{x}) = \hat{\mathbf{k}} \times \nabla \psi(\mathbf{x}),$$

from which it follows that (see (4.7)):

$$\nabla \times \mathbf{u}(\mathbf{x}) = \hat{\mathbf{k}} \nabla^2 \psi(\mathbf{x}).$$

The first equation of (5.1) may be written in terms of the stream function as:

$$\nabla^2 \psi(\mathbf{x}) \hat{\mathbf{k}} \times (\hat{\mathbf{k}} \times \nabla \psi) = -\nabla H(\mathbf{x}).$$

Expanding the vector product, we see that the system (5.1) of PDE's for a steady flow can be expressed in terms of the stream function in two dimensions as:

$$\nabla^2 \psi(\mathbf{x}) \nabla \psi(\mathbf{x}) = -\nabla H(\mathbf{x})$$

for some function $H(\mathbf{x})$. Recall that incompressibility is automatically guaranteed in the vorticity-stream formulation. Now, a vector field is (locally) the gradient of some scalar field precisely when it is curl-free. Thus, the condition for a stream function to represent a steady flow is just:

$$\nabla \times (\nabla^2 \psi(\mathbf{x}) \nabla \psi(\mathbf{x})) = 0. \tag{5.2}$$

257

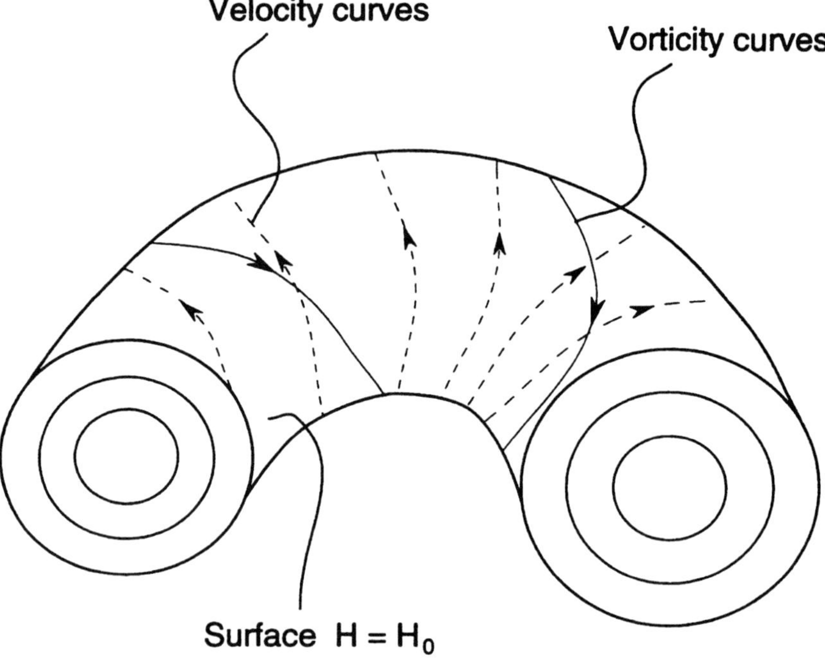

Figure 2. Vortex lines and streamlines winding around a noncritical level set $H(\mathbf{x}) = H_0$

Since

$$\nabla \times (\nabla^2 \psi(\mathbf{x}) \nabla \psi(\mathbf{x})) = \nabla(\nabla^2 \psi(\mathbf{x})) \times \nabla \psi(\mathbf{x}),$$

we see that the steady condition (5.2) is simply the condition that the gradients of the vorticity field $\nabla^2 \psi(\mathbf{x})$ and of the stream function $\psi(\mathbf{x})$ be parallel. This means they share the same level sets, and thus that either scalar field may be locally expressed as a function of the other:

$$\nabla^2 \psi = f(\psi), \text{ or } \psi = g(\nabla^2 \psi), \tag{5.3}$$

where f and g are arbitrary functions. Hence, the problem of determining the 2-D equilibrium states of an ideal fluid is, in general, the problem of solving a generically nonlinear elliptic PDE with appropriate boundary conditions.

5.2. Three dimensional case

Again, the situation is not so straightforward in three dimensions. We shall here consider two extreme cases for the steady equation (5.1). The first class will consist of steady flows for which $\nabla H(\mathbf{x})$ is nonvanishing, while the second class will correspond to head functions which are everywhere constant ($\nabla H(\mathbf{x}) \equiv 0$.)

5.2.1. Integrable flows

We consider a compact level surface $H(\mathbf{x}) = H_0$ which is noncritical, i. e. $\nabla H \neq 0$ on this surface. Arnold [10] showed that such a surface is necessarily diffeomorphic to the 2-D torus. From (5.1) it follows that the streamlines and the vortex lines are each perpendicular to $\nabla H(\mathbf{x})$; thus, these curves wrap around the surface of the torus $H(\mathbf{x}) = H_0$. [see Figure 2]. Note that by the assumption of noncriticality, the vortex lines and streamlines are nowhere parallel. Such a configuration is a model for a vortex ring. Flows $\frac{d\mathbf{X}}{dt} = \mathbf{u}(\mathbf{X}(t))$ for which almost all level sets are noncritical are referred to as integrable. Each streamline has an integral of motion $H(\mathbf{x})$, and the noncriticality condition $\nabla H(\mathbf{x}) \neq 0$ implies that the level sets of $H(\mathbf{x})$ foliate $\mathbb{R}^3$. Thus every streamline is constrained to lie on a smooth 2-D manifold, and it is well-known (by, for example, the Poincaré-Bendixson Theorem [**53**]) that there is no room for chaotic behavior in two dimensions. The topology of vortex surfaces in an integrable flow is trivial; they are nicely foliated.

5.2.2. Chaotic flows

In the case when $\nabla H(\mathbf{x})$ is identically zero, the condition for steady flow (5.1) implies that the vorticity $\nabla \times \mathbf{u}(\mathbf{x})$ is everywhere parallel to $\mathbf{u}(\mathbf{x})$:

$$\nabla \times \mathbf{u}(\mathbf{x}) = \gamma(\mathbf{x})\mathbf{u}(\mathbf{x}), \tag{5.4}$$

$$\nabla \cdot \mathbf{u}(\mathbf{x}) = 0, \tag{5.5}$$

for some scalar function $\gamma(\mathbf{x})$. Taking the divergence of the first equation, and using incompressibility yields the following condition:

$$\mathbf{u}(\mathbf{x}) \cdot \nabla \gamma(\mathbf{x}) = 0. \tag{5.6}$$

When $\gamma(\mathbf{x})$ is nonconstant, the streamlines (and the parallel vortex lines) will again wind around tori corresponding to level sets of $\gamma(\mathbf{x})$. When $\gamma(\mathbf{x}) \equiv \gamma$ is constant, however, there are no evident nontrivial integrals of motion (such as $H(\mathbf{x})$ or $\gamma(\mathbf{x})$ as previously discussed.) The first equation in (5.5) then states that $\mathbf{u}(\mathbf{x})$ is an eigenfunction of the curl operator. Such flows have very nontrivial topology, and are sometimes said to exhibit Lagrangian chaos (see below). One physical way to understand why chaotic flows should correspond to regions of $\nabla H(\mathbf{x}) = 0$ is to recall that $H(\mathbf{x})$ is really the thermodynamic enthalpy of the fluid. Chaotic trajectories would be expected to equilibrate the thermodynamics in the region of space which they "fill," and since they often occupy a manifold of dimension greater than 2, one would expect to see them only in regions where $H(\mathbf{x})$ was locally constant.

A specific example of a (periodic) flow $\mathbf{u}(\mathbf{x})$ which is the eigenfunction of the curl operator is given by the so called "ABC" flow after the mathematicians Arnold, Beltrami, and Childress. In component form,

$$\mathbf{u}(\mathbf{x}) = \begin{bmatrix} u_1(\mathbf{x}) \\ u_2(\mathbf{x}) \\ u_3(\mathbf{x}) \end{bmatrix} ;$$

it may be defined on the 3-D torus as:

$$
\begin{aligned}
u_1(\mathbf{x}) &= A \sin x_3 + C \cos x_2, \\
u_2(\mathbf{x}) &= B \sin x_1 + A \cos x_3, \\
u_3(\mathbf{x}) &= C \sin x_2 + B \cos x_1.
\end{aligned}
\tag{5.7}
$$

The reader may readily verify that (5.7) is an eigenfunction of the curl operator with eigenvalue $\gamma = 1$.

ABC flows have received considerable attention in the literature. For nonzero A,B,C, the system of ODE's given by (5.7) is non-integrable. Numerical evidence [52] indicates that the streamlines appear to densely fill some open domains of the 3-D torus — this phenomenon is referred to as Lagrangian chaos. Dombre et al. [28] showed that for certain parameter values (A, B, C), resonances occur which disrupt the KAM surfaces. Stagnation points (points where the velocity vanishes) may also appear, interconnected by a web of heteroclinic streamlines.

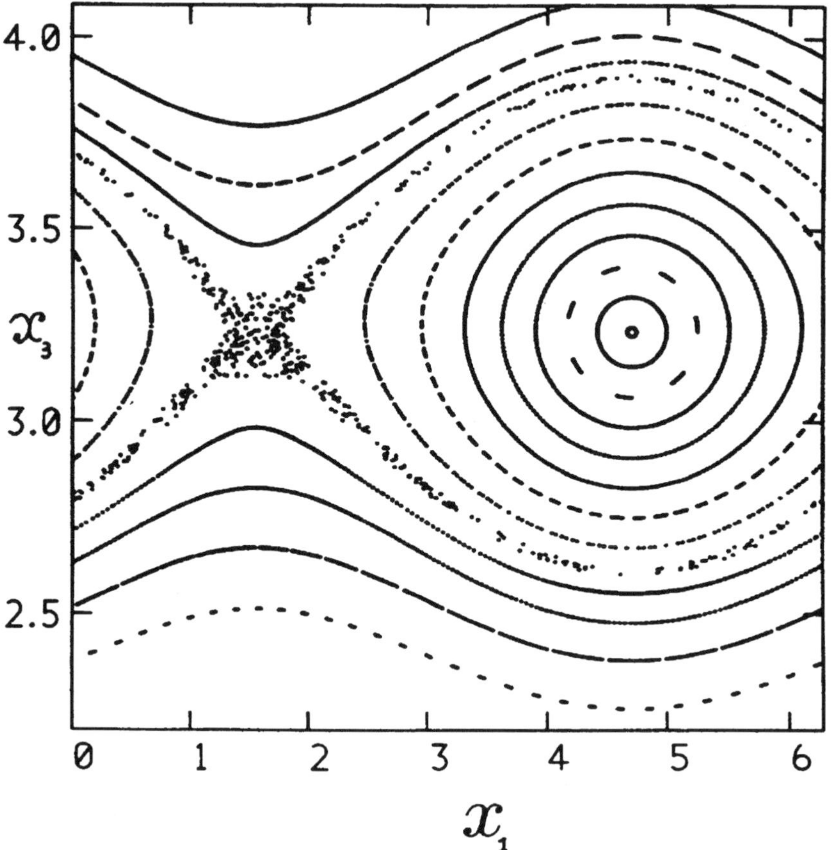

Figure 3. Poincaré section of ABC flow near stagnation points

One can visualize the complex behavior of the streamlines for such a nonintegrable flow by the method of Poincaré sections. One chooses some 2-D surface (the Poincaré section) in such a way that its intersection with the streamlines is for the most part transverse. Then, one encodes a 3-D streamline on this Poincaré section by simply integrating along and marking down the locations of the successive intersections of the particle trajectory with the Poincaré section. By considering the locus of these intersection points, various features of interest such as chaotic regions and resonances can be illuminated. Figure 3 is a Poincaré section in a neighborhood of stagnation points of an ABC flow with 2 small parameters. There appears to be one point about which the flow is not strongly stretching (an *elliptic fixed point*), and another point about which the flow appears to be strongly stretching (a *hyperbolic fixed point*). By stretching, we roughly mean the property of rapid divergence between two streamlines which originally started close to each other; we give a precise definition later in Section 12. Streamlines near the elliptic fixed point appear to remain nearby for quite a while, while the intersection points of some streamlines with the Poincaré plane near the hyperbolic fixed point rapidly run away.

LECTURE 6
Stability/Instability of an Equilibrium State

In this section we develop some notions of *stability* and *instability* for a given steady state solution $\mathbf{u}(\mathbf{x})$ of the Euler equations. There is a very extensive literature connected with the stability of the Euler equations, and different definitions may be found. In fact, arguments have appeared in print with claims and counter-claims about the stability or instability of specific equilibria. Part of the reason for such controversy is that there are a number of definitions of stability that are interrelated but not equivalent. For example, [54] enumerate certain definitions of stability and present a survey of a portion of the literature in connection with the stability of the Euler (and related) equations. Loosely speaking, the question of stability is the following: Consider a steady configuration of a dynamical problem, and make a "small" disturbance. This disturbance may either:

1. die away, in which case the equilibrium state is said to be *asymptotically stable*,
2. persist, but not grow, in which case the equilibrium is *neutrally stable*,
3. grow, in which case the equilibrium is *unstable*.

Observe how the concept of stability/instability will be crucially connected with the norm $\| \cdot \|$ which is chosen to measure the size of the disturbance.

We will adopt the following definitions of linear and nonlinear stability which we think are conceptually appropriate and avoid certain anomalies that may occur under other definitions.

6.1. Linear stability (instability)

We performed the linearization of the Euler equations in Subsection 3.1, and found that the linearized dynamics of a small perturbation $\mathbf{v}(\mathbf{x}, t)$ to a steady state $\mathbf{u}(\mathbf{x})$ satisfies:

$$\frac{\partial \mathbf{v}(\mathbf{x}, t)}{\partial t} = -\mathbf{u}(\mathbf{x}) \cdot \nabla \mathbf{v}(\mathbf{x}, t) - \mathbf{v}(\mathbf{x}, t) \cdot \nabla \mathbf{u}(\mathbf{x}) - \nabla p(\mathbf{x}, t) \equiv L\mathbf{v}$$
$$\nabla \cdot \mathbf{v}(\mathbf{x}, t) = 0. \tag{6.1}$$

The initial perturbation is specified:

$$\mathbf{v}(\mathbf{x}, 0) = \mathbf{v}_0(\mathbf{x}), \tag{6.2}$$

263

and the boundary conditions are the same as for the full nonlinear Euler equations (see Section 2).

The initial value problem given by (6.1) and (6.2) can be attacked using the well-known method of normal modes. We seek a solution in the form of eigenvalues and eigenfunctions, namely:

$$\mathbf{v}(\mathbf{x}, t) = \mathbf{v}(\mathbf{x})e^{\sigma t}. \tag{6.3}$$

We are taking a little notational license in using the same symbol $\mathbf{v}$ in different ways on both sides of the equation, but the appropriate meaning of $\mathbf{v}$ will be clear from the context. In particular, for the spectral problem, where no time dependence enters, the meaning of $\mathbf{v}$ is the purely spatially dependent function appearing on the right hand side. The spectral eigenvalue problem for (6.1) is given by:

$$\begin{aligned}
\sigma\mathbf{v}(\mathbf{x}) &= L\mathbf{v}(\mathbf{x}), \\
\nabla \cdot \mathbf{v}(\mathbf{x}) &= 0,
\end{aligned} \tag{6.4}$$

plus the appropriate boundary conditions. Of course, there remains the issue as to whether the eigenfunctions of (6.4) are complete so that the initial perturbation $\mathbf{v}_0(\mathbf{x})$ may be expanded as a meaningful sum of such eigenfunctions.

We recall that the operator L for the linearized Euler equations is degenerate non-elliptic. This means that in sharp contrast with the Navier-Stokes equations where the operator is elliptic, we cannot assert that the spectrum of L is discrete. In general, the spectrum of L (Spec L) will be continuous, possibly together with some discrete eigenvalues. The spectrum of L depends on the functional space in which we are considering (6.1) and (6.2). For any Banach space Z we use the following classification of spectral points (see, for example, [**83**] for further discussion on the character of spectra):

Definition 1. *A point $z \in$ Spec L is called a point of the* discrete *spectrum if it satisfies the following conditions:*

1. *z is an isolated point of* Spec L, *and*
2. *z has finite multiplicity, and*
3. *the range of $(z - L)$ is closed, which implies that there is a complementary subspace of Z in which $(z - L)$ is invertible.*

On the contrary, if $z \in$ Spec L but does not satisfy all of these three conditions, then it is called a part of the essential *spectrum, which is primarily composed of the continuous spectrum.*

An element of the discrete spectrum is an eigenvalue corresponding to a finite number of eigenfunctions of the problem (6.4), whereas the continuous spectrum may be associated with *generalized* eigenfunctions (not necessarily belonging to Z) satisfying (6.4) with the appropriate boundary conditions.

The possibly most familiar concept of stability is related to the eigenvalues of the discrete spectrum. This is the concept of (discrete) *spectral stability* much discussed in the physical literature (see, for example, [**19**], [**72**], and [**29**]). If all the eigenvalues σ of (6.4) are purely imaginary, then the equilibrium $\mathbf{u}(\mathbf{x})$ is said to be (discretely) spectrally stable. If there exists at least one eigenvalue with positive real part, then the associated eigenfunction will grow exponentially in time, implying instability.

We will adopt the following definition of *linear stability/instability* for the linearized Euler equations, where, in general, the spectrum is a union of a discrete and continuous part: (see [**32**] or [**58**] for terminology)

Definition 2. *Let* $\mathbf{u}(\mathbf{x})$ *be a* C^∞ *smooth equilibrium satisfying equations (5.1) with the appropriate boundary conditions. Consider the linear problem given by (6.1) and (6.2). Let* Z *be a Banach space in which these equations define a* C_0 *semigroup in* Z. *For example,* Z *may be the Sobolev space* $Z = \{\mathbf{v} \in H^s(\mathcal{D}) | \nabla \cdot \mathbf{v} = 0, \mathbf{v} \cdot \mathbf{n}|_{\partial\mathcal{D}} = 0\}$ *for some real* $s \geq 0$. *Then* $\mathbf{u}(\mathbf{x})$ *is called* linearly stable *if:*

$$\lim_{t \to \infty} \frac{1}{t} \log \|e^{tL}\|_Z = 0, \tag{6.5}$$

and linearly unstable if the left hand side is positive. In this notation, the semigroup of operators e^{tL} *is defined:* $e^{tL}\mathbf{v}_0(\mathbf{x}) = \mathbf{v}(\mathbf{x}, t)$, *where* $\mathbf{v}(\mathbf{x}, t)$ *satisfies the linearized Euler equations (6.1) and (6.2). Also, the norm appearing in (6.5) is the natural operator norm associated to the Banach space* Z.

It is clear that if there exists an eigenvalue σ in the discrete spectrum of L with $\Re\sigma > 0$, then the left hand side of (6.5) is positive (i. e. (discrete) spectral instability implies linear instability). However, it is possible that the left hand side of (6.5) is positive due to growth from the continuous spectrum without the existence of discrete unstable eigenvalues.

We reiterate the importance of the choice of Banach space Z and associated norm $\| \cdot \|_Z$ in which the exponential growth is measured for the determination of stability.

6.2. Nonlinear (Lyapunov) stability

This notion of stability relates directly to the full Euler equations, and not its linearization.

Definition 3. *An equilibrium* $\mathbf{u}(\mathbf{x})$ *is called* nonlinearly stable *if for every neighborhood* N_1 *of* $\mathbf{u}$, *there exists another neighborhood* N_2 *of* $\mathbf{u}$ *such that functions* $\mathbf{q}(\mathbf{x}, t)$ *initially in* N_2 *never leave* N_1 *as they evolve under the Euler equations.*

When the topology is taken to be induced by some norm $\| \cdot \|_Z$, this statement translates to the following:

Definition 4. *the equilibrium* $\mathbf{u}(\mathbf{x})$ *is* nonlinearly stable *if for every* $\varepsilon > 0$, *there exists a* $\delta > 0$ *such that if* $\|\mathbf{u} - \mathbf{q}(\mathbf{x}, 0)\| < \delta$, *then* $\|\mathbf{u} - \mathbf{q}(\mathbf{x}, t)\| < \varepsilon$ *for all* $t > 0$, *where* $\mathbf{q}(\mathbf{x}, t)$ *is a solution to the Euler equations (1.2) with initial data* $\mathbf{q}(\mathbf{x}, 0)$.

In the following sections we will discuss these concepts of stability in more detail for some specific classes of steady Euler flows $\mathbf{u}(\mathbf{x})$. We will examine the spectral problem for the eigenvalues of the linearized Euler operator. Even in two dimensions, understanding the spectrum is a highly nontrivial problem, and there are open questions associated with the stability of a flow as relatively simple as plane parallel shear flow. To prove nonlinear stability is an even more difficult task than to prove linear stability. We will discuss the "Arnold criterion" for nonlinear stability which can be applied successfully to certain 2-D, but not to 3-D, equilibria. Finally, we will discuss a criterion which gives an effective sufficient condition for linear instability with respect to the energy norm for very large classes of Euler equilibria. These results are suggestive of the conclusion that there are almost no steady states in three dimensions that are stable with respect to the energy norm.

LECTURE 7
Two-Dimensional Spectral Problem

We shall now begin to investigate the spectral problem (6.4) for the 2-D Euler equations. As we discussed in Subsection 4.1, the divergence equation may be replaced by writing the velocity in terms of a scalar stream function. The vorticity then is naturally expressed in terms of the stream function, according to (4.7). So let us define the stream function for the steady state:

$$\mathbf{u}(\mathbf{x}) = \hat{\mathbf{k}} \times \nabla\psi(\mathbf{x}), \ \nabla\times\mathbf{u}(\mathbf{x}) = \hat{\mathbf{k}}\nabla^2\psi(\mathbf{x}) \tag{7.1}$$

and the stream function for the perturbed velocity field:

$$\mathbf{v}(\mathbf{x},t) = \hat{\mathbf{k}} \times \nabla\phi(\mathbf{x},t), \ \nabla\times\mathbf{v}(\mathbf{x},t) = \hat{\mathbf{k}}\nabla^2\phi(\mathbf{x},t). \tag{7.2}$$

Our spectral problem for the vorticity equation obtained by taking the curl of (6.4) and applying the vector identities (4.1) and (4.4) is:

$$\sigma\omega = \{\mathbf{u},\omega\} + \{\mathbf{v},\Omega\}, \tag{7.3}$$

where:

$$\omega(\mathbf{x},t) = \nabla\times\mathbf{v}(\mathbf{x},t), \ \Omega(\mathbf{x}) = \nabla\times\mathbf{u}(\mathbf{x}), \tag{7.4}$$

and $\{,\}$ denotes the Poisson bracket defined in (4.5). We now insert the expressions for the stream functions:

$$\sigma\nabla^2\phi\,\hat{\mathbf{k}} = \{\hat{\mathbf{k}} \times \nabla\psi, \hat{\mathbf{k}}\nabla^2\phi\} + \{\hat{\mathbf{k}} \times \nabla\phi, \hat{\mathbf{k}}\nabla^2\psi\}. \tag{7.5}$$

Expand the Poisson brackets, and recall that all functions depend only on two coordinates x_1 and x_2 so that $\hat{\mathbf{k}}\cdot\nabla$ is a null operator. The resulting vector on the right hand side is parallel to $\hat{\mathbf{k}}$, and equating this component to that of the right hand side gives:

$$\sigma\nabla^2\phi = \left(\psi_{x_2}\frac{\partial}{\partial x_1} - \psi_{x_1}\frac{\partial}{\partial x_2}\right)\nabla^2\phi + \left(\phi_{x_2}\frac{\partial}{\partial x_1} - \phi_{x_1}\frac{\partial}{\partial x_2}\right)\nabla^2\psi \tag{7.6}$$

with the appropriate boundary conditions. Let ψ satisfy an equation of the form

$$\nabla^2\psi = f(\psi)$$

where f is some smooth function. Recall that stream functions satisfying such an equation always represent steady states, and that all steady states may be locally

267

written in such a form, away from critical points. Equation (7.6) then reduces to:

$$\sigma \nabla^2 \phi = \left(\psi_{x_2} \frac{\partial}{\partial x_1} - \psi_{x_1} \frac{\partial}{\partial x_2} \right) \nabla^2 \phi + \left(\phi_{x_2} \frac{\partial \psi}{\partial x_1} - \phi_{x_1} \frac{\partial \psi}{\partial x_2} \right) f'(\psi) \tag{7.7}$$

$$= \left(\psi_{x_2} \frac{\partial}{\partial x_1} - \psi_{x_1} \frac{\partial}{\partial x_2} \right) \left(\nabla^2 \phi - f'(\psi)\phi \right). \tag{7.8}$$

In the last equality, we used the fact that for any smooth function g,:

$$\left(\psi_{x_2} \frac{\partial}{\partial x_1} - \psi_{x_1} \frac{\partial}{\partial x_2} \right) g(\psi) = \left(\psi_{x_2}\psi_{x_1} - \psi_{x_1}\psi_{x_2} \right) g'(\psi) = 0, \tag{7.9}$$

which implies that $g(\psi)$ commutes with this differential operator.

We will use so-called "energy methods" to analyze the spectral problem:

$$\sigma \nabla^2 \phi = \left(\psi_{x_2} \frac{\partial}{\partial x_1} - \psi_{x_1} \frac{\partial}{\partial x_2} \right) \left(\nabla^2 \phi - f'(\psi)\phi \right). \tag{7.10}$$

We consider this equation in a rigid bounded, domain $\mathcal{D}$, so that the perturbation stream function satisfies the boundary condition:

$$\phi|_{\partial \mathcal{D}} = 0. \tag{7.11}$$

If $\mathcal{D}$ has multiple components, one might wonder whether we shouldn't permit ϕ to attain different constant values on different components. The reason that we demand that ϕ be the same value (zero) on all components of the boundary is that this restricts the perturbed velocity field $\mathbf{q} = \mathbf{u} + \mathbf{v}$ and the unperturbed velocity field $\mathbf{u}$ to have the same circulation about each obstacle in the flow and the same momentum flux through each channel. We define a linear operator D on the set of smooth functions α on $\mathcal{D}$ satisfying the boundary conditions (7.11) by the formula:

$$D\alpha = \left(\psi_{x_2} \frac{\partial}{\partial x_1} - \psi_{x_1} \frac{\partial}{\partial x_2} \right) \alpha. \tag{7.12}$$

We note that D is a skew-symmetric operator in $L^2(\mathcal{D})$, i.e. for all smooth α_1, α_2, such that

$$\alpha_1|_{\partial \mathcal{D}} = 0, \ \alpha_2|_{\partial \mathcal{D}} = 0, \tag{7.13}$$

we may integrate by parts to find:

$$(D\alpha_1, \alpha_2) = -(\alpha_1, D\alpha_2).$$

Let G be the Green's operator for the negative Laplacian with boundary condition (7.11), i.e.

$$-\nabla^2 \phi = \omega \Longrightarrow \phi = G\omega. \tag{7.14}$$

We really won't need to know any properties of G; this operator is only introduced to shift focus from the stream function ϕ to the vorticity $\omega = \nabla^2 \phi$. Just as vorticity is obtained from the stream function by the Laplacian, so the stream function is obtained from the vorticity via G. Now using the symbolism from (7.12) and (7.14), we can rewrite the eigenvalue equation (7.10) in the compact form:

$$\sigma \omega = DT\omega \tag{7.15}$$

where the operator T is defined by:

$$T\omega = \omega + f'(\psi)G\omega. \tag{7.16}$$

We see that even in two dimensions, the spectral problem associated with the linearized Euler operator is highly nontrivial. This operator is the composition of the skew-symmetric operator D and the symmetrizable operator T.

We consider the particular case where f is a smooth monotone function of ψ. Let us first treat the case $f'(\psi) > 0$ for all values of ψ assumed within the domain $\mathcal{D}$. Set

$$f'(\psi) = h^2(\psi)$$

with $h(\psi) > 0$. Also define the a new function $\tilde{\omega}(\mathbf{x})$ by:

$$\omega(\mathbf{x}) = h(\psi(\mathbf{x}))\tilde{\omega}(\mathbf{x}).$$

Substituting these last two equations into (7.15) results in:

$$\sigma h\tilde{\omega} = D(h\tilde{\omega} + h^2 Gh\tilde{\omega}).$$

Since D commutes with functions of ψ (as can be seen from (7.9)), we can rewrite this as:

$$D(\tilde{\omega} + hGh\tilde{\omega}) = \sigma\tilde{\omega}. \tag{7.17}$$

We form an "energy integral" by taking the (complex) inner product of this equation with the function $\tilde{\omega} + hGh\tilde{\omega}$. Then:

$$(D(\tilde{\omega} + hGh\tilde{\omega}), \tilde{\omega} + hGh\tilde{\omega}) = \sigma(\tilde{\omega}, \tilde{\omega} + hGh\tilde{\omega}), \tag{7.18}$$

where we define the inner product to be linear in the first argument and conjugate linear in the second argument. Now the left hand side must be purely imaginary. For, by skew-symmetry of D and the definition of the complex inner product,

$$(D(\tilde{\omega} + hGh\tilde{\omega}), \tilde{\omega} + hGh\tilde{\omega}) = -(\tilde{\omega} + hGh\tilde{\omega}, D(\tilde{\omega} + hGh\tilde{\omega}))$$

$$= -\overline{(D(\tilde{\omega} + hGh\tilde{\omega}), \tilde{\omega} + hGh\tilde{\omega})},$$

where an overbar denotes complex conjugate. So the left hand side of (7.18) is equal to its negative complex conjugate, and must hence be purely imaginary. But the same must then be true of the right hand side, hence:

$$\operatorname{Re}\sigma(\tilde{\omega}, \tilde{\omega} + hGh\tilde{\omega}) = 0. \tag{7.19}$$

The same analysis can be made when $f'(\psi) < 0$ everywhere in $\mathcal{D}$ by setting $f'(\psi) = -h^2(\psi)$ with $h > 0$. In this case, we obtain the result:

$$\operatorname{Re}\sigma(\tilde{\omega}, \tilde{\omega} - hGh\tilde{\omega}) = 0. \tag{7.20}$$

Let us recast these equalities in terms of the original variables. First realize that

$$(\tilde{\omega}, \tilde{\omega}) = (h^{-1}\omega, h^{-1}\omega) = \int_{\mathcal{D}} \frac{|\omega|^2}{h(\psi)^2}\, dx_1\, dx_2$$

and that:

$$(\tilde{\omega}, hGh\tilde{\omega}) = (h\tilde{\omega}, Gh\tilde{\omega}) = (\omega, G\omega) = (\omega, \phi) = -\int_{\mathcal{D}} \nabla^2\phi\,\overline{\phi}\, dx_1\, dx_2$$

$$= \int_{\mathcal{D}} |\nabla\phi|^2\, dx_1\, dx_2,$$

where we have used the boundary condition (7.11) in the last integration by parts. Hence (7.19) and (7.20) expressed in original variables are precisely:

$$\pm \operatorname{Re} \sigma \int_{\mathcal{D}} \left[\frac{|\nabla^2 \phi|^2}{f'(\psi)} + |\nabla \phi|^2 \right] dx_1\, dx_2 = 0.$$

Thus, we have obtained a sufficient condition for spectral stability, namely, the definiteness of the quadratic form:

$$\int_{\mathcal{D}} \left[\frac{|\nabla^2 \phi|^2}{f'(\psi)} + |\nabla \phi|^2 \right] dx_1\, dx_2. \tag{7.21}$$

Hence, any flow for which $f'(\psi) > 0$ is spectrally stable since in this case the quadratic form is automatically positive definite. In the case where $f'(\psi) < 0$, the quadratic form may be definite depending on the magnitude of $f'(\psi)$ in $\mathcal{D}$. For example, we are guaranteed stability when $f' < 0$ and the quadratic form

$$\int_{\mathcal{D}} \left[|\nabla \phi|^2 + \left(\max_{\mathcal{D}} \frac{1}{f'} \right) |\nabla^2 \phi|^2 \right] dx_1\, dx_2$$

is negative definite.

LECTURE 8
"Arnold" Criterion for Nonlinear Stability

In the 1960's, Arnold ([**4**], [**5**], [**7**], [**6**], and [**8**]) exploited the Hamiltonian nature of the Euler equations to study the nonlinear stability of steady flows of an ideal fluid. (See [**11**, App. 2], or [**23**, Sec. 1.4] for how the Euler equations may be put into a Hamiltonian formalism.) Arnold showed that the fluid stability problem could be formulated as an infinite-dimensional analogue of the stability of a rotating top. He used the theory of invariant metrics on Lie groups[1] to obtain a sufficient condition for nonlinear stability. In the stability analysis, the appropriate subspace of perturbation velocities is constrained by the enforcement of Kelvin's circulation theorem, which in particular requires that the perturbation velocities "infinitesimally" preserve the topology of the vortex fields. For example, the perturbations are not allowed to push vortex lines through each other. We define *surfaces of Kelvin type* as the infinite-dimensional surfaces of velocity fields which may be so connected while respecting Kelvin's circulation theorem. These Kelvin surfaces can have a very complicated geometry.

According to the treatment of Arnold ([**4**], [**6**], and [**11**]), two conditions are sufficient to ensure nonlinear stability of the steady flow:

1. the local convexity of the energy

$$\mathcal{H} = \frac{1}{2} \int_{\mathcal{D}} d\mathbf{x} \, \mathbf{q}(\mathbf{x})^2,$$

and the definiteness of its second variation $d^2\mathcal{H}$ when restricted to a Kelvin surface, and

2. the regularity of the infinite-dimensional Kelvin surfaces.

An equilibrium point $\mathbf{u}$ is said to be a *regular* point if a partition of a neighborhood of $\mathbf{u}$ into Kelvin surfaces is diffeomorphic to a partition of Euclidean space into parallel planes. In some treatments of the "Arnold method," the second condition is replaced by the existence of sufficiently many conserved quantities referred to as Casimirs (see, for example, [**54**]). In two dimensions, the regularity condition is easy to check, and as we will discuss later, there exist specific equilibria for which condition (1) holds. However, in three dimensions, not only is the second condition (2) (or the alternative Casimir condition) hard to verify, but $d^2\mathcal{H}$ is in general *not*

---

[1] For the Euler equations, the appropriate group is the infinite-dimensional group SDiff($\mathcal{D}$) of volume preserving diffeomorphisms of the domain $\mathcal{D}$.

271

definite. Recently, Sadun & Vishik [**86**] proved in 3 (or more) dimensions that the second variation is not only never definite, but that it is generally unbounded from above and below. The sole exception in three dimensions is the case of uniformly zero vorticity, where the second variation of the energy is identically zero. Thus, for the Euler equations, the effectiveness of the Arnold criteria for nonlinear stability is essentially restricted to two dimensions, once again emphasizing the crucial differences between two and three dimensions. We note that the Arnold method has indeed been applied with success to other fluid models. (See, for example, [**39**], [**54**], [**1**], and [**13**].)

In two dimensions, the Arnold [**6**] method gives the following sufficient condition for the nonlinear stability of a steady flow with stream function ψ, namely that one of the following two conditions is satisfied:

1. The quadratic form

$$\int_{\mathcal{D}} \left[|\nabla^2 \phi|^2 \frac{\nabla \psi}{\nabla \nabla^2 \psi} + |\nabla \phi|^2 \right] dx_1 \, dx_2$$

   is positive definite with respect to $\nabla \phi$, or

2. the quadratic form

$$\int_{\mathcal{D}} \left[|\nabla^2 \phi|^2 \left(\max_{\mathcal{D}} \frac{\nabla \psi}{\nabla \nabla^2 \psi} \right) + |\nabla \phi|^2 \right] dx_1 \, dx_2$$

   is negative definite with respect to $\nabla \phi$.

The quotient of vectors $\dfrac{\nabla \psi}{\nabla \nabla^2 \psi}$ is actually a scalar because ψ and $\nabla^2 \psi$ are functionally related, and thus have parallel gradients. In fact, $\dfrac{\nabla \nabla^2 \psi}{\nabla \psi} = f'(\psi)$ in the notation of Section 7, so Arnold's conditions for *nonlinear* stability coincide with the sufficient conditions for *spectral* stability derived in that section.

We remark that these conditions are sufficient for stability with respect to the L^2 norm, which measures the energy of the perturbation. (Our derivation of the stability criterion in Section 7 also clearly investigated the growth of the magnitude of the velocity, and not of its gradients.) However, as we will show in Section 10, stability in the energy norm does not preclude growth of a perturbation in a different norm which, for example, might include derivatives of the velocity.

LECTURE 9
Plane Parallel Shear Flow

In the past hundred years, a vast body of literature has appeared on the subject of the stability of one of the most basic fluid equilibria, namely plane parallel shear flow. Here the basic state velocity is of the 2-D form

$$\mathbf{u(x)} = u(x_2)\hat{\mathbf{i}}, \tag{9.1}$$

where $\hat{\mathbf{i}}$ denotes a unit vector along the x_1 direction. Descriptions of some of the results can be found in the texts [**72**], [**19**], and [**29**]. This simple shear flow admits a much more tractable analysis than most flows because it depends on only one variable, and thus the spectral eigenvalue problem is an ODE instead of a PDE. In spite of this long history of study, open questions associated with this most basic of Euler flows remain.

Classically, the spectral problem for the 2-D Euler equation linearized about $u(x_2)\hat{\mathbf{i}}$ is written in the following form: Consider a perturbation stream function:

$$\Phi(x_2)e^{i\alpha(x_1-ct)}, \tag{9.2}$$

where $\Phi(x_2)$ is some complex-valued structure function; an arbitrary perturbation may be written as a Fourier integral over functions of this form. The physical stream function is really to be interpreted as the real part of this complex function. Due to linearity of the equations involved, it is permissible to work with the complex stream function (9.2) and take the real part at the end. In the notation of Section 7, the corresponding time-independent stream function would read:

$$\phi(x_1, x_2) = \Phi(x_2)e^{i\alpha x_1},$$

and the associated eigenvalue of the linearized Euler operator L would be:

$$\sigma = -i\alpha c.$$

Here the complex number c is called the phase speed, since the perturbed stream function (9.2) clearly moves along the $\hat{\mathbf{i}}$ direction at this velocity. The wavenumber along this direction is α, which will be restricted to real values. Linear stability is equivalent to the spectrum of values of c lying only on the *real* axis.

For the perturbation stream function (9.2) and steady profile (9.1), the eigen-value equation in the form (7.3) reads:

$$- i\alpha c \left(-\alpha^2 \Phi(x_2) + \frac{d^2\Phi(x_2)}{dx_2^2} \right) e^{i\alpha x_1}$$

$$= -u(x_2) \frac{d}{dx_1} \left(\left(-\alpha^2 \Phi(x_2) + \frac{d^2\Phi(x_2)}{dx_2^2} \right) e^{i\alpha x_1} \right)$$

$$- e^{i\alpha x_1} \left(-\Phi'(x_2) \frac{\partial}{\partial x_1} + i\alpha \Phi(x_2) \frac{\partial}{\partial x_2} \right) (-u'(x_2)),$$

which reduces to:

$$(u(x_2) - c) \left(\frac{d^2}{dx_2^2} - \alpha^2 \right) \Phi(x_2) - u''(x_2)\Phi(x_2) = 0. \tag{9.3}$$

Appropriate boundary conditions could be one of the following:

1. A channel with rigid walls at $x_1 = a, x_2 = b$. Then the boundary conditions on the perturbation stream function are, according to (7.11),

$$\Phi(a) = 0, \Phi(b) = 0. \tag{9.4}$$

2. Periodic boundary conditions:

$$\Phi(b) = \Phi(a).$$

Equation (9.3) is referred to as the Rayleigh equation (alternatively the inviscid Orr-Sommerfeld equation). It is clear that this differential equation is not self-adjoint. The Rayleigh equation exhibits critical point behavior at any point $x_2 = X_c$ where $u(X_c) = c$, because the coefficient of the leading derivative vanishes at such a point. There will, in general, be multiple solutions which join each other at a branch point located at X_c.

In this example, one can construct an energy integral analogous (but not identical) to (7.21). One simply multiplies both sides of (9.3) by the complex conjugate $\bar{\Phi}$, divides by $u(x_2) - c$, and integrates from $x_2 = a$ to $x_2 = b$. The imaginary part of the resulting equation reads:

$$\mathrm{Im}\, c \int_a^b \frac{u'' |\Phi|^2}{|u - c|^2} \, dx_2 = 0. \tag{9.5}$$

Hence, we obtain the Rayleigh's celebrated *sufficient* condition for spectral stability, namely $u''(x_2) \neq 0$ for $a < x_2 < b$. For, then the integral must be sign-definite and the imaginary part of c must vanish. From the Arnold criterion, it also follows that the flow is *nonlinearly* stable when $u''(x_2) \neq 0$. (At first sight, the Arnold criterion requires that $\frac{u''}{u}$ be positive definite. But since stability is a concept independent of the Galilean frame of reference in which we view the flow, we may at will replace $u(x_2)$ by any $u(x_2) + U$, for any constant U. Thus, if u'' is sign-definite, we may add a uniform flow onto the given shear profile $u(x_2)$ to assume that u and u'' both have the same sign.) It is, however, harder to show the existence of instability when this inflection point condition is violated, and the problem is open for general profiles.

Here is a brief list of some of the results known to hold for the Rayleigh equation with rigid boundary conditions. The stability results are with respect to the energy norm.

1. There exists a continuous spectrum for the boundary value problem defined by (9.3) and (9.4). All values of c within the range of $u(x_2)$ lie in the continuous spectrum, which is associated with the existence of critical points X_c where $(u(X_c) - c)$ vanishes. The continuous spectrum is always stable for this problem. A proof of this result was sketched by Case. Faddeev [34] constructed the "generalized eigenfunctions" for the continuous spectrum.

2. Due to the preceding result, if any instability exists, it must come from the discrete spectrum. By Rayleigh's criterion, the discrete spectrum can only be unstable for profiles where $u''(x_2) = 0$ somewhere in $[a, b]$. Howard [55] proved a more general theorem that for certain classes of profiles, there are no more unstable modes than inflection points of u. Lin [71] gave heuristic arguments and later Faddeev [34] proved the existence of an unstable mode for *monotonic* profiles with an inflection point. Friedlander and Howard [36] used the method of continued fractions to construct an unstable mode in the case of a sinusoidal profile with sufficiently many oscillations.

These results suggest that any profile with an inflection point ought to be unstable (in the energy norm), but such an instability result has not yet been rigorously proved. In fact, the functional analytic content of the spectral problem given by (9.3) and (9.4) is not well understood even though shear flows are among the simplest examples of fluid motion. Clearly the spectral problem for more general flows in two or three dimensions will be very difficult indeed, particularly since the differential equations involved will generally be PDE's.

LECTURE 10
Instability in a Vorticity Norm

The following example due to Yudovich [95], [97] illustrates that the matter of stability is dependent on spatial dimension and the norm. Consider the parallel shear flow

$$\mathbf{u}(\mathbf{x}) = u(x_2)\hat{\mathbf{i}}$$

as a flow in 3-D space with, for simplicity, 2π periodic boundary conditions. It is easy to check that the following velocity, in component form,

$$\mathbf{q}(\mathbf{x}, t) = \begin{bmatrix} u(x_2) \\ 0 \\ w(x_1 - tu(x_2)) \end{bmatrix} \tag{10.1}$$

is an exact solution of the full nonlinear Euler equations (1.2). Here u and w are arbitrary, smooth 2π periodic functions; the third component of the velocity is simply carried by the flow. The vorticity corresponding to (10.1) is:

$$\nabla \times \mathbf{q}(\mathbf{x}, t) = - \begin{bmatrix} tu'(x_2)w'(x_1 - tu(x_2)) \\ w'(x_1 - tu(x_2)) \\ u'(x_2) \end{bmatrix}. \tag{10.2}$$

We consider $\mathbf{u}(\mathbf{x}) = u(x_2)\hat{\mathbf{i}}$ as the basic state and take a perturbation $w(x_1 - tu(x_2))\hat{\mathbf{k}}$ which is small in a norm including the magnitude of the vorticity, for example:

$$\|\mathbf{q}\| \equiv \max_{\mathbf{x} \in \mathcal{D}} |\nabla \times \mathbf{q}| + \|\mathbf{q}\|_1,$$

where $\|\cdot\|_1$ is some other norm depending only on the magnitude of the velocity, and none of its derivatives. Clearly the size of the initial perturbation $\|\mathbf{q}(\mathbf{x}, 0) - \mathbf{u}(\mathbf{x})\|$ can be chosen arbitrarily small by an appropriate choice of w. In the norm $\|\cdot\|_1$ depending only on the magnitude of the velocity, such as

$$\|\mathbf{q}\|_1 = \max_{\mathbf{x} \in \mathcal{D}} |\mathbf{q}|,$$

it is clear that such a perturbation is neutrally stable; the norm of the perturbation $\|\mathbf{q}(\mathbf{x}, t) - \mathbf{u}(\mathbf{x})\|_1$ remains constant in time. However, it follows from (10.2) that the norm incorporating a measure of the vorticity magnitude,

$$\|\mathbf{q}(\mathbf{x}, t) - \mathbf{u}(\mathbf{x})\|$$

277

grows linearly with time. Thus, any plane parallel shear flow, with $u(x_2)$ nonconstant, is *nonlinearly unstable* as a 3-D flow in a norm that includes the magnitude of the vorticity. This is in contrast with the result in Section 9 that such a flow is nonlinearly stable in the *energy* norm against 2-D perturbations, provided that the shear profile satisfies the condition $u''(x_2) \neq 0$.

Yudovich [**97**] points out that there are many 2-D flows that are stable to 2-D perturbations in a uniform metric for the vorticity. (This is essentially a consequence of the fact that vorticity is conserved along particle trajectories in 2-D flow.) He proves, however, that any 2-D steady flow in a strip, at least one boundary of which is a line, is Lyapunov *unstable* in any metric that includes the maximum of the *gradient* of the vorticity. One can easily visualize a mechanism leading to growth of the *derivatives* of the vorticity in two dimensions, while the vorticity itself remains the same for each fluid particle. It is probable that, in a given flow, there will exist many pairs of particles which approach each other at large times. If they possess different values of vorticity, the magnitude of the derivative of the vorticity becomes infinitely large as $t \to \infty$. This observation has interesting implications for the loss of smoothness (regularity) for solutions evolving under the Euler equations [**95**].

Here is a summary of stability results in the context of shear flows that shows the importance of both the spatial dimension in which perturbations are allowed and the norm in which the size of the perturbation is measured. These results hold for shear flows with rigid boundary conditions of the form (9.4).

1. Any shear flow with $u'' \neq 0$ is nonlinearly stable with respect to 2-D perturbations in the energy norm.
2. Certain shear flows with an inflection point (such as monotonic or sinusoidal profiles with sufficiently many oscillations) are linearly unstable in the energy norm. This holds in two or three dimensions. As we will discuss in Section 15, these shear flows are, in fact, nonlinearly unstable.
3. Yudovich's arguments [**95**] show that there exist shear flows that are stable to 2-D perturbations in a uniform metric for vorticity. These flows are, however, nonlinearly *unstable* to 2-D perturbations in any metric which includes the maximum of the *vorticity gradient*.
4. Any shear flow with u nonconstant is nonlinearly unstable to *3-D* perturbations in a metric that includes the magnitude of the *vorticity*.

LECTURE 11
Sufficient Condition for Instability

As we remarked in Section 8, the Arnold criteria for nonlinear stability are almost never applicable to the 3-D Euler equations. This suggests that most Euler equilibria might be unstable in three dimensions. In [9], Arnold remarks, in the context of steady ideal fluid flows, "there appears to be an infinitely great number of unstable configurations." In this section we will discuss a sufficient condition for instability which is an effective criterion which can be used to prove the linear instability of many classes of Euler equilibria.

The techniques used to obtain the instability criteria are similar to the method of geometrical optics, where the evolution of high frequency waves of sufficiently small amplitude are described by a system of ODE's (see for example, [92], [63], [56]). This geometric approach to problems governed by hyperbolic PDE's was first applied to the equations of *compressible* fluids (which are hyperbolic) by Friedlander [35], Ludwig [73], and later Eckhoff [31]. Recent results of Vishik and Friedlander [91] show that this technique, using WKB type asymptotics (see [16]), can also be successfully applied to the Euler equations, even though they are not exactly hyperbolic. The main tool is the full asymptotic expansion of the evolution operator (Green's function) for the linearized Euler equation. This approach has an advantage over the more traditional analysis of the spectrum. We have already pointed out that the spectral problem associated with even the simplest Euler equilibrium, namely plane parallel shear flow, involves deep questions in non-self-adjoint spectral theory. It seems very hard indeed to obtain a detailed spectral analysis of more general Euler equilibria. The geometrical optics approach circumvents this difficulty by producing a geometric quantity which gives a *lower* bound on the maximum growth rate of the linearized Euler operator L (see equation 6.1). This geometric quantity can be considered as a Lyapunov type exponent for ideal fluid flow. Positivity of this exponent implies that the *unstable* spectrum of L is nonempty, i.e. the corresponding steady Euler flow is linearly unstable. In the final section, we will discuss the implications of this result for nonlinear instability.

The following arguments motivate the instability criterion that is proved in Vishik & Friedlander [91].

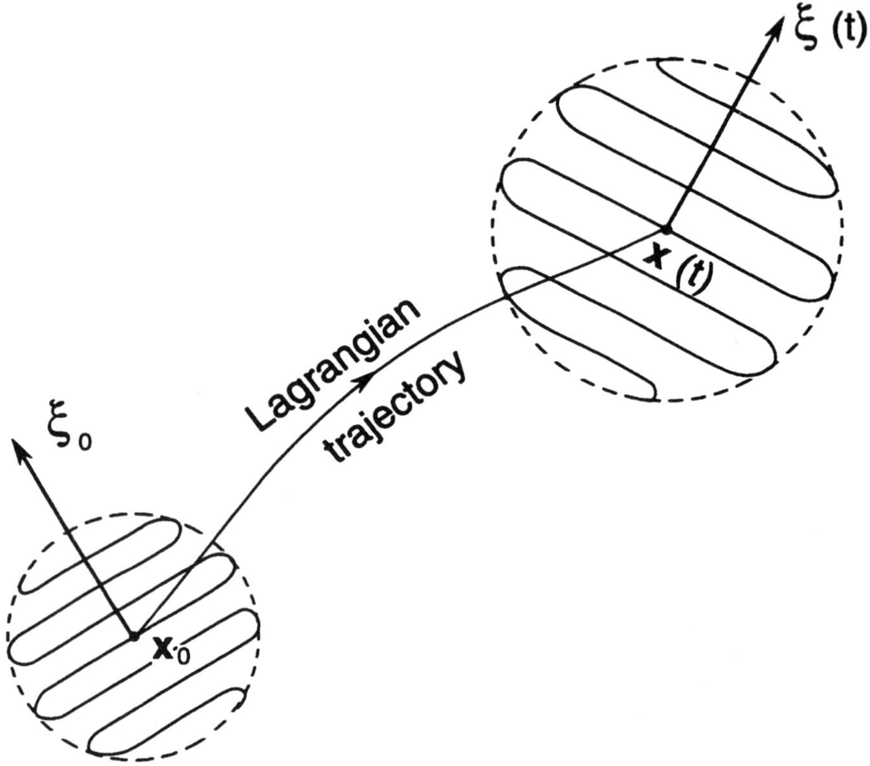

Figure 4. Evolution of high-frequency wavelet disturbance

Recall the linearized Euler equations (6.1):

$$\frac{\partial \mathbf{v}(\mathbf{x},t)}{\partial t} = -(\mathbf{u}(\mathbf{x})\cdot\nabla)\mathbf{v}(\mathbf{x},t) - (\mathbf{v}(\mathbf{x},t)\cdot\nabla)\mathbf{u}(\mathbf{x},t) - \nabla p(\mathbf{x}) \equiv L\mathbf{v}(\mathbf{x},t),$$
$$\nabla \cdot \mathbf{v}(\mathbf{x}) = 0, \tag{11.1}$$

where $\mathbf{u}(\mathbf{x})$ is a smooth solution of the steady flow equation (5.1). We consider periodic or free space boundary conditions. Let the initial disturbance be a high frequency wavelet localized at a point $\mathbf{x}_0$ with wavevector $\delta^{-1}\xi_0$, i.e.:

$$\mathbf{v}(\mathbf{x},t=0) = \mathbf{b}_0(\mathbf{x},\xi_0)e^{i(\mathbf{x}\cdot\xi_0)/\delta} \tag{11.2}$$

where $\delta \ll 1$ and $\mathbf{b}_0$ denotes the amplitude of the localized wavelet. $\mathbf{b}_0(\mathbf{x},\xi_0)$ is a function sharply peaked at $\mathbf{x} = \mathbf{x}_0$. This initial disturbance is advected with the flow so that at time t, the fluid particle at $\mathbf{x}_0$ has moved to a point $\mathbf{X}(\mathbf{x}_0,t)$, while the dominant wavevector of the wavelet has evolved to $\delta^{-1}\xi(t)$ and the amplitude to $\mathbf{b}(t)$. [See Figure 4]. To solve for the evolution equations for these quantities "following a fluid particle," we will first consider the linearized Euler equations for the full field $\mathbf{v}(\mathbf{x},t)$. Subsequently, we will be able to read off the appropriate laws for the various quantities evaluated at a particular fluid particle as it moves along. We seek a WKB type asymptotic expansion in powers of the small parameter δ for

$\mathbf{v}(\mathbf{x}, t)$ satisfying (11.1):

$$\mathbf{v} = \mathbf{b}(\mathbf{x}, \xi_0, t)e^{iS(\mathbf{x}, \xi_0, t)/\delta} + \dots ,$$
$$p = \delta d(\mathbf{x}, \xi_0, t)e^{iS(\mathbf{x}, \xi_0, t)/\delta} + \dots , \tag{11.3}$$

where the initial conditions are given by (11.2):

$$\mathbf{b}(\mathbf{x}, \xi_0, t = 0) = \mathbf{b}_0(\mathbf{x}, \xi_0), \; S(\mathbf{x}, \xi_0, t = 0) = \mathbf{x}\cdot\xi_0. \tag{11.4}$$

We have written the asymptotic expansion for the pressure without a zero order term. This is because pressure only serves to enforce incompressibility, and from the following analysis it will be clear that no zero order pressure term is necessary.

The asymptotic expansions (11.3) are now substituted into the linearized Euler equations (11.1), giving a hierarchy of equations determined by balancing terms multiplied by the same power of δ. The highest order equations in powers of δ are known in geometric optics language as the Eikonal equation for the phase function S and the transport equation for the amplitude $\mathbf{b}$:

$$\frac{\partial S}{\partial t} = -(\mathbf{u}\cdot\nabla)S, \tag{11.5}$$

$$\frac{\partial \mathbf{b}}{\partial t} = -(\mathbf{u}\cdot\nabla)\mathbf{b} - \left(\frac{\partial \mathbf{u}}{\partial \mathbf{x}}\right)\mathbf{b} - id\nabla S. \tag{11.6}$$

We have written the operator $(\mathbf{b}\cdot\nabla)\mathbf{u}$ in the form $\left(\frac{\partial \mathbf{u}}{\partial \mathbf{x}}\right)\mathbf{b}$, where $\left(\frac{\partial \mathbf{u}}{\partial \mathbf{x}}\right)$ denotes the 3×3 Jacobian matrix with entries:

$$\left(\frac{\partial \mathbf{u}}{\partial \mathbf{x}}\right)_{ij} = \frac{\partial u_i}{\partial x_j}.$$

The local wavevector of a wavelet of the form (11.3) is given by:

$$\xi(\mathbf{x}, \xi_0, t) = \nabla S(\mathbf{x}, \xi_0, t);$$

taking the gradient of (11.5) gives the evolution equation for the field ξ :

$$\frac{\partial \xi}{\partial t} = -(\mathbf{u}\cdot\nabla)\xi - \left(\frac{\partial \mathbf{u}}{\partial \mathbf{x}}\right)^T \xi. \tag{11.7}$$

The equations (11.5), (11.6), and (11.7) are equations for the fields

$$S(\mathbf{x}, \xi_0, t), \mathbf{b}(\mathbf{x}, \xi_0, t), \text{ and } \xi(\mathbf{x}, \xi_0, t).$$

We are really primarily interested in the values of these fields at a particular fluid particle as it moves along *the unperturbed flow*. That is, the perturbation velocity field does not contribute to the advection of our observation point. To transform to this *Lagrangian* point of view, define $\mathbf{X}(\mathbf{x}, t)$ to be the position at time t of a fluid particle initially located at $\mathbf{x}_0$:

$$\frac{d\mathbf{X}(\mathbf{x}_0, t)}{dt} = \mathbf{u}(\mathbf{X}(\mathbf{x}_0, t)), \; \mathbf{X}(\mathbf{x}_0, t = 0) = \mathbf{x}_0. \tag{11.8}$$

Then the evolution of physical quantities evaluated at this moving point can be determined by the evolution of the fields via the relation (1.4), with the unperturbed velocity field $\mathbf{u}(\mathbf{x}, t)$ playing the role of the advecting velocity here:

$$\frac{d}{dt} = \frac{\partial}{\partial t} + \mathbf{u}(\mathbf{x})\cdot\nabla.$$

(We don't use the notation for the material derivative since we are not strictly differentiating with respect to a particle moving with the actual flow, but rather with respect to a particle moving with the *unperturbed basic flow*.)

So, from the Lagrangian point of view, the following functions evaluated at a given fluid particle centered at the disturbance:

$$S = S(\mathbf{X}(\mathbf{x}_0, t), \xi_0, t),$$
$$\xi = \xi(\mathbf{X}(\mathbf{x}_0, t), \xi_0, t),$$
$$\mathbf{b} = \mathbf{b}(\mathbf{X}(\mathbf{x}_0, t), \xi_0, t),$$

evolve in time according to the following system of ordinary differential equations:

$$\frac{dS}{dt} = 0,$$
$$\frac{d\mathbf{b}}{dt} = -\left(\frac{\partial \mathbf{u}}{\partial \mathbf{x}}\right)\mathbf{b} - id\xi, \tag{11.9}$$
$$\frac{d\xi}{dt} = -\left(\frac{\partial \mathbf{u}}{\partial \mathbf{x}}\right)^T \xi.$$

Now we seek to eliminate the pressure (entering through the function d) from this system of ODE's using the fact that $\mathbf{v}$ is divergence-free. By computing $\nabla \cdot \mathbf{v}$ to leading order in δ, we find that this requires:

$$\mathbf{b} \cdot \xi = 0, \tag{11.10}$$

which is the well-known fact that for an incompressible fluid, the amplitude of a wave must always be perpendicular to the wavevector.

We will now see how the pressure term proportional to δ is determined by the incompressibility condition (11.10). We can substitute the ODE's from (11.9) into:

$$\frac{d}{dt}(\mathbf{b} \cdot \xi) = 0 \tag{11.11}$$

to obtain:

$$\xi \cdot \left(-\left(\frac{\partial \mathbf{u}}{\partial \mathbf{x}}\right)\mathbf{b} - id\xi\right) + \mathbf{b} \cdot \left(-\left(\frac{\partial \mathbf{u}}{\partial \mathbf{x}}\right)^T \xi\right) = 0.$$

Solving for d, we have:

$$-id = 2\left\{\left[\left(\frac{\partial \mathbf{u}}{\partial \mathbf{x}}\right)\mathbf{b}\right] \cdot \xi\right\} |\xi|^{-2}. \tag{11.12}$$

Thus, we have the following system of *ODE's* obtained from the leading order terms in an asymptotic expansion of δ, namely:

$$\frac{d\mathbf{X}}{dt} = \mathbf{u}, \tag{11.13a}$$
$$\frac{d\xi}{dt} = -\left(\frac{\partial \mathbf{u}}{\partial \mathbf{x}}\right)^T \xi, \tag{11.13b}$$
$$\frac{d\mathbf{b}}{dt} = -\left(\frac{\partial \mathbf{u}}{\partial \mathbf{x}}\right)\mathbf{b} + 2\left\{\left[\left(\frac{\partial \mathbf{u}}{\partial \mathbf{x}}\right)\mathbf{b}\right] \cdot \xi\right\} \xi|\xi|^{-2}. \tag{11.13c}$$

The first equation describes the streamlines (Lagrangian trajectories) of the flow. The second equation is called the *co-tangent* equation for a *co-vector* ξ. (See [30] or [**11**, App. 2].) The third equation is for the *bicharacteristic amplitude* **b**. The initial conditions for this system of ODE's read:

$$\mathbf{X}(\mathbf{x}_0, 0) = \mathbf{x}_0,$$

$$\xi(\mathbf{x}_0, \xi_0, 0) = \xi_0, \tag{11.14}$$

$$\mathbf{b}(\mathbf{x}_0, \xi_0, 0) = \mathbf{b}_0(\mathbf{x}_0, \xi_0).$$

One can apply knowledge of behavior of solutions to this ODE system to infer a lower limit on the growth rate of perturbations obeying the linearized Euler equations, as the following theorem states:

Theorem 11.1 (First Instability Theorem). *Let* $\mathbf{u}(\mathbf{x})$ *be a* C^∞ *steady solution of the Euler equations in a* 2π- *periodic domain (i.e. ,* $\mathcal{D} = \mathbb{R}^n/2\pi\mathbb{Z}^n$; *the theorem is valid for any integer* n *but only* $n = 2, 3$ *are physically interesting). Let* λ *be the exact growth rate of the evolution operator (or Green's function)* $G(t)$ *for the linearized Euler equation (6.1), i.e.*

$$\lambda = \lim_{t \to \infty} \frac{1}{t} \log \|G(t)\|_{L^2}$$

where $\mathbf{v}(t) = G(t)\mathbf{v}_0$ *is a solution to the linearized Euler's equations with initial data* $\mathbf{v}(\mathbf{x}, t = 0) = \mathbf{v}_0$. *Then* λ *is bounded from below by:*

$$\lim_{t \to \infty} \frac{1}{t} \log \sup_{\substack{\mathbf{x}_0, \xi_0, \mathbf{b}_0 \\ \xi_0 \cdot \mathbf{b}_0 = 0, |\xi_0| = 1, |\mathbf{b}_0| = 1}} |\mathbf{b}(\mathbf{x}_0, \xi_0, t)| \le \lambda \tag{11.15}$$

where **b** *and* ξ *satisfy the system of ODE's (11.13) with initial conditions (11.14).*

Remark. This theorem is also valid when the domain is $\mathbb{R}^n$ with free-space boundary conditions.

We note that the fact that the lower bound involves the supremum over all initial conditions $\mathbf{x}_0, \xi_0, \mathbf{b}_0$ means that the instability criterion is effective. It is not necessary to solve the system (11.13) for all initial conditions to apply the criteria. Only one set of initial conditions needs to be found so that the Lyapunov exponent type quantity on the left hand side of (11.15) is positive, and linear instability in the energy norm of the ideal fluid flow $\mathbf{u}(\mathbf{x})$ will be demonstrated. (Recall the definition of linear instability given in Section 6.1.)

The proof of the theorem is given in Vishik & Friedlander [**91**]. Simultaneously, another instability theorem is proved, stating that a necessary condition for instability in the Navier-Stokes equations (2.3) *in the limit of vanishing viscosity* is an instability in the underlying Euler equation. Examples of fluid flow are known which are unstable under the Navier-Stokes dynamics *with finite viscosity*, but stable under evolution by the Euler equations. The theorem mentioned above shows that any such instability only persists for a range of viscosities bounded below by some *positive* number. These theorems are proved by developing an asymptotic expansion (in δ) for the exact evolution operator $G(t)$, and bounding the correction to the leading order term by a quantity which vanishes with δ, uniformly with respect to $t \in [0, T]$ where $T > 0$ is fixed. Recently, Vishik [**90**] has improved the instability theorem for the Euler equations by proving that the Lyapunov exponent type quantity on the left hand side of (11.15) is in fact the logarithm of the *essential spectral radius* for the operator L. Let us explain this more carefully since spectral radius is sometimes defined a little differently than what we mean here. Consider the exponentiated spectrum. The points on the unit circle correspond to the stable spectrum. The "radius" we refer to is the largest magnitude of any complex value in this *exponentiated* spectrum. The word "essential" refers to the contribution from the continuous spectrum; discrete eigenvalues are not to be considered in the calculation of the essential spectral radius. Vishik's theorem tells us that when the left hand side of (11.15) is positive, then the radius of the exponentiated continuous spectrum is greater than one. The fact that this approach detects instability due to the continuous spectrum, as opposed to the discrete spectrum, is a consequence of the assumption that the perturbations have high (spatial) frequency. Possible instabilities corresponding to low frequency perturbations are not detected by this asymptotic analysis.

LECTURE 12

Exponential Stretching

We first recall the definition of the Lyapunov exponent for a flow (dynamical system):

$$\frac{d\mathbf{X}(\mathbf{x}_0, t)}{dt} = \mathbf{u}(\mathbf{X}(\mathbf{x}_0, t)),$$

$$\mathbf{X}(\mathbf{x}_0, t = 0) = \mathbf{x}_0.$$

Let $\eta(\mathbf{x}_0, \eta_0, t)$ be a tangent vector advected with the flow (see [**30**]), meaning that it satisfies the ODE:

$$\frac{d\eta}{dt} = (\eta \cdot \nabla \mathbf{u}(\mathbf{X}(\mathbf{x}_0, t))) = \left(\frac{\partial \mathbf{u}(\mathbf{x})}{\partial \mathbf{x}}\right)\Bigg|_{\mathbf{x} = \mathbf{X}(\mathbf{x}_0, t)} \eta, \qquad (12.1)$$

with the initial conditions:

$$\eta(\mathbf{x}_0, \eta_0, t) = \eta_0.$$

The Lyapunov exponent Λ for the flow $\mathbf{u}(\mathbf{x})$ is defined (for each initial position) to be the maximum exponential growth rate of the length of such tangent vectors η, i.e.

$$\Lambda(\mathbf{x}_0) = \lim_{t \to \infty} \frac{1}{t} \log \sup_{|\eta_0| = 1} |\eta(\mathbf{x}_0, \eta_0, t)|. \qquad (12.2)$$

We say that $\Lambda > 0$ implies *exponential stretching* in the flow. The displacement between infinitesimally close fluid particles behaves like a tangent vector; so an exponential growth rate of tangent vectors implies that certain fluid particles are being separated at an exponential rate. This separation can be visualized as a stretching in the flow field. If $\Lambda(\mathbf{x}_0) = 0$ for all values of $\mathbf{x}_0 \in \mathcal{D}$, there is at most polynomial growth in the separation of trajectories in the flow $\mathbf{u}(\mathbf{x})$. Incompressibility requires that $\Lambda(\mathbf{x}_0)$ be nonnegative, so we have covered the only two possible cases.

A simple example of a flow in which $\Lambda > 0$ is a flow with a hyperbolic stagnation point, i.e. a point $\mathbf{x}_0$ where $\mathbf{u}(\mathbf{x}_0) = 0$ and the matrix $\left(\frac{\partial \mathbf{u}}{\partial \mathbf{x}}\right)$ has at least one positive real eigenvalue. (Note that the condition $\nabla \cdot \mathbf{u} = 0$ implies that $\mathrm{Tr}\left(\frac{\partial \mathbf{u}}{\partial \mathbf{x}}\right) = 0$.) Clearly, the vector

$$\eta(\mathbf{x}_0, \eta_0, t) = \eta_0 e^{\lambda t},$$

285

where η_0 is an eigenvector of $\left(\frac{\partial \mathbf{u}}{\partial \mathbf{x}}\right)$ with eigenvalue λ, is a solution to (12.1). Hence, λ is a positive lower bound for Λ.

The "chaotic" ABC flow discussed in Subsection 5.2.2 is an example of a steady solution of the Euler equations which exhibits exponential stretching. Without loss of generality, we may normalize the constants so that:

$$1 \geq A \geq B \geq C \geq 0.$$

(There is an obvious permutation symmetry of the coefficients A, B, and C, and dilation of each coefficient by the same number only rescales time. We can further choose A, B, and C positive since a sign reversal only amounts to a rigid translation of the flow.) In the parameter domain $B^2 + C^2 \geq 1$, the ABC flow (5.7) has hyperbolic stagnation points, and thus exponential stretching. Outside this domain, there are *no* stagnation points. In the case B and C are small, Friedlander, Gilbert, & Vishik [37] prove the existence of exponential stretching by constructing a hyperbolic closed trajectory in the flow. (See [82] for the definition of a hyperbolic trajectory; it is a natural extension of the notion of a hyperbolic fixed point.) This special closed trajectory lies on the rational torus corresponding to the lowest resonance (the ratio between the winding numbers of the orbits on this torus is 0:1). The "hyperbolic point" behavior of the trajectories is illustrated via the method of Poincare sections in Figure 3. Chicone [22] has extended this result to show the existence of exponential stretching in ABC flows with only one small parameter. It is conjectured that all ABC flows (except the degenerate integrable case where $B = C = 0$) have positive Lyapunov exponent. Quite possibly, all non-degenerate flows which are eigenfunctions of the curl operator exhibit exponential stretching.

Clearly, flows with exponential stretching are of special interest. They possess strong stochastic properties: the Lagrangian trajectories do not admit globally defined first integrals. Since the ODE for such a flow is not integrable, only a limited class of explicit examples are known. A classical example of a somewhat artificial flow is an Anosov flow [3], which by construction has exponential stretching everywhere. However, there are certain topological obstructions to having exponential stretching everywhere in, say, a sphere, so these flows have limited physical significance.

The idea that exponential stretching could imply fluid instability is originally due to Arnold [9]. In this seminal paper, he examined an example of a model irrotational flow on a compact manifold $\mathcal{M}$ with a Riemannian metric. Every fluid element moving in this flow stretches in one direction and contracts in another direction. Such a flow is unstable. We will now use the instability theorem in Section 11 to prove that all flows of an ideal fluid with exponential stretching are, in fact, hydrodynamically unstable.

Theorem 12.1. *Let the 3-D flow $\frac{d\mathbf{X}}{dt} = \mathbf{u}(\mathbf{X})$ have a positive Lyapunov exponent at some point $\mathbf{x}_0$. Then $\mathbf{u}(\mathbf{x})$ is linearly unstable as a steady flow of an ideal fluid.*

Proof. Let $\mathbf{b}_1(\mathbf{x}_0, \xi_0, t)$ and $\mathbf{b}_2(\mathbf{x}_0, \xi_0, t)$ be two linearly independent solutions of (11.13) for some co-vector $\xi(\mathbf{x}_0, \xi_0, t)$ with $\mathbf{b}_1(\mathbf{x}_0, \xi_0, 0)\cdot\xi_0$ and $\mathbf{b}_2(\mathbf{x}_0, \xi_0, 0)\cdot\xi_0 = 0$. (Note that the evolution of the covector ξ is independent of $\mathbf{b}$.)

The volume of the parallelpiped spanned by $\mathbf{b}_1, \mathbf{b}_2$, and ξ is given by the determinant of these vectors. Consider the time derivative of this volume:

$$\frac{d\,\mathrm{Det}\,(\mathbf{b}_1, \mathbf{b}_2, \xi)}{dt} = \mathrm{Det}\,(\dot{\mathbf{b}}_1, \mathbf{b}_2, \xi) + \mathrm{Det}\,(\mathbf{b}_1, \dot{\mathbf{b}}_2, \xi) + \mathrm{Det}\,(\mathbf{b}_1, \mathbf{b}_2, \dot{\xi}),$$

by multilinearity of the determinant. Substituting for the time derivatives from (11.13), we obtain:

$$\frac{d \operatorname{Det}\,(\mathbf{b}_1, \mathbf{b}_2, \xi)}{dt} = - \operatorname{Det}\left(\left(\frac{\partial \mathbf{u}}{\partial \mathbf{x}}\right)\mathbf{b}_1, \mathbf{b}_2, \xi\right) - \operatorname{Det}\left(\mathbf{b}_1, \left(\frac{\partial \mathbf{u}}{\partial \mathbf{x}}\right)\mathbf{b}_2, \xi\right)$$
$$- \operatorname{Det}\left(\mathbf{b}_1, \mathbf{b}_2, \left(\frac{\partial \mathbf{u}}{\partial \mathbf{x}}\right)^T \xi\right),$$

where we have already discarded two terms which would have had two vectors in the determinant parallel to ξ. We now invoke the following identity, valid for any matrix A and set of vectors $\mathbf{v}_1, \mathbf{v}_2$, and $\mathbf{v}_3$:

$$\operatorname{Det}\,(A\mathbf{v}_1, \mathbf{v}_2, \mathbf{v}_3) + \operatorname{Det}\,(\mathbf{v}_1, A\mathbf{v}_2, \mathbf{v}_3) + \operatorname{Det}\,(\mathbf{v}_1, \mathbf{v}_2, A\mathbf{v}_3) = (\operatorname{Tr} A)\operatorname{Det}\,(\mathbf{v}_1, \mathbf{v}_2, \mathbf{v}_3).$$

To verify this, one could transform the three vectors to the canonical unit vectors, and note that both sides of the equation transform like pseudoscalars under co-ordinate changes. Hence, one need only directly check this equation for the case where the three vectors are the basic orthonormal triad $\hat{\mathbf{i}}, \hat{\mathbf{j}}, \hat{\mathbf{k}}$. (In case the vectors are linearly dependent, the equality is immediate.) Returning to the differential equation, we have:

$$\frac{d \operatorname{Det}\,(\mathbf{b}_1, \mathbf{b}_2, \xi)}{dt} = - \operatorname{Tr}\left(\frac{\partial \mathbf{u}}{\partial \mathbf{x}}\right) \operatorname{Det}\,(\mathbf{b}_1, \mathbf{b}_2, \xi)$$
$$+ \operatorname{Det}\left(\mathbf{b}_1, \mathbf{b}_2, \left(\left(\frac{\partial \mathbf{u}}{\partial \mathbf{x}}\right) - \left(\frac{\partial \mathbf{u}}{\partial \mathbf{x}}\right)^T\right)\xi\right).$$

It is an easy exercise to check that:

$$\left(\left(\frac{\partial \mathbf{u}}{\partial \mathbf{x}}\right) - \left(\frac{\partial \mathbf{u}}{\partial \mathbf{x}}\right)^T\right)\xi = (\nabla\times\mathbf{u}) \times \xi,$$

so we finally have:

$$\frac{d \operatorname{Det}\,(\mathbf{b}_1, \mathbf{b}_2, \xi)}{dt} = - \operatorname{Tr}\left(\frac{\partial \mathbf{u}}{\partial \mathbf{x}}\right) \operatorname{Det}\,(\mathbf{b}_1, \mathbf{b}_2, \xi) + \operatorname{Det}\,(\mathbf{b}_1, \mathbf{b}_2, (\nabla\times\mathbf{u}) \times \xi). \quad (12.3)$$

Since $\nabla \cdot \mathbf{u} = 0$, it follows that $\operatorname{Tr}\left(\frac{\partial \mathbf{u}}{\partial \mathbf{x}}\right) = 0$. Also, $(\nabla\times\mathbf{u}) \times \xi$ is orthogonal to ξ. By incompressibility, the vectors $\mathbf{b}_1, \mathbf{b}_2$ remain always orthogonal to ξ as well. We deduce that $(\nabla\times\mathbf{u}) \times \xi$, $\mathbf{b}_1$, and $\mathbf{b}_2$, span at most a 2-D subspace orthogonal to ξ, so that the second term on the right hand side of (12.3) is also zero. Hence, the volume of the triad is exactly conserved by the flow:

$$\frac{d \operatorname{Det}\,(\mathbf{b}_1, \mathbf{b}_2, \xi)}{dt} = 0. \quad (12.4)$$

Since the flow has positive Lyapunov exponent, there exists a tangent vector η that increases exponentially with time. As the flow is volume preserving ($\nabla \cdot \mathbf{u} = 0$), the existence of an exponentially increasing tangent vector implies the existence of an exponentially decreasing co-tangent vector which we choose to be our $\xi(t)$. Now take two linearly independent solutions $\mathbf{b}_1, \mathbf{b}_2$ of (11.13) which are orthogonal to this co-tangent vector ξ; then by (12.4), one of the vectors $\mathbf{b}_1$ or $\mathbf{b}_2$ must be exponentially growing. From Theorem 11.1, this implies that the flow $\mathbf{u}(\mathbf{x})$ is (linearly) unstable as an ideal fluid flow.

This theorem, in particular, proves that many ABC flows are hydrodynamically unstable. We conjecture that any fluid flow that is an eigenfunction of curl is unstable.

The equivalent result in two dimensions is even easier to prove, since exponential stretching can only occur in the plane if there is a hyperbolic stagnation point $\mathbf{x}_0$. (See [**38**] for a proof.) In this case, the matrix $\left(\frac{\partial \mathbf{u}(\mathbf{x}_0)}{\partial \mathbf{x}}\right)$ has eigenvalues $\pm\lambda$, with $\lambda > 0$. Change coordinates so that the matrix is diagonalized with a left eigenvector a_- corresponding to eigenvalue $-\lambda$ and a right eigenvector a_+ corresponding to eigenvalue λ. This pair of vectors is orthogonal. (See [**60**]) The expressions:

$$\mathbf{b}(\mathbf{x}_0, a_-, a_+, t) = a_+ e^{\lambda t}, \ \xi(\mathbf{x}_0, a_-, t) = a_- e^{-\lambda t}$$

are then solutions to the system of ODE given by (11.13). Hence, the flow is unstable by Theorem 11.1.

LECTURE 13
Integrable Flows

In Section 5 we discussed the two "extreme" classes of steady Euler flows, namely, flows where $\mathbf{u}$ and $\nabla\times\mathbf{u}$ are everywhere parallel (chaotic flows) and flows where $\mathbf{u}$ and $\nabla\times\mathbf{u}$ are nonzero and nowhere parallel (integrable flows). We have shown how the instability criterion given by Theorem 11.1 proves that at least certain classes of chaotic flows are unstable as fluid flows because of the existence of exponential stretching. The question arises whether this criterion, which is detecting instability in the continuous spectrum, can demonstrate instability for an integrable flow without exponential stretching. Friedlander & Vishik [42] use Theorem 11.1 to obtain a sufficient condition for instability of a general smooth axisymmetric flow:

$$\mathbf{u} \times (\nabla\times\mathbf{u}) = \nabla H, \ \nabla \cdot \mathbf{u} = 0,$$

with $\nabla H \neq 0$ almost everywhere. Taking the case where the surfaces $H = H_0$ are noncritical ($\nabla H \neq 0$ everywhere on H_0) and compact, a sufficient condition for instability of a toroidal vortex ring can be given in terms of geometric quantities characterizing the ring. This condition reads:

$$\int_0^T \kappa\mathbf{n}\cdot\nabla H - \frac{\tau_g\mathbf{u}\cdot(\nabla\times\mathbf{u})}{|\nabla H|^2} \, dt \geq 0, \tag{13.1}$$

where all quantities in the integrand are evaluated at the location of a particle $\mathbf{X}(t)$ moving along a streamline which wraps once around the torus $H = H_0$ with period T. (The streamlines and vortex lines are not closed; the period here refers to the time required to return to a given circle on the torus.) The symbol κ denotes the curvature of the streamline, τ_g is the geodesic torsion (see [27]), and $\mathbf{n}$ is the unit principal normal. We note that this condition is not sharp, and a more general sufficient condition for instability can be obtained from Theorem 11.1. But this more general condition is not readily expressed in terms of fundamental quantities from differential geometry.

The basic philosophy for using Theorem 11.1 to obtain explicit conditions for instability, such as condition (13.1), is to seek a covector $\xi(t)$ satisfying (11.13a) and (11.13b) that does not grow in time. This approach is suggested by the result expressed in equation (12.4), showing that growth in $\xi(t)$ must be associated with decay for at least some vectors $\mathbf{b}(t)$. In Friedlander & Vishik [42], it is proved

289

that for the vortex ring model, any solution $\mathbf{b}(t)$ to (11.13c) could grow at most algebraically when $\xi(t)$ grows unboundedly. Thus, the only possibility of finding an exponentially growing $\mathbf{b}(t)$ satisfying (11.13c) is to choose a co-tangent vector $\xi(t)$ that is bounded. Using a perturbation approach with

$$\xi(\mathbf{x}_0, t) = \nabla H(\mathbf{X}(t)) + \varepsilon \Psi(t), \quad \varepsilon \ll 1,$$

it is possible to construct a solution $\mathbf{b}(t)$

$$\mathbf{b}(t) = \mathbf{b}_0(t) + \varepsilon \mathbf{b}_1(t) + \varepsilon^2 \mathbf{b}_2(t) + \dots,$$

which, under certain conditions, exhibits exponential growth. An equation is obtained for the Floquet exponents associated with the monodromy operator for equation (11.13c) (see [50]). A sufficient condition for exponential growth of $\mathbf{b}(t)$ (which in turn implies fluid instability) is the existence of one Floquet exponent with modulus greater than unity. The inequality (13.1) is such a condition.

LECTURE 14
Baroclinic Instability

We now turn to an application of Theorem 11.1 to a problem in geophysical fluid dynamics. The fundamental forces that play a vital role in the fluid motion of the ocean and the atmosphere are the Coriolis force arising from the Earth's rotation and the force of gravity acting on a fluid with density stratification. The Boussinesq approximation for an incompressible fluid with gradual density variations yields the following modification of the Euler equations ([67]):

$$\frac{\partial \mathbf{q}(\mathbf{x},t)}{\partial t} + \mathbf{q}(\mathbf{x},t)\cdot\nabla\mathbf{q}(\mathbf{x},t) = -\frac{1}{\bar{\rho}}\nabla p(\mathbf{x},t) - \rho(\mathbf{x},t)\nabla\Phi(\mathbf{x}), \tag{14.1a}$$

$$\nabla \cdot \mathbf{q}(\mathbf{x},t) = 0, \tag{14.1b}$$

$$\frac{\partial \rho(\mathbf{x},t)}{\partial t} + (\mathbf{q}(\mathbf{x},t)\cdot\nabla)\rho(\mathbf{x},t) = 0. \tag{14.1c}$$

Here $\rho(\mathbf{x},t)$ is the density field of the fluid with average value $\bar{\rho}$, which, by a proper choice of units, may be taken as unity. $\mathbf{q}(\mathbf{x},t)$ and $p(\mathbf{x},t)$ are the velocity and pressure fields, like before. The final new quantity in these equations is the external gravitational potential $\Phi(\mathbf{x})$, the negative gradient of which gives the gravitational force. So the momentum equation (14.1a) is altered by the addition of an external gravitational body force acting on the fluid. The incompressibility condition (14.1b) is just as before. The new equation (14.1c) describes how the density field reacts to the velocity field; it is derived from the conservation of mass of a fluid element, coupled with the incompressibility condition (see for example [67]). Since the additional variable $\rho(\mathbf{x},t)$ in the system is a scalar field, the analysis used to derive the instability criterion in Theorem 11.1 can be modified in a straightforward way to apply to the system (14.1). The characteristics for the geometric optics equations are unchanged by the coupling of the scalar density field. Incidentally, this is not the case for magnetohydrodynamics (MHD) where the interaction of the two *vector* fields (velocity and magnetic field) produce characteristics of variable multiplicity for the linearized MHD equations. This makes it necessary to introduce a new ansatz for the asymptotics which differs from the classical WKB treatment that is appropriate for the Euler equations or the stratified system (14.1). For MHD, it is possible to obtain an instability criterion from a system of *local, hyperbolic PDE* as opposed to the system of ODE derived for the Euler equations [44].

291

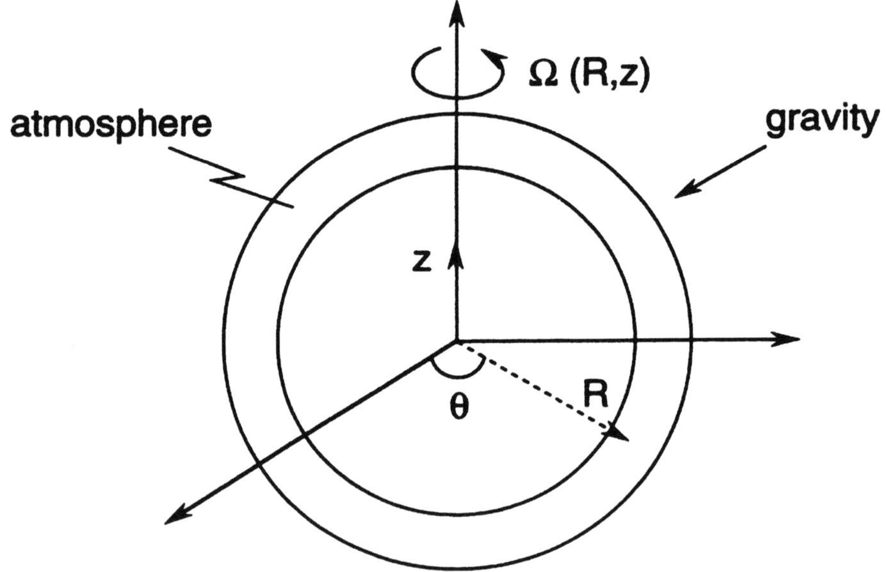

Figure 5. Model of axisymmetric flow about a sphere

For a density-stratified (Boussinesq) fluid, the statement of Theorem 11.1 given by (11.15) is valid with $\mathbf{b}$ and ξ satisfying instead the extended system [**43**]:

$$\frac{d\mathbf{X}}{dt} = \mathbf{u}, \tag{14.2a}$$

$$\frac{d\xi}{dt} = -\left(\frac{\partial \mathbf{u}}{\partial \mathbf{x}}\right)^T \xi, \tag{14.2b}$$

$$\frac{d\mathbf{b}}{dt} = -\left(\frac{\partial \mathbf{u}}{\partial \mathbf{x}}\right)\mathbf{b} + 2\left\{\left[\left(\frac{\partial \mathbf{u}}{\partial \mathbf{x}}\right)\mathbf{b}\right]\cdot\xi\right\}\xi|\xi|^{-2} + r\left[\nabla\rho_0 - (\xi\cdot\nabla\rho_0)\xi|\xi|^{-2}\right] \tag{14.2c}$$

$$\frac{dr}{dt} = -\mathbf{b}\cdot\nabla\Phi, \tag{14.2d}$$

with initial conditions:

$$\mathbf{X}(\mathbf{x}_0, t = 0) = \mathbf{x}_0, \tag{14.3a}$$

$$\xi(\mathbf{x}_0, t = 0) = \xi_0, \tag{14.3b}$$

$$\mathbf{b}(\mathbf{x}_0, \xi_0, r_0, t = 0) = \mathbf{b}_0, \tag{14.3c}$$

$$r(\mathbf{x}_0, \xi_0, r_0, t = 0) = r_0. \tag{14.3d}$$

Here r corresponds to the amplitude of the disturbance to the equilibrium density field $\rho_0(\mathbf{x})$.

We consider a basic steady state that models the large scale features of a global model for the oceans or the atmosphere, namely:

$$\mathbf{u}(\mathbf{x}) = R\Omega(R, z)\hat{\theta}, \quad \rho_0(\mathbf{x}) = \rho_0(R, z) \tag{14.4}$$

where (R, θ, z) are cylindrical polar coordinates. Figure 5 illustrates this basic state of an axisymmetric velocity given by a rotation (with shear) about the axis of the Earth at a local angular velocity Ω, and an axisymmetric density stratification induced by a spherically directed gravitational field $\nabla\Phi$. The equilibrium condition for such a steady configuration is:

$$-\hat{\theta}R\frac{\partial\Omega^2}{\partial z} + (\nabla\Phi \times \nabla\rho_0) = 0, \tag{14.5}$$

which is obtained by taking the curl of the momentum equation (14.1a). Meteorologists refer to this as the "thermal wind" equation [46], [81], which shows how the velocity must adjust so that the pressure gradients induced by the density variation are balanced by the Coriolis force gradients of the rotating fluid.

We seek an exponentially growing $\mathbf{b}(t), r(t)$ satisfying (14.2) for some choice of covector $\xi(t)$ when $\mathbf{u}$ and ρ_0 are given by (14.4), under the relation (14.5). We note that there is no exponential stretching in this equilibrium flow. Let us see if we can find a constant covector ξ satisfying (14.2b) which is of unit length and has the form (in cylindrical coordinates):

$$\xi = \begin{bmatrix} \cos\alpha \\ 0 \\ \sin\alpha \end{bmatrix}. \tag{14.6}$$

In these coordinates, the Jacobian matrix for the flow field reads:

$$\left(\frac{\partial\mathbf{u}}{\partial\mathbf{x}}\right) = \begin{bmatrix} 0 & -\Omega & 0 \\ \Omega + R\Omega_R & 0 & R\Omega_z \\ 0 & 0 & 0 \end{bmatrix}. \tag{14.7}$$

Let us look for solutions to (14.2c) and (14.2d) of the form:

$$\mathbf{b} = \begin{pmatrix} b_1 \\ b_2 \\ b_3 \end{pmatrix} e^{\lambda t}, \; r = r_1 e^{\lambda t}$$

where b_1, b_2, b_3, r_1 are constants such that

$$\mathbf{b}\cdot\xi = b_1 \cos\alpha + b_3 \sin\alpha = 0.$$

Substituting this choice of functions into (14.2) gives an algebraic equation for the eigenvalue λ namely

$$\lambda^2 \left[\lambda^2 + \sin^2\alpha \left(4\Omega^2 + 2R\Omega\Omega_R + \Phi_R\rho_{0R}\right) + \cos^2\alpha \, \Phi_z\rho_{0z} - 2\cos\alpha\sin\alpha \, \Phi_z\rho_{0R}\right] = 0 \tag{14.8}$$

The R and z suffixes denote partial differentiation with respect to these variables, and we have used the equilibrium condition (14.5) to simplify this expression.

The choice of angle α is exactly the choice of the direction of the (constant) wavevector ξ with respect to the equatorial plane. If for some choice of angle α, there exists a solution λ to (14.8) that has a positive real part, then the vector $\mathbf{b}(t)$ grows exponentially with time, implying fluid instability. Clearly this happens if *any* of the following conditions occur:

1. $4\Omega^2 + 2R\Omega\Omega_R + \Phi_R\rho_{0R} < 0$. This is the analogue of the Rayleigh condition for centrifugal instability for the case of a density-stratified fluid (see [2]).
2. $\Phi_z\rho_{0z} < 0$. This is pure gravitational instability when heavy fluid lies over light fluid.

3.

$$R\Omega\Omega_z\Phi_z\rho_{0R} - (2\Omega^2 + R\Omega\Omega_R)\Phi_z\rho_{0z} > 0.$$

This condition can be considered a generalized "baroclinic instability" criterion (see [81]). This instability is an important feature governing the global behavior of the atmosphere which occurs when the gravitational potential energy due to the horizontal density gradient ρ_{0R} is large enough to overcome the stabilizing effects of rotation $2\Omega^2 + R\Omega\Omega_R$ and buoyancy $\Phi_z\rho_{0z}$. (To see how rotation can have a stabilizing effect on a flow, see for example [67].)

In this geophysical example, the techniques we have been discussing have given a precise mathematical proof of localized linear instability results. Thus, the existence of instabilities in a geophysical model can be shown without the traditional approximations, i.e. the so called β plane approximation for a thin shell of rotating fluid. This demonstrates the effectiveness of Theorem 11.1 which allows a difficult problem in the spectral theory of PDE to be answered by construction of particular solutions of a system of ODE's. It is an example of the power of the technique of geometrical optics.

LECTURE 15
Nonlinear Instability

We have discussed an effective sufficient condition for instability which shows that large classes of steady flows of an ideal fluid are linearly unstable. We now ask whether these flows are also *nonlinearly* unstable, as defined in Subsection 6.2. It is well known that linear instability implies nonlinear instability for ODE's (see [68]), but there is no such general result for PDE's. Friedlander, Strauss, and Vishik [38] prove such a result under certain conditions. They consider a general evolution equation:

$$\frac{\partial w}{\partial t} = Lw + N(w), \quad w(0) = w_0 \tag{15.1}$$

within two fixed Banach spaces $X \hookrightarrow Z$. L is the generator of a C_0 group of operators in $\mathcal{L}(Z)$ and N is a nonlinear operator mapping X into Z. As a concrete example, we can think of L as the linearized operator of the Euler equations, while N is the difference between the full nonlinear differential operator and its linearization about some steady state. The spectrum of the linearized operator L is studied in the "large" space Z; X is a "small" space in which a *local* existence theorem for the nonlinear equation (15.1) can be proved. Under certain conditions on the spectrum of L, Friedlander, Strauss, and Vishik [38] prove an abstract theorem which states that spectral instability for L implies nonlinear instability for the general evolution equation (15.1). It is quite possible that the specific conditions on Spec L are an artifact of the method of proof of the theorem, and it is conjectured (but not proved) that the result holds in greater generality. The specific conditions under which the nonlinear instability theorem is proved include the requirement that Spec L satisfies either of the following conditions:

1. There is a "gap" in the spectrum, with all the modes on one side being growing modes, or
2. there is an *unstable* discrete eigenvalue of magnitude larger than, or almost as large as the maximum growth rate due to the continuous unstable spectrum.

Under these conditions, the result that spectral instability proves nonlinear instability is proved by contradiction.

The abstract nonlinear instability theorem can be applied to the Euler equations for an arbitrary smooth Euler equilibrium $\mathbf{u}(\mathbf{x})$ and for the usual functional

295

spaces: Z is taken to be the subspace of L^2 consisting of divergence-free vecto fields in $\mathbb{R}^n$, and X is the space of divergence-free vector fields with components in the Sobolev space $H^s, s > \frac{n}{2} + 1$. The flow domain $\mathcal{D}$ is either bounded or periodic. With this choice of X and Z, the nonlinear term satisfies the inequality:

$$\|N(w)\|_Z \leq c\|w\|_X\|w\|_Z \text{ for } w \in X,$$

for some constant c, and this is one of the conditions under which the instability theorem is proved. Furthermore, a local existence theorem for the Euler equations in $H^s, s > \frac{n}{2} + 1$ is well known [see for example [**88**] or [**77**, Ch. 2]].

The difficulty in applying the abstract nonlinear instability theorem to the Euler equations comes from attempting to check the spectral conditions (1) and (2). If there were no continuous unstable spectrum, i. e., the unstable modes are all classical eigenfunctions with discrete eigenvalues, then both conditions (1) and (2) would hold, and the abstract theorem could be easily applied. However the spectrum of the linearization of the Euler equation has a nonempty continuous part in general. The recent result of Vishik [**90**] that we mentioned in Section 11 shows that the Lyapunov-type quantity on the left hand side of (11.15) is in fact the logarithm of the essential spectral radius. Hence all the examples of flows we discussed in Sections 12 through 14 where this quantity was shown to be positive are associated to an unstable spectrum of L with a non-empty continuous part. There are possible discrete eigenvalues in addition to this unstable continuous spectrum. Thus, the application of the abstract nonlinear instability theorem requires information about the structure of Spec L. As we have pointed out in earlier sections, analysis of the details of the spectrum of the Euler equation linearized about a specific equilibrium is difficult.

One example of a steady Euler flow to which one may straightforwardly apply the nonlinear instability theorem is 2-D plane parallel shear flow $\mathbf{u}(\mathbf{x}) = u(x_2)\hat{\mathbf{i}}$ discussed in Section 9. In this case, it is easy to show that the left hand side of (11.15) is zero for any profile $u(x_2)$. (One observes that tangent vectors can grow at most linearly in time, and deduces from the equations (11.13) that $\mathbf{b}$ can grow at most polynomially in time.) Hence, the logarithm of the essential spectral radius is zero, implying that any unstable spectrum, if it exists, must be purely discrete. In Section 9, we described examples of profiles where the existence of unstable eigenvalues has been demonstrated, i. e. $u(x_2)$ is monotonic with an inflection point [**34**] or $u(x_2)$ is sinusoidal with sufficiently many oscillations [**36**]. These are therefore examples of *nonlinearly* unstable ideal fluid flows.

Conclusion

We have considered a subset of the problems that arise in the theory of stability/instability of the Euler equations which describe the motion of an ideal fluid. After giving the definitions of linear and nonlinear stability, we explained how the concept of stability is crucially dependent on the norm in which the growth rate of a disturbance is measured. The stability of one of the most basic "simple" flows, namely plane parallel flow, was then discussed in some detail. We used this example as an illustration of a flow that can exhibit linear stability, linear instability, nonlinear stability, and nonlinear instability, depending on the profile of the flow, the spatial dimension of the perturbations, and the norm in which growth is measured. The spectral problem for the linearized Euler equations is very hard to analyze in detail (in fact there are open questions connected with the spectrum of the operator linearized about even the "simple" flow mentioned above). We described how asymptotic techniques of geometric optics can be used to circumvent the need for detailed spectral analysis and to prove linear instability, in the energy norm, of large classes of steady flows. Loosely speaking, "most" steady flows of an ideal fluid in three dimensions are unstable. These results follow from a theorem that gives a lower bound on the growth rate of the linearized Euler operator in terms of a Lyapunov type exponent for fluid flow. Finally, we discussed the applicability to the Euler equations of a theorem that proves for rather general nonlinear evolution PDE's that, under certain circumstances, linear instability implies nonlinear instability.

References

[1] H. D. I. Abarbanel, D. D. Holm, J. E. Marsden, and T. S. Ratiu. Nonlinear stability analysis of stratified fluid equilibria. *Philosophical Transactions of the Royal Society of London, Series A*, 318:349, 1986.

[2] D. J. Acheson. *Elementary Fluid Dynamics*, chapters 1.2,1.3,2,5.1-5.3,9.4. Oxford Applied Mathematics and Computing Science Series. Clarendon Press, Oxford, 1990.

[3] D. V. Anosov. Geodesic flows on compact Riemannian manifolds of negative curvature. *Publ. Steklov Inst. Ac-Sc. USSR*, 90:210, 1967.

[4] V. I. Arnold. Conditions for nonlinear stability of stationary plane curvilinear flows of an ideal fluid. *Dokl. Acad. Nauk SSSR*, 162(5):975–978, 1965. Translated in Sov. Math. **6**, 773-777.

[5] V. I. Arnold. Sur la topologie des écoulements stationnaires des fluides parfaits. *C. R. Acad. Sci. Paris*, 261:17–20, 1965.

[6] V. I. Arnold. On an a priori estimate in the theory of hydrodynamic stability. *Izv. Vyssh. Ucheb. Zaved. Mathematika*, 54:3–5, 1966. Translated in Amer. Math. Soc. Trans. Ser. 2 **19**, 267-269.

[7] V. I. Arnold. Sur la géometrie differentielle des groupes de Lie de dimension infinie et ses applications a l'hydrodynamique des fluides parfaits. *Annales de l'Institut Fourier, Grenoble*, 16:316–361, 1966.

[8] V. I. Arnold. The Hamiltonian nature of the Euler equations in the dynamics of a rigid body and of an ideal fluid. *Usp. Math. Nauk.*, 24(3):225–226, 1969. Russian.

[9] V. I. Arnold. Notes on the three-dimensional flow pattern of a perfect fluid in the presence of a small perturbation of the initial velocity field. *Applied Mathematics & Mechanics*, 36(2):236–242, 1972.

[10] V. I. Arnold. *Selecta Mathematica Sovietica*, 5:327, 1986. English Translation.

[11] V. I. Arnold. *Mathematical Methods of Classical Mechanics*. Number 60 in Graduate Texts in Mathematics. Springer-Verlag, New York, second edition, 1989.

[12] V. I. Arnold. *Ordinary Differential Equations*, appendix 2. Springer-Verlag, Berlin, third edition, 1992.

[13] V. I. Arnold and B. A. Khesin. *Annual Review of Fluid Mechanics*, 24:145–166, 1992.

[14] G. K. Batchelor. *An Introduction to Fluid Dynamics*, section 1.1. Cambridge University Press, Cambridge, 1967.

[15] J. T. Beale, T. Kato, and A. J. Majda. Remarks on the breakdown of smooth solutions for the 3-D Euler equations. *Communications in Mathematical Physics*, 94:61–66, 1984.

[16] C. M. Bender and S. A. Orszag. *Advanced Mathematical Methods for Scientists and Engineers*, chapter 10. International Series in Pure and Applied Mathematics. McGraw-Hill, New York, 1978.

[17] A. Bertozzi and P. Constantin. Global regularity for vortex patches. *Communications in Mathematical Physics*, 152:19–28, 1993.

[18] K. M. Case. Stability of inviscid plane Couette flow. *Physics of Fluids*, 3:143–148, 1960.

[19] S. Chandrasekhar. *Hydrodynamics and Hydromagnetic Stability*. Dover, New York, 1961.

[20] J.-Y. Chemin. Sur le mouvement des particules d'un fluide parfait incompressible bidimensionnel. *Inventiones Mathematicae*, 103:599, 1991.

[21] J.-Y. Chemin. Régularité de la trajectoire des particules d'un fluide parfait incompressible remplissant l'espace. *Journal de Mathematiques Pures et Appliquees*, 71:407, 1992.

[22] C. Chicone. A geometric approach to regular perturbation theory with application to hydrodynamics. To appear in Transactions of the AMS.

[23] A. J. Chorin. *Vorticity and Turbulence*, volume 103 of *Applied Mathematical Sciences*. Springer-Verlag, New York, 1994.

[24] A. J. Chorin and J. E. Marsden. *A Mathematical Introduction to Fluid Mechanics*, volume 4 of *Texts in Applied Mathematics*. Springer-Verlag, New York, third edition, 1993.

[25] P. Constantin. A few results and open problems regarding incompressible fluids. *Notices of the AMS*, 42(6):658–663, June 1995.

[26] P. Constantin and C. Foias. *Navier-Stokes Equations*. University of Chicago Press, Chicago, 1988.

[27] M. P. do Carmo. *Differential Geometry of Curves and Surfaces*, section 3.2, page 153. Prentice-Hall, Englewood Cliffs, NJ, 1976.

[28] T. Dombre, U. Frisch, J. M. Greene, M. Hénon, A. Mehr, and A. M. Soward. Chaotic streamlines in the ABC flows. *Journal of Fluid Mechanics*, 167:353–391, 1986.

[29] P. G. Drazin and W. H. Reid. *Hydrodynamic Stability*. Cambridge Monographs on Mechanics and Applied Mathematics. Cambridge University Press, Cambridge, 1982.

[30] B. A. Dubrovin, A. T. Fomenko, and S. P. Novikov. *Modern Geometry – Methods and Applications, Vol. 1*, volume 93 of *Graduate Texts in Mathematics*, sections 2.2, 23.3. Springer-Verlag, New York, second edition, 1992.

[31] K. S. Eckhoff. On stability for symmetric hyperbolic systems. *Journal of Differential Equations*, 40:94–115, 1981.

[32] S. N. Ethier and T. G. Kurtz. *Markov Processes: Characterization and Convergence*, chapter 1. John Wiley & Sons, New York, 1986.

[33] L. Euler. Principes généraux du mouvement des fluides. *Hist. de L'Acad. de Berlin*, 1755.

[34] L. D. Faddeev. On the theory of the stability of stationary plane-parallel flows of an ideal fluid. *Zapiski Nauchnykh Seminarov LOMI*, 21:164–172, 1971.

[35] F. G. Friedlander. *Sound Pulses*. Cambridge University Press, Cambridge, 1958.

[36] S. Friedlander and L. N. Howard. Instability in parallel flows revisited. To appear, Studies in Applied Math, 1998.

[37] S. Friedlander, A. D. Gilbert, and M. M. Vishik. Hydrodynamic instability for certain ABC flows. *Geophysical and Astrophysical Fluid Dynamics*, 73:97–107, 1993.

[38] S. Friedlander, W. Strauss, and M. M. Vishik. Nonlinear instability in an ideal fluid. Annales I.H.P., J. Nonlineaire, 14 no 2:187-209, 1997.

[39] S. Friedlander and M. M. Vishik. Nonlinear stability for stratified magneto-hydrodynamics. *Geophysical and Astrophysical Fluid Dynamics*, 55:19–45, 1990.

[40] S. Friedlander and M. M. Vishik. Dynamo theory, vorticity generation, and exponential stretching. *Chaos*, 1(2):198–205, 1991.

[41] S. Friedlander and M. M. Vishik. Instability criteria for the flow of an inviscid incompressible fluid. *Physical Review Letters*, 66(17):2204–2206, Apr. 29 1991.

[42] S. Friedlander and M. M. Vishik. Instability criteria for steady flows of a perfect fluid. *Chaos*, 2(3):455–460, 1992.

[43] S. Friedlander and M. M. Vishik. Instability criteria in fluid dynamics. In H. K. Moffatt, editor, *Proceedings of the Workshop on Topological Methods in Fluid Mechanics*, volume 128 of *NATO ASI Series*, pages 535–549. Kluwer, 1992.

[44] S. Friedlander and M. M. Vishik. On stability and instability criteria for magnetohydrodynamics. *Chaos*, 5(2):416–423, 1995.

[45] P. Gérard. Résultats récents sur les fluides parfaits incompressibles bidimensionnels. *Seminaire Bourbaki*, 44ème année(757):411–444, 1991-92.

[46] A. E. Gill. *Atmosphere-Ocean Dynamics*, volume 30 of *International Geophysics Series*, section 7.7. Academic Press, San Diego, 1982.

[47] N. Gunther. On the motion of fluid in the moving container. *Izvestiya Akademii Nauk SSSR Seriya Fiziko-Matemicheskikh*, 20:1323–1348,1503–1532, 1926.

[48] N. Gunther. On the motion of fluid in the moving container. *Izvestiya Akademii Nauk SSSR Seriya Fiziko-Matemicheskikh*, 21:621-656,735–756, 1139-1162, 1927.

[49] N. Gunther. On the motion of fluid in the moving container. *Izvestiya Akademii Nauk SSSR Seriya Fiziko-Matemicheskikh*, 22:9–30, 1928.

[50] J. K. Hale. *Ordinary Differential Equations*, section III.7. Krieger, Malabar, Florida, second edition, 1980.

[51] H. Helmholtz. On the integrals of the hydrodynamical equations which express vortex-motion. *Philosophical Magazine, Series 4*, 33:485–512, 1867. Translation.

[52] M. Hénon. Sur la topologie des lignes de courant dans un eas particulier. *Comptes Rendus de L'Académie de Sciences Paris*, 262:312, 1966.

[53] M. W. Hirsch and S. Smale. *Differential Equations, Dynamical Systems, and Linear Algebra*, volume 60 of *Pure and Applied Mathematics*, chapter 11. Academic Press, Boston, 1974.

[54] D. D. Holm, J. E. Marsden, T. Ratiu, and A. Weinstein. Nonlinear stability of fluid and plasma equilibria. *Physics Reports*, 123(1 & 2):1–116, 1985.

[55] L. N. Howard. Note on a paper of Miles. *Journal of Fluid Mechanics*, 10:502–512, 1961.

[56] E. Infeld and G. Rowlands. *Nonlinear Waves, Solitons and Chaos*, section 8.1.3. Cambridge University Press, 1990.

[57] F. John. *Partial Differential Equations*, volume 1 of *Applied Mathematical Sciences*, section 3.2. Springer-Verlag, New York, fourth edition, 1982.

[58] S. Karlin and H. M. Taylor. *A Second Course in Stochastic Processes*, section 15.11. Academic Press, Boston, 1981.

[59] T. Kato. On classical solutions of the two-dimensional non-stationary Euler equations. *Archive for Rational Mechanics & Analysis*, 25:188–200, 1967.

[60] J. P. Keener. *Principles of Applied Mathematics*, chapter 1, page 49. Addison-Wesley, Reading, MA, 1988.

[61] L. Kelvin. *Math. and Phys. Papers*, 4:13–66, 1869.

[62] H. Lamb. *Hydrodynamics*. Dover, New York, sixth edition, 1945.

[63] L. D. Landau and E. M. Lifshitz. *Fluid Mechanics*, volume 6 of *Course of Theoretical Physics*, sections 13,67. Pergamon Press, Oxford, second edition, 1987.

[64] J. Leray. Etude de diverses équations intégrales nonlinéaires et de quelques problémes que pose l'hydrodynamique. *Journal de Mathematiques Pures et Appliquees*, 12:1–82, 1933.

[65] J. Leray. Essai sur les mouvements plans d'un liquide visqueux que limitent des parois. *Journal de Mathematiques Pures et Appliquees*, 13:331–418, 1934.

[66] J. Leray. Sur le mouvement d'un liquide visqueux emplissant l'espace. *Acta Math.*, 63:193–248, 1934.

[67] M. Lesieur. *Turbulence in Fluids*, volume 1 of *Fluid Mechanics and its Applications*, sections II.2,8, III.5. Kluwer, Dordrecht, second revised edition, 1990.

[68] A. J. Lichtenberg and M. A. Lieberman. *Regular and Chaotic Dynamics*, volume 38 of *Applied Mathematical Sciences*, section 3.3d. Springer-Verlag, New York, second edition, 1992.

[69] L. Lichtenstein. Über Einige Existenzprobleme der Hydrodynamik homogener unzusammendrückbarer reibunglosser Flüssigkeiten und die Helmholtzschen Wirbelsatze, zweite Abhandlung. *Mathematische Zeitschrift*, 26:196–323,387–415,725, 1927.

[70] L. Lichtenstein. Über Einige Existenzprobleme der Hydrodynamik homogener unzusammendrückbarer reibunglosser Flüssigkeiten und die Helmholtzschen Wirbelsatze, vierte Abhandlung. *Mathematische Zeitschrift*, 32:608, 1930.

[71] C. C. Lin. On the stability of a 2-dimensional parallel flows I. *Quart. Appl. Math.*, 3:117–142, 1945.

[72] C. C. Lin. *The Theory of Hydrodynamic Stability*. Cambridge University Press, Cambridge, 1966.

[73] D. Ludwig. Exact and asymptotic solutions of the Cauchy problem. *Communications on Pure and Applied Mathematics*, 13:473–508, 1960.

[74] A. J. Majda. *Compressible Fluid Flow and Systems of Conservation Laws in Several Space Variables*, volume 53 of *Applied Mathematical Sciences*, section 2.4. Springer-Verlag, New York, 1984.

[75] A. J. Majda. Vorticity and the mathematical theory of incompressible fluid flow. *Communications on Pure and Applied Mathematics*, 39:S187–S220, 1986.

[76] A. J. Majda. Vorticity, turbulence, and acoustics in fluid flow. *SIAM Review*, 33(3):349–388, Sept. 1991.

[77] C. Marchioro and M. Pulvirenti. *Mathematical Theory of Incompressible Nonviscous Fluids*, volume 96 of *Applied Mathematical Sciences*. Springer-Verlag, New York, 1994.

[78] J. E. Marsden, D. G. Ebin, and A. E. Fisher. Diffeomorhism groups, hydrodynamics, and relativity. In J. R. Vanstone, editor, *Proceedings of the 13th Biennial Seminar of the Canadian Mathematical Congress*, volume 1, pages 135–279. Canadian Mathematical Congress, 1972.

[79] L. Meshalkin and Y. Sinai. Investigation of stability for a system of equations describing plane motion of a viscous incompressible fluid. *Applied Mathematics and Mechanics*, 25:1140–1143, 1961.

[80] R. H. Morf, S. A. Orszag, and U. Frisch. Spontaneous singularity in three-dimensional, inviscid, incompressible flow. *Physical Review Letters*, 44:572, 1980.

[81] J. Pedlosky. *Geophysical Fluid Dynamics*, chapter 2.9,7. Springer-Verlag, New York, second edition, 1987.

[82] L. Perko. *Differential Equations and Dynamical Systems*, volume 7 of *Texts in Applied Mathematics*, section 3.5. Springer-Verlag, New York, 1991.

[83] M. Reed and B. Simon. *Functional Analysis*, volume 1 of *Methods of Modern Mathematical Physics*, chapter VI.3. Academic Press, San Diego, revised and enlarged edition, 1980.

[84] S. I. Rosencrans and D. M. Sattinger. On the spectrum of an operator occuring in the theory of hydrodynamic stability. *Journal of Mathematical and Physical Sciences*, 45:289–300, 1966.

[85] W. Rudin. *Principles of Mathematical Analysis*, chapter 10, pages 275–280. International Series in Pure and Applied Mathematics. McGraw-Hill, New York, third edition, 1976.

[86] L. Sadun and M. Vishik. The spectrum of the second variation of the energy for an ideal incompressible fluid. *Physics Letters A*, 182:394–398, 1993.

[87] G. I. Taylor and A. E. Green. Mechanism of the production of small eddies from large ones. *Proceedings of the Royal Society of London, Series A*, 158:499–521, 1937.

[88] R. Temam. *Navier-Stokes Equations*, volume 2 of *Studies in Mathematics and its Applications*. North-Holland, New York, third revised edition, 1984.

[89] M. Van Dyke. *An Album of Fluid Motion*. Parabolic Press, Stanford, 1982.

[90] M. M. Vishik. Spectrum of small oscillations of an ideal fluid and lyapunov exponents. J. Math. Pures Appl., 75: 531-557, 1996.

[91] M. M. Vishik and S. Friedlander. Dynamo theory methods for hydrodynamic stability. *Journal de Mathematiques Pures et Appliquees*, 72:145–180, 1993.

[92] G. B. Whitham. *Linear and Nonlinear Waves*, section 7.7. Pure and Applied Mathematics. John Wiley & Sons, New York, 1974.

[93] W. Wolibner. Un théorème sur l'existence du mouvement plan d'un fluide parfait, homogéne, incompressible, pendant un temps infiniment long. *Math. Zam.*, 37:698–726, 1933.

[94] V. I. Yudovich. Non-stationary flow of an ideal incompressible fluid. *Akademiya Nauk SSSR. Zhurnal Vychislitelnol Matematiki I Matematicheskoi Fiziki*, 3:1032–1066, 1963. Russian.

[95] V. I. Yudovich. On loss of smoothness for the Euler equations. *Dinamika Sploshoy Sredy (Dynamics of Continuum Media)*, 16:71–78, 1974.

[96] V. I. Yudovich. Uniqueness theorem for the basic nonstationary problem in the dynamics of an ideal incompressible fluid. *Math. Res. Letters*, 2:27–38, 1995.

[97] V. I. Yudovich. Private communication.

Waves and Transport

George Papanicolaou
Leonid Ryzhik

Waves and Transport

G. Papanicolaou
L. Ryzhik

LECTURE 1
Introduction

1.1. The geophysical problem

Many propagation phenomena that arise from waves in randomly inhomogeneous environments or media can be described very effectively by radiative transport equations. In these notes we shall explain in some detail when this is possible and why.

Radiative transport theory was introduced phenomenologically in order to describe the propagation of light energy through the atmosphere [16]. It remained a subject without any connection to wave motion until the nineteen sixties [54, 58, 40, 59, 7] when it it was realized that it arises rather naturally in wave propagation through random media. It was applied successfully to underwater sound propagation [11] and in 1984 R.S. Wu used it to analyze and interpret seismic data [66]. Since then phenomenological transport theory has been applied frequently in seismology [68, 69, 49] but its connections to the underlying elastic wave equations have not been analyzed. We will explain this here.

As in astronomical problems that motivated the introduction of radiative transport theory, it is seismic events that propagate over long distances that are best suited to a transport description. Long here means compared to typical wavelengths and, moreover, the seismic activity must not be confined near the surface where the geological structure is highly anisotropic, as in exploration seismology. Typical features that favor a transport approximation arise, for example, in the following situation:

---

[1]Department of Mathematics, Stanford University, Stanford CA 94305.
E-mail address: papanico@math.stanford.edu, ryzhik@math.stanford.edu.

- Frequency content of the source is in the 1-10Hz range. This means that with a propagation speed of 3 km/sec, say, typical wavelengths are in the 300m-3km range.
- Propagation distances are in the 100-1000km range
- The random inhomogeneities are weak, not very anisotropic and have typical correlation lengths that are comparable to the wavelengths.

If we ignore the random inhomogeneities then with the above parameters we are in the high frequency regime and traditional seismology deals almost exclusively with this approximation: description of the seismic data through arrival times of main events and their amplitude. The structure of the coda, or part of the seismogram that follows a main event, has a lot of information but is difficult to use because it is not understood well how to extract information from it except in rather empirical ways [2].

Deciding how to model the random inhomogeneities is not easy because there is so much variability in the geology of the earth's crust. A simple and reasonably good way to proceed is to assume that the waves interact fully with the inhomogeneities but the fluctuations of the inhomogeneities are weak, as stated in item three above. Full interaction means that typical wavelengths and typical correlation lengths of the inhomogeneities are comparable.

Transport theory has been used successfully to interpret and classify seismic data [66]. So far, however, only infinite, isotropic media are considered, for both steady and time dependent problems. Moreover, wave polarization is not taken into account and, as we will see later, it does play an important role. The seismic waves are treated as if they were scalar rather than vector waves.

In its simplest form transport theory is as follows. Let $a(\mathbf{x}, \mathbf{k})$ be the angularly resolved wave energy density, that is, the energy density at $\mathbf{x}$ propagating in the direction $\mathbf{k}$, which is a unit vector $|\mathbf{k}| = 1$. Then a satisfies the transport equation

$$(1.1) \qquad \mathbf{k} \cdot \nabla_{\mathbf{x}} a = \frac{1}{4\pi} \int_{|\mathbf{k}'|=1} a(\mathbf{x}, \mathbf{k}) d\mathbf{k}' - \alpha a + S(x)$$

where $\alpha \geq 1$ is called the total cross section of the scattering and measures the relative effect of absorption to scattering ($\alpha = 1$ means no absorption) and $S(\mathbf{x})$ is the source of wave energy, assumed isotropic. Since there are no boundaries and the equation is spatially homogeneous we also get an equation for spatial energy density

$$(1.2) \qquad J(\mathbf{x}) = \int_{|\mathbf{k}|=1} a(\mathbf{x}, \mathbf{k}) d\mathbf{k}$$

which is a convolution equation

$$(1.3) \qquad J(\mathbf{x}) = \gamma \int_{\mathbf{R}^3} J(\mathbf{y}) \frac{e^{-|\mathbf{x}-\mathbf{y}|}}{|\mathbf{x}-\mathbf{y}|^2} d\mathbf{y} + S(x).$$

When $|\mathbf{x}|$ is large and S has compact support the energy density J has *diffusive* behavior

$$(1.4) \qquad J(\mathbf{x}) \approx S_0 \frac{e^{-\gamma_0 |\mathbf{x}|}}{|\mathbf{x}|}$$

where $\gamma_0 = \gamma_0(\alpha)$ and tends to zero as $\alpha \to 1$. The constant S_0 is characteristic of the source strength.

Much of the geophysical work is based on making good use of this last formula (1.4) to classify seismic data sets by identifying the decay parameter γ_0, hence the total cross section α, and the source strength S_0. More accurate results can be obtained by using the time dependent transport equation

$$(1.5) \qquad \frac{1}{c}\frac{\partial a}{\partial t} + \mathbf{k}\cdot\nabla_{\mathbf{x}}a = \frac{1}{4\pi}\int_{|\mathbf{k}'|=1} a(t,\mathbf{x},\mathbf{k})d\mathbf{k}' - \alpha a + S(t,x)$$

where c is the propagation speed, $|\mathbf{k}| = 1$ and the initial value of the energy density a is given.

It is important to keep in mind that in some situations radiative transport theory cannot be used because wave energy gets *localized*. This is typically the case in exploration seismology where the propagation features have the following form:

- The source is near the surface where the geological structure is highly anisotropic (nearly layered) down to ten or more kilometers.
- The frequency content of the source is in the 20-50Hz regime so that wavelengths are typically 150 meters or so, with a propagation speed of 3km/sec.
- Propagation distances of interest are around 10km in depth over a surface of 1-3 square kilometers.
- The correlation length of the inhomogeneities is in the 3-10m range, as has been estimated from diverse well log data [64].
- The fluctuations in the medium parameters are not small, with standard deviations around 15%.

It is not really obvious that the correct conclusion from the mathematical articulation of the above specifications is absence of energy transport and reflection back to the surface of all the energy released by the source. This, however, has been done in detail [4, 38], where it is shown that we do have wave localization. Note that we are in the high frequency regime here also (ratio of wavelength to propagation distance is small) so it is really because of stochastic effects that we do not get transport behavior.

1.2. Radiative transport equations

As we have noted already, the theory of radiative transport was originally developed to describe how light energy propagates through a turbulent atmosphere. We shall show how this theory can be derived from the governing equations for light and for other waves of any type, in a randomly inhomogeneous medium. Our results take into account nonuniformity of the background medium, scattering by random inhomogeneities, the effect of polarization, the coupling of different types of waves, etc. The main new application is to elastic waves, in which shear waves exhibit polarization effects while the compressional waves do not, and the two types of waves are coupled. We also analyze solutions of the transport equations at long times and long distances and show that they have diffusive behavior.

Transport equations arise because a wave with wave vector $\mathbf{k}'$ at a point $\mathbf{x}$ in a randomly inhomogeneous medium can be scattered into any direction $\hat{\mathbf{k}}$ with wave vector $\mathbf{k}$, where $\hat{\mathbf{k}} = \mathbf{k}/|\mathbf{k}|$. Therefore one must consider, as above, the angularly resolved, wave vector dependent, scalar energy density $a(t,\mathbf{x},\mathbf{k})$ defined for all $\mathbf{k}$ at each point $\mathbf{x}$ and time t. For scalar waves, energy conservation is expressed by the general transport equation

$$(1.6) \quad \frac{\partial a(t,\mathbf{x},\mathbf{k})}{\partial t} + \nabla_\mathbf{k}\omega(\mathbf{x},\mathbf{k})\cdot\nabla_\mathbf{x}a(t,\mathbf{x},\mathbf{k}) - \nabla_\mathbf{x}\omega(\mathbf{x},\mathbf{k})\cdot\nabla_\mathbf{k}a(t,\mathbf{x},\mathbf{k})$$

$$= \int_{R^3}\sigma(\mathbf{x},\mathbf{k},\mathbf{k}')a(t,\mathbf{x},\mathbf{k}')d\mathbf{k}' - \Sigma(\mathbf{x},\mathbf{k})a(t,\mathbf{x},\mathbf{k}).$$

Here $\omega(\mathbf{x},\mathbf{k})$ is the frequency at $\mathbf{x}$ of the wave with wave vector $\mathbf{k}$, $\sigma(\mathbf{x},\mathbf{k},\mathbf{k}')$ is the differential scattering cross-section -the rate at which energy with wave vector $\mathbf{k}'$ is converted to wave energy with wave vector $\mathbf{k}$ at position $\mathbf{x}$- and

$$(1.7) \quad \int \sigma(\mathbf{x},\mathbf{k}',\mathbf{k})d\mathbf{k}' = \Sigma(\mathbf{x},\mathbf{k})$$

is the total scattering cross-section. Both σ and Σ are nonnegative and σ is usually symmetric in $\mathbf{k}$ and $\mathbf{k}'$. For acoustic waves $\omega(\mathbf{x},\mathbf{k}) = v(\mathbf{x})|\mathbf{k}|$, with v the sound speed (3.36), and the differential scattering cross-section is given by

$$(1.8) \quad \sigma(\mathbf{x},\mathbf{k},\mathbf{k}') = \frac{\pi v^2(\mathbf{x})|\mathbf{k}|^2}{2}\ \{\ (\hat{\mathbf{k}}\cdot\hat{\mathbf{k}}')^2\hat{R}_{\rho\rho}(\mathbf{k}-\mathbf{k}') + 2(\hat{\mathbf{k}}\cdot\hat{\mathbf{k}}')\hat{R}_{\rho\kappa}(\mathbf{k}-\mathbf{k}') +$$

$$\hat{R}_{\kappa\kappa}(\mathbf{k}-\mathbf{k}')\}\ \cdot\ \delta(v(\mathbf{x})|\mathbf{k}| - v(\mathbf{x})|\mathbf{k}'|).$$

Here $\hat{R}_{\rho\rho}$, $\hat{R}_{\rho\kappa}$ and $\hat{R}_{\kappa\kappa}$ are the power spectra of the fluctuations of the density ρ and compressibility κ defined by (4.22) and (4.53) below. The left side of (1.6) is the total time derivative of $a(t,\mathbf{x},\mathbf{k})$ at a point moving along a ray in phase space $(\mathbf{x},\mathbf{k})$, with the frequency adjusting to the appropriate local value. The right side of (1.6) represents the effects of scattering.

The transport equation (1.6) is conservative when (1.7) holds because then

$$\iint a(t,\mathbf{x},\mathbf{k})d\mathbf{x}d\mathbf{k} = \text{const}$$

independent of time. For simplicity we will assume that there is no intrinsic attenuation. However, it is accounted for easily by letting the total scattering cross-section be the sum of two terms

$$\Sigma(\mathbf{x},\mathbf{k}) = \Sigma_{sc}(\mathbf{x},\mathbf{k}) + \Sigma_{ab}(\mathbf{x},\mathbf{k})$$

where $\Sigma_{sc}(\mathbf{x},\mathbf{k})$ is the total scattering cross-section given by (1.7) and $\Sigma_{ab}(\mathbf{x},\mathbf{k})$ is the intrinsic attenuation rate.

The reason that the power spectral densities of the inhomogeneities determine the scattering cross-section (1.8) is seen most easily from a Born expansion of the wave solution for weak inhomogeneities. The single scattering approximate solution of (1.6) and the second moments of the single scattering approximate solution for the underlying wave equation must be the same. The latter are determined by the power spectra of the inhomogeneities. The same considerations explain the appearance of the delta function in the scattering cross-section (1.8) when the random inhomogeneities do not depend on time, for then the frequency is unchanged by scattering. The transport equation (1.6) arises also when the waves are scattered by discrete scatterers that are randomly distributed in the medium. In this case the scattering cross-section (1.8) is the same as the cross-section of a single scatterer times the density of scatterers. We will deal only with continuous random media.

Equation (1.6) has been derived from equations governing particular wave motions by various authors, such as Stott [54], Watson et.al. [60], [58], [40], [59], Barabanenkov et.al. [7], Besieris and Tappert [10], Howe [32], Ishimaru [34] and

Besieris et. al. [11] with a recent survey presented in [8]. These derivations also determine the functions $\omega(\mathbf{x}, \mathbf{k})$ and $\sigma(\mathbf{x}, \mathbf{k}, \mathbf{k}')$ and show how a is related to the wave field. We shall derive (1.6) and these functions as a special case of our more general theory.

As we explained in the seismological context, we expect that radiative transport equations will provide a good description of wave energy transport when (i) typical wavelengths are short compared to macroscopic features of the medium (high frequency case), (ii) correlation lengths of the inhomogeneities are comparable to wavelengths and (iii) the fluctuations of the inhomogeneities are weak. Condition (ii) is important because it allows strong interaction between the waves and the inhomogeneities, which is the most interesting and difficult case to analyze. In addition to these three conditions, the inhomogeneities must not be too anisotropic because in layered random media wave localization occurs even with weak fluctuations, instead of transport [4]. When the fluctuations are strong, wave localization can occur even when the inhomogeneities are isotropic [24], [52].

We shall also analyze the diffusive behavior of solutions of (1.6) which emerges at times and distances that are long compared to a typical transport mean free time $1/\Sigma$ and a typical transport mean free path $|\nabla_{\mathbf{k}}\omega|/\Sigma$. In this regime the phase space energy density $a(t, \mathbf{x}, \mathbf{k})$ is approximately independent of the direction of the wave vector $\mathbf{k}$, $a(t, \mathbf{x}, \mathbf{k}) \sim \bar{a}(t, \mathbf{x}, |\mathbf{k}|)$. In the simplest, spatially homogeneous case, $\bar{a}$ satisfies the diffusion equation

$$(1.9) \qquad \frac{\partial \bar{a}}{\partial t} = \nabla_{\mathbf{x}} \cdot (D\nabla_{\mathbf{x}}\bar{a})$$

with a constant diffusion coefficient $D = D(|\mathbf{k}|)$, (5.13, 5.14), determined by the differential scattering cross-section σ. Diffusion approximations for scalar transport equations are well known [15], including their behavior near boundaries [41], [9]. Our results show that diffusion approximations are also valid for the more general transport equations that arise for electromagnetic and elastic waves.

1.3. Transport theory for electromagnetic waves

To describe electromagnetic waves in isotropic media we must know their state of polarization. Therefore the radiative transport theory of electromagnetic waves must account for energy transport in different states of polarization. Such transport equations were first proposed by Chandrasekhar [16]. They are a coupled system of transport equations for the Stokes parameters I, Q, U, V as functions of time, position and wave number [12]. The Stokes vector is related to the coherence matrix $W(t, \mathbf{x}, \mathbf{k})$ by

$$(1.10) \qquad W(t, \mathbf{x}, \mathbf{k}) = \frac{1}{2}\begin{pmatrix} I + Q & U + iV \\ U - iV & I - Q \end{pmatrix}.$$

In terms of W, which is Hermitian and positive definite, Chandrasekhar's transport equation is

$$\frac{\partial W}{\partial t} + \nabla_{\mathbf{k}}\omega(\mathbf{x}, \mathbf{k}) \cdot \nabla_{\mathbf{x}}W - \nabla_{\mathbf{x}}\omega(\mathbf{x}, \mathbf{k}) \cdot \nabla_{\mathbf{k}}W + WN(\mathbf{x}, \mathbf{k}) - N(\mathbf{x}, \mathbf{k})W$$

$$(1.11) \qquad = \int_{R^3} \sigma(\mathbf{x}, \mathbf{k}, \mathbf{k}')[W(t, \mathbf{x}, \mathbf{k}')]d\mathbf{k}' - \Sigma(\mathbf{x}, \mathbf{k})W(t, \mathbf{x}, \mathbf{k}).$$

Here $\omega(\mathbf{x}, \mathbf{k}) = v(\mathbf{x})|\mathbf{k}|$ is the local frequency and $v(\mathbf{x}) = (\epsilon(\mathbf{x})\mu(\mathbf{x}))^{-1/2}$ is the local speed of light. The differential scattering cross-section $\sigma(\mathbf{x}, \mathbf{k}, \mathbf{k}')$ is a tensor. In the simplest case of isotropic random inhomogeneities, without fluctuations in the magnetic permeability μ, it has the form

$$\sigma(\mathbf{x}, \mathbf{k}, \mathbf{k}')[W(t, \mathbf{x}, \mathbf{k}')] = \frac{\pi v^2(\mathbf{x})|\mathbf{k}|^2 \hat{R}_{\epsilon\epsilon}(|\mathbf{k} - \mathbf{k}'|)}{2} \; T\,(\mathbf{k}, \mathbf{k}')W(t, \mathbf{x}, \mathbf{k}')T(\mathbf{k}', \mathbf{k})$$

$$(1.12) \qquad\qquad\qquad\qquad\qquad\qquad\qquad \cdot \; \delta(v(\mathbf{x})|\mathbf{k}| - v(\mathbf{x})|\mathbf{k}'|).$$

Here $\hat{R}_{\varepsilon\varepsilon}(\mathbf{k})$ is the power spectrum of the dimensionless fluctuations of the relative dielectric permittivity. The total scattering cross-section $\Sigma(\mathbf{x}, \mathbf{k})$ is given by

$$(1.13) \qquad \Sigma(\mathbf{x}, \mathbf{k}) = \frac{\pi^2 |\mathbf{k}|^4 v(\mathbf{x})}{2} \int_{-1}^{1} \hat{R}_{\epsilon\epsilon}(|\mathbf{k}| \sqrt{2 - 2\eta})(1 + \eta^2)d\eta.$$

The differential scattering cross-section σ and the total scattering cross-section Σ are related by the matrix analog of (1.7)

$$(1.14) \qquad \int_{R^3} \sigma(\mathbf{k}', \mathbf{k})[I]d\mathbf{k}' = \Sigma(\mathbf{k})I,$$

where I is 2×2 identity matrix.

To define the matrices T and N, which occur in (1.12) and (1.11), respectively, we let $(\hat{\mathbf{k}}, \mathbf{z}^{(1)}(\mathbf{k}), \mathbf{z}^{(2)}(\mathbf{k}))$ be the orthonormal propagation triple consisting of the direction of propagation $\hat{\mathbf{k}}$ and two transverse unit vectors $\mathbf{z}^{(1)}(\mathbf{k}), \mathbf{z}^{(2)}(\mathbf{k})$. In polar coordinates they are

$$(1.15)\; \hat{\mathbf{k}} = \begin{pmatrix} \sin\theta\cos\phi \\ \sin\theta\sin\phi \\ \cos\theta \end{pmatrix}, \; \mathbf{z}^{(1)}(\mathbf{k}) = \begin{pmatrix} \cos\theta\cos\phi \\ \cos\theta\sin\phi \\ -\sin\theta \end{pmatrix}, \; \mathbf{z}^{(2)}(\mathbf{k}) = \begin{pmatrix} -\sin\phi \\ \cos\phi \\ 0 \end{pmatrix}.$$

Then the 2×2 matrix T is given by

$$(1.16) \qquad\qquad T_{ij}(\mathbf{k}, \mathbf{k}') = \mathbf{z}^{(i)}(\mathbf{k}) \cdot \mathbf{z}^{(j)}(\mathbf{k}')$$

and in polar coordinates it has the form

$$T(\mathbf{k}, \mathbf{k}') = \begin{pmatrix} \cos\theta\cos\theta'\cos(\phi - \phi') + \sin\theta\sin\theta' & \cos\theta\sin(\phi - \phi') \\ \cos\theta'\sin(\phi' - \phi) & \cos(\phi - \phi') \end{pmatrix}.$$

The coupling matrix N is given by

$$(1.17) \qquad N(\mathbf{x}, \mathbf{k}) = \sum_{i=1}^{3} \frac{\partial v(\mathbf{x})}{\partial x^i}|\mathbf{k}|\mathbf{z}^{(1)}(\mathbf{k}) \cdot \frac{\partial \mathbf{z}^{(2)}(\mathbf{k})}{\partial k_i} \begin{pmatrix} 0 & 1 \\ -1 & 0 \end{pmatrix}.$$

Chandrasekhar considered a homogeneous background only, in which case the speed of light v is a constant so that $\nabla_\mathbf{x}\omega = 0$ and $N = 0$. Law and Watson [40] derived (1.11) in general, from Maxwell's equations in a random medium, as was done also in [14].

We will now explain the physical meaning of the matrices T and N, which do not appear in the scalar transport equation (1.6). The 2×2 matrix $T(\mathbf{k}, \mathbf{k}')$ involves the angles between the directions transverse to $\mathbf{k}'$ before the scattering and the directions transverse to $\mathbf{k}$ after the scattering. Thus T_{ij} is the fraction of wave amplitude going in the direction $\mathbf{k}'$ and polarized along the transverse direction $\mathbf{z}^{(j)}(\mathbf{k}')$ that scatters into a wave going in the direction $\mathbf{k}$ and polarized along the transverse direction $\mathbf{z}^{(i)}(\mathbf{k})$. Since the coherence matrix W is related to the mean square of the wave amplitudes (see sections 3.3 and 4.4), the transformation matrix T acts on W twice in (1.12). The coupling matrix $N(\mathbf{x}, \mathbf{k})$, defined by (1.17), arises from the slow variations of the background because the rays in inhomogeneous media are curved, and this leads to rotation of the polarization vector around the ray as the wave propagates (Lewis [42]). This rotation corresponds to parallel transport along the ray in the metric $v^{-1}(\mathbf{x})ds$ where $v(\mathbf{x})$ is the propagation speed. The coherence matrix $W(t, \mathbf{x}, \mathbf{k})$ captures this behavior of polarization for quantities quadratic in the electromagnetic field through the matrix N. In the absence of scattering, so that the right side of (1.11) is zero, the solution of (1.11) with geometrical optics initial conditions (see (2.10) and (3.57)) is the coherence matrix of Lewis' solution.

When the transport mean free path is small compared to the overall propagation distance, there is a diffusion approximation for Chandrasekhar's equation (1.11). The coherence matrix W is approximated by $\bar{\phi}(t, \mathbf{x}, |\mathbf{k}|)I$ with I the 2×2 identity matrix and $\bar{\phi}$ the solution of a diffusion equation (see section 5.2). In this approximation the coherence matrix is independent of the direction of the wave vector $\mathbf{k}$ and is completely depolarized since it is proportional to the identity. In section 5.2 we give an explicit formula (5.30) for the diffusion coefficient $D(|\mathbf{k}|)$.

1.4. Transport theory for elastic waves

Radiative transport theory was used in seismology by Wesley [63], Nakamura [46], Dainty and Toksöz [19], Wu [66] and others. The stationary, scalar transport equation was used to successfully assess scattering and intrinsic attenuation (the albedo) [67], [57], [51], [44], [45], [23] and the time dependent scalar transport equation was used by Zeng, Su and Aki [68], Zeng [69] and Hoshiba [31]. In all these papers the vector nature of the underlying elastic wave motion was not taken into consideration. Mode conversion for surface waves was considered in a phenomenological way by Chen and Aki [17] and general mode conversion between longitudinal compressional or P waves and transverse shear or S waves was considered by Sato [49] and by Zeng [70]. However, the transport equations proposed phenomenologically in [49], [70] do not account for polarization of the shear waves. Starting from the elastic wave equations in a random medium we derive a system of transport equations that accounts correctly for P to S mode conversion and for polarization effects.

Longitudinal or P waves propagate with local speed

$$v_P(\mathbf{x}) = \sqrt{(2\mu(\mathbf{x}) + \lambda(\mathbf{x}))/\rho(\mathbf{x})}$$

and transverse shear or S waves propagate with local speed

$$v_S(\mathbf{x}) = \sqrt{\mu(\mathbf{x})/\rho(\mathbf{x})}.$$

The corresponding dispersion relations are $\omega_P = v_P|\mathbf{k}|$ and $\omega_S = v_S|\mathbf{k}|$, respectively. Here λ and μ are the Lame parameters. The P and S wave modes interact in an inhomogeneous medium because a P wave with wavenumber $\mathbf{k}$ can scatter into an S wave with wavenumber $\mathbf{p}$ with the same frequency; that is, with $v_P(\mathbf{x})|\mathbf{k}| = v_S(\mathbf{x})|\mathbf{p}|$, and vice versa. Therefore the transport equations for P and S wave energy densities must be coupled. The transport equation for the P wave should be a scalar equation similar to (1.6) with an additional term that accounts for S to P conversion. Similarly, the transport equation for the S wave coherence matrix should be like

Chandrasekhar's equation (1.11) with an additional term that accounts for P to S conversion. We show in section 4.5 that this is indeed the case and we determine explicitly the form of the scattering cross-sections in terms of the power spectral densities of the material inhomogeneities.

The coupled radiative transport equations for the P wave energy density $a^P(t, \mathbf{x}, \mathbf{k})$ and the 2×2 coherence matrix $W^S(t, \mathbf{x}, \mathbf{k})$ for the S waves have the forms

$$(1.18) \qquad \frac{\partial a^P}{\partial t} + \nabla_{\mathbf{k}}\omega^P \cdot \nabla_{\mathbf{x}}a^P - \nabla_{\mathbf{x}}\omega^P \cdot \nabla_{\mathbf{k}}a^P$$

$$= \int \sigma^{PP}(\mathbf{k}, \mathbf{k}')a^P(\mathbf{k}')d\mathbf{k}' - \Sigma^{PP}(\mathbf{k})a^P(\mathbf{k})$$

$$+ \int \sigma^{PS}(\mathbf{k}, \mathbf{k}')[W^S(\mathbf{k}')]d\mathbf{k}' - \Sigma^{PS}(\mathbf{k})a^P(\mathbf{k})$$

and

$$(1.19) \qquad \frac{\partial W^S}{\partial t} + \nabla_{\mathbf{k}}\omega^S \cdot \nabla_{\mathbf{x}}W^S - \nabla_{\mathbf{x}}\omega^S \cdot \nabla_{\mathbf{k}}W^S + WN - NW$$

$$= \int \sigma^{SS}(\mathbf{k}, \mathbf{k}')[W^S(\mathbf{k}')]d\mathbf{k}' - \Sigma^{SS}(\mathbf{k})W^S(\mathbf{k})$$

$$+ \int \sigma^{SP}(\mathbf{k}, \mathbf{k}')[a^P(\mathbf{k}')]d\mathbf{k}' - \Sigma^{SP}(\mathbf{k})W^S(\mathbf{k}).$$

The coupling matrix N is the same as (1.17) for electromagnetic waves except that the speed v is now the shear speed $v_S(\mathbf{x}) = \sqrt{\mu(\mathbf{x})/\rho(\mathbf{x})}$. The differential scattering cross-section $\sigma^{PP}(\mathbf{k}, \mathbf{k}')$ for P to P scattering is similar to (1.8) for scattering of scalar waves and the differential scattering tensor $\sigma^{SS}(\mathbf{k}, \mathbf{k}')$ is similar to Chandrasekhar's tensor (1.12). They have the forms

$$(1.20) \qquad \sigma^{PP}(\mathbf{k}, \mathbf{k}') = \sigma_{pp}(\mathbf{k}, \mathbf{k}')\delta(v_P|\mathbf{k}| - v_P|\mathbf{k}'|)$$

and

$$\sigma^{SS}(\mathbf{k}, \mathbf{k}')[W(\mathbf{k}')] = \{\ \sigma_{ss}^{TT}T(\mathbf{k}, \mathbf{k}')W(\mathbf{k}')T(\mathbf{k}', \mathbf{k}) + \sigma_{ss}^{\Gamma\Gamma}\Gamma(\mathbf{k}, \mathbf{k}')W(\mathbf{k}')\Gamma(\mathbf{k}', \mathbf{k})$$

$$+ \sigma_{ss}^{\Gamma T}[T(\mathbf{k}, \mathbf{k}')W(\mathbf{k}')\Gamma(\mathbf{k}', \mathbf{k}) + \Gamma(\mathbf{k}, \mathbf{k}')W(\mathbf{k}')T(\mathbf{k}', \mathbf{k})]\}$$

$$(1.21) \qquad \cdot\ \delta(v_S|\mathbf{k}| - v_S|\mathbf{k}'|).$$

The 2×2 matrix $\Gamma(\mathbf{k}, \mathbf{k}')$ is similar to T and is defined by

$$(1.22) \quad \Gamma_{ij}(\mathbf{k}, \mathbf{k}') = (\hat{\mathbf{k}} \cdot \hat{\mathbf{k}}')(\mathbf{z}^{(i)}(\mathbf{k}) \cdot \mathbf{z}^{(j)}(\mathbf{k}')) + (\hat{\mathbf{k}} \cdot \mathbf{z}^{(j)}(\mathbf{k}'))(\hat{\mathbf{k}}' \cdot \mathbf{z}^{(i)}(\mathbf{k}))$$

while σ_{pp} and σ_{ss} are scalar functions given in terms of power spectral densities of the inhomogeneities by (4.68) and (4.79). The total scattering cross-sections Σ^{PP} and Σ^{SS} are the integrals of the corresponding differential scattering cross-sections, as in (1.7) and (1.14).

The scattering cross-sections for the S to P and P to S coupling terms, σ^{PS} and σ^{SP}, respectively, have the forms

$$(1.23) \quad \sigma^{PS}(\mathbf{k}, \mathbf{k}')[W^S(\mathbf{k}')] = \text{Tr}[\sigma_{ps}(\mathbf{k}, \mathbf{k}')G(\mathbf{k}, \mathbf{k}')W^S(\mathbf{k}')]\delta(v_P|\mathbf{k}| - v_S|\mathbf{k}'|)$$

$$(1.24) \quad \sigma^{SP}(\mathbf{k}, \mathbf{k}')[a^P(\mathbf{k}')] = \sigma_{ps}(\mathbf{k}', \mathbf{k})G(\mathbf{k}', \mathbf{k})a^P(\mathbf{k}')\delta(v_S|\mathbf{k}| - v_P|\mathbf{k}'|)$$

with the 2×2 matrix G given by

$$(1.25) \qquad G_{ij}(\mathbf{k}, \mathbf{k}') = (\hat{\mathbf{k}} \cdot \mathbf{z}^{(i)}(\mathbf{k}'))(\hat{\mathbf{k}} \cdot \mathbf{z}^{(j)}(\mathbf{k}')).$$

The scalar function σ_{ps} is given explicitly in terms of power spectral densities of the inhomogeneities by (4.80). The scattering operator on the right side of (4.69) and (4.70) is symmetric in a^P, W^S and conservative. This implies in particular that

$$(1.26) \qquad \Sigma^{SP}(\mathbf{k}) = \int \sigma_{ps}(\mathbf{k}', \mathbf{k}) G(\mathbf{k}', \mathbf{k}) \delta(v_S|\mathbf{k}| - v_P|\mathbf{k}'|) d\mathbf{k}'$$

with

$$(1.27) \qquad \Sigma^{PS}(\mathbf{k}) = \int \sigma_{ps}(\mathbf{k}, \mathbf{k}') \mathrm{Tr} G(\mathbf{k}, \mathbf{k}') \delta(v_S|\mathbf{k}'| - v_P|\mathbf{k}|) d\mathbf{k}'.$$

The geometrical meaning of the 2×2 matrices T, Γ and G that appear in the differential scattering cross-sections (4.72) and (4.74) is similar to that of T in the electromagnetic case (1.12). They arise from a single scattering event of P or S waves with wave vector $\mathbf{k}'$ that scatter to P or S waves with wave vector $\mathbf{k}$, and from the fact that the transport equations deal with quadratic field quantities. In the analysis given in sections 3.4 and 4.5 this is captured in the structure of the eigenvalues and eigenvectors of the dispersion matrix L (3.81) for the elastic wave equations.

As for the scalar transport equation (1.6) and Chandrasekhar's equation (1.11), the elastic transport equations (4.69) and (4.70) simplify considerably in the regime where the diffusion approximation is valid. This occurs when the scattering mean free path is small compared to the propagation distance. In this regime the P wave energy density $a^P(t, \mathbf{x}, \mathbf{k})$ and the S wave coherence matrix $W^S(t, \mathbf{x}, \mathbf{k})$ are independent of the direction of the wave vector $\mathbf{k}$. Furthermore, W^S is proportional to the identity matrix

$$(1.28) \qquad a^P(t, \mathbf{x}, \mathbf{k}) \sim \phi(t, \mathbf{x}, |\mathbf{k}|), \ \ W^S(t, \mathbf{x}, \mathbf{k}) \sim w(t, \mathbf{x}, |\mathbf{k}|)I$$

and the equipartition relation

$$(1.29) \qquad \phi(t, \mathbf{x}, |\mathbf{k}|) = w(t, \mathbf{x}, \frac{v_P|\mathbf{k}|}{v_S})$$

holds with ϕ satisfying the diffusion equation (1.9). The diffusion coefficient $D(|\mathbf{k}|)$ is given by (5.46).

When integrated over $\mathbf{k}$, the equipartition relation (1.29) is

$$(1.30) \qquad \mathcal{E}_P(t, \mathbf{x}) = \frac{v_S^3}{2v_P^3} \mathcal{E}_S(t, \mathbf{x})$$

where $\mathcal{E}_P$ and $\mathcal{E}_S$ are the P and S wave spatial energy densities. They are related to a^P and W^S by

$$\mathcal{E}_P(t, \mathbf{x}) = \int a^P(t, \mathbf{x}, \mathbf{k}) d\mathbf{k}$$

and

$$\mathcal{E}_S(t, \mathbf{x}) = \int \mathrm{Tr} W^S(t, \mathbf{x}, \mathbf{k}) d\mathbf{k},$$

respectively. From the point of view of seismological applications of transport theory, relation (1.30) is important because it predicts universal behavior of the P to S wave energy ratio in the diffusive regime. This ratio is independent of the details of the multiple scattering process and of the source distribution. When we use the typical S to P wave speed ratio of 1 to 1.7, relation (1.30) predicts $\mathcal{E}_S/\mathcal{E}_P \sim 10$.

This is in general agreement with seismological data and it would be interesting to identify cases where $\mathcal{E}_S/\mathcal{E}_P$ stabilizes. This stabilization, which is derived here from first principles, is reminiscent of the important empirical observation of Hansen, Ringdal and Richards [29] regarding the stabilization of crustal waveguide mode energy ratios.

1.5. Brief outline

In Lecture 2, we analyze the Schrödinger equation in order to motivate the phase space setup in a relatively simple context. We introduce the Wigner distribution and show how it can be used effectively for energy transport calculations. In Lecture 3 we present the high frequency approximation for general symmetric hyperbolic systems and in particular the equations of acoustic, electromagnetic and elastic waves. We do this in phase space using the Wigner distribution, and explain the connections with the standard high frequency approximation. In Lecture 4 we derive the transport equations, first for the Schrodinger equation and then for general symmetric hyperbolic systems. We apply this general formalism to the equations of acoustic, electromagnetic and elastic waves in the second half of this lecture. The diffusion approximation is analyzed in detail in Lecture 5. Lecture 6, which is based on [47], is a self-contained description of the geophysical aspects and applications of the radiative transport equations. It may be read independently of the previous material and is oriented towards a reader interested more in applied problems than in mathematical questions.

Acknowledgement

We are grateful to Joseph B. Keller with whom we collaborated on most of the work reviewed here. We also thank A. Fannjiang with whom we have discussed at length mathematical issues that come up in the passage to transport theory from random waves.

LECTURE 2
The Schrödinger Equation

2.1. The Schrödinger equation

It is convenient to analyze high frequency asymptotics in a simple setup, that of the Schrödinger equation where waves go in only one direction. The Schrödinger equation arises in many contexts in wave propagation, not only in quantum mechanics. We review briefly the parabolic or paraxial approximation for forward propagation time harmonic waves.

The wave equation

$$(2.1) \qquad \frac{1}{c^2}\frac{\partial^2 u}{\partial t^2} - \Delta u = 0,$$

where $c(x, y, z)$ is the propagation speed, has time harmonic solutions of the form $e^{i\omega t}u(x, y, z)$ with the complex wave function u satisfying the Helmholtz or reduced wave equation

$$(2.2) \qquad \Delta u + \frac{\omega^2}{c^2(x, y, z)}u = 0.$$

Let c_0 be a uniform reference speed, let $k = \omega/c_0$ be the wave number and let $c_0/c(x, y, z) = n(x, y, z)$ be the index of refraction. The reduced wave equation has then the form

$$(2.3) \qquad \Delta u + k^2 n^2(x, y, z)u = 0.$$

When waves are approximately plane and move in one direction primarily, say the z direction, we look for solutions of the form

$$(2.4) \qquad u = e^{ikz}\phi(z, \mathbf{x})$$

where $\mathbf{x} = (x, y)$ denotes the transverse variables. We insert this into the reduced wave equation and drop the ϕ_{zz} term to get the Schrödinger equation

$$(2.5) \qquad 2ik\phi_z + \Delta_\perp\phi + k^2\mu(z, \mathbf{x})\phi = 0$$

where $\Delta_\perp$ is the Laplacian in the transverse variables and $\mu = n^2 - 1$ is the fluctuation in the refractive index. Note that the direction of propagation z plays the role of time and $-k^2\mu(z, \mathbf{x})$ is the ('time' dependent) potential. Of course (2.5) is only an approximation to the full reduced wave equation and it is valid when the variations in the index of refraction are smooth and the bulk of the wave energy

317

is away from boundaries. We have then an important and very useful approximation that is well suited for numerical computations since we now have an initial value problem for ϕ, assuming that $\phi(0, \mathbf{x})$ is known, rather than a boundary value problem for u.

We shall now analyze high frequency solutions to the Schrödinger equation without regard to its origin through the parabolic approximation. So we write (2.5) in standard (dimensionless) form

$$(2.6) \qquad i\phi_t + \frac{1}{2}\Delta\phi - V(\mathbf{x})\phi = 0$$

with $\mathbf{x} \in R^d$ and initial conditions

$$(2.7) \qquad \phi(0, \mathbf{x}) = \phi_0(\mathbf{x}).$$

Note that for simplicity the potential is taken to be time independent but all the analysis extends to the time dependent case as well.

If the potential and the initial data are smooth and decaying at infinity then the solution will be smooth and decaying. The evolution is unitary and preserves the L^2 norm

$$(2.8) \qquad \int_{R^d} |\phi(t, \mathbf{x})|^2 d\mathbf{x} = \int_{R^d} |\phi(\mathbf{x})|^2 d\mathbf{x}.$$

2.2. Standard high frequency asymptotics

We consider high frequency asymptotics which concerns approximate solutions of (2.6) that are good approximations to oscillatory solutions. For such solutions the propagation distance is long compared to the wavelength, the propagation time is large compared to the period and the potential $V(\mathbf{x})$ varies slowly. To make this precise we introduce slow time and space variables $t \to t/\varepsilon$, $\mathbf{x} \to \mathbf{x}/\varepsilon$ and the scaled wave function $\phi^\varepsilon(t, \mathbf{x}) = \phi(t/\varepsilon, \mathbf{x}/\varepsilon)$ which satisfies the scaled Schrödinger equation

$$(2.9) \qquad i\varepsilon\phi_t^\varepsilon + \frac{\varepsilon^2}{2}\Delta\phi^\varepsilon - V(\mathbf{x})\phi^\varepsilon = 0.$$

In the standard high frequency approximation [37] we consider initial data of the form

$$(2.10) \qquad \phi^\varepsilon(0, \mathbf{x}) = e^{iS_0(\mathbf{x})/\varepsilon} A_0(\mathbf{x})$$

with a smooth, real valued initial phase function $S_0(\mathbf{x})$ and a smooth compactly supported complex valued initial amplitude $A_0(\mathbf{x})$. We then look for an asymptotic solution of (2.9) in the same form as the initial data (2.10), with evolved phase and amplitude

$$(2.11) \qquad \phi^\varepsilon(t, \mathbf{x}) \sim e^{iS(t,\mathbf{x})/\varepsilon} A(t, \mathbf{x}).$$

Inserting this form into (2.9) and equating powers of ε we get evolution equations for the phase and amplitude

$$(2.12) \qquad S_t + \frac{1}{2}|\nabla S|^2 + V(\mathbf{x}) = 0, \;\; S(0, \mathbf{x}) = S_0(\mathbf{x})$$

and

$$(2.13) \qquad (|A|^2)_t + \nabla \cdot (|A|^2\nabla S) = 0, \;\; |A(0, \mathbf{x})|^2 = |A_0(\mathbf{x})|^2.$$

The phase equation (2.12) is called the *eiconal* and the amplitude equation (2.13) the *transport* equation. The terminology for the latter is standard in the high frequency approximation but should not be confused with the radiative transport equation that will be derived later. The eiconal equation that evolves the phase is nonlinear and, in general, it will have a solution only up to some finite time t^* that depends on the initial phase. The solution can be constructed by characteristics, the rays, and singularities in the phase arise when the rays form envelopes which are the caustics. They are called caustics because the amplitude is singular there, which means that there is wave energy concentration. Let

$$(2.14) \qquad\qquad H(\mathbf{x}, \mathbf{k}) = \frac{1}{2}|\mathbf{k}|^2 + V(\mathbf{x})$$

be the Hamiltonian so that the eiconal equation (2.12) is

$$S_t + H(\mathbf{x}, \mathbf{k}) = 0,$$

and

$$(|A|^2)_t + \nabla \cdot (|A|^2 \nabla_{\mathbf{k}} H(\mathbf{x}, \mathbf{k})|_{\mathbf{k}=\nabla S}) = 0.$$

This form of the eiconal and transport equations is general and remains valid in the case of symmetric hyperbolic systems. The Lagrangian is given by

$$(2.15) \qquad \mathcal{L}(\mathbf{x}, \dot{\mathbf{x}}) = \sup_{\mathbf{k}}\{\mathbf{k} \cdot \dot{\mathbf{x}} - H(\mathbf{x}, \mathbf{k})\} = \frac{1}{2}|\dot{\mathbf{x}}|^2 - V(\mathbf{x})$$

and then the solution of the eiconal equation is given by the principle of least action

$$(2.16) \qquad S(t, \mathbf{x}) = \inf_{\mathbf{x}(\cdot),\ \mathbf{x}(t)=\mathbf{x}} \{S_0(\mathbf{x}(0)) + \int_0^t \mathcal{L}(\mathbf{x}(s), \dot{\mathbf{x}}(s))ds\}.$$

The Euler equations for the minimizing path are Lagrange's equations

$$(2.17) \qquad \frac{d^2\mathbf{x}}{dt^2} + \nabla V(\mathbf{x}) = 0\ ,\ \mathbf{x}(t) = \mathbf{x}\ ,\ \nabla S_0(\mathbf{x}(0)) = \frac{d}{dt}\mathbf{x}(0)$$

and the solutions of this two point boundary value problem are the rays.

The transport equation (2.13) can be interpreted as follows. At two time points $0 \le t_1 < t_2$ consider the level surfaces $S(t_1, \mathbf{x}) = S^*$ and $S(t_2, \mathbf{x}) = S^*$, which are wavefronts at these two times. Let $d\Sigma_1$ be an element of area on the wavefront at t_1 around the point $\mathbf{x}_1$ and let $d\Sigma_2$ be the element of surface area carried by a tube of rays from $\mathbf{x}_1$ to the wavefront at t_2. Then

$$(2.18) \qquad\qquad |A(t_2, \mathbf{x}_2)|^2 d\Sigma_2 = |A(t_1, \mathbf{x}_1)|^2 d\Sigma_1$$

where $\mathbf{x}_2$ is the point around $d\Sigma_2$ on the wavefront at t_2. This relation is obtained from the transport equation (2.13) by integrating it over the tube of rays and using the divergence theorem. The relation (2.18) is the intensity law of geometrical optics: The ratio of the intensity at $(t_2, \mathbf{x}_2)$ to that at $(t_1, \mathbf{x}_1)$ is inversely proportional to the ratio of the ray tube areas, which is the Jacobian of the ray map. Thus, converging rays increase intensities as is intuitively clear.

The fact that expressions of the form $e^{iS/\varepsilon}A$ are asymptotic solutions is easy to see when everything is smooth since the L^2 norm is preserved by the Schrödinger equation. The main issue in more detailed asymptotics is to extend this elementary theory so that diffraction effects are included. This was done by J.B.Keller [36] and a lot of the work from that period is in the 1964 review of Keller and Lewis that is contained in [37]. An exposition of the Hamilton-Jacobi theory can be found in

[25]. Microlocal asymptotic analysis is the modern mathematical version of these developments.

2.3. The Wigner distribution

An essential step in our approach to deriving radiative transport equations from wave equations is the introduction of the Wigner distribution [65][1]. For any smooth function ϕ, rapidly decaying at infinity, the Wigner distribution is defined by

$$(2.19) \qquad W(\mathbf{x}, \mathbf{k}) = \left(\frac{1}{2\pi}\right)^d \int_{R^d} e^{i\mathbf{k}\cdot\mathbf{y}} \phi(\mathbf{x} - \frac{\mathbf{y}}{2})\overline{\phi}(\mathbf{x} + \frac{\mathbf{y}}{2}) d\mathbf{y}$$

where $\overline{\phi}$ is the complex conjugate of ϕ and the dimension $d = 2$ or 3. The Wigner distribution is defined on phase space and has many important properties. It is real and its $\mathbf{k}$-integral is the modulus square of the function ϕ,

$$(2.20) \qquad \int_{R^d} W(\mathbf{x}, \mathbf{k}) d\mathbf{k} = |\phi(\mathbf{x})|^2,$$

so we may think of $W(\mathbf{x}, \mathbf{k})$ as wave number resolved energy density. This is not quite right though because $W(\mathbf{x}, \mathbf{k})$ is not always positive but it does become positive in the high frequency limit. The energy flux is expressed through $W(\mathbf{x}, \mathbf{k})$ by

$$(2.21) \qquad \mathcal{F} = \frac{1}{2i}(\phi\nabla\overline{\phi} - \overline{\phi}\nabla\phi) = \int_{R^d} \mathbf{k} W(\mathbf{x}, \mathbf{k}) d\mathbf{k}$$

and its second moment in $\mathbf{k}$ is

$$(2.22) \qquad \iint |\mathbf{k}|^2 W(\mathbf{x}, \mathbf{k}) d\mathbf{k} d\mathbf{x} = \int |\nabla\phi(\mathbf{x})|^2 d\mathbf{x}.$$

The Wigner distribution possesses an important $\mathbf{x}$-to-$\mathbf{k}$ duality given by the alternative definition

$$(2.23) \qquad W(\mathbf{x}, \mathbf{k}) = \int e^{i\mathbf{p}\cdot\mathbf{x}} \hat{\phi}(-\mathbf{k} - \frac{\mathbf{p}}{2})\overline{\hat{\phi}(-\mathbf{k} + \frac{\mathbf{p}}{2})} d\mathbf{p}$$

where $\hat{\phi}$ is the Fourier transform of ϕ

$$\hat{\phi}(\mathbf{k}) = \frac{1}{(2\pi)^d} \int e^{i\mathbf{k}\cdot\mathbf{x}} \phi(\mathbf{x}) d\mathbf{x}.$$

These properties make the Wigner distribution a good candidate for analyzing the evolution of wave energy in phase space.

We begin our study of the Wigner distribution by calculating it for wave functions of the form (2.11) that enter as data in high frequency calculations

$$(2.24) \qquad \phi^\varepsilon(\mathbf{x}) = e^{iS(\mathbf{x})/\varepsilon} A(\mathbf{x}),$$

with S smooth and bounded and A smooth and of compact support. The Wigner distribution is

$$(2.25) \quad W(\mathbf{x}, \mathbf{k}) = \frac{1}{(2\pi)^d} \int_{R^d} e^{i\mathbf{k}\cdot\mathbf{y}} e^{iS(\mathbf{x}-\frac{\mathbf{y}}{2})/\varepsilon} A(\mathbf{x} - \frac{\mathbf{y}}{2}) e^{-iS(\mathbf{x}+\frac{\mathbf{y}}{2})/\varepsilon} \overline{A}(\mathbf{x} + \frac{\mathbf{y}}{2}) d\mathbf{y}$$

---

[1] Actually in [65] Wigner gives credit to Szillard for introducing this function.

and we see that the rescaled version

$$(2.26) \qquad W^\varepsilon(\mathbf{x}, \mathbf{k}) = \frac{1}{\varepsilon^d} W(\mathbf{x}, \frac{\mathbf{k}}{\varepsilon})$$

$$= \frac{1}{(2\pi)^d} \int_{R^d} e^{i\mathbf{k}\cdot\mathbf{y}} e^{iS(\mathbf{x}-\frac{\varepsilon\mathbf{y}}{2})/\varepsilon} A(\mathbf{x} - \frac{\varepsilon\mathbf{y}}{2}) e^{-iS(\mathbf{x}+\frac{\varepsilon\mathbf{y}}{2})/\varepsilon} \bar{A}(\mathbf{x} + \frac{\varepsilon\mathbf{y}}{2}) d\mathbf{y}$$

converges weakly (as a distribution in S') to

$$(2.27) \qquad W(\mathbf{x}, \mathbf{k}) = \delta(\mathbf{k} - \nabla S(\mathbf{x}))|A(\mathbf{x})|^2$$

or, if Z is any test function in $S(R^{2d})$

$$(2.28) \qquad \int_{R^{2d}} W^\varepsilon Z d\mathbf{x}d\mathbf{k} \rightarrow \int_{R^d} Z(\mathbf{x}, \nabla S(\mathbf{x}))|A(\mathbf{x})|^2 d\mathbf{x}$$

as $\varepsilon \rightarrow 0$. This calculation tells us two things. One is the correct scaling of the Wigner distribution for the high frequency limit

$$(2.29) \qquad W^\varepsilon(\mathbf{x}, \mathbf{k}) = \frac{1}{(2\pi)^d} \int_{R^d} e^{i\mathbf{k}\cdot\mathbf{y}} \phi^\varepsilon(\mathbf{x} - \frac{\varepsilon\mathbf{y}}{2}) \bar{\phi}^\varepsilon(\mathbf{x} + \frac{\varepsilon\mathbf{y}}{2}) d\mathbf{y},$$

or that for the *evolved* Wigner density

$$W^\varepsilon(t, \mathbf{x}, \mathbf{k}) = \frac{1}{(2\pi)^d} \int_{R^d} e^{i\mathbf{k}\cdot\mathbf{y}} \phi^\varepsilon(t, \mathbf{x} - \frac{\varepsilon\mathbf{y}}{2}) \bar{\phi}^\varepsilon(t, \mathbf{x} + \frac{\varepsilon\mathbf{y}}{2}) d\mathbf{y}$$

of the solution $\phi^\varepsilon(t, \mathbf{x})$ of the scaled Schrödinger equation (2.9). The second is that in the high frequency limit, and for the kind of inhomogeneous, near-plane waves that we are considering, the limit Wigner function is a measure, a delta function supported on the surface $\mathbf{k} = \nabla S(\mathbf{x})$ in phase space.

From (2.27) we conclude that as $\varepsilon \rightarrow 0$ the scaled Wigner distribution of the solution $\phi^\varepsilon(t, \mathbf{x})$ of (2.9) with initial data (2.10) is given by

$$(2.30) \qquad W(t, \mathbf{x}, \mathbf{k}) = |A(t, \mathbf{x})|^2 \delta(\mathbf{k} - \nabla S(t, \mathbf{x})),$$

where $S(t, \mathbf{x})$ and $A(t, \mathbf{x})$ are solutions of the eiconal and transport equations (2.12) and (2.13), respectively.

We will now derive the high frequency approximation of the scaled Wigner distribution directly from the differential equations. Let us assume that the initial Wigner distribution $W_0^\varepsilon(\mathbf{x}, \mathbf{k})$ tends to a smooth function $W_0(\mathbf{x}, \mathbf{k})$ that decays at infinity. Note that this is not the case with the Wigner function corresponding to $\phi^\varepsilon(0, \mathbf{x})$ given by (2.10) but it is the case for random initial wave functions. The evolution equation for $W^\varepsilon(t, \mathbf{x}, \mathbf{k})$ corresponding to the Schrödinger equation (2.9) is the Wigner equation

$$(2.31) \qquad W_t^\varepsilon + \mathbf{k} \cdot \nabla_{\mathbf{x}} W^\varepsilon + \mathcal{L}^\varepsilon W^\varepsilon = 0.$$

Here the operator $\mathcal{L}^\varepsilon$ is defined by

$$(2.32) \quad \mathcal{L}^\varepsilon Z(\mathbf{x}, \mathbf{k}) = i \int_{R^d} e^{-i\mathbf{p}\cdot\mathbf{x}} \hat{V}(\mathbf{p}) \frac{1}{\varepsilon} \left[Z(\mathbf{x}, \mathbf{k} + \frac{\varepsilon\mathbf{p}}{2}) - Z(\mathbf{x}, \mathbf{k} - \frac{\varepsilon\mathbf{p}}{2}) \right] d\mathbf{p}$$

on any smooth function Z in phase space. The Fourier transform is denoted by a hat

$$(2.33) \qquad \hat{V}(\mathbf{p}) = \frac{1}{(2\pi)^d} \int e^{i\mathbf{p}\cdot\mathbf{x}} V(\mathbf{x}) d\mathbf{x}.$$

From (2.32) we can find easily the limit of the operator $\mathcal{L}^\varepsilon$ as $\varepsilon \to 0$. For any smooth and decaying function $Z(\mathbf{x}, \mathbf{k})$ we have

$$(2.34) \qquad \mathcal{L}^\varepsilon Z(\mathbf{x}, \mathbf{k}) \to -\nabla_{\mathbf{x}} V \cdot \nabla_{\mathbf{k}} Z.$$

Thus, the limit Wigner equation is the Liouville equation in phase space

$$(2.35) \qquad W_t + \mathbf{k} \cdot \nabla_{\mathbf{x}} W - \nabla V \cdot \nabla_{\mathbf{k}} W = 0$$

with the initial condition $W(0, \mathbf{x}, \mathbf{k}) = W_0(\mathbf{x}, \mathbf{k})$. This is a linear partial differential equation that can be solved by characteristics. When the initial Wigner distribution has the high frequency form

$$(2.36) \qquad W_0(\mathbf{x}, \mathbf{k}) = |A_0(\mathbf{x})|^2 \delta(\mathbf{k} - \nabla S_0(\mathbf{x}))$$

then it is easy to see that the solution of (2.35) is given by

$$(2.37) \qquad W(t, \mathbf{x}, \mathbf{k}) = |A(t, \mathbf{x})|^2 \delta(\mathbf{k} - \nabla S(t, \mathbf{x}))$$

where $S(t, \mathbf{x})$ and $A(t, \mathbf{x})$ are solutions of the eiconal and transport equations (2.12) and (2.13), respectively. We see, therefore, that from the Wigner distribution we can recover all the information in the standard high frequency approximation. In addition, it provides flexibility to deal with initial data that are not of the form (2.36).

2.4. General properties of the Wigner distribution

We want to consider the Wigner distributions of high frequency waves, i.e. of functions $f_\varepsilon(\mathbf{x})$ which are oscillating on a scale ε as $\varepsilon \to 0$, not necessarily of the form (2.11). Our exposition follows the ideas of L.Tartar and P.Gerard [**55, 26**]. We consider the rescaled Wigner distribution

$$(2.38) \qquad W_\varepsilon(\mathbf{x}, \mathbf{k}) = \int \frac{d\mathbf{y}}{(2\pi)^d} e^{i\mathbf{k} \cdot \mathbf{y}} f_\varepsilon(\mathbf{x} - \frac{\varepsilon \mathbf{y}}{2}) f_\varepsilon^*(\mathbf{x} + \frac{\varepsilon \mathbf{y}}{2}).$$

In terms of the Fourier transform $\hat{f}$ of f, the scaled form of W is

$$(2.39) \qquad W_\varepsilon(\mathbf{x}, \mathbf{k}) = \int \frac{d\mathbf{p}}{(2\pi\sqrt{\varepsilon})^{2d}} e^{i\mathbf{p} \cdot \mathbf{x}} \hat{f}_\varepsilon(\frac{\mathbf{k}}{\varepsilon} + \frac{\mathbf{p}}{2}) \hat{f}_\varepsilon^*(\frac{\mathbf{k}}{\varepsilon} - \frac{\mathbf{p}}{2}).$$

Here $\hat{f}$ is

$$(2.40) \qquad \hat{f}(\mathbf{k}) = \int d\mathbf{x} e^{-i\mathbf{k} \cdot \mathbf{x}} f(\mathbf{x})$$

and not normalized like (2.33). We do this in section 2.4 only.

The duality between (2.38) and (2.39) can be expressed succinctly by

$$(2.41) \qquad W_\varepsilon[f_\varepsilon(\cdot)](\mathbf{x}, \mathbf{k}) = W_\varepsilon[\frac{1}{(2\pi\varepsilon)^{d/2}} \hat{f}_\varepsilon\left(\frac{\cdot}{\varepsilon}\right)](\mathbf{k}, -\mathbf{x}).$$

We are interested in the weak limit of W_ε as $\varepsilon \to 0$. We first define W_ε weakly for $f_\varepsilon \in \mathcal{S}'$, the space of tempered distributions. Let $a(\mathbf{x}, \mathbf{k})$ be a test function in $\mathcal{S}(R^d \times R^d)$. Then

$$(2.42) \qquad < a, W_\varepsilon >= (a^w(\mathbf{x}, \varepsilon D) f_\varepsilon, f_\varepsilon),$$

where $<,>$ is the usual inner product on $R^d \times R^d$, $(,)$ is inner product on R^d, and the Weyl operator $a^w(\mathbf{x}, \varepsilon D)$ is defined by

$$(2.43) \qquad a^w(\mathbf{x}, \varepsilon D)f(\mathbf{x}) = \iint \frac{d\mathbf{y}\,d\mathbf{k}}{(2\pi)^d} e^{i(\mathbf{x}-\mathbf{y})\cdot\mathbf{k}} a\left(\frac{\mathbf{x}+\mathbf{y}}{2}, \varepsilon\mathbf{k}\right) f(\mathbf{y})$$

$$= \int \frac{d\mathbf{y}}{(2\pi\varepsilon)^d} \hat{a}\left(\frac{\mathbf{x}+\mathbf{y}}{2}, \frac{\mathbf{y}-\mathbf{x}}{\varepsilon}\right) f(\mathbf{y}).$$

Here $\hat{a}$ is the Fourier transform of $a(\mathbf{x}, \mathbf{k})$ in the variable $\mathbf{k}$ only. An alternative way to get to limits of Wigner distributions is to introduce

$$(2.44) \qquad \widetilde{W}_\varepsilon(\mathbf{x}, \mathbf{k}) = \int \frac{d\mathbf{y}}{(2\pi)^d} e^{i\mathbf{k}\cdot\mathbf{y}} f_\varepsilon(\mathbf{x} - \varepsilon\mathbf{y}) f_\varepsilon^*(\mathbf{x}) = \frac{e^{i\mathbf{k}\cdot\mathbf{x}/\varepsilon}}{(2\pi\varepsilon)^d} \hat{f}_\varepsilon\left(\frac{\mathbf{k}}{\varepsilon}\right) f_\varepsilon^*(\mathbf{x})$$

and weakly, for $a(\mathbf{x}, \mathbf{k}) \in \mathcal{S}(R^d \times R^d)$,

$$(2.45) \qquad\qquad < a, \widetilde{W}_\varepsilon > = (a(\mathbf{x}, \varepsilon D)f_\varepsilon, f_\varepsilon).$$

Now the semiclassical operators $a(\mathbf{x}, \varepsilon D)$ are defined by

$$(2.46)\ \ a(\mathbf{x}, \varepsilon D)f(\mathbf{x}) = \int \frac{d\mathbf{k}}{(2\pi)^d} e^{i\mathbf{k}\cdot\mathbf{x}} a(\mathbf{x}, \varepsilon\mathbf{k}) \hat{f}(\mathbf{k}) = \int \frac{d\mathbf{y}}{(2\pi\varepsilon)^d} \hat{a}\left(\mathbf{x}, \frac{\mathbf{y}-\mathbf{x}}{\varepsilon}\right) f(\mathbf{y})$$

and they are frequently more convenient for calculations than (2.43). The main fact is that (2.46) and (2.43) are asymptotically equivalent as $\varepsilon \to 0$

$$(2.47) \qquad\qquad ||a^w(\mathbf{x}, \varepsilon D) - a(\mathbf{x}, \varepsilon D)||_{L^2 \to L^2} \to 0.$$

To prove (2.47) we recall first that if $K = K(\mathbf{x}, \mathbf{y})$ is a continuous function on $R^d \times R^d$ and there exists C such that

$$\int_{R^d} |K(\mathbf{x}, \mathbf{y})|d\mathbf{y} \leq C, \quad \int_{R^d} |K(\mathbf{x}, \mathbf{y})|d\mathbf{x} \leq C,$$

then the operator

$$Af(\mathbf{x}) = \int_{R^d} K(\mathbf{x}, \mathbf{y})f(y)dy$$

is bounded on L^2 and

$$(2.48) \qquad\qquad ||A||_{L^2} \leq C.$$

Then given a function $f \in L^2$ we have

$$(2.49) \qquad\qquad a(\mathbf{x}, \varepsilon D)f = \int \frac{d\mathbf{y}}{\varepsilon^d} \tilde{a}\left(\mathbf{x}, \frac{\mathbf{x}-\mathbf{y}}{\varepsilon}\right) f(\mathbf{y})$$

$$a^w(\mathbf{x}, \varepsilon D)f = \int \frac{d\mathbf{y}}{\varepsilon^d} \tilde{a}\left(\frac{\mathbf{x}+\mathbf{y}}{2}, \frac{\mathbf{x}-\mathbf{y}}{\varepsilon}\right) f(\mathbf{y})$$

and thus

$$(a^w(\mathbf{x}, \varepsilon D) - a(\mathbf{x}, \varepsilon D))f = \int \frac{d\mathbf{y}}{\varepsilon^d} \left\{ \tilde{a}\left(\frac{\mathbf{x}+\mathbf{y}}{2}, \frac{\mathbf{x}-\mathbf{y}}{\varepsilon}\right) - \tilde{a}\left(\mathbf{x}, \frac{\mathbf{x}-\mathbf{y}}{\varepsilon}\right) \right\} f(\mathbf{y}).$$

Then (2.47) follows by (2.48) and the dominated convergence theorem. The semiclassical operators $a(\mathbf{x}, \varepsilon D)$ are bounded on L^2, uniformly in ε:

$$(2.50) \qquad\qquad ||a(\mathbf{x}, \varepsilon D)||_{L^2 \to L^2} \leq C(a).$$

To show this we note that

$$(2.51) \qquad\qquad a(\mathbf{x}, \varepsilon D)f = \int \frac{d\mathbf{y}}{(2\pi\varepsilon)^d} \hat{a}\left(\mathbf{x}, \frac{\mathbf{y}-\mathbf{x}}{\varepsilon}\right) f(\mathbf{y}),$$

where hat denotes the Fourier transform in $\mathbf{k}$. Since

$$\int \frac{d\mathbf{y}}{(2\pi\varepsilon)^d}\left|\hat{a}\left(\mathbf{x}, \frac{\mathbf{y}-\mathbf{x}}{\varepsilon}\right)\right| = \int \frac{d\mathbf{y}}{(2\pi)^d}\,|\hat{a}(\mathbf{x}, \mathbf{y})|$$

and

$$\int \frac{d\mathbf{x}}{(2\pi\varepsilon)^d}\left|\hat{a}\left(\mathbf{x}, \frac{\mathbf{y}-\mathbf{x}}{\varepsilon}\right)\right| = \int \frac{d\mathbf{z}}{(2\pi)^d}\,|\hat{a}(\mathbf{y}+\varepsilon\mathbf{z}, \mathbf{z})| \le \int \frac{d\mathbf{z}}{(2\pi)^d}\,\sup_{\mathbf{x}}|\hat{a}(\mathbf{x}, \mathbf{z})|\,,$$

the estimate (2.50) follows by (2.48). We note that if the functions f_ε are uniformly bounded in L^2, then there is a μ in $\mathcal{S}'(R^d \times R^d)$ and a subsequence of their Wigner distributions W_ε that converges weakly to it. This follows from the general theory in Chapter VII of [50]. The family $\widetilde{W}_\varepsilon$ has the same weak limits as W_ε, as $\varepsilon \to 0$, and if μ is a limit Wigner distribution of f_ε then

$$(2.52) \qquad \lim_{\varepsilon \to 0}(a(\mathbf{x}, \varepsilon D)f_\varepsilon, f_\varepsilon) =<a, \mu>= \mathrm{Tr}\int a(\mathbf{x}, \mathbf{k})\mu(d\mathbf{x}d\mathbf{k})$$

with the limit taken along the subsequence ε_j corresponding to the subsequence W_{ε_j} converging weakly to μ.

A basic tool in Fourier integral operator calculus is the product formula

$$(2.53) \quad b(\mathbf{x}, \varepsilon D)a(\mathbf{x}, \varepsilon D) = (ba)(\mathbf{x}, \varepsilon D) + \frac{\varepsilon}{i}(\nabla_{\mathbf{k}}b \cdot \nabla_{\mathbf{x}}a)(\mathbf{x}, \varepsilon D) + \varepsilon^2 Q_\varepsilon,$$

where the operators Q_ε are uniformly bounded on L^2. The adjoint operators $a(\mathbf{x}, \varepsilon D)^*$ are given by

$$(2.54) \qquad\qquad a(\mathbf{x}, \varepsilon D)^* = a^*(\mathbf{x}, \varepsilon D) + \varepsilon R_\varepsilon$$

with R_ε uniformly bounded on L^2.

The formalism of semiclassical pseudodifferential operators developed provides a fast and elegant way to derive the Wigner transport equation (2.35). Let ϕ^ε satisfy the Schrödinger equation

$$(2.55) \qquad\qquad i\varepsilon\frac{\partial\phi^\varepsilon}{\partial t} + \frac{\varepsilon^2}{2}\Delta\phi^\varepsilon - V(\mathbf{x})\phi^\varepsilon = 0$$

with the initial data $\phi_0^\varepsilon(\mathbf{x})$ being ε-oscillatory (Definition 2.3) with the corresponding Wigner measure $W_0(\mathbf{x}, \mathbf{k})$. Then for any test function $a(\mathbf{x}, \mathbf{k}) \in \mathcal{S}(\mathbb{R}^d \times \mathbb{R}^d)$ we have, using (2.55) and the product formula (2.53):

$$<a, \frac{\partial W}{\partial t}> = \lim_{\varepsilon \to 0}\left[(a(\mathbf{x}, \varepsilon D)\phi^\varepsilon, \frac{\partial\phi^\varepsilon}{\partial t}) + (a(\mathbf{x}, \varepsilon D)\frac{\partial\phi^\varepsilon}{\partial t}, \phi^\varepsilon)\right]$$
$$= \lim_{\varepsilon \to 0}\frac{1}{i\varepsilon}([\varepsilon^2\Delta - V(\mathbf{x}), a(\mathbf{x}, \varepsilon D)]\phi^\varepsilon . \phi^\varepsilon)$$
$$=<a, -\mathbf{k}\cdot\nabla_{\mathbf{x}}W + \nabla_{\mathbf{x}}V \cdot \nabla_{\mathbf{k}}W >,$$

which is the same as (2.35).

A basic property of Wigner distributions is that their high frequency limits are nonnegative distributions, that is, measures. It is sufficient to verify that

$$<a, \mu >\ge 0$$

for all test functions $a(\mathbf{x}, \mathbf{k})$ of the form $a(\mathbf{x}, \mathbf{k}) = |b(\mathbf{x}, \mathbf{k})|^2$ because those are dense in the set of positive test functions. Then (2.54) and the product rule (2.53) implies that for such test functions

$$(2.56) \quad <a, \mu >= \lim_{\varepsilon \to 0}(a^w(\mathbf{x}, \varepsilon D)f_\varepsilon, f_\varepsilon) = \lim_{\varepsilon \to 0}(b^w(\mathbf{x}, \varepsilon D)f_\varepsilon, b^w(\mathbf{x}, \varepsilon D)f_\varepsilon) \ge 0,$$

where the limit is along a convergent subsequence. Another important property is that the Wigner measure is a local notion in space.

Property 2.1 (Localization). *Let $f_\varepsilon(\mathbf{x})$ be a family of uniformly bounded functions in L^2 and let $\mu_f(\mathbf{x},\mathbf{k})$ be any limit Wigner measure. Let $\phi(\mathbf{x})$ be a smooth function. Then the Wigner measure of the family $g_\varepsilon(\mathbf{x}) = \phi(\mathbf{x})f_\varepsilon(\mathbf{x})$ is $|\phi(\mathbf{x})|^2\mu_f(\mathbf{x},\mathbf{k})$. Moreover, let f_ε, g_ε be two uniformly bounded families of L^2 functions which coincide in an open neighborhood of a point $\mathbf{x}_0$. Then any limit Wigner measures μ_f and μ_g coincide in this neighborhood.*

Let $a(\mathbf{x},\mathbf{k})$ be a test function compactly supported in $\mathbf{x}$, such that its Fourier transform in $\mathbf{k}$ is compactly supported. Then

$$(a(\mathbf{x},\varepsilon D)g_\varepsilon, g_\varepsilon) - (|\phi(\mathbf{x})|^2 a(\mathbf{x},\varepsilon D)f_\varepsilon, f_\varepsilon)$$

$$= \int d\mathbf{x}d\mathbf{z} f_\varepsilon^*(\mathbf{x})\ \phi^*(\mathbf{x})\hat{a}(\mathbf{x},\mathbf{z})(\phi(\mathbf{x}+\varepsilon\mathbf{z}) - \phi(\mathbf{x}))f_\varepsilon(\mathbf{x}+\varepsilon\mathbf{z}) \to 0$$

as $\varepsilon \to 0$. This proves the first statement which implies in turn the second statement.

The localization property is quite useful because it allows the consideration of Wigner measures for families of functions f_ε that are uniformly bounded in L^2_{loc}.

Another useful and intuitively clear property is that the Wigner measure of waves going in different directions is the sum of the individual Wigner measures.

Property 2.2 (Orthogonality). *Let f_ε, g_ε be two families of functions with Wigner measures μ_f and μ_g, which are mutually singular. Then the Wigner measure of the sum $f_\varepsilon + g_\varepsilon$ is $\mu_f + \mu_g$.*

Proof. Let $a(\mathbf{x},\mathbf{k})$ be a positive test function of the form $a(\mathbf{x},\mathbf{k}) = |b(\mathbf{x},\mathbf{k})|^2$. Then by the product rule (2.53) and the Schwartz inequality we have

$$|(a(\mathbf{x},\varepsilon D)f_\varepsilon, g_\varepsilon)| = |(b(\mathbf{x},\varepsilon D)f_\varepsilon, b(\mathbf{x},\varepsilon D)g_\varepsilon)| + O(\varepsilon)$$

$$\leq |(b(\mathbf{x},\varepsilon D)f_\varepsilon, b(\mathbf{x},\varepsilon D)f_\varepsilon)|^{1/2}|(b(\mathbf{x},\varepsilon D)g_\varepsilon, b(\mathbf{x},\varepsilon D)g_\varepsilon)|^{1/2} + O(\varepsilon)$$

$$(2.57) \qquad \to\ < a, \mu_f >^{1/2} < a, \mu_g >^{1/2}.$$

Since μ_f and μ_g are mutually singular, we can split $a(\mathbf{x},\mathbf{k})$ into a sum of three terms:

$$(2.58) \qquad\qquad a = a_1 + a_2 + a_3,$$

so that a_1 is orthogonal to μ_f, a_2 is orthogonal to μ_g, and the integral of a_3 with respect to both measures can be made arbitrarily small. Then the right side of (2.57) is arbitrarily small and thus its left side goes to zero in the limit $\varepsilon \to 0$.

2.5. Convergence of energy

The Wigner distribution is well suited for studying high frequency limits and, in particular, families of functions that depend on a small parameter in an oscillatory manner, the ε-oscillatory families of [26]. The oscillatory property is conveniently characterized by the following definition.

Definition 2.3. A family of functions f_ε that is bounded in L^2_{loc} is said to be ε-oscillatory if for every smooth and compactly supported function $\phi(\mathbf{x})$

$$(2.59) \qquad \limsup_{\varepsilon \to 0} \int_{|\boldsymbol{\xi}| \geq R/\varepsilon} |\widehat{\phi f_\varepsilon}(\boldsymbol{\xi})|^2 d\boldsymbol{\xi} \to 0 \text{ as } R \to +\infty.$$

A simple and intuitive sufficient condition for (2.59) is that there exist a positive integer j and a constant C independent of ε such that

$$(2.60) \qquad \varepsilon^j \left\| \frac{\partial^j f_\varepsilon}{\partial x^j} \right\|_{L^2_{loc}} \leq C.$$

This condition holds, for instance, for high frequency plane waves

$$(2.61) \qquad f_\varepsilon(\mathbf{x}) = A e^{i \boldsymbol{\xi} \cdot \mathbf{x}/\varepsilon}.$$

Another natural example of ε-oscillatory functions is

$$(2.62) \qquad g_\varepsilon(\mathbf{x}) = g(\frac{\mathbf{x}}{\varepsilon}),$$

where $g(\mathbf{x})$ is a periodic function with bounded gradient.

The main reason for introducing ε-oscillatory functions is the following theorem concerning weak convergence of energy, i.e. of the integral of the square of the wave function.

Proposition 2.4. *Let f_ε be a bounded family in L^2_{loc} with limit Wigner measure $\mu(\mathbf{x}, \mathbf{k})$. Then for any smooth function of compact support $\theta(\mathbf{x})$*

$$(2.63) \qquad \iint |\theta(\mathbf{x})|^2 \mu(d\mathbf{x}, d\mathbf{k}) \leq \limsup_{\varepsilon \to 0} \int_{R^d} |\theta(\mathbf{x}) f_\varepsilon(\mathbf{x})|^2 d\mathbf{x}$$

with equality holding if and only if f_ε is ε-oscillatory. In this case $\limsup$ can be replaced by $\lim$ on the right side of (2.63).

The proof can be found in [**26**].

With this proposition and the positivity property we can interpret $\mu(\mathbf{x}, \mathbf{k})$ as a phase space energy density, that is, energy density resolved over directions and wavenumbers.

LECTURE 3
Symmetric Hyperbolic Systems

3.1. General symmetric hyperbolic systems

We will use the Wigner distribution to get the high frequency approximation of symmetric hyperbolic systems [18] in phase space. As we will see in sections 3.2-3.4, many wave equations arising from physical problems can be written as symmetric hyperbolic systems of the form[1]

$$(3.1) \qquad A(\mathbf{x})\frac{\partial \mathbf{u}}{\partial t} + D^i \frac{\partial \mathbf{u}}{\partial x^i} = 0,$$
$$\mathbf{u}(0,\mathbf{x}) = \mathbf{u}_0(\mathbf{x}),$$

where $\mathbf{u}$ is a complex valued N-vector and $\mathbf{x} \in R^3$. We assume that the matrix $A(\mathbf{x})$ is symmetric and positive definite and that the matrices D^j are symmetric and independent of $\mathbf{x}$ and t.

The energy density $\mathcal{E}(t,\mathbf{x})$ for solutions of (3.1) is given by the inner product

$$(3.2) \qquad \mathcal{E}(t,\mathbf{x}) = \frac{1}{2}(A(\mathbf{x})\mathbf{u}(t,\mathbf{x}),\mathbf{u}(t,\mathbf{x})) = \frac{1}{2}\sum_{i,j=1}^{N} A_{ij}(\mathbf{x})u_i(t,\mathbf{x})\bar{u}_j(t,\mathbf{x})$$

and the flux $\mathcal{F}(\mathbf{x})$ by

$$(3.3) \qquad \mathcal{F}_i(t,\mathbf{x}) = \frac{1}{2}(D^i\mathbf{u}(t,\mathbf{x}),\mathbf{u}(t,\mathbf{x})).$$

Taking the inner product of (3.1) with $\mathbf{u}(t,\mathbf{x})$ yields the energy conservation law

$$(3.4) \qquad \frac{\partial \mathcal{E}}{\partial t} + \nabla \cdot \mathcal{F} = 0.$$

Integration of (3.4) shows that the total energy is conserved:

$$(3.5) \qquad \frac{d}{dt}\int \mathcal{E}(t,\mathbf{x})d\mathbf{x} = 0.$$

It is convenient to introduce the new inner product

$$(3.6) \qquad < \mathbf{u},\mathbf{v} >_A = (A\mathbf{u},\mathbf{v}).$$

---

[1]We use the summation convention as follows: repeated Latin indices are summed, while repeated Greek indices are not summed.

327

Then the energy density is $\mathcal{E} = \frac{1}{2} < \mathbf{u}, \mathbf{u} >_A$. This inner product is the natural one for the system (3.1).

For N-vector functions we define the Wigner distribution an $N \times N$ matrix,

$$(3.7) \qquad W(t, \mathbf{x}, \mathbf{k}) = \left(\frac{1}{2\pi}\right)^d \int e^{i\mathbf{k}\cdot\mathbf{y}} \mathbf{u}(t, \mathbf{x} - \mathbf{y}/2)\mathbf{u}^*(t, \mathbf{x} + \mathbf{y}/2)d\mathbf{y},$$

where $\mathbf{u}^* = \bar{\mathbf{u}}^t$ is the conjugate transpose of $\mathbf{u}$. The matrix $W(t, \mathbf{x}, \mathbf{k})$ is Hermitian but not necessarily positive definite. As in the scalar case, $W(t, \mathbf{x}, \mathbf{k})$ has the properties

$$\int W(t, \mathbf{x}, \mathbf{k})d\mathbf{k} = \mathbf{u}(t, \mathbf{x})\mathbf{u}^*(t, \mathbf{x})$$

and

$$\left(\frac{1}{2\pi}\right)^d \int W(t, \mathbf{x}, \mathbf{k})d\mathbf{x} = \hat{\mathbf{u}}(-\mathbf{k}, t)\widehat{\mathbf{u}^*}(-\mathbf{k}, t).$$

It follows that the energy density is expressible in terms of $W(t, \mathbf{x}, \mathbf{k})$ by

$$(3.8) \qquad \mathcal{E}(t, \mathbf{x}) = \frac{1}{2} < \mathbf{u}(t, \mathbf{x}), \mathbf{u}(t, \mathbf{x}) >_A = \frac{1}{2} A_{ij}(\mathbf{x})\, u_i(t, \mathbf{x})\bar{u}_j(t, \mathbf{x})$$

$$= \frac{1}{2} A_{ij}(\mathbf{x}) \int W_{ij}(t, \mathbf{x}, \mathbf{k})d\mathbf{k} = \frac{1}{2} \int \mathrm{Tr}(A(\mathbf{x})W(t, \mathbf{x}, \mathbf{k}))d\mathbf{k}.$$

The flux $\mathcal{F}(t, \mathbf{x})$ can be expressed via $W(t, \mathbf{x}, \mathbf{k})$ by

$$(3.9) \qquad \mathcal{F}_i(t, \mathbf{x}, \mathbf{k}) = \frac{1}{2} D^i_{nm} u_n(t, \mathbf{x})\bar{u}_m(t, \mathbf{x}) = \frac{1}{2} \int \mathrm{Tr}(D^i W(t, \mathbf{x}, \mathbf{k}))d\mathbf{k}.$$

To study the high frequency approximation of solutions of (3.1), we assume that the coefficients of the matrix $A(\mathbf{x})$ vary on a scale much longer than the scale on which the initial data vary. Let ε be the ratio of these two scales. We rescale space and time coordinates $(\mathbf{x}, t)$ by $\mathbf{x} \to \varepsilon\mathbf{x}$, $t \to \varepsilon t$ as in (2.9). In scaled coordinates (3.1) has the form

$$(3.10) \qquad A(\mathbf{x})\frac{\partial \mathbf{u}_\varepsilon}{\partial t} + D^j \frac{\partial \mathbf{u}_\varepsilon}{\partial x^j} = 0$$

$$(3.11) \qquad \mathbf{u}_\varepsilon(0, \mathbf{x}) = \mathbf{u}_0^\varepsilon(\mathbf{x}).$$

The initial data $\mathbf{u}_0^\varepsilon(\mathbf{x})$ is ε-oscillatory and uniformly bounded in L^2, so that the limit Wigner distribution of the initial data exists and has the correct total mass. Note that the parameter ε does not appear explicitly in (3.10). It enters through the initial conditions (3.11), which may be of the standard geometrical optics form (2.10). The scaled Wigner distribution matrix W^ε is defined, as in the scalar case, by

$$(3.12) \qquad W^\varepsilon(t, \mathbf{x}, \mathbf{k}) = \left(\frac{1}{2\pi}\right)^d \int e^{i\mathbf{k}\cdot\mathbf{y}} \mathbf{u}_\varepsilon(t, \mathbf{x} - \varepsilon\mathbf{y}/2)\mathbf{u}_\varepsilon^*(\mathbf{x} + \varepsilon\mathbf{y}/2)d\mathbf{y}.$$

Although W^ε is not positive definite, it becomes so in the high frequency limit $\varepsilon \to 0$. The proof of that is exactly the same as in the scalar case.

The transport equation for the Wigner distribution may be derived as in the case of the Schrödinger equation using the formalism of psudodifferential operators. This was done in [**27**]. This method does not simplify the calculations considerably and so we shall resort to the method of perturbation expansions which may be easily modified in the case of a random medium.

As in (2.31), W^ε satisfies the evolution equation

$$(3.13) \qquad \frac{\partial W^\varepsilon}{\partial t} + \mathcal{Q}_1^\varepsilon W^\varepsilon + \frac{1}{\varepsilon}\mathcal{Q}_2^\varepsilon W^\varepsilon = 0$$
$$W^\varepsilon(0, \mathbf{x}, \mathbf{k}) = W_0^\varepsilon(\mathbf{x}, \mathbf{k}).$$

Here the operators $\mathcal{Q}_1^\varepsilon$ and $\mathcal{Q}_2^\varepsilon$ are given by

$$\mathcal{Q}_1^\varepsilon W^\varepsilon = \frac{1}{2}\int e^{-i\mathbf{p}\cdot\mathbf{x}}\{\widehat{A^{-1}}(\mathbf{p})D^j\frac{\partial W^\varepsilon(t, \mathbf{x}, \mathbf{k} + \varepsilon\mathbf{p}/2)}{\partial x^j} +$$
$$\frac{\partial W^\varepsilon(t, \mathbf{x}, \mathbf{k} - \varepsilon\mathbf{p}/2)}{\partial x^j}D^j\widehat{A^{-1}}(\mathbf{p})$$
$$+i\widehat{A^{-1}}(\mathbf{p})p_j D^j W^\varepsilon(t, \mathbf{x}, \mathbf{k} + \varepsilon\mathbf{p}/2)$$
$$(3.14) \qquad +W^\varepsilon(\mathbf{k} - \varepsilon\mathbf{p}/2)ip_j D^j\widehat{A^{-1}}(\mathbf{p})\}d\mathbf{p}$$

and

$$\mathcal{Q}_2^\varepsilon W^\varepsilon = \int e^{-i\mathbf{p}\cdot\mathbf{x}}\{i\widehat{A^{-1}}(\mathbf{p})k_j D^j W^\varepsilon(t, \mathbf{x}, \mathbf{k} + \varepsilon\mathbf{p}/2)$$
$$(3.15) \qquad -iW^\varepsilon(t, \mathbf{x}, \mathbf{k} - \varepsilon\mathbf{p}/2)k_j D^j\widehat{A^{-1}}(\mathbf{p})\}d\mathbf{p}.$$

The hat denotes the Fourier transform as in (2.33). The initial condition for (3.13) is obtained by inserting (3.11) into (3.12).

A new feature of (3.13), not found in the scalar case (2.31), is the appearance of the factor $1/\varepsilon$ in front of the term $\mathcal{Q}_2^\varepsilon W^\varepsilon$. There is no other term in the equation to balance it. This means that the limiting Wigner distribution $W(t, \mathbf{x}, \mathbf{k})$ ($W^\varepsilon \to W$ as $\varepsilon \to 0$) must belong to the null space of the limit operator $\mathcal{Q}_2$, where $\mathcal{Q}_2^\varepsilon \to \mathcal{Q}_2$ as $\varepsilon \to 0$. From (3.15) this operator acting on a matrix $Z(\mathbf{x}, \mathbf{k})$ has the form

$$(3.16) \qquad \mathcal{Q}_2 Z(\mathbf{x}, \mathbf{k}) = iA^{-1}k_j D^j Z(\mathbf{x}, \mathbf{k}) - iZ(\mathbf{x}, \mathbf{k})k_j D^j A^{-1}.$$

The next term in the expansion of $\mathcal{Q}_2^\varepsilon$ in ε, $\mathcal{Q}_2^\varepsilon = \mathcal{Q}_2 + \varepsilon\mathcal{Q}_{21} + \dots$, is given by

$$(3.17) \qquad \mathcal{Q}_{21} Z(\mathbf{x}, \mathbf{k}) = -\frac{1}{2}\frac{\partial A^{-1}}{\partial x^i}k_j D^j\frac{\partial Z}{\partial k_i} - \frac{1}{2}\frac{\partial Z}{\partial k_i}k_j D^j\frac{\partial A^{-1}}{\partial x^i}.$$

This introduces the term with the gradient with respect to $\mathbf{k}$ into the transport equation, as we shall see. Similarly, the limit operator $\mathcal{Q}_1$, $\mathcal{Q}_1^\varepsilon \to \mathcal{Q}_1$ as $\varepsilon \to 0$ is given by

$$(3.18) \quad \mathcal{Q}_1 Z(\mathbf{x}, \mathbf{k}) = \frac{1}{2}A^{-1}D^j\frac{\partial Z}{\partial x^j} + \frac{1}{2}\frac{\partial Z}{\partial x^j}D^j A^{-1} - \frac{1}{2}\frac{\partial A^{-1}}{\partial x^j}D^j Z - \frac{1}{2}ZD^j\frac{\partial A^{-1}}{\partial x^j}.$$

This operator introduces the term with the **x**-gradient. The undifferentiated terms in $\mathcal{Q}_1$ also contribute to the transport equation, as we explain below. With the expansions of the $\mathcal{Q}$'s given by (3.16)-(3.18) equation (3.13) becomes

$$(3.19) \qquad \frac{\partial W^\varepsilon}{\partial t} + \frac{1}{\varepsilon}\mathcal{Q}_2 W^\varepsilon + (\mathcal{Q}_{21} + \mathcal{Q}_1)W^\varepsilon + O(\varepsilon) = 0.$$

We analyze (3.19) by expanding W^ε

$$W^\varepsilon(t, \mathbf{x}, \mathbf{k}) = W^{(0)}(t, \mathbf{x}, \mathbf{k}) + \varepsilon W^{(1)}(t, \mathbf{x}, \mathbf{k}) + \dots$$

This leads to the following equations for $W^{(0)}$ and $W^{(1)}$

$$(3.20) \qquad\qquad \mathcal{Q}_2 W^{(0)} = 0$$

and

$$(3.21) \qquad\qquad \mathcal{Q}_2 W^{(1)} = -\{\frac{\partial W^{(0)}}{\partial t} + (\mathcal{Q}_{21} + \mathcal{Q}_1) W^{(0)}\}.$$

We introduce the *dispersion* matrix $L(\mathbf{x}, \mathbf{k})$, defined by

$$(3.22) \qquad\qquad L(\mathbf{x}, \mathbf{k}) = A^{-1}(\mathbf{x}) k_i D^i.$$

It is self-adjoint with respect to the inner product $<, >_A$:

$$
\begin{aligned}
< L\mathbf{u}, \mathbf{v} >_A &= (AL\mathbf{u}, \mathbf{v}) = (k_j D^j \mathbf{u}, \mathbf{v}) \\
&= (\mathbf{u}, k_j D^j \mathbf{v}) = (A\mathbf{u}, A^{-1} k_j D^j \mathbf{v}) = < \mathbf{u}, L\mathbf{v} >_A .
\end{aligned}
$$

Therefore, all its eigenvalues ω_τ are real and the corresponding eigenvectors $\mathbf{b}^\tau$ can be chosen to be orthonormal with respect to $<, >_A$:

$$L(\mathbf{x}, \mathbf{k})\mathbf{b}^\tau(\mathbf{x}, \mathbf{k}) = \omega_\tau(\mathbf{x}, \mathbf{k})\mathbf{b}^\tau(\mathbf{x}, \mathbf{k}) , \quad < \mathbf{b}^\tau, \mathbf{b}^\beta >_A = \delta_{\tau\beta}.$$

We assume throughout that the eigenvalues have constant multiplicity independent of $\mathbf{x}$ and $\mathbf{k}$. This hypothesis is satisfied in the case of acoustic, electromagnetic and elastic waves. In terms of the dispersion matrix L, (3.20) becomes

$$\mathcal{Q}_2 W^{(0)}(t, \mathbf{x}, \mathbf{k}) = iL(\mathbf{x}, \mathbf{k})W^{(0)}(t, \mathbf{x}, \mathbf{k}) - iW^{(0)}(t, \mathbf{x}, \mathbf{k})L^*(\mathbf{x}, \mathbf{k}) = 0.$$

The structure of this null space when all the eigenvalues of $L(\mathbf{x}, \mathbf{k})$ are distinct is different from that when there are some multiple eigenvalues.

We assume first that all the eigenvalues $\omega_\tau(\mathbf{x}, \mathbf{k})$ are simple. Define the matrices $B^\tau(\mathbf{x}, \mathbf{k})$ by

$$(3.23) \qquad\qquad B^\tau(\mathbf{x}, \mathbf{k}) = \mathbf{b}^\tau(\mathbf{x}, \mathbf{k})\mathbf{b}^{\tau*}(\mathbf{x}, \mathbf{k}).$$

They span the null space of $\mathcal{Q}_2$, so the limit Wigner matrix $W^{(0)}(t, \mathbf{x}, \mathbf{k})$ has the form

$$(3.24) \qquad\qquad W^{(0)}(t, \mathbf{x}, \mathbf{k}) = \sum_{\tau=1}^{N} a^\tau(t, \mathbf{x}, \mathbf{k}) B^\tau(\mathbf{x}, \mathbf{k}).$$

The $a^\tau(t, \mathbf{x}, \mathbf{k})$ are scalar functions determined by projection

$$a^\tau = \mathrm{Tr}(AW^{(0)*} AB^\tau).$$

We now insert (3.24) into equation (3.21) for $W^{(1)}$, which is an inhomogeneous form of (3.20). The operator $\frac{1}{i}\mathcal{Q}_2$ is self-adjoint with respect to the matrix inner product $<< X, Y >> = \mathrm{Tr}(AX^* AY)$. Since the null space of $\mathcal{Q}_2$ is spanned by the matrices B^τ given by (3.23), the solvability condition for (3.21) is that its right hand side be orthogonal to these matrices, relative to the $<<, >>$ inner product. This leads to the following equations for the functions a^τ:

$$(3.25) \qquad\qquad \frac{\partial a^\tau}{\partial t} + \nabla_\mathbf{k}\omega_\tau \cdot \nabla_\mathbf{x} a^\tau - \nabla_\mathbf{x}\omega_\tau \cdot \nabla_\mathbf{k} a^\tau = 0.$$

These are Liouville equations in phase space. We leave the details of the derivation to the reader.

We see, therefore, that in the absence of polarization (simple eigenvalues of the dispersion matrix) the amplitudes a^τ decouple from each other and each satisfies the Liouville equation with Hamiltonian equal to the corresponding eigenvalue ω_τ.

We see also that the Liouville equation is not satisfied by the limiting Wigner distribution but by its projections on the eigenspaces generated by the matrices B^τ given by (3.23). Moreover, we do not have a single Liouville equation but several decoupled ones. When small random perturbations are present the Liouville equations are coupled (section 4.1).

Consider now the case where the dispersion matrix $L(\mathbf{x}, \mathbf{k})$ has multiple eigenvalues. Let $\omega_\tau(\mathbf{x}, \mathbf{k})$ be an eigenvalue of multiplicity r and let the corresponding eigenvectors $\mathbf{b}^{\tau,i}$, $i = 1, \ldots, r$ be orthonormal with respect to $<, >_A$. Given a pair of eigenvectors $\mathbf{b}^{\tau,i}$, $\mathbf{b}^{\tau,j}$ we define the $N \times N$ matrix

$$(3.26) \qquad B^{\tau,ij} = \mathbf{b}^{\tau,i}\mathbf{b}^{\tau,j*},$$

with $i, j = 1 \ldots r$. These matrices span the null space of the operator $\mathcal{Q}_2$ and so the limiting Wigner matrix $W^{(0)}(t, \mathbf{x}, \mathbf{k})$ has the representation

$$(3.27) \qquad W^{(0)}(t, \mathbf{x}, \mathbf{k}) = \sum_{\tau, i, j} a_{ij}^\tau(t, \mathbf{x}, \mathbf{k}) B^{\tau,ij}(\mathbf{x}, \mathbf{k}),$$

where $a_{ij}^\tau(t, \mathbf{x}, \mathbf{k})$ are scalar functions. Define the $r \times r$ *coherence* matrices $W^\tau(t, \mathbf{x}, \mathbf{k})$ by

$$(3.28) \qquad W_{ij}^\tau(t, \mathbf{x}, \mathbf{k}) = a_{ij}^\tau(t, \mathbf{x}, \mathbf{k}) \ , \ i, j = 1 \ldots r.$$

The multiplicity r of the eigenvalue ω_τ depends on τ but we do not indicate this explicitly. The functions a_{ij}^τ are given by

$$a_{ij}^\tau(t, \mathbf{x}, \mathbf{k}) = << W^{(0)}(t, \mathbf{x}, \mathbf{k}), B^{\tau,ij}(\mathbf{x}, \mathbf{k}) >> .$$

Then, by applying the solvability condition for (3.21) as before, we find that each of the coherence matrices $W^\tau(t, \mathbf{x}, \mathbf{k})$ satisfies the transport equation

$$(3.29) \qquad \frac{\partial W^\tau}{\partial t} + \nabla_\mathbf{k}\omega_\tau \cdot \nabla_\mathbf{x}W^\tau - \nabla_\mathbf{x}\omega_\tau \cdot \nabla_\mathbf{k}W^\tau + W^\tau N^\tau - N^\tau W^\tau = 0.$$

The skew-symmetric *coupling* matrices $N^\tau(\mathbf{x}, \mathbf{k})$ are given by

$$(3.30) \quad N_{mn}^\tau(\mathbf{x}, \mathbf{k}) = (\mathbf{b}^{\tau,n}, D^i \frac{\partial \mathbf{b}^{\tau,m}}{\partial x^i}) - \frac{\partial \omega_\tau}{\partial x^i}(A(\mathbf{x})\mathbf{b}^{\tau,n}, \frac{\partial \mathbf{b}^{\tau,m}}{\partial k_i}) - \frac{1}{2}\frac{\partial^2 \omega_\tau}{\partial x^i \partial k_i}\delta_{nm}.$$

The last term in (3.30) is retained to make the coupling matrices N skew symmetric even though it cancels in the transport equation (3.29).

The coherence matrices $W^\tau(t, \mathbf{x}, \mathbf{k})$ are Hermitian and positive definite because they are projections of the limiting Wigner matrix $W^{(0)}(t, \mathbf{x}, \mathbf{k})$ which is Hermitian and positive definite. Equations (3.29) preserve both of these properties: if the initial conditions for W^τ are Hermitian and positive definite then the solution is Hermitian and positive definite for all t. The fact that the coupling matrices N are skew-symmetric is important for these properties.

We see that in the case of polarized waves, i.e. waves for which the eigenvalues of the dispersion matrix have multiplicity larger than one, the quantities satisfying the transport equations are not scalars but matrices. Their sizes are equal to the degeneracies of the corresponding wave modes. However, modes corresponding to different eigenvalues still decouple from each other. Random inhomogeneities will couple them in general (section 4.2).

The reason we call the $W^\tau(t, \mathbf{x}, \mathbf{k})$ coherence matrices is because their off-diagonal terms capture cross-polarization effects. Their diagonal terms represent the phase space energy density in each state of polarization. That is, since $\mathrm{Tr}(AB^{\tau,ij}) = \delta_{ij}$, the energy density (3.8) is given by

$$(3.31) \quad \mathcal{E}(t,\mathbf{x}) = \frac{1}{2}\int \mathrm{Tr}(A(\mathbf{x})W(t,\mathbf{x},\mathbf{k}))d\mathbf{k} = \frac{1}{2}\int \sum_\tau \mathrm{Tr}W^\tau(t,\mathbf{x},\mathbf{k})d\mathbf{k}$$

and the flux (3.9) is given by

$$\begin{aligned}
\mathcal{F}_i(t,\mathbf{x}) &= \frac{1}{2}\mathrm{Tr}\int D^i W(t,\mathbf{x},\mathbf{k})d\mathbf{k} \\
(3.32) \qquad &= \frac{1}{2}\int \sum_\tau \frac{\partial\omega_\tau}{\partial k_i}\mathrm{Tr}W^\tau(t,\mathbf{x},\mathbf{k})d\mathbf{k} \ , \ i = 1,2,3.
\end{aligned}$$

These relations hold because

$$\begin{aligned}
\mathrm{Tr}D^i W(t,\mathbf{x},\mathbf{k}) &= \sum_{\tau,n,m} a^\tau_{nm}(t,\mathbf{x},\mathbf{k})\mathrm{Tr}\{D^i b^{\tau,n}(\mathbf{x},\mathbf{k})b^{\tau,m*}(\mathbf{x},\mathbf{k})\} \\
&= \sum_{\tau,n,m} a^\tau_{nm}(t,\mathbf{x},\mathbf{k})(D^i b^{\tau,n}(\mathbf{x},\mathbf{k}),b^{\tau,m}(\mathbf{x},\mathbf{k})) \\
&= \sum_{\tau,n,m} a^\tau_{nm}(t,\mathbf{x},\mathbf{k})(\frac{\partial\omega_\tau}{\partial k_i}Ab^{\tau,n} + \omega_\tau A\frac{\partial b^{\tau,n}}{\partial k_i} - k_j D^j\frac{\partial b^{\tau,n}}{\partial k_i},b^{\tau,m}) \\
&= \sum_{\tau,n,m} a^\tau_{nm}(t,\mathbf{x},\mathbf{k})\frac{\partial\omega_\tau}{\partial k_i}(Ab^{\tau,n},b^{\tau,m}) = \sum_\tau \frac{\partial\omega_\tau}{\partial k_i}\mathrm{Tr}W^\tau(t,\mathbf{x},\mathbf{k}).
\end{aligned}$$

Here we have used the fact that $L\mathbf{b}^\tau = A^{-1}k_i D^i\mathbf{b}^\tau = \omega_\tau\mathbf{b}^\tau$, which implies after differentiation with respect to k_i, that

$$D^i\mathbf{b}^\tau = \frac{\partial\omega_\tau}{\partial k_i}A\mathbf{b}^\tau + \omega_\tau A\frac{\partial\mathbf{b}^\tau}{\partial k_i} - k_j D^j\frac{\partial\mathbf{b}^\tau}{\partial k_i}.$$

The energy equation (3.4) follows from (3.29) when $\mathcal{E}$ and $\mathcal{F}$ are defined by (3.31) and (3.32), respectively. Thus, the total energy

$$\int \mathcal{E}(t,\mathbf{x})d\mathbf{x}$$

is conserved by the transport equations (3.29).

Expressions (3.31) and (3.32) for the energy and flux are similar to (2.20) and (2.21) because $\mathbf{k}W$, which is the flux density for the Schrödinger equation, can be written as $\nabla_\mathbf{k}H(\mathbf{x},\mathbf{k})W(t,\mathbf{x},\mathbf{k})$ where $H(\mathbf{x},\mathbf{k})$ is the Hamiltonian (2.14).

In the case of multiple eigenvalues, there is a basis of eigenvectors $\mathbf{b}^{\tau,i}(\mathbf{x},\mathbf{k})$ such that the transport equations (3.29) for the coherence matrices have the form (3.25); that is, we can eliminate the matrices N^τ from (3.29) by a rotation of the basis. Small random perturbations couple the components of the coherence matrices, and to keep the coupling explicit we do not use a basis which eliminates the N's.

3.2. High frequency approximation for acoustic waves

We will now apply the results of the previous section to acoustic waves. We will also review the usual form of the high frequency approximation and make explicit the relation between the phase space form of the high frequency approximation and the usual one.

The acoustic equations for the velocity and pressure disturbances $\mathbf{u}$ and p are

$$\rho \frac{\partial \mathbf{u}}{\partial t} + \nabla p = 0$$

(3.33)
$$\kappa \frac{\partial p}{\partial t} + \operatorname{div} \mathbf{u} = 0.$$

Here $\rho = \rho(\mathbf{x})$ is the density and $\kappa = \kappa(\mathbf{x})$ is the compressibility. Equations (3.33) can be put in the form of a symmetric hyperbolic system

$$A(\mathbf{x}) \frac{\partial}{\partial t} \begin{pmatrix} \mathbf{u} \\ p \end{pmatrix} + \sum_{i=1}^{3} D^i \frac{\partial}{\partial x^i} \begin{pmatrix} \mathbf{u} \\ p \end{pmatrix} = 0.$$

The matrix $A(\mathbf{x}) = \operatorname{diag}(\rho(\mathbf{x}), \rho(\mathbf{x}), \rho(\mathbf{x}), \kappa(\mathbf{x}))$ and each of the matrices D^i has all zero entries except for D^i_{i4} and D^i_{4i} which are equal to one. Then the dispersion matrix $L(\mathbf{x}, \mathbf{k})$, defined by (3.22), is

(3.34)
$$L = \begin{pmatrix} 0 & 0 & 0 & k_1/\rho \\ 0 & 0 & 0 & k_2/\rho \\ 0 & 0 & 0 & k_3/\rho \\ k_1/\kappa & k_2/\kappa & k_3/\kappa & 0 \end{pmatrix}.$$

It has one double eigenvalue $\omega_1 = \omega_2 = 0$ and two simple eigenvalues

(3.35)
$$\omega_+ = v(\mathbf{x})|\mathbf{k}| \ , \quad \omega_- = -v(\mathbf{x})|\mathbf{k}| \ ,$$

where $|\mathbf{k}| = \sqrt{k_1^2 + k_2^2 + k_3^2}$ and v is the sound speed

(3.36)
$$v(\mathbf{x}) = \frac{1}{\sqrt{\kappa(\mathbf{x})\rho(\mathbf{x})}}.$$

The corresponding basis of eigenvectors orthonormal with respect to the inner product $<,>_A$ is

$$\mathbf{b}^1 = \frac{1}{\sqrt{\rho}} (\mathbf{z}^{(1)}(\mathbf{k}), 0)^t,$$

$$\mathbf{b}^2 = \frac{1}{\sqrt{\rho}} (\mathbf{z}^{(2)}(\mathbf{k}), 0)^t,$$

(3.37)
$$\mathbf{b}^+ = (\frac{\hat{\mathbf{k}}}{\sqrt{2\rho}}, \frac{1}{\sqrt{2\kappa}})^t,$$

$$\mathbf{b}^- = (\frac{\hat{\mathbf{k}}}{\sqrt{2\rho}}, -\frac{1}{\sqrt{2\kappa}})^t,$$

the vectors $\hat{\mathbf{k}}$, $\mathbf{z}^{(1)}(\mathbf{k})$, $\mathbf{z}^{(2)}(\mathbf{k})$, which form an orthonormal triplet, are

(3.38)
$$\hat{\mathbf{k}} = \begin{pmatrix} \sin\theta\cos\phi \\ \sin\theta\sin\phi \\ \cos\theta \end{pmatrix}, \mathbf{z}^{(1)} = \begin{pmatrix} \cos\theta\cos\phi \\ \cos\theta\sin\phi \\ -\sin\theta \end{pmatrix}, \mathbf{z}^{(2)} = \begin{pmatrix} -\sin\phi \\ \cos\phi \\ 0 \end{pmatrix}.$$

The physical meaning of the eigenvectors is as follows. The eigenvectors $\mathbf{b}^1(\mathbf{x}, \mathbf{k})$ and $\mathbf{b}^2(\mathbf{x}, \mathbf{k})$ correspond to transverse advection modes, orthogonal to the direction of propagation. These modes do not propagate because $\omega_{1,2} = 0$. The eigenvectors $\mathbf{b}^+(\mathbf{x}, \mathbf{k})$ and $\mathbf{b}^-(\mathbf{x}, \mathbf{k})$ represent acoustic waves, which are longitudinal , and which propagate with the sound speed $\pm v(\mathbf{x})$ given by (3.36).

The energy density (3.2) for acoustic waves is given by

$$(3.39) \qquad \mathcal{E}(t,\mathbf{x}) = \frac{1}{2}\rho(\mathbf{x})|\mathbf{u}(t,\mathbf{x})|^2 + \frac{1}{2}\kappa(\mathbf{x})p^2(t,\mathbf{x})$$

and the flux (3.3) by

$$(3.40) \qquad \mathcal{F}(t,\mathbf{x}) = p(t,\mathbf{x})\mathbf{u}(t,\mathbf{x}).$$

We now express the *unscaled* amplitudes $a^j(t,\mathbf{x},\mathbf{k})$, in terms of the acoustic velocity and pressure fields $\underline{\mathbf{u}} = (\mathbf{u},p)^t$. The amplitudes $a^{\pm}(t,\mathbf{x},\mathbf{k})$ are given by

$$(3.41) \quad a^{\pm}(t,\mathbf{x},\mathbf{k}) = \frac{1}{(2\pi)^3}\int d\mathbf{y}\, e^{i\mathbf{k}\cdot\mathbf{y}} f_{\pm}(t,\mathbf{x},\mathbf{x}-\mathbf{y}/2,\mathbf{k})f_{\pm}^*(t,\mathbf{x},\mathbf{x}+\mathbf{y}/2,\mathbf{k}),$$

where

$$(3.42) \quad f_{\pm}(t,\mathbf{x},\mathbf{z},\mathbf{k}) = <\underline{\mathbf{u}}(t,\mathbf{z}),b^{\pm}(\mathbf{x},\mathbf{k})>_A = \sqrt{\frac{\rho(\mathbf{x})}{2}}(\mathbf{u}(t,\mathbf{z})\cdot\hat{\mathbf{k}}) \pm \sqrt{\frac{\kappa(\mathbf{x})}{2}}p(t,\mathbf{z}).$$

This shows that

$$(3.43) \qquad a^+(t,\mathbf{x},\mathbf{k}) = a^-(t,\mathbf{x},-\mathbf{k})$$

and therefore we need only keep track of $a^+(t,\mathbf{x},\mathbf{k})$. The advective mode amplitudes are given by

$$a^0_{ij}(t,\mathbf{x},\mathbf{k}) = \frac{1}{(2\pi)^3}\int d\mathbf{y}\, e^{i\mathbf{k}\cdot\mathbf{y}}\frac{\rho(\mathbf{x})}{2}(\mathbf{u}(t,\mathbf{x}-\mathbf{y}/2)\cdot\mathbf{z}^{(i)}(\mathbf{k}))\overline{(\mathbf{u}(t,\mathbf{x}+\mathbf{y}/2)\cdot\mathbf{z}^{(j)}(\mathbf{k}))}.$$

By direct computation we verify that

$$(3.44) \qquad \int a^+(t,\mathbf{x},\mathbf{k})d\mathbf{k} + \frac{1}{2}\int \{a^0_{11}(t,\mathbf{x},\mathbf{k}) + a^0_{22}(t,\mathbf{x},\mathbf{k})\}d\mathbf{k}$$

$$= \frac{1}{2}\rho(\mathbf{x})|\mathbf{u}(t,\mathbf{x})|^2 + \frac{1}{2}\kappa(\mathbf{x})p^2(t,\mathbf{x}) = \mathcal{E}(t,\mathbf{x})$$

and

$$(3.45) \qquad \int \hat{\mathbf{k}}v(\mathbf{x})a^+(t,\mathbf{x},\mathbf{k})d\mathbf{k} = p(t,\mathbf{x})\mathbf{u}(t,\mathbf{x}) = \mathcal{F}(t,\mathbf{x}).$$

The first integral in (3.44) represents the part of the energy density which is propagating with speed v. The second integral gives the energy of the non-propagating waves.

Equation (3.29) for W^0 is of the form $\dfrac{\partial W^0}{\partial t} = 0$ and so $W^0(t,\mathbf{x},\mathbf{k}) = 0$ if it is zero initially. Then from (3.44)

$$(3.46) \qquad \int a^+(t,\mathbf{x},\mathbf{k})d\mathbf{k} = \frac{1}{2}\rho(\mathbf{x})|\mathbf{u}(t,\mathbf{x})|^2 + \frac{1}{2}\kappa(\mathbf{x})p^2(t,\mathbf{x}).$$

This shows that when $W^0 = 0$, the amplitude $a^+(t,\mathbf{x},\mathbf{k})$ is the phase space energy density. In the high frequency limit it satisfies the Liouville equation (3.25)

$$(3.47) \qquad \frac{\partial a^+}{\partial t} + v(\mathbf{x})\hat{\mathbf{k}}\cdot\nabla_{\mathbf{x}}a^+ - |\mathbf{k}|\nabla_{\mathbf{x}}v(\mathbf{x})\cdot\nabla_{\mathbf{k}}a^+ = 0$$

with the initial condition

$$(3.48) \qquad a^+(0,\mathbf{x},\mathbf{k}) = a_0(\mathbf{x},\mathbf{k}).$$

Next we establish the connection with the usual high frequency approximation. We consider (3.33) with initial data of the form

$$\mathbf{\underline{u}}(0, \mathbf{x}) = \mathbf{\underline{u}}_0(\mathbf{x})e^{iS_0(\mathbf{x})/\varepsilon}, \tag{3.49}$$

where $\mathbf{\underline{u}} = (\mathbf{u}, p)$ and S_0 is the real valued initial phase function. We look for a solution of (3.33) in the form

$$\mathbf{\underline{u}}(t, \mathbf{x}) = (\underline{A}_0(t, \mathbf{x}) + \varepsilon\underline{A}_1 + \dots)e^{iS(t,\mathbf{x})/\varepsilon}, \tag{3.50}$$

where $\underline{A}_0 = (\mathbf{u}_0, p_0)$. We insert (3.50) into (3.33) to get to leading order in ε

$$\begin{pmatrix} \rho S_t & \nabla S \\ \nabla S\cdot & \kappa S_t \end{pmatrix} \begin{pmatrix} \mathbf{u}_0 \\ p_0 \end{pmatrix} = 0. \tag{3.51}$$

The next term in the expansion yields

$$-i \begin{pmatrix} \rho S_t & \nabla S \\ \nabla S\cdot & \kappa S_t \end{pmatrix} \begin{pmatrix} \mathbf{u}_1 \\ p_1 \end{pmatrix} = \begin{pmatrix} \rho\partial_t & \nabla \\ \nabla\cdot & \kappa\partial_t \end{pmatrix} \begin{pmatrix} \mathbf{u}_0 \\ p_0 \end{pmatrix}. \tag{3.52}$$

Equation (3.51) gives the eiconal equation for the phase S

$$\frac{1}{v^2}S_t^2 - (\nabla S)^2 = 0. \tag{3.53}$$

Then assuming that $S_t = +v|\nabla S|$ we have

$$\begin{pmatrix} \mathbf{u}_0 \\ p_0 \end{pmatrix} = \mathcal{A}(\mathbf{x})\mathbf{b}^+(\mathbf{x}, \nabla S(t, \mathbf{x})), \tag{3.54}$$

where $\mathbf{b}^+$ is given by (3.37). The amplitude $\mathcal{A}(t, \mathbf{x})$ is determined by the solvability condition for (3.52), which gives the transport equation

$$\frac{\partial}{\partial t}|\mathcal{A}|^2 + \nabla\cdot(|\mathcal{A}|^2 v\frac{\nabla S}{|\nabla S|}) = 0. \tag{3.55}$$

The terminology 'transport equation' is standard in high frequency asymptotics for this equation and should not be confused with the radiative transport equations which are defined in phase space. As expected, equation (3.55) is the same as (3.4), to principal order in ε when $\mathbf{\underline{u}}$ is of the form (3.50) and (3.54). It is also the same as the transport equation (2.13) for the Schrödinger equation and both can be written in the form

$$\frac{\partial}{\partial t}|\mathcal{A}|^2 + \nabla\cdot(|\mathcal{A}|^2\nabla_{\mathbf{k}}H(\mathbf{x}, \nabla S)) = 0. \tag{3.56}$$

The Hamiltonian for the acoustic waves is the eigenvalue $H(\mathbf{x}, \mathbf{k}) = v(\mathbf{x})|\mathbf{k}|$ and for the Schrödinger equation it is given by (2.14).

The eiconal and transport equations (3.53) and (3.55) can also be derived from (3.47) as follows. In the high frequency limit, initial conditions of the form (3.49) imply that

$$a^+(0, \mathbf{x}, \mathbf{k}) = |\mathcal{A}_0(\mathbf{x})|^2\delta(\mathbf{k} - \nabla S_0(\mathbf{x})). \tag{3.57}$$

Let the functions $S(t, \mathbf{x})$ and $|\mathcal{A}(t, \mathbf{x})|^2$ be the solutions of the eiconal and transport equations (3.53) and (3.55), respectively, with the initial conditions $S(0, \mathbf{x}) = S_0(\mathbf{x})$ and $|\mathcal{A}(0, \mathbf{x})|^2 = |\mathcal{A}_0(\mathbf{x})|^2$. Then the solution of equation (3.47) is

$$a^+(t, \mathbf{x}, \mathbf{k}) = |\mathcal{A}(t, \mathbf{x})|^2\delta(\mathbf{k} - \nabla S(t, \mathbf{x})). \tag{3.58}$$

Conversely, given initial conditions of the form (3.57) for (3.47) and a^+ given by (3.58), then S and $\mathcal{A}$ must satisfy the eiconal and transport equations (3.53) and

(3.55), respectively. This is because the eiconal equation follows by integrating (3.47) with respect to $\mathbf{k}$ while the transport equation follows by multiplying it by $\mathbf{k}$ and then integrating with respect to $\mathbf{k}$. This shows that we can recover from the Liouville equation (3.25) the usual high frequency approximation.

3.3. Geometrical optics for electromagnetic waves

Maxwell's equations in an isotropic medium and in suitable units are

$$(3.59) \qquad \frac{\partial \mathbf{E}}{\partial t} = \frac{1}{\epsilon}\mathrm{curl}\mathbf{H}$$

$$\frac{\partial \mathbf{H}}{\partial t} = -\frac{1}{\mu}\mathrm{curl}\mathbf{E}$$

where the dielectric permittivity [2] is $\epsilon(\mathbf{x})$ and the relative magnetic permeability is $\mu(\mathbf{x})$. As a symmetric hyperbolic system they are

$$(3.60) \qquad \begin{pmatrix} \epsilon & 0 \\ 0 & \mu \end{pmatrix} \frac{\partial}{\partial t} \begin{pmatrix} \mathbf{E} \\ \mathbf{H} \end{pmatrix} + \begin{pmatrix} 0 & -\nabla\times \\ \nabla\times & 0 \end{pmatrix} \begin{pmatrix} \mathbf{E} \\ \mathbf{H} \end{pmatrix} = 0.$$

These equations imply that if at some initial time we have

$$(3.61) \qquad \mathrm{div}(\epsilon\mathbf{E}) = 0$$

$$\mathrm{div}(\mu\mathbf{H}) = 0$$

then these equations hold for all time. We assume (3.3.3) holds. The 6×6 dispersion matrix L defined by (3.22) is

$$(3.62) \qquad L = - \begin{pmatrix} 0 & 0 & 0 & 0 & -k_3/\epsilon & k_2/\epsilon \\ 0 & 0 & 0 & k_3/\epsilon & 0 & -k_1/\epsilon \\ 0 & 0 & 0 & -k_2/\epsilon & k_1/\epsilon & 0 \\ 0 & k_3/\mu & -k_2/\mu & 0 & 0 & 0 \\ -k_3/\mu & 0 & k_1/\mu & 0 & 0 & 0 \\ k_2/\mu & -k_1/\mu & 0 & 0 & 0 & 0 \end{pmatrix}$$

or in block form

$$L = \begin{pmatrix} 0 & -\frac{1}{\epsilon}P \\ \frac{1}{\mu}P & 0 \end{pmatrix}.$$

The matrix $P(\mathbf{k})\mathbf{p} = \mathbf{k}\times\mathbf{p}$ or

$$(3.63) \qquad P(\mathbf{k}) = \begin{pmatrix} 0 & -k_3 & k_2 \\ k_3 & 0 & -k_1 \\ -k_2 & k_1 & 0 \end{pmatrix}.$$

The dispersion matrix L has three eigenvalues, each with multiplicity two. They are $\omega_0 = 0$, $\omega_+ = v|\mathbf{k}|$, $\omega_- = -v|\mathbf{k}|$ with the speed of propagation v given by

$$(3.64) \qquad v(\mathbf{x}) = \frac{1}{\sqrt{\epsilon(\mathbf{x})\mu(\mathbf{x})}}.$$

---

[2]Throughout this section and when we consider electromagnetic waves ϵ denotes the dielectric permittivity while the small parameter is denoted by ε.

The basis formed by the corresponding eigenvectors is

$$\mathbf{b}^{(01)} = \frac{1}{\sqrt{\epsilon}}(\hat{\mathbf{k}}, 0), \ \ \mathbf{b}^{(02)} = \frac{1}{\sqrt{\mu}}(0, \hat{\mathbf{k}}),$$

$$\mathbf{b}^{(+,1)} = (\sqrt{\frac{1}{2\epsilon}}\mathbf{z}^{(1)}, \sqrt{\frac{1}{2\mu}}\mathbf{z}^{(2)}), \ \ \mathbf{b}^{(+,2)} = (\sqrt{\frac{1}{2\epsilon}}\mathbf{z}^2, -\sqrt{\frac{1}{2\mu}}\mathbf{z}^{(1)}),$$

$$(3.65) \quad \mathbf{b}^{(-,1)} = (\sqrt{\frac{1}{2\epsilon}}\mathbf{z}^{(1)}, -\sqrt{\frac{1}{2\mu}}\mathbf{z}^{(2)}), \ \ \mathbf{b}^{(-,2)} = (\sqrt{\frac{1}{2\epsilon}}\mathbf{z}^{(2)}, \sqrt{\frac{1}{2\mu}}\mathbf{z}^{(1)}),$$

where the vectors $\mathbf{z}^{(1)}(\mathbf{k})$ and $\mathbf{z}^{(2)}(\mathbf{k})$ are given by (3.38). The eigenvectors $\mathbf{b}^{(01)}$ and $\mathbf{b}^{(02)}$ represent the non-propagating longitudinal modes and do not satisfy (3.61) so they will be assumed to be absent from the solution. The other eigenvectors correspond to transverse modes propagating with the speed $v(\mathbf{x})$. As in the acoustic case, we need only consider the eigenspace corresponding to ω_+. With this choice for the basis of eigenvectors, the skew symmetric coupling matrix $N(\mathbf{x}, \mathbf{k})$, given by (3.30), is

$$(3.66) \qquad N = \frac{\partial v}{\partial x^i}|\mathbf{k}|\mathbf{z}^{(1)} \cdot \frac{\partial \mathbf{z}^{(2)}}{\partial k_i} \begin{pmatrix} 0 & 1 \\ -1 & 0 \end{pmatrix}.$$

Note that the vector $\mathbf{z}^{(2)}(\mathbf{k})$ does not depend on k_3. From (3.66) we conclude that if the medium is layered, so that $v = v(x_3)$, then the coupling matrix N vanishes. This means that in the case of a layered medium there is no coupling between the two polarizations of the electromagnetic field, a well known fact. We note also that there is a choice of the vectors $\mathbf{z}^{(1)}(\mathbf{k})$, $\mathbf{z}^{(2)}(\mathbf{k})$, different from (3.38), which eliminates the coupling terms [13]. As explained earlier, we will use (3.38) because they are convenient for the analysis of random effects.

The transport equation (3.29) for the matrix W^+ is

$$(3.67) \quad \frac{\partial W^+}{\partial t} + v(\mathbf{x})\hat{\mathbf{k}} \cdot \nabla_{\mathbf{x}} W^+ - |\mathbf{k}|\nabla_{\mathbf{x}} v(\mathbf{x}) \cdot \nabla_{\mathbf{k}} W^+ + W^+ N - NW^+ = 0.$$

The energy density (3.2) for the electromagnetic waves is given by

$$(3.68) \qquad \mathcal{E}(t, \mathbf{x}) = \frac{1}{2}\epsilon(\mathbf{x})|\mathbf{E}(t, \mathbf{x})|^2 + \frac{1}{2}\mu(\mathbf{x})|\mathbf{H}(t, \mathbf{x})|^2$$

while the energy flux (3.3) is the Poynting vector

$$(3.69) \qquad \mathcal{F}(t, \mathbf{x}) = \mathbf{E}(t, \mathbf{x}) \times \mathbf{H}(t, \mathbf{x}).$$

Let $\underline{\mathbf{u}}(t, \mathbf{x}) = (\mathbf{E}, \mathbf{H})$. Then, as in the case of acoustic waves, we will consider the *unscaled* amplitudes $a_{ij}^{\pm}(t, \mathbf{x}, \mathbf{k})$

$$a_{ij}^{\pm}(t, \mathbf{x}, \mathbf{k}) = \frac{1}{(2\pi)^3} \int e^{i\mathbf{k}\cdot\mathbf{y}} f_i^{\pm}(t, \mathbf{x}, \mathbf{x} - \mathbf{y}/2, \mathbf{k}) f_j^{\pm *}(t, \mathbf{x}, \mathbf{x} + \mathbf{y}/2, \mathbf{k}) d\mathbf{y},$$

where

$$f_i(t, \mathbf{x}, \mathbf{z}, \mathbf{k}) = \ < \underline{\mathbf{u}}(t, \mathbf{z}), \mathbf{b}^{\pm i}(\mathbf{x}, \mathbf{k}) >_A = \sqrt{\frac{\epsilon(\mathbf{x})}{2}}(\mathbf{E}(t, \mathbf{z}) \cdot \mathbf{z}^{(i)}(\mathbf{k}))$$

$$(3.70) \qquad \pm \sqrt{\frac{\mu(\mathbf{x})}{2}}(\mathbf{H}(t, \mathbf{z}) \cdot (\hat{\mathbf{k}} \times \mathbf{z}^{(i)}(\mathbf{k}))).$$

The amplitudes of the longitudinal, nonpropagating modes are

$$a_{11}^0(t,\mathbf{x},\mathbf{k})) = \frac{1}{(2\pi)^3}\int e^{i\mathbf{k}\cdot\mathbf{y}}\,\epsilon(\mathbf{x})(\mathbf{E}(t,\mathbf{x}-\mathbf{y}/2)\cdot\hat{\mathbf{k}})\overline{(\mathbf{E}(t,\mathbf{x}+\mathbf{y}/2)\cdot\hat{\mathbf{k}})}d\mathbf{y}$$

$$a_{12}^0(t,\mathbf{x},\mathbf{k}) = \frac{1}{(2\pi)^3}\int e^{i\mathbf{k}\cdot\mathbf{y}}\,\sqrt{\epsilon(\mathbf{x})\mu(\mathbf{x})}(\mathbf{E}(t,\mathbf{x}-\mathbf{y}/2)\cdot\hat{\mathbf{k}})\overline{(\mathbf{H}(t,\mathbf{x}+\mathbf{y}/2)\cdot\hat{\mathbf{k}})}d\mathbf{y}$$

$$a_{21}^0(t,\mathbf{x},\mathbf{k}) = \frac{1}{(2\pi)^3}\int e^{i\mathbf{k}\cdot\mathbf{y}}\,\sqrt{\epsilon(\mathbf{x})\mu(\mathbf{x})}(\mathbf{H}(t,\mathbf{x}-\mathbf{y}/2)\cdot\hat{\mathbf{k}})\overline{(\mathbf{E}(t,\mathbf{x}+\mathbf{y}/2)\cdot\hat{\mathbf{k}})}d\mathbf{y}$$

$$a_{22}^0(t,\mathbf{x},\mathbf{k}) = \frac{1}{(2\pi)^3}\int e^{i\mathbf{k}\cdot\mathbf{y}}\,\mu(\mathbf{x})(\mathbf{H}(t,\mathbf{x}-\mathbf{y}/2)\cdot\hat{\mathbf{k}})\overline{(\mathbf{H}(t,\mathbf{x}+\mathbf{y}/2)\cdot\hat{\mathbf{k}})}d\mathbf{y}.$$

As before we denote the coherence matrices by $W^\pm = (a_{ij}^\pm)$ and $W^0 = (a_{ij}^0)$. The latter is zero since there are no longitudinal modes. Moreover, as in the acoustic case, we have the symmetry

$$(3.71)\qquad W^-(t,\mathbf{x},-\mathbf{k}) = \begin{pmatrix} W_{11}^+(\mathbf{k}) & -W_{12}^+(\mathbf{k}) \\ -W_{21}^+(\mathbf{k}) & W_{22}^+(\mathbf{k}) \end{pmatrix}.$$

Hence, by direct computation, we get the energy relation

$$(3.72)\qquad \int \mathrm{Tr}W^+(t,\mathbf{x},\mathbf{k})d\mathbf{k} = \frac{1}{2}\epsilon(\mathbf{x})|\mathbf{E}(t,\mathbf{x})|^2 + \frac{1}{2}\mu(\mathbf{x})|\mathbf{H}(t,\mathbf{x})|^2 = \mathcal{E}(t,\mathbf{x}).$$

Thus, $\mathrm{Tr}W^+(t,\mathbf{x},\mathbf{k})$ is the phase space energy density. By a similar calculation using (3.70) we find that the Poynting vector (3.69) is

$$(3.73)\qquad \mathcal{F}(t,\mathbf{x}) = \mathbf{E}(t,\mathbf{x})\times\mathbf{H}(t,\mathbf{x}) = v(\mathbf{x})\int \hat{\mathbf{k}}\mathrm{Tr}W^+(t,\mathbf{x},\mathbf{k})d\mathbf{k}.$$

The coherence matrix $W^+(t,\mathbf{x},\mathbf{k})$ is related to the four Stokes parameters [**16, 12**], which are commonly used for the description of polarized light because they are directly measurable. Let l and r be two directions orthogonal to the direction of propagation and let $I = I_l + I_r$ be the the total intensity of light, with I_l and I_r denoting the intensities in the directions l and r, respectively. Let $Q = I_l - I_r$ be the difference between the two intensities. Also let $U = 2 < E_l E_r \cos\delta >$ and $V = 2 < E_l E_r \sin\delta >$ denote the intensity coherence, with fixed phase shift δ, between the amplitude of light in the directions l and r, respectively. Light is unpolarized if $U = V = Q = 0$. If the directions l and r are chosen to be $\mathbf{z}^{(1)}(\mathbf{k})$ and $\mathbf{z}^{(2)}(\mathbf{k})$, given by (3.38), then the coherence matrix $W^+(t,\mathbf{x},\mathbf{k})$ is related to the Stokes parameters (I,Q,U,V) by

$$(3.74)\qquad W^+(t,\mathbf{x},\mathbf{k}) = \frac{1}{2}\begin{pmatrix} I+Q & U+iV \\ U-iV & I-Q \end{pmatrix}.$$

When the light is unpolarized, then the coherence matrix W^+ is proportional to the 2×2 identity matrix I.

3.4. High frequency approximation for elastic waves

The equations of motion for small displacements $u_i(t,\mathbf{x})$, $i = 1,2,3$ of an elastic medium are

$$(3.75)\qquad \rho\frac{d^2 u_i}{dt^2} = \frac{\partial\tau_{ij}}{\partial x^j}, \quad i = 1,2,3.$$

Here $\rho(\mathbf{x})$ is the density, $\tau_{ij}(t, \mathbf{x})$ is the stress tensor, which, in an isotropic medium is

$$(3.76) \qquad \tau_{ij} = \lambda(\mathbf{x})\frac{\partial u_k}{\partial x^k}\delta_{ij} + \mu(\mathbf{x})\left(\frac{\partial u_i}{\partial x^j} + \frac{\partial u_j}{\partial x^i}\right),$$

and $\lambda(\mathbf{x})$ and $\mu(\mathbf{x})$ are the Lame parameters. Equation (3.75) is then

$$(3.77) \qquad \rho\frac{d^2 u_i}{dt^2} = \frac{\partial}{\partial x^i}(\lambda\mathrm{div}\mathbf{u}) + \frac{\partial}{\partial x^j}\left(\mu\frac{\partial u_j}{\partial x^i} + \mu\frac{\partial u_i}{\partial x^j}\right).$$

We now write these equations as a symmetric hyperbolic system (3.1) and apply the high frequency analysis to them.

We introduce new dependent variables by

$$(3.78) \qquad p = \lambda\mathrm{div}\mathbf{u}, \ \ \xi_i = \dot{u}_i, \ \ \varepsilon_{ij} = \mu\left(\frac{\partial u_i}{\partial x^j} + \frac{\partial u^j}{\partial x^i}\right),$$

where dot stands for derivative with respect to time. Clearly p is similar to pressure, $\boldsymbol{\xi}$ is the velocity of the medium and ε_{ij} is part of the stress tensor. Equations (3.77) are equivalent to

$$(3.79) \qquad \begin{aligned} \rho\dot{\xi}_i &= \frac{\partial p}{\partial x^i} + \frac{\partial \varepsilon_{ij}}{\partial x^j} \\ \dot{\varepsilon}_{ij} &= \mu\left(\frac{\partial \xi_i}{\partial x^j} + \frac{\partial \xi_j}{\partial x^i}\right) \\ \dot{p} &= \lambda\mathrm{div}\boldsymbol{\xi}. \end{aligned}$$

Note that if the shear modulus μ is zero in (3.79) then $\varepsilon_{ij} = 0$ and we have the acoustic equations (3.33) for the velocity $\boldsymbol{\xi}$ and pressure p. From these variables we form the 10-vector $\mathbf{w} = (\xi_1, \xi_2, \xi_3, \varepsilon_{11}, \varepsilon_{22}, \varepsilon_{33}, \varepsilon_{23}, \varepsilon_{13}, \varepsilon_{12}, p)$ and rewrite (3.79) as a system

$$(3.80) \qquad A(\mathbf{x})\frac{\partial \mathbf{w}}{\partial t} + D^i\frac{\partial \mathbf{w}}{\partial x^i} = 0,$$

with the 10×10 matrix $A(\mathbf{x}) = \mathrm{diag}(\rho, \rho, \rho, 1/2\mu, 1/2\mu, 1/2\mu, 1/\mu, 1/\mu, 1/\mu, 1/\lambda)$. The 10×10 matrices D^i are constant and symmetric and the dispersion matrix $L(\mathbf{x}, \mathbf{k})$ defined by (3.22) is

$$L = -\begin{pmatrix} 0 & 0 & 0 & k_1/\rho & 0 & 0 & 0 & k_3/\rho & k_2/\rho & k_1/\rho \\ 0 & 0 & 0 & 0 & k_2/\rho & 0 & k_3/\rho & 0 & k_1/\rho & k_2/\rho \\ 0 & 0 & 0 & 0 & 0 & k_3/\rho & k_2/\rho & k_1/\rho & 0 & k_3/\rho \\ 2\mu k_1 & 0 & 0 & 0 & 0 & 0 & 0 & 0 & 0 & 0 \\ 0 & 2\mu k_2 & 0 & 0 & 0 & 0 & 0 & 0 & 0 & 0 \\ 0 & 0 & 2\mu k_3 & 0 & 0 & 0 & 0 & 0 & 0 & 0 \\ 0 & \mu k_3 & \mu k_2 & 0 & 0 & 0 & 0 & 0 & 0 & 0 \\ \mu k_3 & 0 & \mu k_1 & 0 & 0 & 0 & 0 & 0 & 0 & 0 \\ \mu k_2 & \mu k_1 & 0 & 0 & 0 & 0 & 0 & 0 & 0 & 0 \\ \lambda k_1 & \lambda k_2 & \lambda k_3 & 0 & 0 & 0 & 0 & 0 & 0 & 0 \end{pmatrix}.$$

In block form

$$(3.81) \qquad L = -\begin{pmatrix} 0 & K(\mathbf{k})/\rho & M(\mathbf{k})/\rho & \frac{1}{\rho}\mathbf{k} \\ 2\mu K(\mathbf{k}) & 0 & 0 & 0 \\ \mu M(\mathbf{k}) & 0 & 0 & 0 \\ \lambda\mathbf{k}^t & 0 & 0 & 0 \end{pmatrix},$$

where the matrix $K(\mathbf{k}) = \mathrm{diag}(k_1, k_2, k_3)$ and

$$(3.82) \qquad M(\mathbf{k}) = \begin{pmatrix} 0 & k_3 & k_2 \\ k_3 & 0 & k_1 \\ k_2 & k_1 & 0 \end{pmatrix}.$$

The matrix $M(\mathbf{k})$ is a symmetrized version of the matrix $P(\mathbf{k})$ in (3.63) that appears in Maxwell's equations.

The eigenvalues of the dispersion matrix L are

$$(3.83) \qquad \begin{aligned} \omega_0 &= 0 \text{ with multiplicity four,} \\ \omega_{\pm}^P &= \pm v_P |\mathbf{k}| \text{ each with multiplicity one,} \\ \omega_{\pm}^S &= \pm v_S |\mathbf{k}| \text{ each with multiplicity two,} \end{aligned}$$

with the corresponding compressional and shear speeds given by

$$(3.84) \qquad v_P = \sqrt{(2\mu + \lambda)/\rho}\ , \quad v_S = \sqrt{\mu/\rho}.$$

The eigenvectors of the dispersion matrix are orthonormal with respect to the inner product $<,>_A$, defined in (3.6), and are given by

$$\mathbf{b}_{\pm}^P = (\frac{\hat{\mathbf{k}}}{\sqrt{2\rho}}, \mp \frac{2\mu K(\hat{\mathbf{k}})\hat{\mathbf{k}}}{\sqrt{2(2\mu + \lambda)}}, \mp \frac{\mu M(\hat{\mathbf{k}})\hat{\mathbf{k}}}{\sqrt{2(2\mu + \lambda)}}, \mp \frac{\lambda}{\sqrt{2(2\mu + \lambda)}})$$

$$\mathbf{b}_{\pm}^{Sj} = (\frac{\mathbf{z}^{(j)}}{\sqrt{2\rho}}, \mp \frac{2\sqrt{\mu} K(\hat{\mathbf{k}})\mathbf{z}^{(j)}}{\sqrt{2}}, \mp \frac{\sqrt{\mu} M(\hat{\mathbf{k}})\mathbf{z}^{(j)}}{\sqrt{2}}, 0), \ j = 1, 2$$

$$(3.85) \qquad \mathbf{b}^{0j} = (0, \sqrt{2\mu} K(\mathbf{z}^{(j)})\mathbf{z}^{(j)}, \sqrt{\frac{\mu}{2}} M(\mathbf{z}^{(j)})\mathbf{z}^{(j)}, 0), \ j = 1, 2$$

$$\mathbf{b}^{03} = (0, 2\sqrt{\mu} K(\mathbf{z}^{(1)})\mathbf{z}^{(2)}, \sqrt{\mu} M(\mathbf{z}^{(1)})\mathbf{z}^{(2)}, 0)$$

$$\mathbf{b}^{04} = (0, \frac{2\sqrt{\lambda\mu} K(\hat{\mathbf{k}})\hat{\mathbf{k}}}{\sqrt{2(\lambda + 2\mu)}}, \sqrt{\frac{\lambda\mu}{2(\lambda + 2\mu)}} M(\hat{\mathbf{k}})\hat{\mathbf{k}}, -\frac{2\sqrt{\lambda\mu}}{\sqrt{2(\lambda + 2\mu)}}).$$

The orthonormal triple $\hat{\mathbf{k}}, \mathbf{z}^{(1)}(\mathbf{k}), \mathbf{z}^{(2)}(\mathbf{k})$ is defined by (3.38). The eigenvectors $\mathbf{b}_{\pm}^P$ represent longitudinal or compressional modes, the P waves. They are similar to the acoustic longitudinal modes and if $\mu = 0$ then $\mathbf{b}_{\pm}^P$ is equivalent to the vector $\mathbf{b}^{\mp}$ for acoustics (3.37). The eigenvectors $\mathbf{b}_{\pm}^{Sj}$ represent transverse or shear waves, the S waves. They are similar to the eigenvectors (3.65) in Maxwell's equations, because they correspond to transverse waves admitting two states of polarization. The eigenvectors $\mathbf{b}^{0j}$, $j = 1, \ldots 4$ correspond to non-propagating modes.

The energy density for elastic waves is given by

$$\mathcal{E}(t, \mathbf{x}) = \frac{1}{2}\rho(\mathbf{x})|\dot{\mathbf{u}}(t, \mathbf{x})|^2 + \frac{1}{2}\lambda(\mathbf{x})(\mathrm{div}\mathbf{u}(\mathbf{x}))^2 + \frac{1}{2}\mu(\mathbf{x})\mathrm{Tr}(\nabla\mathbf{u}(t, \mathbf{x}) + \nabla^t\mathbf{u}(t, \mathbf{x}))^2.$$
$$(3.86)$$

The first term is the kinetic energy and the sum of the last two terms is the strain energy. The energy flux of the elastic waves is

$$(3.87) \qquad \mathcal{F}(t, \mathbf{x}) = \{\lambda \mathrm{div}\mathbf{u}(\mathbf{x})) + \mu(\mathbf{x})(\nabla\mathbf{u}(t, \mathbf{x}) + \nabla^t\mathbf{u}(t, \mathbf{x}))\}\dot{\mathbf{u}}(t, \mathbf{x}),$$

which in view of (3.76) is also

$$\mathcal{F}(t, \mathbf{x}) = \tau(t, \mathbf{x})\dot{\mathbf{u}}(t, \mathbf{x}).$$

The *unscaled* amplitudes $a_\pm^P(t,\mathbf{x},\mathbf{k})$ are

$$(3.88) \quad a_\pm^P = \left(\frac{1}{2\pi}\right)^3 \int e^{i\mathbf{k}\cdot\mathbf{y}} f_\pm^P(t,\mathbf{x},\mathbf{x}-\mathbf{y}/2,\mathbf{k}) \bar{f}_\pm^P(t,\mathbf{x},\mathbf{x}+\mathbf{y}/2,\mathbf{k})d\mathbf{y},$$

where

$$f_\pm^P(t,\mathbf{x},\mathbf{z},\mathbf{k}) \;=\; <\underline{\mathbf{u}}(t,\mathbf{z}),\mathbf{b}^{P\pm}(\mathbf{x},\mathbf{k})>_A = \sqrt{\frac{\rho(\mathbf{x})}{2}}(\hat{\mathbf{k}}\cdot\dot{\mathbf{u}}(t,\mathbf{z})) \mp$$

$$\frac{\mu(\mathbf{x})}{\sqrt{2(2\mu(\mathbf{x})+\lambda(\mathbf{x}))}}\,(\hat{\mathbf{k}}\cdot(\nabla\mathbf{u}(t,\mathbf{z})+\nabla^t\mathbf{u}(t,\mathbf{z}))\hat{\mathbf{k}}) \mp \frac{\lambda(\mathbf{x})\mathrm{div}\mathbf{u}(t,\mathbf{z})}{\sqrt{2(2\mu(\mathbf{x})+\lambda(\mathbf{x}))}}.$$

The 2×2 coherence matrices $W_\pm^S$ for the S waves are

$$W_{\pm ij}^S(t,\mathbf{x},\mathbf{k}) = \left(\frac{1}{2\pi}\right)^3 \int e^{i\mathbf{k}\cdot\mathbf{y}} f_i^{S\pm}(t,\mathbf{x},\mathbf{x}-\mathbf{y}/2,\mathbf{k})\bar{f}_j^{S\pm}(t,\mathbf{x},\mathbf{x}+\mathbf{y}/2,\mathbf{k})d\mathbf{y},$$

$$(3.89)$$

where

$$f_i^{S\pm}(t,\mathbf{x},\mathbf{z},\mathbf{k}) = \sqrt{\frac{\rho(\mathbf{x})}{2}}(\mathbf{z}^{(i)}(\mathbf{k})\cdot\dot{\mathbf{u}}(t,\mathbf{z})) \mp \sqrt{\frac{\mu(\mathbf{x})}{2}}(\hat{\mathbf{k}}\cdot(\nabla\mathbf{u}(\mathbf{z})+\nabla^t\mathbf{u}(\mathbf{z}))\mathbf{z}^{(i)}(\mathbf{k})).$$

The entries of the 4×4 coherence matrix for the nonpropagating modes are

$$(3.90)\quad a_{ij}^0(t,\mathbf{x},\mathbf{k}) = \left(\frac{1}{2\pi}\right)^3 \int e^{i\mathbf{k}\cdot\mathbf{y}} f_i^0(t,\mathbf{x},\mathbf{x}-\mathbf{y}/2,\mathbf{k})\bar{f}_j^0(t,\mathbf{x},\mathbf{x}+\mathbf{y}/2,\mathbf{k})d\mathbf{y},$$

where

$$f_j^0(t,\mathbf{x},\mathbf{z},\mathbf{k}) = \sqrt{\frac{\mu(\mathbf{x})}{2}}(\mathbf{z}^{(j)}(\mathbf{k})\cdot(\nabla\mathbf{u}(t,\mathbf{z})+\nabla^t\mathbf{u}(t,\mathbf{z}))\mathbf{z}^{(j)}(\mathbf{k})),\; j=1,2$$

$$f_3^0(t,\mathbf{x},\mathbf{z},\mathbf{k}) = \sqrt{\mu(\mathbf{x})}(\mathbf{z}^{(1)}(\mathbf{k})\cdot(\nabla\mathbf{u}(t,\mathbf{z})+\nabla^t\mathbf{u}(t,\mathbf{z}))\mathbf{z}^{(2)}(\mathbf{k}))$$

$$f_4^0(t,\mathbf{x},\mathbf{z},\mathbf{k}) = \sqrt{\frac{\lambda(\mathbf{x})\mu(x)}{2(\lambda(\mathbf{x})+2\mu(\mathbf{x}))}}(\hat{\mathbf{k}}\cdot(\nabla\mathbf{u}(t,\mathbf{z})+\nabla^t\mathbf{u}(t,\mathbf{z}))\hat{\mathbf{k}})$$

$$-\frac{2\sqrt{\lambda(\mathbf{x})\mu(\mathbf{x})}\mathrm{div}\mathbf{u}(t,\mathbf{z})}{\sqrt{2(2\mu(\mathbf{x})+\lambda(\mathbf{x}))}}.$$

Note that (3.88) implies that the amplitudes a_+^P and a_-^P are related by

$$(3.91)\qquad\qquad a_+^P(t,\mathbf{x},\mathbf{k}) = a_-^P(t,\mathbf{x},-\mathbf{k}),$$

which is analogous to (3.43), while the coherence matrices W_+^S and W_-^S are related by the analog of (3.71) and

$$(3.92)\qquad\qquad \mathrm{Tr}W_+^S(t,\mathbf{x},\mathbf{k}) = \mathrm{Tr}W_-^S(t,\mathbf{x},-\mathbf{k}).$$

A direct calculation using (3.88-3.90) shows that the energy density (3.86) is

$$(3.93)\qquad \mathcal{E}(t,\mathbf{x}) = \int (a_+^P + \mathrm{Tr}W_+^S)d\mathbf{k} + \frac{1}{2}\int \sum_{i=1}^{4} a_{ii}^0 d\mathbf{k}.$$

The first term is the energy density of the P and S waves while the second is the energy of the zero velocity waves. The flux (3.87) is

$$(3.94)\qquad \mathcal{F}(t,\mathbf{x}) = \int \hat{\mathbf{k}}[v_P a_+^P(t,\mathbf{x},\mathbf{k}) + v_S \mathrm{Tr}W_+^S(t,\mathbf{x},\mathbf{k})]d\mathbf{k}.$$

Using the eigenvalues (3.83) and (3.84) in (3.25) and (3.29) we obtain the transport equations for the scalar amplitude a_+^P and the coherence matrix W_+^S:

$$(3.95) \qquad \frac{\partial a_+^P}{\partial t} + v_P(\mathbf{x})\hat{\mathbf{k}} \cdot \nabla_{\mathbf{x}} a_+^P - |\mathbf{k}|\nabla_{\mathbf{x}} v_P(\mathbf{x}) \cdot \nabla_{\mathbf{k}} a_+^P = 0$$

$$(3.96) \qquad \frac{\partial W_+^S}{\partial t} + v_S(\mathbf{x})\hat{\mathbf{k}} \cdot \nabla_{\mathbf{x}} W_+^S - |\mathbf{k}|\nabla_{\mathbf{x}} v_S(\mathbf{x}) \cdot \nabla_{\mathbf{k}} W_+^S + W_+^S N - N W_+^S = 0.$$

The coupling matrix $N(\mathbf{x}, \mathbf{k})$ is exactly the same as in the case of Maxwell's equations (3.66) with the speed $v = v_S$. In the high frequency limit the longitudinal P waves behave exactly like acoustic waves. This is because in both cases the waves correspond to a simple eigenvalue of the dispersion matrix. The S waves behave exactly like electromagnetic waves. The same results were obtained in [13] by ray methods.

LECTURE 4
Waves in Random Media

4.1. The Schrödinger equation

We now consider small random perturbations of the potential $V(\mathbf{x})$ in the scaled Schrödinger equation (2.9). It is well known that in one space dimension, waves in a random medium get localized even when the random perturbations are small [52], so our analysis is restricted to three dimensions. We could treat two-dimensional problems with time dependent perturbations. There is in fact quite a lot of work done on limit theorems with randomly time dependent potentials that decorrelate rapidly or are Markovian [5, 6, 22]. We want to consider only spatial randomness here so $V(\mathbf{x})$ does not depend on time.

There is not a lot of mathematical work on the transport limit for the Schrödinger equation with random potential. We cite here the work of Martin and Emch [43], of Spohn [53], of Dell'Antonio [20] and the recent extensive study of Ho, Landau and Wilkins [30]. They treat only spatially homogeneous problems but it is known how to extend the analysis to the spatially inhomogeneous case (x-dependent initial data) [21]. A really satisfactory mathematical treatment of radiative transport asymptotics from random wave equations is lacking at present.

We now give a formal analysis of the transport limit.

We assume that the correlation length of the random perturbation is of the same order as the wavelength, so the potential has the form

$$(4.1) \qquad V(\mathbf{x}) = V_0(\mathbf{x}) + V_1(\frac{\mathbf{x}}{\varepsilon}).$$

Here $V_0(\mathbf{x})$ is the slowly varying background and $V_1(\mathbf{y})$ is a mean zero, stationary random function with correlation length of order one. This scaling allows the random potential to interact fully with the waves. We shall also assume that the fluctuations are space-homogeneous and isotropic so that

$$(4.2) \qquad < V_1(\mathbf{x})V_1(\mathbf{y}) >= R(|\mathbf{x} - \mathbf{y}|),$$

where $<, >$ denotes statistical averaging and $R(|\mathbf{x}|)$ is the covariance of random the fluctuations. The power spectrum of the fluctuations is defined by

$$(4.3) \qquad \hat{R}(\mathbf{k}) = \left(\frac{1}{2\pi}\right)^d \int e^{i\mathbf{k}\cdot\mathbf{y}} R(\mathbf{x})d\mathbf{k}.$$

343

When (4.2) holds the fluctuations are isotropic and $\hat{R}$ is a function of $|\mathbf{k}|$ only. Moreover,

$$(4.4) \qquad < \hat{V}(\mathbf{p})\hat{V}(\mathbf{q}) >= \hat{R}(\mathbf{p})\delta(\mathbf{p}+\mathbf{q}).$$

If the amplitude of these fluctuations is strong then scattering will dominate and waves will be localized [**24**]. This means that we cannot assume that the fluctuations in the random potential $V_1(\mathbf{y})$ are large. If the random fluctuations are too weak they will not affect energy transport at all. In order that the scattering produced by the random potential and the influence of the slowly varying background affect energy transport in comparable ways the fluctuations in the random potential must be of order $\sqrt{\varepsilon}$. Then equation (2.9) becomes

$$i\varepsilon\frac{\partial\phi^\varepsilon}{\partial t} + \frac{\varepsilon^2}{2}\Delta\phi^\varepsilon - (V_0(\mathbf{x}) + \sqrt{\varepsilon}V_1(\frac{\mathbf{x}}{\varepsilon}))\phi^\varepsilon = 0$$
$$(4.5) \qquad \phi^\varepsilon(0,\mathbf{x}) = \phi_0(\frac{\mathbf{x}}{\varepsilon},\mathbf{x}).$$

To describe the passage from (4.5) to the transport equation in its simplest form we will set $V_0(\mathbf{x}) = 0$ and drop the subscript one from $V_1(\mathbf{x})$. A $V_0(\mathbf{x})$ that is not zero will not change the scattering terms in the radiative transport equation. Now (2.31) for W^ε has the form

$$(4.6) \qquad \frac{\partial W^\varepsilon}{\partial t} + \mathbf{k}\cdot\nabla_\mathbf{x}W^\varepsilon + \frac{1}{\sqrt{\varepsilon}}\mathcal{L}_{\frac{\mathbf{x}}{\varepsilon}}W^\varepsilon = 0$$

where the operator $\mathcal{L}_{\frac{\mathbf{x}}{\varepsilon}}$, a rescaled form of (2.32), is given by

$$(4.7) \quad \mathcal{L}_{\frac{\mathbf{x}}{\varepsilon}}Z(\mathbf{x},\mathbf{k}) = i\int e^{-i\mathbf{p}\cdot\mathbf{x}/\varepsilon}\hat{V}(\mathbf{p})\left(Z(\mathbf{x},\mathbf{k}+\frac{\mathbf{p}}{2}) - Z(\mathbf{x},\mathbf{k}-\frac{\mathbf{p}}{2})\right)d\mathbf{p}.$$

The behavior of this operator as $\varepsilon \to 0$ is very different from (2.34) when V is slowly varying. We can find the correct results by a multiscale analysis as follows.

Let $\boldsymbol{\xi} = \mathbf{x}/\varepsilon$ be a fast space variable (on the scale of the wavelength) and introduce an expansion of W^ε of the form

$$(4.8)\ W^\varepsilon(t,\mathbf{x},\mathbf{k}) = W^{(0)}(t,\mathbf{x},\mathbf{k}) + \varepsilon^{1/2}W^{(1)}(t,\mathbf{x},\boldsymbol{\xi},\mathbf{k}) + \varepsilon W^{(2)}(t,\mathbf{x},\boldsymbol{\xi},\mathbf{k}) + \ldots .$$

We assume that the leading term does not depend on the fast scale and that the initial Wigner distribution $W^\varepsilon(0,\mathbf{x},\mathbf{k})$ tends to a smooth function $W_0(\mathbf{x},\mathbf{k})$ which is decaying fast enough at infinity. Then the average of the Wigner distribution, $< W^\varepsilon >$, is close to $W^{(0)}$ which satisfies the transport equation

$$\frac{\partial W}{\partial t} + \mathbf{k}\cdot\nabla_\mathbf{x}W = \overline{\mathcal{L}}W$$
$$(4.9) \qquad W(0,\mathbf{x},\mathbf{k}) = W_0(\mathbf{x},\mathbf{k}),$$

where we have dropped the superscript zero. The operator $\overline{\mathcal{L}}$ is given by

$$(4.10) \quad \overline{\mathcal{L}}W(\mathbf{x},\mathbf{k}) = 4\pi\int \hat{R}(\mathbf{p}-\mathbf{k})\delta(\mathbf{k}^2-\mathbf{p}^2)(W(\mathbf{x},\mathbf{p}) - W(\mathbf{x},\mathbf{k}))d\mathbf{p}.$$

Equation (4.9) has precisely the form (1.6). From (2.14)

$$\omega = \frac{\mathbf{k}^2}{2},$$

since the background potential V_0 is zero. The differential scattering cross-section $\sigma(\mathbf{k}, \mathbf{k}')$ is given by

$$(4.11) \qquad \sigma(\mathbf{k}, \mathbf{p}) = 4\pi \hat{R}(\mathbf{p} - \mathbf{k})\delta(\mathbf{k}^2 - \mathbf{p}^2)$$

and the total scattering cross-section $\Sigma(\mathbf{k})$ is given by

$$(4.12) \qquad \Sigma(\mathbf{k}) = 4\pi \int \hat{R}(\mathbf{k} - \mathbf{p})\delta(\mathbf{k}^2 - \mathbf{p}^2)d\mathbf{p}.$$

Note also that the transport equation (4.9) has two important properties. First, the total energy

$$(4.13) \qquad E(t) = \iint W(t, \mathbf{x}, \mathbf{k})d\mathbf{k}d\mathbf{x}$$

is conserved and second, the positivity of the solution $W(t, \mathbf{x}, \mathbf{k})$ is preserved, that is, if the initial Wigner distribution $W_0(\mathbf{x}, \mathbf{k})$ is non-negative then $W(t, \mathbf{x}, \mathbf{k}) \geq 0$ for $t > 0$.

We proceed now with the derivation of the radiative transport equation (4.9) from the multiple scales expansion (4.8) of the solution of (4.6). Assume formally that the leading term $W^{(0)}$ does not depend on the fast scale $\boldsymbol{\xi} = \mathbf{x}/\varepsilon$ and is deterministic. We replace

$$\nabla_{\mathbf{x}} \rightarrow \frac{1}{\varepsilon}\nabla_{\boldsymbol{\xi}} + \nabla_{\mathbf{x}}$$

in (4.6) and insert expansion (4.8) into (4.6). The term $W^{(1)}$ satisfies the equation

$$(4.14) \quad \mathbf{k} \cdot \nabla_{\boldsymbol{\xi}} W^{(1)} + \theta W^{(1)} = i \int e^{-i\mathbf{p}\cdot\boldsymbol{\xi}}\hat{V}(\mathbf{p})\{W^{(0)}(\mathbf{k} - \frac{\mathbf{p}}{2}) - W^{(0)}(\mathbf{k} + \frac{\mathbf{p}}{2})\}d\mathbf{p},$$

where θ is a regularization parameter which will be set to zero later. This equation can be solved explicitly and the Fourier transform in $\boldsymbol{\xi}$ of $W^{(1)}$ is given by

$$(4.15) \qquad \frac{\hat{V}(\mathbf{p})[W^{(0)}(\mathbf{k} + \frac{\mathbf{p}}{2}) - W^{(0)}(\mathbf{k} - \frac{\mathbf{p}}{2})]}{\mathbf{k} \cdot \mathbf{p} + i\theta}.$$

The next term $W^{(2)}$ satisfies the equation

$$(4.16) \qquad \begin{aligned} &\frac{\partial W^{(0)}}{\partial t} + \mathbf{k} \cdot \nabla_{\mathbf{x}} W^{(0)} + \mathbf{k} \cdot \nabla_{\boldsymbol{\xi}} W^{(2)} \\ &+ i \int e^{-i\mathbf{p}\cdot\boldsymbol{\xi}}\hat{V}(\mathbf{p})[W^{(1)}(\mathbf{k} + \frac{\mathbf{p}}{2}) - W^{(1)}(\mathbf{k} - \frac{\mathbf{p}}{2})]d\mathbf{p} = 0. \end{aligned}$$

Note that

$$(4.17) \qquad < \frac{\partial W^{(2)}}{\partial \boldsymbol{\xi}} >= 0$$

and so after averaging (4.16) has the form

$$(4.18) \qquad \begin{aligned} &\frac{\partial W^{(0)}}{\partial t} + \mathbf{k} \cdot \nabla_{\mathbf{x}} W^{(0)} \\ &+ < i \int e^{-i\mathbf{p}\cdot\boldsymbol{\xi}}\hat{V}(\mathbf{p})[W^{(1)}(\mathbf{k} + \frac{\mathbf{p}}{2}) - W^{(1)}(\mathbf{k} - \frac{\mathbf{k}}{2})]d\mathbf{p} >= 0. \end{aligned}$$

We insert the Fourier transform (4.15) in (4.18) and use (4.4) to obtain as $\theta \to 0$

$$
(4.19) \qquad < i \int e^{-i\mathbf{p}\cdot\boldsymbol{\xi}} \hat{V}(\mathbf{p})[W^{(1)}(\mathbf{k} + \frac{\mathbf{p}}{2}) - W^{(1)}(\mathbf{k} - \frac{\mathbf{p}}{2})]d\mathbf{p} >
$$

$$
= \int \hat{R}(\mathbf{p} - \mathbf{k})[W^{(0)}(\mathbf{k}) - W^{(0)}(\mathbf{p})]\frac{2\theta}{\frac{1}{4}(\mathbf{k}^2 - \mathbf{p}^2)^2 + \theta^2}d\mathbf{p}
$$

$$
\to 4\pi \int \hat{R}(\mathbf{p} - \mathbf{k})[W^{(0)}(\mathbf{k}) - W^{(0)}(\mathbf{p})]\delta(\mathbf{k}^2 - \mathbf{p}^2)d\mathbf{p}.
$$

This holds because

$$
\frac{\theta}{x^2 + \theta^2} \to \pi\delta(x)
$$

as $\theta \to 0$. We insert (4.19) in (4.18) and find that $W^{(0)}$ satisfies the transport equation (4.9).

In the rest of this lecture we extend the analysis of this section to symmetric hyperbolic systems of partial differential equations. The main steps are (i) developing the high frequency approximation in phase space using the Wigner distribution and (ii) getting the scattering cross-sections from the random inhomogeneities of the medium.

4.2. Transport equations without polarization

We now consider wave propagation in a slowly varying background with small random perturbations of the systems of the form (3.1). The symmetric hyperbolic system (3.1) is

$$
(4.20) \qquad A(\mathbf{x})\{I + \varepsilon^{1/2}V(\frac{\mathbf{x}}{\varepsilon})\}\frac{\partial \mathbf{u}}{\partial t} + D^j \frac{\partial \mathbf{u}}{\partial x^j} = 0,
$$

where $V(\mathbf{x})$ is a statistically homogeneous matrix-valued random process with mean zero that models the parameter fluctuations. The scale of variation of the fluctuations is of order ε and therefore comparable to the wave length so that the random inhomogeneities can interact fully with the propagating waves. The magnitude $\sqrt{\varepsilon}$ of the fluctuations is chosen, as in the case of the Schrödinger equation (4.5), so that the effect of scattering by the inhomogeneities be comparable to the effect of the slowly varying background. In order that the system (4.20) remain symmetric hyperbolic the random inhomogeneities must satisfy the condition

$$
(4.21) \qquad A(\mathbf{x})V(\mathbf{y}) = V^*(\mathbf{y})A(\mathbf{x}).
$$

for all $\mathbf{x}$ and $\mathbf{y}$, which implies conservation of energy. The matrices A and D^j are symmetric and A is positive definite. In all three cases considered here – acoustic, electromagnetic and elastic waves – condition (4.21) is satisfied. In this section we will assume that the dispersion matrix (3.22) for the deterministic background has simple eigenvalues. The case of polarization (multiple eigenvalues) is considered in the next section.

The covariance functions $R_{ijkl}(\mathbf{x})$ and the power spectral densities $\hat{R}_{ijkl}(\mathbf{k})$ are defined by

$$
(4.22) \qquad R_{ijkl}(\mathbf{x}) = \langle V_{ij}(\mathbf{y})V_{kl}(\mathbf{x} + \mathbf{y})\rangle = \int e^{-i\mathbf{p}\cdot\mathbf{x}}\hat{R}_{ijkl}(\mathbf{p})d\mathbf{p},
$$

where $<,>$ denotes statistical average. Spatial homogeneity implies

$$(4.23) \qquad \langle \widehat{V}_{ij}(\mathbf{p})\widehat{V}_{kl}(\mathbf{q})\rangle = \hat{R}_{ijkl}(\mathbf{p})\delta(\mathbf{p}+\mathbf{q})$$

and

$$(4.24) \qquad \hat{R}_{ijkl}(\mathbf{p}) = \hat{R}_{klij}(-\mathbf{p}).$$

We assume that the power spectral densities $\hat{R}_{ijkl}(\mathbf{p})$ are real, which is equivalent to

$$(4.25) \qquad \hat{R}_{ijkl}(\mathbf{p}) = \hat{R}_{ijkl}(-\mathbf{p})$$

and holds when the covariance functions $R_{ijkl}(\mathbf{x})$ are even . This is the case when the fluctuations are isotropic in space, that is

$$(4.26) \qquad R_{ijkl}(\mathbf{x}) = R_{ijkl}(|\mathbf{x}|).$$

The symmetry condition (4.21) implies that the matrix A and the covariance tensor R_{ijkl} satisfy the relations

$$(4.27) \qquad A_{ni}A_{mk}R_{ijkl} = A_{ji}A_{mk}R_{inkl} = A_{ji}A_{lk}R_{inkm}.$$

When (4.20) holds, the evolution equation (3.13) for W^ε has the form

$$(4.28) \qquad \frac{\partial W^\varepsilon}{\partial t} + \mathcal{Q}_1^\varepsilon W^\varepsilon + \frac{1}{\varepsilon}\mathcal{Q}_2^\varepsilon W^\varepsilon - \frac{1}{\sqrt{\varepsilon}}\mathcal{P}_2^\varepsilon W^\varepsilon - \sqrt{\varepsilon}\mathcal{P}_1^\varepsilon W^\varepsilon = 0,$$

where the operators $\mathcal{Q}_1^\varepsilon$ and $\mathcal{Q}_2^\varepsilon$ are defined by (3.14) and (3.15). The operators $\mathcal{P}_1^\varepsilon$ and $\mathcal{P}_2^\varepsilon$ come from the random inhomogeneities and are given by

$$(4.29) \quad \mathcal{P}_1^\varepsilon W^\varepsilon = \frac{1}{2}\iint \frac{e^{i\mathbf{q}\cdot\mathbf{y}}\,d\mathbf{y}\,d\mathbf{q}}{(2\pi)^d}\{V(\frac{\mathbf{x}}{\varepsilon}+\mathbf{y})A^{-1}(\mathbf{x}+\varepsilon\mathbf{y})D^j\frac{\partial W^\varepsilon(\mathbf{k}+\mathbf{p}/2)}{\partial x^j}$$

$$+\frac{\partial W^\varepsilon(\mathbf{k}-\mathbf{p}/2)}{\partial x^j}D^j A^{-1}(\mathbf{x}+\varepsilon\mathbf{y})V^*(\frac{\mathbf{x}}{\varepsilon}+\mathbf{y})\}$$

and

$$(4.30) \; \mathcal{P}_2^\varepsilon W^\varepsilon = i\iint \frac{e^{i\mathbf{q}\cdot\mathbf{y}}\,d\mathbf{y}\,d\mathbf{q}}{(2\pi)^d}\{(k_j+\frac{q_j}{2})V(\frac{\mathbf{x}}{\varepsilon}+\mathbf{y})A^{-1}(\mathbf{x}+\varepsilon\mathbf{y})W^\varepsilon(\mathbf{k}+\mathbf{q}/2)$$

$$-W^\varepsilon(\mathbf{k}-\mathbf{q}/2)(k_j-\frac{q_j}{2})D^j A^{-1}(\mathbf{x}+\varepsilon\mathbf{y})V^*(\frac{\mathbf{x}}{\varepsilon}+\mathbf{y})\}.$$

The double integrals enter in (4.29) and (4.30) because we inserted the Fourier transform $\hat{V}$ into (3.14) and (3.15). The operator $\mathcal{P}_1^\varepsilon$ corresponds to the terms in (3.14) involving the $\mathbf{x}$-gradient of W^ε, while the undifferentiated terms in (3.14) and (3.15) combine to produce the operator $\mathcal{P}_2^\varepsilon$.

We analyze equation (4.28) by a multiple scales expansion, following our analysis for the Schrödinger equation. We introduce the fast space variable $\boldsymbol{\xi} = \mathbf{x}/\varepsilon$ and the expansion

$$(4.31) \; W^\varepsilon(t,\mathbf{x},\boldsymbol{\xi},\mathbf{k}) = W^{(0)}(t,\mathbf{x},\mathbf{k}) + \varepsilon^{1/2}W^{(1)}(t,\mathbf{x},\boldsymbol{\xi},\mathbf{k}) + \varepsilon W^{(2)}(t,\mathbf{x},\boldsymbol{\xi},\mathbf{k}) + \cdots$$

We replace $\dfrac{\partial}{\partial x^i}$ by

$$(4.32) \qquad \frac{\partial}{\partial x^i} + \frac{1}{\varepsilon}\frac{\partial}{\partial \xi^i}$$

and expand the $\mathcal{Q}$ and $\mathcal{P}$ operators in powers of ε:

$$
\begin{aligned}
\mathcal{Q}_1^\varepsilon &= \frac{1}{\varepsilon}\tilde{\mathcal{Q}}_1 + \mathcal{Q}_1 + \tilde{\mathcal{Q}}_{11} + \dots \\
\mathcal{Q}_2^\varepsilon &= \mathcal{Q}_2 + \varepsilon \mathcal{Q}_{21} + \dots \\
\mathcal{P}_1^\varepsilon &= \frac{1}{\varepsilon}\mathcal{P}_1(\frac{\partial}{\partial\boldsymbol{\xi}}) + \mathcal{P}_1(\frac{\partial}{\partial\mathbf{x}}) + \dots \\
\mathcal{P}_2^\varepsilon &= \mathcal{P}_2 + \dots
\end{aligned}
$$

The operator $\tilde{\mathcal{Q}}_1$ is

$$
(4.33) \qquad \tilde{\mathcal{Q}}_1 Z = \frac{1}{2}A^{-1}D^j\frac{\partial Z}{\partial\xi^j} + \frac{1}{2}\frac{\partial Z}{\partial\xi^j}D^j A^{-1}
$$

and the operators $\mathcal{P}_1$ and $\mathcal{P}_2$ are

$$
\begin{aligned}
\mathcal{P}_1 Z(\mathbf{x},\boldsymbol{\xi},\mathbf{k}) = \frac{1}{2}\int dq e^{-i\mathbf{q}\cdot\boldsymbol{\xi}}\Big\{ &\hat{V}(\mathbf{q})A^{-1}(\mathbf{x})D^j\frac{\partial Z(\mathbf{k}+\mathbf{q}/2)}{\partial x^j} \\
&+ \frac{\partial Z(\mathbf{k}-\mathbf{q}/2)}{\partial x^j}D^j A^{-1}(\mathbf{x})\widehat{V^*}(\mathbf{q})\Big\}
\end{aligned}
$$

and

$$
\begin{aligned}
\mathcal{P}_2 Z(\mathbf{x},\boldsymbol{\xi},\mathbf{k}) = i\int dq e^{-i\mathbf{q}\cdot\boldsymbol{\xi}}\Big\{ &\hat{V}(\mathbf{p})A^{-1}(\mathbf{x})(k_j + q_j/2)D^j Z(\mathbf{k}+\mathbf{q}/2) \\
&- Z(\mathbf{k}-\mathbf{q}/2)(k_j - p_j/2)D^j A^{-1}(\mathbf{x})\widehat{V^*}(\mathbf{q})\Big\}.
\end{aligned}
$$

We do not give an explicit expression for $\tilde{\mathcal{Q}}_{11}$ since we shall not need it. It is the first order term in the expansion in ε of the part involving the $\boldsymbol{\xi}$-gradient of the operator $\mathcal{Q}_1(\frac{\partial}{\partial\boldsymbol{\xi}})$. With these definitions, (4.28) becomes

$$
\frac{\partial W^\varepsilon}{\partial t} + \Big\{\frac{1}{\varepsilon}\mathcal{Q}_2 + \mathcal{Q}_{21} + \frac{1}{\varepsilon}\tilde{\mathcal{Q}}_1 + \mathcal{Q}_1 + \tilde{\mathcal{Q}}_{11} - \frac{1}{\sqrt{\varepsilon}}\mathcal{P}_2 - \frac{1}{\sqrt{\varepsilon}}\mathcal{P}_1(\frac{\partial}{\partial\boldsymbol{\xi}}) + O(\varepsilon)\Big\}W^\varepsilon = 0.
$$

We assume that the average of the leading term $W^{(0)}$ in the expansion (4.28) depends only on the slow space variable $\mathbf{x}$. To simplify the presentation we will assume that $W^{(0)}$ itself is independent of $\boldsymbol{\xi}$. We insert expansion (4.31) into (4.28) and find that $W^{(0)}$ satisfies

$$
(4.34) \qquad \mathcal{Q}_2 W^{(0)} = 0
$$

as in (3.20). We assume in this section that all the eigenvalues of the dispersion matrix $L(\mathbf{x},\mathbf{k})$ in (3.22) are simple. The case of multiple eigenvalues is considered in the next section. Then the Wigner matrix $W^{(0)}$ has the form

$$
(4.35) \qquad W^{(0)}(t,\mathbf{x},\mathbf{k}) = \sum_{\tau=1}^{N} a^\tau(t,\mathbf{x},\mathbf{k})B^\tau(\mathbf{x},\mathbf{k}),
$$

where the matrices $B^\tau(\mathbf{x},\mathbf{k})$ are defined by (3.23), as in (3.24).

The term $W^{(1)}$ satisfies

$$
(4.36) \qquad \mathcal{Q}_2 W^{(1)} + \tilde{\mathcal{Q}}_1 W^{(1)} = \mathcal{P}_2 W^{(0)}.
$$

We insert (4.35) into (4.36) and solve this equation explicitly for $F^{(1)}(t.\mathbf{x}, \mathbf{p}, \mathbf{k})$, the Fourier transform in $\boldsymbol{\xi}$ of $W^{(1)}$:

$$F^{(1)} = \frac{1}{\omega_j(\mathbf{k} + \frac{\mathbf{P}}{2}) - \omega_i(\mathbf{k} - \frac{\mathbf{P}}{2}) - i\nu}$$

$$\left\{ \omega_i(\mathbf{k} - \frac{\mathbf{P}}{2}) a^i(\mathbf{k} - \frac{\mathbf{P}}{2}) c_m^j(\mathbf{k} + \frac{\mathbf{P}}{2}) \hat{V}_{ml}(\mathbf{p}) b_l^i(\mathbf{k} - \frac{\mathbf{P}}{2}) \right.$$

$$\left. - \omega_j(\mathbf{k} + \frac{\mathbf{P}}{2}) a^j(\mathbf{k} + \frac{\mathbf{P}}{2}) c_m^i(\mathbf{k} - \frac{\mathbf{P}}{2}) \hat{V}_{ml}(\mathbf{p}) b_l^j(\mathbf{k} + \frac{\mathbf{P}}{2}) \right\} \mathbf{b}^i(\mathbf{k} - \frac{\mathbf{P}}{2}) \mathbf{b}^{j*}(\mathbf{k} + \frac{\mathbf{P}}{2}).$$

(4.37)

Here the vectors $\mathbf{b}^j(\mathbf{x}, \mathbf{k})$ are the right eigenvectors of the dispersion matrix $L(\mathbf{x}, \mathbf{k})$, orthonormal with respect to the inner product $<,>_A$, and the vectors $\mathbf{c}^i(\mathbf{x}, \mathbf{k})$ are the left eigenvectors of the dispersion matrix, given by

$$(4.38) \qquad \mathbf{c}^i(\mathbf{x}, \mathbf{k}) = A(\mathbf{x}) \mathbf{b}^i(\mathbf{x}, \mathbf{k}).$$

The second order term $W^{(2)}$ satisfies the equation

$$(4.39) \qquad \mathcal{Q}_2 W^{(2)} + \tilde{\mathcal{Q}}_1 W^{(2)} = -\frac{\partial W^{(0)}}{\partial t} - \mathcal{Q}_{21} W^{(0)} - \mathcal{Q}_1 W^{(0)} +$$

$$\mathcal{P}_2 W^{(1)} + \mathcal{P}_1 (\frac{\partial}{\partial \boldsymbol{\xi}}) W^{(1)},$$

because $\tilde{\mathcal{Q}}_{11} W^{(0)} = 0$ since $W^{(0)}$ is independent of $\boldsymbol{\xi}$. As discussed in Appendix for the analogous situation for the Schrödinger equation, the average

$$< \tilde{\mathcal{Q}}_1 W^{(2)} >= 0$$

and so the average of the right side of (4.39) is orthogonal to the null space of $\mathcal{Q}_2$. We insert expression (4.37) for $W^{(1)}$ into (4.39), average it and obtain from the orthogonality condition that the amplitudes a^τ satisfy the radiative transport equations

$$(4.40) \quad \frac{\partial a^\tau}{\partial t} + \nabla_{\mathbf{k}} \omega_\tau \cdot \nabla_{\mathbf{x}} a^\tau - \nabla_{\mathbf{x}} \omega_\tau \cdot \nabla_{\mathbf{k}} a^\tau = \int \sigma_{\tau i}(\mathbf{k}, \mathbf{k}') a^i(\mathbf{k}') d\mathbf{k}' - \Sigma_\tau(\mathbf{k}) a^\tau(\mathbf{k}).$$

The differential scattering cross-sections $\sigma_{\tau i}(\mathbf{k}, \mathbf{k}')$ and the total scattering cross-sections $\Sigma_\tau(\mathbf{k})$ are given by

$$(4.41) \quad \sigma_{\tau i}(\mathbf{k}, \mathbf{k}') = 2\pi \omega_\tau^2(\mathbf{k}) c_s^\tau(\mathbf{k}) c_l^\tau(\mathbf{k}) b_v^i(\mathbf{k}') b_w^i(\mathbf{k}') \hat{R}_{svlw}(\mathbf{k} - \mathbf{k}') \delta(\omega_\tau(\mathbf{k}) - \omega_i(\mathbf{k}'))$$

and

$$(4.42) \qquad \Sigma_\tau(\mathbf{k}) = \sum_i \int \sigma_{\tau i}(\mathbf{k}, \mathbf{k}') d\mathbf{k}'.$$

Equation (4.40) has the form (1.6). The scattering cross-sections $\sigma_{\tau i}(\mathbf{k}, \mathbf{k}')$ defined by (4.41) are always positive because the power spectral densities $\hat{R}_{ijkl}(\mathbf{k})$ are positive definite matrices with respect to the pairs of indices ik and jl, by Bochner's theorem [28]. Two modes generated by the eigenvalues ω_i and ω_j are coupled only if ω_i and ω_j coincide for some values of the wave vectors $\mathbf{k}$, $\mathbf{k}'$, that is if for a fixed $\mathbf{k}$ there exists a hypersurface of solutions $\mathbf{k}'$ to the equation

$$(4.43) \qquad \omega_\tau(\mathbf{k}) = \omega_i(\mathbf{k}').$$

If there is scattering between two modes then the symmetries (4.24), (4.25) and (4.21), and (4.41) imply that the differential scattering cross-sections of the direct and reverse scattering processes are the same, i.e.,

$$(4.44) \qquad \sigma_{\tau i}(\mathbf{k}, \mathbf{k}') = \sigma_{i\tau}(\mathbf{k}', \mathbf{k}).$$

This implies that the total energy

$$(4.45) \qquad E(t) = \iint \sum_{j=1}^{N} a^j(t, \mathbf{x}, \mathbf{k}) d\mathbf{x} d\mathbf{k}$$

is conserved.

4.3. Transport equations with polarization

When the eigenvalues of the dispersion matrix $L(\mathbf{x}, \mathbf{k})$ have multiplicities greater than one the perturbation analysis of the previous section must be modified. Equation (4.34) implies that the Wigner matrix $W^{(0)}$ has the form

$$(4.46) \qquad W^{(0)}(t, \mathbf{x}, \mathbf{k}) = \sum_{\tau, i, j} a_{ij}^{\tau}(t, \mathbf{x}, \mathbf{k}) B^{\tau, ij}(\mathbf{x}, \mathbf{k})$$

where the matrices $B^{\tau, ij}$ are defined by (3.26), as in (3.27). We define the coherence matrices $W^{\tau}(t, \mathbf{x}, \mathbf{k})$ as in (3.28) by

$$(4.47) \qquad W_{ij}^{\tau} = a_{ij}^{\tau}.$$

We express $W^{(1)}$ through the coherence matrix using (4.36) and insert it into (4.39). We average (4.39) and use the orthogonality conditions to obtain the radiative transport equations for the coherence matrices

$$(4.48) \qquad \frac{\partial W^{\tau}}{\partial t} + \nabla_{\mathbf{k}} \omega_{\tau} \cdot \nabla_{\mathbf{x}} W^{\tau} - \nabla_{\mathbf{x}} \omega_{\tau} \cdot \nabla_{\mathbf{k}} W^{\tau} + W^{\tau} N^{\tau} - N^{\tau} W^{\tau}$$

$$= \int \sigma^{\tau i}(\mathbf{k}, \mathbf{k}')[W^i(\mathbf{k}')]\delta(\omega_i(\mathbf{k}') - \omega_{\tau}(\mathbf{k})) d\mathbf{k}' - \Sigma^{\tau}(\mathbf{k}) W^{\tau}(\mathbf{k}) - W^{\tau}(\mathbf{k}) \Sigma^{\tau *}(\mathbf{k}).$$

The differential scattering cross-section matrix is

$$\left(\sigma^{\tau i}(\mathbf{k}, \mathbf{k}')[W^i(\mathbf{k}')]\right)_{mj}$$

$$(4.49) \qquad = 2\pi \omega_{\tau}^2(\mathbf{k}) b_v^{i,q}(\mathbf{k}') b_w^{i,r}(\mathbf{k}') c_l^{\tau, j}(\mathbf{k}) c_s^{\tau, m}(\mathbf{k}) \hat{R}_{svlw}(\mathbf{k} - \mathbf{k}') W_{qr}^i(\mathbf{k}')$$

and the total scattering cross-section matrix Σ^{τ} is

$$\Sigma^{\tau} = \frac{1}{2} \sum_{j} \int \sigma^{\tau j}(\mathbf{k}, \mathbf{k}')[I]\delta(\omega_{\tau}(\mathbf{k}) - \omega_i(\mathbf{k}')) d\mathbf{k}'$$

$$(4.50) \qquad - \frac{i}{2} \int \frac{1}{\omega_{\tau}(\mathbf{k}) - \omega_i(\mathbf{k}')} \sigma^{\tau j}(\mathbf{k}, \mathbf{k}')[I] d\mathbf{k}'.$$

The singular integrals in (4.50) should be interpreted in the principal value sense. The imaginary terms in (4.50) are related to the anisotropy of the random perturbations. We will see in particular examples that they are absent when the random perturbations are isotropic.

The radiative transport equations (4.48) preserve W^j as positive definite Hermitian matrices; that is if all the $W^j(0, \mathbf{x}, \mathbf{k})$ are Hermitian and positive definite then $W^j(t, \mathbf{x}, \mathbf{k})$ is Hermitian and positive definite for $t > 0$ and all j. Another important property of equations (4.48) is that they conserve the total energy

$$(4.51) \qquad E(t) = \sum_j \iint \mathrm{Tr} W^j(t, \mathbf{x}, \mathbf{k}) d\mathbf{x} d\mathbf{k} = \text{const.}$$

4.4. Transport equations for acoustic waves

We will now apply the results derived above to the acoustic equations (3.33). The symmetric hyperbolic system for acoustic waves has simple structure because all the non-zero speeds of propagation are distinct and there is no scattering between different modes, even in the presence of random inhomogeneities. This is because the frequency (3.35) $\omega_+(\mathbf{k})$ is always positive and the frequency $\omega_-(\mathbf{k})$ is negative for all $\mathbf{k} \neq 0$ and so the radiative transport equations (4.40) for the amplitudes a^+ and a^- are decoupled from each other. Moreover, these amplitudes are related by (3.43) and so we consider only $a^+(t, \mathbf{x}, \mathbf{k})$, which we denote by $a(t, \mathbf{x}, \mathbf{k})$.

The perturbed matrix A of the symmetric hyperbolic system (3.33) is

$$(4.52) \qquad \begin{pmatrix} \rho I & 0 \\ 0 & \kappa \end{pmatrix} \left[\begin{pmatrix} I & 0 \\ 0 & 1 \end{pmatrix} + \sqrt{\varepsilon} \begin{pmatrix} \tilde{\rho} I & 0 \\ 0 & \tilde{\kappa} \end{pmatrix} \right]$$

where I is the 3×3 identity matrix and $\tilde{\rho}$ and $\tilde{\kappa}$ are the fluctuations in the density and compressibility, respectively. Therefore the power spectral densities $\hat{R}_{svlw}(\mathbf{p})$ in (4.22) have therefore the form

$$(4.53) \qquad \hat{R}_{svlw}(\mathbf{p}) = \delta_{sv}\delta_{lw}\delta_{s\leq3}\delta_{l\leq3}\hat{R}_{\rho\rho}(\mathbf{p}) + \delta_{sv}\delta_{s\leq3}\delta_{lw}\delta_{l,4}\hat{R}_{\rho\kappa}(\mathbf{p})$$
$$+ \; \delta_{sv}\delta_{s,4}\delta_{lw}\delta_{l,4}\hat{R}_{\kappa\kappa}(\mathbf{p}) + \delta_{sv}\delta_{s,4}\delta_{lw}\delta_{l\leq3}\hat{R}_{\rho\kappa}(\mathbf{p}).$$

Here $\hat{R}_{\rho\rho}$, $\hat{R}_{\rho\kappa}$, $\hat{R}_{\kappa\kappa}$ are the power spectral densities of the fluctuations of the density ρ and compressibility κ. The indices go from 1 to 4 and we use the notation $\delta_{l\leq3}$ which is equal to one if $l \leq 3$ and to zero otherwise.

We insert into (4.41) the expression (4.53) for the power spectral densities, the eigenvalues (3.35) and the eigenvectors (3.37) and obtain for the phase space energy density $a(t, \mathbf{x}, \mathbf{k})$ the radiative transport equation (4.40) in the form

$$\frac{\partial a}{\partial t} + v\hat{\mathbf{k}} \cdot \nabla_{\mathbf{x}} a - |\mathbf{k}| \nabla_{\mathbf{x}} v \cdot \nabla_{\mathbf{k}} a = \frac{\pi v^2 |\mathbf{k}|^2}{2} \int \delta(v|\mathbf{k}| - v|\mathbf{k}'|)[a(\mathbf{k}') - a(\mathbf{k})]$$

$$(4.54) \qquad \cdot \left\{ (\hat{\mathbf{k}} \cdot \hat{\mathbf{k}}')^2 \hat{R}_{\rho\rho}(\mathbf{k} - \mathbf{k}') + 2(\hat{\mathbf{k}} \cdot \hat{\mathbf{k}}')\hat{R}_{\rho\kappa}(\mathbf{k} - \mathbf{k}') + \hat{R}_{\kappa\kappa}(\mathbf{k} - \mathbf{k}') \right\} d\mathbf{k}'.$$

This is equation (1.6) with the scattering cross-section as in (1.8). It is also similar to the radiative transport equation (4.9) for the Schrödinger equation but the scattering cross-sections differ.

4.5. Transport equations for electromagnetic waves

Electromagnetic waves are polarized so propagation of wave energy is described by the coherence matrices $W^+(t, \mathbf{x}, \mathbf{k})$ and $W^-(t, \mathbf{x}, \mathbf{k})$ that satisfy the relation (3.71). Note that the frequency $\omega_+(\mathbf{x}, \mathbf{k}) = v(\mathbf{x})|\mathbf{k}|$, with v given by (3.64), is always positive while the frequency $\omega_-(\mathbf{x}, \mathbf{k}) = -v(\mathbf{x})|\mathbf{k}|$ is always negative. According to (4.48) this implies that the radiative transport equations for the coherence matrices W^+ and W^- are not coupled so we consider only the radiative transport equation for W^+ and drop the superscript $+$.

We assume that the random fluctuations of the medium properties are isotropic with perturbed A matrix in (3.60) given by

$$\begin{pmatrix} \epsilon I & 0 \\ 0 & \mu I \end{pmatrix} \left[\begin{pmatrix} I & 0 \\ 0 & I \end{pmatrix} + \sqrt{\varepsilon} \begin{pmatrix} \tilde{\epsilon} I & 0 \\ 0 & \tilde{\mu} I \end{pmatrix} \right].$$

Here I is the 3×3 identity matrix and $\tilde{\epsilon}$ and $\tilde{\mu}$ are the fluctuations in the dielectric permittivity and the magnetic permeability, respectively. The power spectral densities of the fluctuations (4.22), $\hat{R}_{svlw}(\mathbf{k})$, have the form

$$(4.55) \quad \hat{R}_{svlw}(\mathbf{k}) = \delta_{\ sv}\delta_{lw}\delta_{s\leq 3}\delta_{w\leq 3}\hat{R}_{\epsilon\epsilon}(|\mathbf{k}|) + \delta_{sv}\delta_{lw}\delta_{s\leq 3}\delta_{w\geq 4}\hat{R}_{\epsilon\mu}(|\mathbf{k}|) +$$
$$\delta_{\ sv}\delta_{lw}\delta_{s\geq 4}\delta_{w\leq 3}\hat{R}_{\epsilon\mu}(|\mathbf{k}|) + \delta_{sv}\delta_{lw}\delta_{s\geq 4}\delta_{w\geq 4}\hat{R}_{\mu\mu}(|\mathbf{k}|),$$

where $\hat{R}_{ij}(\mathbf{k})$, $i,j = \epsilon,\mu$ are the power spectral densities of the fluctuations of ϵ and μ. In (4.55) the indices run from 1 to 6 and we use the delta notation as in (4.53).

We introduce the 2×2 matrices $T(\mathbf{k},\mathbf{k}')$ and $X(\mathbf{k},\mathbf{k}')$ by

$$(4.56) \qquad\qquad T_{ij}(\mathbf{k},\mathbf{p}) = \mathbf{z}^{(i)}(\mathbf{k}) \cdot \mathbf{z}^{(j)}(\mathbf{p})$$

and

$$(4.57) \qquad\qquad X_{ij} = \tilde{\mathbf{z}}^{(i)}(\mathbf{k}) \cdot \tilde{\mathbf{z}}^{(j)}(\mathbf{k}),$$

where the vectors $\mathbf{z}^{(i)}(\mathbf{k})$ are given by (3.38), and $\tilde{\mathbf{z}}^{(1)}(\mathbf{k}) = -\mathbf{z}^{(2)}(\mathbf{k})$ and $\tilde{\mathbf{z}}^{(2)}(\mathbf{k}) = \mathbf{z}^{(1)}(\mathbf{k})$. These matrices are related by

$$(4.58) \qquad\qquad T(\mathbf{k},\mathbf{p})X^*(\mathbf{k},\mathbf{p}) = (\hat{\mathbf{k}} \cdot \hat{\mathbf{p}})I$$

where I denotes 2×2 matrix. Moreover

$$(4.59) \qquad\qquad \begin{aligned} T^*(\mathbf{k},\mathbf{p}) &= T(\mathbf{p},\mathbf{k}) \\ X^*(\mathbf{k},\mathbf{p}) &= X(\mathbf{p},\mathbf{k}). \end{aligned}$$

We now calculate the scattering cross-sections in terms of the matrices T and X and the power spectral densities by using in the general formulas (4.49) and (4.50), the eigenvalues and eigenvectors (3.65) and the power spectral densities (4.55). The power spectral density tensor (4.55) has four terms and each one generates a term in the differential scattering cross-section. The one with $\hat{R}_{\epsilon\epsilon}$ is

$$\sigma_1(\mathbf{k},\mathbf{k}')[W(\mathbf{k}')]_{mj}$$
$$= 2\pi v^2 |\mathbf{k}|^2 \sqrt{\frac{1}{2\epsilon}} z_v^{(q)}(\mathbf{k}')\sqrt{\frac{1}{2\epsilon}} z_w^{(r)}(\mathbf{k}')\sqrt{\frac{\epsilon}{2}} z_w^{(j)}(\mathbf{k})\sqrt{\frac{\epsilon}{2}} z_v^{(m)}(\mathbf{k})W_{qr}(\mathbf{k}')$$
$$\cdot \hat{R}_{\epsilon\epsilon}(\mathbf{k}-\mathbf{k}')$$
$$(4.60) \qquad = \frac{\pi v^2 |\mathbf{k}|^2}{2}\hat{R}_{\epsilon\epsilon}(\mathbf{k}-\mathbf{k}')T_{mq}(\mathbf{k},\mathbf{k}')W_{qr}(\mathbf{k}')T_{rj}(\mathbf{k}',\mathbf{k}).$$

The other terms in the scattering cross-section are calculated in the same way and they yield

$$\sigma[W](\mathbf{k},\mathbf{k}') = \frac{\pi v^2 |\mathbf{k}|^2}{2}\{\hat{R}_{\epsilon\epsilon}(|\mathbf{k}-\mathbf{k}'|)T(\mathbf{k},\mathbf{k}')W(\mathbf{k}')T(\mathbf{k}',\mathbf{k})$$
$$(4.61) \qquad + \hat{R}_{\mu\mu}(|\mathbf{k}-\mathbf{k}'|)X(\mathbf{k},\mathbf{k}')W(\mathbf{k}')X(\mathbf{k}',\mathbf{k})$$
$$+ \hat{R}_{\epsilon\mu}(|\mathbf{k}-\mathbf{k}'|)[T(\mathbf{k},\mathbf{k}')W(\mathbf{k}')X(\mathbf{k}',\mathbf{k}) + X(\mathbf{k},\mathbf{k}')W(\mathbf{k}')T(\mathbf{k}',\mathbf{k})]\}.$$

This differential scattering cross-section has the correct structure so that the radiative transport equation (4.62) below conserves the Hermitian and positive definite properties of the coherence matrix W.

By direct calculation we find that $\int \sigma(\mathbf{k}, \mathbf{k}')[I]d\Omega(\hat{\mathbf{p}})$ is proportional to the identity matrix and the imaginary terms in (4.50) vanish. The total scattering cross-section $\Sigma(\mathbf{k})$ is therefore

$$\Sigma(|\mathbf{k}|) = \frac{\pi^2 |\mathbf{k}|^4}{2\sqrt{\epsilon\mu}} \int_{-1}^{1} [(\hat{R}_{\epsilon\epsilon}(|\mathbf{k}|\sqrt{2-2\eta}) + \hat{R}_{\mu\mu}(|\mathbf{k}|\sqrt{2-2\eta}))(1+\eta^2)$$
$$+ 4\eta\hat{R}_{\epsilon\mu}(|\mathbf{k}|\sqrt{2-2\eta})]d\eta.$$

Thus the radiative transport equation (4.48) for the coherence matrix W is

$$\frac{\partial W}{\partial t} + v\hat{\mathbf{k}} \cdot \nabla_{\mathbf{x}}W - |\mathbf{k}|\nabla_{\mathbf{x}}v \cdot \nabla_{\mathbf{k}}W + WN - NW$$

(4.62)
$$= \frac{\pi|\mathbf{k}|^4}{2\sqrt{\epsilon\mu}} \int_{|\mathbf{k}'|=|\mathbf{k}|} [\hat{R}_{\epsilon\epsilon}(|\mathbf{k}-\mathbf{k}'|)T(\mathbf{k},\mathbf{k}')W(\mathbf{k}')T(\mathbf{k}',\mathbf{k})$$
$$+ \hat{R}_{\epsilon\mu}(|\mathbf{k}-\mathbf{k}'|)(T(\mathbf{k},\mathbf{k}')W(\mathbf{k}')X(\mathbf{k}',\mathbf{k}) + X(\mathbf{k},\mathbf{k}')W(\mathbf{k}')T(\mathbf{k}',\mathbf{k}))$$
$$+ \hat{R}_{\mu\mu}(|\mathbf{k}-\mathbf{k}'|)X(\mathbf{k},\mathbf{p})W(\mathbf{p})X(\mathbf{p},\mathbf{k})]d\Omega(\hat{\mathbf{p}}) - \Sigma(|\mathbf{k}|)W(\mathbf{k}).$$

The coupling matrix N is given by (3.66).

When the power spectral densities of the fluctuations $\hat{R}_{ij}$ are constants, the scattering cross-sections are proportional to $|\mathbf{k}|^4$, which corresponds to Rayleigh scattering. If, in addition, the magnetic permittivity has no fluctuations then the radiative transport equation (4.62) in a uniform background medium coincides, up to a normalization constant, with Chandrasekhar's equation of radiative transfer (equation (212) in [1]).

In the transport equations corresponding to Maxwell's equations, there is scattering only between modes propagating with the same speed. This is not true in general, as we saw in section 4.2.

4.6. Transport equations for elastic waves

The elastic wave equations in a random medium are given by the symmetric hyperbolic system (3.80) with the perturbed A matrix

(4.63)
$$\begin{pmatrix} \rho I & 0 & 0 & 0 \\ 0 & \frac{1}{2\mu}I & 0 & 0 \\ 0 & 0 & \frac{1}{\mu}I & 0 \\ 0 & 0 & 0 & \frac{1}{\lambda} \end{pmatrix} \left[\begin{pmatrix} I & 0 & 0 & 0 \\ 0 & I & 0 & 0 \\ 0 & 0 & I & 0 \\ 0 & 0 & 0 & 1 \end{pmatrix} + \sqrt{\varepsilon} \begin{pmatrix} \tilde{\rho}I & 0 & 0 & 0 \\ 0 & \tilde{\theta}I & 0 & 0 \\ 0 & 0 & \tilde{\theta}I & 0 \\ 0 & 0 & 0 & \tilde{\psi} \end{pmatrix} \right].$$

Here I is the 3×3 identity matrix and $\tilde{\theta}$ and $\tilde{\psi}$ are the fluctuations of $\frac{1}{\mu}$ and $\frac{1}{\lambda}$, respectively. The power spectral densities of the fluctuations $\hat{R}_{svlw}(\mathbf{k})$ have the form

$$
\begin{aligned}
\hat{R}_{svlw}(\mathbf{k}) = {}& \delta_{sv}\delta_{lw}\{\delta_{s\leq3}\delta_{l\leq3}\hat{R}_{\rho\rho}(|\mathbf{k}|) + \delta_{4\leq s\leq6}\delta_{l\leq3}\hat{R}_{\mu\rho}(|\mathbf{k}|) \\
& + \delta_{s\leq3}\delta_{4\leq l\leq6}\hat{R}_{\mu\rho}(|\mathbf{k}|) + \delta_{7\leq s\leq9}\delta_{l\leq3}\hat{R}_{\mu\rho}(|\mathbf{k}|) + \delta_{s\leq3}\delta_{7\leq l\leq9}\hat{R}_{\mu\rho}(|\mathbf{k}|) \\
& + \delta_{s,10}\delta_{l\leq3}\hat{R}_{\rho\lambda}(|\mathbf{k}|) + \delta_{s\leq3}\delta_{l,10}\hat{R}_{\rho\lambda}(|\mathbf{k}|) + \delta_{4\leq s\leq6}\delta_{4\leq l\leq6}\hat{R}_{\mu\mu}(|\mathbf{k}|) \\
& + \delta_{4\leq s\leq6}\delta_{7\leq l\leq9}\hat{R}_{\mu\mu}(|\mathbf{k}|) + \delta_{7\leq s\leq9}\delta_{4\leq s\leq6}\hat{R}_{\mu\mu}(|\mathbf{k}|) + \delta_{4\leq s\leq6}\delta_{l,10}\hat{R}_{\mu\lambda}(|\mathbf{k}|) \\
& + \delta_{s,10}\delta_{4\leq l\leq6}\hat{R}_{\mu\lambda}(|\mathbf{k}|) + \delta_{7\leq s\leq9}\delta_{7\leq l\leq9}\hat{R}_{\mu\mu}(|\mathbf{k}|) + \delta_{7\leq s\leq9}\delta_{l,10}\hat{R}_{\mu\lambda}(|\mathbf{k}|) \\
& + \delta_{s,10}\delta_{7\leq l\leq9}\hat{R}_{\mu\lambda}(|\mathbf{k}|) + \delta_{s,10}\delta_{l,10}\hat{R}_{\lambda\lambda}(|\mathbf{k}|)\}.
\end{aligned}
\tag{4.64}
$$

The subscripts μ and λ refer to the fluctuations of $1/\mu$ and $1/\lambda$ and the subscript ρ corresponds to the fluctuations of the density ρ. The indices in (4.64) run from 1 to 10 and the delta's are as in (4.53).

The P to S wave resonance condition (4.43) is

$$
\omega_+^S(\mathbf{k}) = \omega_+^P(\mathbf{k}')
$$

with the P and S wave frequencies given by (3.83). For a fixed S wave vector $\mathbf{k}$ there is a sphere of resonant P wave vectors $|\mathbf{k}'| = \sqrt{\mu/(2\mu+\lambda)}|\mathbf{k}|$, so the transport equation (4.48) for the P wave energy density $a_+^P(t,\mathbf{x},\mathbf{k}))$ and the transport equation for the S wave coherence matrix $W_+^S(t,\mathbf{x},\mathbf{k})$ are coupled. Moreover, as in the electromagnetic case, there is no coupling to backward traveling waves so it is enough to consider the two forward modes and to omit the subscript $+$. As we noted earlier, the P wave energy transport is similar to that of acoustic waves, and the S wave energy transport is similar to that of electromagnetic waves. Therefore the system of transport equations for elastic waves will have the form (4.54) for the P waves coupled to a system of the form (4.62) for the S waves. They are given by (4.69) and (4.70).

We now outline the calculation of the scattering cross-sections. We present two calculations: the part of σ^{SS} in (4.72) that involves $\hat{R}_{\rho\rho}$ and the part containing $\hat{R}_{\mu\mu}$. Using the eigenvalues (3.83) and eigenvectors (3.85) of the dispersion matrix (3.81) and the power spectral densities (4.64) in (4.49) we have

$$
\begin{aligned}
& 2\pi v_S^2|\mathbf{k}|^2\sqrt{\frac{1}{2\rho}}z_s^{(q)}(\mathbf{k}')\sqrt{\frac{1}{2\rho}}z_n^{(r)}(\mathbf{k}')\sqrt{\frac{\rho}{2}}z_n^{(j)}(\mathbf{k})\sqrt{\frac{\rho}{2}}z_s^{(m)}(\mathbf{k})W_{qr}^S(\mathbf{k}') \\
& \cdot\hat{R}_{\rho\rho}(|\mathbf{k}-\mathbf{k}'|) \\
& = \frac{\pi v_S^2|\mathbf{k}|^2}{2}\hat{R}_{\rho\rho}(|\mathbf{k}-\mathbf{k}'|)\{T(\mathbf{k},\mathbf{k}')W^S(\mathbf{k}')T(\mathbf{k}',\mathbf{k})\}_{mj}.
\end{aligned}
\tag{4.65}
$$

We show next that the differential scattering cross-section for the S-to-S scattering (4.72) differs slightly from the differential scattering cross-section for electromagnetic waves (4.61). The part of the differential scattering cross-section σ^{SS}

involving the power spectral density $\hat{R}_{\mu\mu}$ is given by

$$\frac{\pi\mu\hat{R}_{\mu\mu}(|\mathbf{k}-\mathbf{k}'|)}{2\rho|\mathbf{k}|^2}(2K(\mathbf{k})K(\mathbf{k}')\mathbf{z}^{(r)}(\mathbf{k}')+M(\mathbf{k})M(\mathbf{k}')\mathbf{z}^{(r)}(\mathbf{k}'),\mathbf{z}^{(j)}(\mathbf{k}))$$

$$\cdot(2K(\mathbf{k})K(\mathbf{k}')\mathbf{z}^{(q)}(\mathbf{k}')+M(\mathbf{k})M(\mathbf{k}')\mathbf{z}^{(q)}(\mathbf{k}'),\mathbf{z}^{(m)}(\mathbf{k}))W_{qr}^S(\mathbf{k}')$$

$$=\frac{\pi\mu\hat{R}_{\mu\mu}(|\mathbf{k}-\mathbf{k}'|)|\mathbf{k}'|^2}{2\rho}\Gamma_{mq}(\mathbf{k},\mathbf{k}')W_{qr}^S(\mathbf{k}')\Gamma_{rj}(\mathbf{k}',\mathbf{k})$$

$$(4.66)\qquad=\frac{\pi v_S^2|\mathbf{k}'|^2\hat{R}_{\mu\mu}(|\mathbf{k}-\mathbf{k}'|)}{2}\{\Gamma(\mathbf{k},\mathbf{k}')W^S(\mathbf{k}')\Gamma(\mathbf{k}',\mathbf{k})\}_{mj}.$$

The matrix Γ is given by (4.73) or equivalently by

$$(4.67)\qquad\Gamma_{mq}(\mathbf{k},\mathbf{k}')=(2K(\hat{\mathbf{k}})K(\hat{\mathbf{k}}')\mathbf{z}^{(q)}(\mathbf{k}')+M(\hat{\mathbf{k}})M(\hat{\mathbf{k}}')\mathbf{z}^{(q)}(\mathbf{k}'),\mathbf{z}^{(m)}(\mathbf{k}))$$

with M defined by (3.82) and $K=\mathrm{diag}(k_1,k_2,k_3)$. The differential scattering cross-section has the form (4.72) with

$$(4.68)\qquad\begin{aligned}\sigma_{ss}^{TT}&=\frac{\pi v_S^2|\mathbf{k}|^2}{2}\hat{R}_{\rho\rho}(|\mathbf{k}-\mathbf{k}'|)\\[4pt]\sigma_{ss}^{\Gamma\Gamma}&=\frac{\pi v_S^2|\mathbf{k}|^2}{2}\hat{R}_{\mu\mu}(|\mathbf{k}-\mathbf{k}'|)\\[4pt]\sigma_{ss}^{\Gamma T}&=\frac{\pi v_S^2|\mathbf{k}|^2}{2}\hat{R}_{\rho\mu}(|\mathbf{k}-\mathbf{k}'|).\end{aligned}$$

This is the same as (4.61) in the electromagnetic case with the matrix X replaced by Γ. A direct calculation shows that the imaginary terms in (4.50) vanish in this case.

The rest of the calculations are similar and we omit them. The coupled radiative transport equations for the P wave energy density $a^P(t,\mathbf{x},\mathbf{k})$ and the 2×2 coherence matrix $W^S(t,\mathbf{x},\mathbf{k})$ for the S waves have the forms

$$(4.69)\qquad\frac{\partial a^P}{\partial t}+\nabla_{\mathbf{k}}\omega^P\cdot\nabla_{\mathbf{x}}a^P-\nabla_{\mathbf{x}}\omega^P\cdot\nabla_{\mathbf{k}}a^P$$

$$=\int\sigma^{PP}(\mathbf{k},\mathbf{k}')a^P(\mathbf{k}')d\mathbf{k}'-\Sigma^{PP}(\mathbf{k})a^P(\mathbf{k})$$

$$+\int\sigma^{PS}(\mathbf{k},\mathbf{k}')[W^S(\mathbf{k}')]d\mathbf{k}'-\Sigma^{PS}(\mathbf{k})a^P(\mathbf{k})$$

and

$$(4.70)\qquad\frac{\partial W^S}{\partial t}+\nabla_{\mathbf{k}}\omega^S\cdot\nabla_{\mathbf{x}}W^S-\nabla_{\mathbf{x}}\omega^S\cdot\nabla_{\mathbf{k}}W^S+WN-NW$$

$$=\int\sigma^{SS}(\mathbf{k},\mathbf{k}')[W^S(\mathbf{k}')]d\mathbf{k}'-\Sigma^{SS}(\mathbf{k})W^S(\mathbf{k})$$

$$+\int\sigma^{SP}(\mathbf{k},\mathbf{k}')[a^P(\mathbf{k}')]d\mathbf{k}'-\Sigma^{SP}(\mathbf{k})W^S(\mathbf{k}).$$

The coupling matrix N is the same as (1.17) for electromagnetic waves except that the speed v is now the shear speed $v_S(\mathbf{x})=\sqrt{\mu(\mathbf{x})/\rho(\mathbf{x})}$. The differential scattering cross-section $\sigma^{PP}(\mathbf{k},\mathbf{k}')$ for P to P scattering is similar to (1.8) for scattering of scalar waves and the differential scattering tensor $\sigma^{SS}(\mathbf{k},\mathbf{k}')$ is similar to

Chandrasekhar's tensor (1.12). They have the forms

$$(4.71) \qquad \sigma^{PP}(\mathbf{k},\mathbf{k}') = \sigma_{pp}(\mathbf{k},\mathbf{k}')\delta(v_P|\mathbf{k}| - v_P|\mathbf{k}'|)$$

and

$$\begin{aligned}
\sigma^{SS}(\mathbf{k},\mathbf{k}')[W(\mathbf{k}')] = \{\ & \sigma_{ss}^{TT}T(\mathbf{k},\mathbf{k}')W(\mathbf{k}')T(\mathbf{k}',\mathbf{k}) + \sigma_{ss}^{\Gamma\Gamma}\Gamma(\mathbf{k},\mathbf{k}')W(\mathbf{k}')\Gamma(\mathbf{k}',\mathbf{k}) \\
& + \sigma_{ss}^{\Gamma T}[T(\mathbf{k},\mathbf{k}')W(\mathbf{k}')\Gamma(\mathbf{k}',\mathbf{k}) + \Gamma(\mathbf{k},\mathbf{k}')W(\mathbf{k}')T(\mathbf{k}',\mathbf{k})]\} \\
(4.72) \qquad & \cdot\ \delta(v_S|\mathbf{k}| - v_S|\mathbf{k}'|).
\end{aligned}$$

The 2×2 matrix $\Gamma(\mathbf{k},\mathbf{k}')$ is similar to T and is defined by

$$(4.73) \quad \Gamma_{ij}(\mathbf{k},\mathbf{k}') = (\hat{\mathbf{k}} \cdot \hat{\mathbf{k}}')(\mathbf{z}^{(i)}(\mathbf{k}) \cdot \mathbf{z}^{(j)}(\mathbf{k}')) + (\hat{\mathbf{k}} \cdot \mathbf{z}^{(j)}(\mathbf{k}'))(\hat{\mathbf{k}}' \cdot \mathbf{z}^{(i)}(\mathbf{k}))$$

while σ_{pp} and σ_{ss} are scalar functions given in terms of power spectral densities of the inhomogeneities by (4.68) and (4.79). The total scattering cross-sections Σ^{PP} and Σ^{SS} are the integrals of the corresponding differential scattering cross-sections, as in (1.7) and (1.14).

The scattering cross-sections for the S to P and P to S coupling terms, σ^{PS} and σ^{SP}, respectively, have the forms

$$(4.74) \quad \sigma^{PS}(\mathbf{k},\mathbf{k}')[W^S(\mathbf{k}')] = \mathrm{Tr}[\sigma_{ps}(\mathbf{k},\mathbf{k}')G(\mathbf{k},\mathbf{k}')W^S(\mathbf{k}')]\delta(v_P|\mathbf{k}| - v_S|\mathbf{k}'|)$$

$$(4.75) \quad \sigma^{SP}(\mathbf{k},\mathbf{k}')[a^P(\mathbf{k}')] = \sigma_{ps}(\mathbf{k}',\mathbf{k})G(\mathbf{k}',\mathbf{k})a^P(\mathbf{k}')\delta(v_S|\mathbf{k}| - v_P|\mathbf{k}'|)$$

with the 2×2 matrix G given by

$$(4.76) \qquad G_{ij}(\mathbf{k},\mathbf{k}') = (\hat{\mathbf{k}} \cdot \mathbf{z}^{(i)}(\mathbf{k}'))(\hat{\mathbf{k}} \cdot \mathbf{z}^{(j)}(\mathbf{k}')).$$

The scalar function σ_{ps} is given explicitly in terms of power spectral densities of the inhomogeneities by (4.80). The scattering operator on the right side of (4.69) and (4.70) is symmetric in a^P, W^S and conservative. This implies in particular that

$$(4.77) \qquad \Sigma^{SP}(\mathbf{k}) = \int \sigma_{ps}(\mathbf{k}',\mathbf{k})G(\mathbf{k}',\mathbf{k})\delta(v_S|\mathbf{k}| - v_P|\mathbf{k}'|)d\mathbf{k}'$$

with

$$(4.78) \qquad \Sigma^{PS}(\mathbf{k}) = \int \sigma_{ps}(\mathbf{k},\mathbf{k}')\mathrm{Tr}G(\mathbf{k},\mathbf{k}')\delta(v_S|\mathbf{k}'| - v_P|\mathbf{k}|)d\mathbf{k}'.$$

The functions σ_{pp} and σ_{ps} are given by

$$\begin{aligned}
\sigma_{pp}(\mathbf{k},\mathbf{k}') = {}& \frac{\pi|\mathbf{k}|^2(2\mu + \lambda)}{2\rho}\left\{ \frac{\lambda^2}{(2\mu + \lambda)^2}\hat{R}_{\lambda\lambda}(|\mathbf{k} - \mathbf{k}'|) \right. \\
& + \frac{4\lambda\mu}{(2\mu + \lambda)^2}(\hat{\mathbf{k}},\hat{\mathbf{k}}')^2 \hat{R}_{\lambda\mu}(|\mathbf{k} - \mathbf{k}'|) \\
(4.79) \qquad & + \frac{4\mu^2}{(2\mu + \lambda)^2}(\hat{\mathbf{k}},\hat{\mathbf{k}}')^4 \hat{R}_{\mu\mu}(|\mathbf{k} - \mathbf{k}'|) + (\hat{\mathbf{k}},\hat{\mathbf{k}}')^2 \hat{R}_{\rho\rho}(|\mathbf{k} - \mathbf{k}'|) \\
& \left. + \frac{2\lambda}{2\mu + \lambda}(\hat{\mathbf{k}},\hat{\mathbf{k}}')\hat{R}_{\lambda\rho}(|\mathbf{k} - \mathbf{k}'|) + \frac{4\mu}{2\mu + \lambda}(\hat{\mathbf{k}},\hat{\mathbf{k}}')^3 \hat{R}_{\rho\mu}(|\mathbf{k} - \mathbf{k}'|) \right\}
\end{aligned}$$

and

$$\begin{aligned}
(4.80) \quad \sigma_{ps}(\mathbf{k},\mathbf{k}') = {}& \frac{\pi\mu}{2\rho}\{|\mathbf{k}'|^2 \hat{R}_{\rho\rho}(|\mathbf{k} - \mathbf{k}'|) + 4|\mathbf{k}|^2(\hat{\mathbf{k}},\hat{\mathbf{k}}')^2 \hat{R}_{\mu\mu}(|\mathbf{k} - \mathbf{k}'|) \\
& + 4|\mathbf{k}||\mathbf{k}'|(\hat{\mathbf{k}},\hat{\mathbf{k}}')\hat{R}_{\mu\rho}(|\mathbf{k} - \mathbf{k}'|)\}.
\end{aligned}$$

The P-to-P part of (4.69) coincides with the transport equation for the acoustic waves when $\mu = 0$. The S-to-S part coincides with the electromagnetic case with the replacement of Γ by X. The scattering operator on the right side of the transport equations (4.69) and (4.70) is symmetric in a^P and W^S. This is an important property that is used in the analysis of the transport equations in the diffusion regime (section 5.3). The applications of the elastic radiative transport equations are discussed in detail in the last lecture.

LECTURE 5
The Diffusion Approximation

5.1. Diffusion approximation for acoustic waves

The diffusion approximation for transport equations like (1.6) is valid at propagation distances much longer than the transport mean free path $|\nabla_{\mathbf{k}}\omega|/\Sigma$ [15]. We show in sections 5.2 and 5.3 that solutions of transport equations for polarized waves also exhibit diffusive behavior and that the waves become approximately depolarized in this regime. For simplicity we will consider only the case when the background is homogeneous and isotropic, in which case the eigenvalues of the dispersion matrix (3.22) are given by $\omega_i(\mathbf{k}) = v_i|\mathbf{k}|$ with the speeds v_i independent of $\mathbf{x}$. We shall consider only conservative transport equations so that (1.7) or (1.14) holds. The results, however, can be generalized to variable backgrounds and to weakly dissipative scattering provided that the background variations and the dissipation are on the scale of the propagation distance.

To derive the diffusion approximation we introduce a dimensionless small parameter ε, not related to the small parameter used in the previous sections. It is ratio of the mean free path to the propagation distance. Then, by rescaling time and space variables by $t \to \varepsilon^2 t$ and $\mathbf{x} \to \varepsilon\mathbf{x}$, we can write (1.6) as

$$\varepsilon^2\frac{\partial a}{\partial t} + \varepsilon v\hat{\mathbf{k}} \cdot \nabla_{\mathbf{x}}a = \int_{|\mathbf{k}|=|\mathbf{k}'|} \sigma(|\mathbf{k}|,\hat{\mathbf{k}},\hat{\mathbf{k}}')a(\mathbf{k}')d\Omega(\hat{\mathbf{k}}') - \Sigma(|\mathbf{k}|)a(|\mathbf{k}|)$$

$$(5.1) \qquad a(0,\mathbf{x},\mathbf{k}) = a_0(\mathbf{x},\mathbf{k}).$$

The total scattering cross-section Σ is

$$(5.2) \qquad \Sigma(|\mathbf{k}|) = \int_{|\mathbf{k}|=|\mathbf{k}'|} \sigma(|\mathbf{k}|,\hat{\mathbf{k}},\hat{\mathbf{k}}')d\Omega(\hat{\mathbf{k}}')$$

and $d\Omega$ denotes the surface element on the unit sphere. We shall consider only rotationally invariant scattering so that the differential scattering cross-section $\sigma(\mathbf{k},\mathbf{k}')$ is a non-negative function that depends only on $|\mathbf{k}|$ and $\mu = \hat{\mathbf{k}} \cdot \hat{\mathbf{k}}'$.

We expand the solution of (5.1) in powers of ε

$$(5.3) \qquad a(t,\mathbf{x},\mathbf{k}) = a^{(0)}(t,\mathbf{x},\mathbf{k}) + \varepsilon a^{(1)}(t,\mathbf{x},\mathbf{k}) + \varepsilon^2 a^{(2)}(t,\mathbf{x},\mathbf{k}) + \dots$$

359

and insert this expansion into (5.1). We find that the leading term $a^{(0)}(t, \mathbf{x}, \mathbf{k})$ satisfies

$$(5.4) \qquad \int_{|\mathbf{k}|=|\mathbf{k}'|} \sigma(|\mathbf{k}|, \hat{\mathbf{k}} \cdot \hat{\mathbf{k}}') a^{(0)}(t, \mathbf{x}, \mathbf{k}') d\Omega(\hat{\mathbf{k}}') = \Sigma(|\mathbf{k}|) a^{(0)}(t, \mathbf{x}, \mathbf{k}).$$

This is an eigenfunction equation for $a^{(0)}$ involving the integral operator $\mathcal{A}$, defined by the left side. The kernel of $\mathcal{A}$ is the scattering cross-section and it is positive. From the general theory of such operators it follows that they have the following properties [**35**]:

(i) the eigenvalue with the largest absolute value is simple,

(ii) the eigenfunction corresponding to this eigenvalue is non-negative,

(iii) this eigenfunction is the only non-negative eigenfunction of this operator.

From (5.2) we see that if $a^{(0)}$ is independent of the direction $\hat{\mathbf{k}}$ it is a solution of (5.4). This fact and properties (i-iii) show that

$$(5.5) \qquad a^{(0)}(t, \mathbf{x}, \mathbf{k}) = a^{(0)}(t, \mathbf{x}, |\mathbf{k}|).$$

This means that $a(t, \mathbf{x}, \mathbf{k})$ is approximately independent of the direction $\hat{\mathbf{k}}$ of the wave vector $\mathbf{k}$.

The first order term $a^{(1)}$ satisfies the equation

$$(5.6) \qquad v\hat{\mathbf{k}} \cdot \nabla_{\mathbf{x}} a^{(0)} = \int_{|\mathbf{k}|=|\mathbf{k}'|} \sigma(|\mathbf{k}|, \hat{\mathbf{k}} \cdot \hat{\mathbf{k}}') a^{(1)}(\mathbf{k}') d\Omega(\hat{\mathbf{k}}') - \Sigma(|\mathbf{k}|) a^{(1)}(\mathbf{k}).$$

To solve (5.6) we note that the function $u(\mathbf{x}, \mathbf{k}) = \hat{\mathbf{k}} \cdot \nabla_{\mathbf{x}} a^{(0)}(t, \mathbf{x}, |\mathbf{k}|)$ is an eigenfunction of the operator $\mathcal{A}$ corresponding to the eigenvalue

$$\lambda = 2\pi \int_{-1}^{1} \sigma(|\mathbf{k}|, \mu) \mu \, d\mu,$$

where $\mu = \hat{\mathbf{k}} \cdot \hat{\mathbf{k}}'$. To show this we let Q be an orthogonal transformation such that $Q\hat{\mathbf{k}} = (0, 0, 1)^t$. Then

$$(5.7) \qquad \begin{aligned} (\mathcal{A}u)(\hat{\mathbf{k}}) &= \int_{|\mathbf{k}|=|\mathbf{k}'|} \sigma(|\mathbf{k}|, \hat{\mathbf{k}} \cdot \hat{\mathbf{k}}')(\hat{\mathbf{k}}' \cdot \nabla_{\mathbf{x}} a^{(0)}) d\Omega(\hat{\mathbf{k}}') \\ &= \int_{|\mathbf{k}|=|\mathbf{k}'|} \sigma(|\mathbf{k}|, \hat{k}_3')(\hat{\mathbf{k}}' \cdot Q\nabla_{\mathbf{x}} a^{(0)}) d\Omega(\hat{\mathbf{k}}') \\ &= 2\pi \int_{-1}^{1} \sigma(|\mathbf{k}|, \mu) \mu \, d\mu (Q\nabla_{\mathbf{x}} a^{(0)})_3 \\ &= 2\pi \int_{-1}^{1} \sigma(|\mathbf{k}|, \mu) \mu \, d\mu (\hat{\mathbf{k}} \cdot \nabla_{\mathbf{x}} a^{(0)}) = \lambda u(\hat{\mathbf{k}}). \end{aligned}$$

Now we write $a^{(1)} = C(|\mathbf{k}|) u$, substitute into (5.6) and use (5.7). Then we can solve for C and u and obtain

$$(5.8) \qquad a^{(1)}(t, \mathbf{x}, \mathbf{k}) = -\frac{v}{\Sigma(|\mathbf{k}|) - \lambda(|\mathbf{k}|)} \hat{\mathbf{k}} \cdot \nabla_{\mathbf{x}} a^{(0)}(t, \mathbf{x}, |\mathbf{k}|).$$

The equation for $a^{(2)}$ is

$$(5.9) \qquad \frac{\partial a^{(0)}}{\partial t} - v\hat{\mathbf{k}} \cdot \nabla_{\mathbf{x}} \left(\frac{v}{\Sigma(|\mathbf{k}|) - \lambda(|\mathbf{k}|)} \hat{\mathbf{k}} \cdot \nabla_{\mathbf{x}} a^{(0)} \right) = \mathcal{A}a^{(2)} - \Sigma(|\mathbf{k}|) a^{(2)}.$$

We integrate (5.9) with respect to direction $\hat{\mathbf{k}}$. The integral of the right side vanishes and we get the solvability condition

$$(5.10) \qquad \int_{|\mathbf{k}|=|\mathbf{k}'|} \left(\frac{\partial a^{(0)}}{\partial t} - v\hat{\mathbf{k}} \cdot \nabla_{\mathbf{x}} \left(\frac{v}{\Sigma(|\mathbf{k}|) - \lambda(|\mathbf{k}|)} \hat{\mathbf{k}} \cdot \nabla_{\mathbf{x}} a^{(0)} \right) \right) d\Omega(\hat{\mathbf{k}}) = 0.$$

After performing the integration over $\hat{\mathbf{k}}$ in (5.10) we obtain the diffusion equation

$$(5.11) \qquad \frac{\partial a^{(0)}(t, \mathbf{x}, |\mathbf{k}|)}{\partial t} = \nabla_{\mathbf{x}} \cdot [D(|\mathbf{k}|)\nabla_{\mathbf{x}} a^{(0)}(t, \mathbf{x}, |\mathbf{k}|)].$$

This equation determines the principal term $a^{(0)}$ in the expansion (5.3). We find that the diffusion coefficient $D(|\mathbf{k}|)$ in (5.11) is given by the well known formula [3].

$$(5.12) \qquad D(|\mathbf{k}|) = \frac{v^2}{3(\Sigma(|\mathbf{k}|) - \lambda(|\mathbf{k}|))}.$$

Note that $D > 0$ because $\Sigma(|\mathbf{k}|)$ is the largest eigenvalue of $\mathcal{A}$ so it is larger than $\lambda(|\mathbf{k}|)$, which is another eigenvalue. The diffusion coefficient can also be written in the form

$$(5.13) \qquad D(\mathbf{k}) = \frac{vl^*(|\mathbf{k}|)}{3},$$

where the diffusion mean free path $l^*(|\mathbf{k}|)$ is given by

$$(5.14) \qquad l^*(|\mathbf{k}|) = v \left(2\pi \int_{-1}^{1} \sigma(|\mathbf{k}|, \mu)(1 - \mu)d\mu \right)^{-1}.$$

The diffusion equation (5.11) cannot accommodate the initial condition $a(0, \mathbf{x}, \mathbf{k}) = a_0(\mathbf{x}, \mathbf{k})$ unless the function $a_0(\mathbf{x}, \mathbf{k})$ is independent of the angular direction $\hat{\mathbf{k}}$. To obtain the correct initial conditions for the diffusion equation (5.11) we must consider the initial layer problem as in [41]. We write a in the form

$$(5.15) \qquad a = a^i + a^{il},$$

where a^i is the solution given by the asymptotic expansion (5.3) and a^{il} is the initial layer solution which decays exponentially in time. The initial layer solution a^{il} depends on the fast time $\tau = t/\varepsilon^2$ and satisfies the equation

$$(5.16) \qquad \frac{\partial a^{il}}{\partial \tau} = \int_{|\mathbf{k}|=|\mathbf{k}'|} \sigma(|\mathbf{k}|, \hat{\mathbf{k}} \cdot \hat{\mathbf{k}}')a^{il}(\mathbf{k}')d\hat{\mathbf{k}}' - \Sigma(|\mathbf{k}|)a^{il}(\mathbf{k}).$$

The solution a^{il} decays exponentially in time if we take as an initial condition for (5.16)

$$(5.17) \qquad a^{il}(0, \mathbf{x}, \mathbf{k}) = a_0(\mathbf{x}, \mathbf{k}) - \frac{1}{4\pi} \int a_0(\mathbf{x}, \mathbf{k}')d\Omega(\hat{\mathbf{k}}').$$

This implies that the initial condition for the diffusion equation (5.11) is the average of a_0,

$$(5.18) \qquad a^{(0)}(0, \mathbf{x}, |\mathbf{k}|) = \frac{1}{4\pi} \int a_0(\mathbf{x}, \mathbf{k})d\Omega(\hat{\mathbf{k}}),$$

as might have been expected from physical considerations.

The perturbation expansion (5.3) may be justified in the spirit of [**9**]. Expansion (5.3) should be replaced by

$$a(t, \mathbf{x}, \mathbf{k}) = a^{(0)}(t, \mathbf{x}, \mathbf{k}) + \varepsilon a^{(1)}(t, \mathbf{x}, \mathbf{k}) + \varepsilon^2 a_\varepsilon^{(2)}(t, \mathbf{x}, \mathbf{k})$$

with $a^{(0)}$ and $a^{(1)}$ computed above. Then it is easy to check that $a_\varepsilon^{(2)} \to 0$ for finite t as $\varepsilon \to 0$.

5.2. Diffusion approximation for electromagnetic waves

We now apply the analysis of the previous section to the transport equation (4.62) for electromagnetic waves. We rewrite this equation in the form

$$(5.19) \qquad \frac{\partial W}{\partial t} + v\hat{\mathbf{k}} \cdot \nabla_\mathbf{x} W = \mathcal{A}W - \Sigma(|\mathbf{k}|)W.$$

Here the integral operator $\mathcal{A}$ acts on matrix valued functions, and is defined by

$$\mathcal{A}f(\mathbf{k}) = \frac{\pi v |\mathbf{k}|^4}{2} \int_{|\mathbf{k}|=|\mathbf{k}'|} \{\hat{R}_{\epsilon\epsilon}(|\mathbf{k} - \mathbf{k}'|)T(\mathbf{k}, \mathbf{k}')f(\mathbf{k}')T(\mathbf{k}', \mathbf{k})$$
$$+ \hat{R}_{\epsilon\mu}(|\mathbf{k} - \mathbf{k}'|)(T(\mathbf{k}, \mathbf{k}')f(\mathbf{k}')X(\mathbf{k}', \mathbf{k}) + X(\mathbf{k}, \mathbf{k}')f(\mathbf{k}')T(\mathbf{k}', \mathbf{k}))$$
$$(5.20) \qquad + \hat{R}_{\mu\mu}(|\mathbf{k} - \mathbf{k}'|)X(\mathbf{k}, \mathbf{k}')f(\mathbf{k}')X(\mathbf{k}', \mathbf{k})\}d\Omega(\hat{\mathbf{k}}'),$$

where $v = 1/\sqrt{\epsilon\mu}$. We assume that the transport mean free path is small compared to the propagation distance and we scale space and time variables $(\mathbf{x}, t)$ by $t \to \varepsilon^2 t$, $\mathbf{x} \to \varepsilon \mathbf{x}$ as in section 5.1. The scaled transport equation (5.19) is

$$(5.21) \qquad \varepsilon^2 \frac{\partial W}{\partial t} + \varepsilon v\hat{\mathbf{k}} \cdot \nabla_\mathbf{x} W = \mathcal{A}W - \Sigma(|\mathbf{k}|)W.$$

We expand the solution of (5.21) in powers of the small parameter ε

$$(5.22) \qquad W = W^{(0)} + \varepsilon W^{(1)} + \varepsilon^2 W^{(2)} + \dots$$

Inserting this into (5.21), we find that the leading term $W^{(0)}$ satisfies the eigenfunction equation

$$(5.23) \qquad \mathcal{A}W^{(0)}(t, \mathbf{x}, \mathbf{k}) = \Sigma(|\mathbf{k}|)W^{(0)}(t, \mathbf{x}, \mathbf{k}),$$

which is analogous to (5.4). The general theory of positive operators [**35**] applies to $\mathcal{A}$ and hence $W^{(0)}(t, \mathbf{x}, \mathbf{k})$ has the form

$$(5.24) \qquad W^{(0)}(t, \mathbf{x}, \mathbf{k}) = \phi(t, \mathbf{x}, |\mathbf{k}|)I,$$

where $\phi(t, \mathbf{x}, |\mathbf{k}|)$ is an unknown scalar function to be determined. Thus, the leading approximation for the coherence matrix is a scalar multiple of the identity and is independent of the direction $\hat{\mathbf{k}}$. This shows that electromagnetic waves are depolarized in the diffusion approximation.

The first order term $W^{(1)}$ satisfies the equation

$$(5.25) \qquad v\hat{\mathbf{k}} \cdot \nabla_\mathbf{x}\phi I = \mathcal{A}W^{(1)} - \Sigma(|\mathbf{k}|)W^{(1)}.$$

The matrix function $u(\hat{\mathbf{k}}) = \hat{\mathbf{k}} \cdot \nabla_\mathbf{x}\phi I$ is an eigenfunction of $\mathcal{A}$, defined by (5.20), corresponding to the eigenvalue

$$\lambda(|\mathbf{k}|) = \frac{\pi v |\mathbf{k}|^4}{2} \int_{-1}^{1} \{\pi(\hat{R}_{\epsilon\epsilon}(|\mathbf{k}|\sqrt{2 - 2\eta}) + \hat{R}_{\mu\mu}(|\mathbf{k}|\sqrt{2 - 2\eta}))(\eta + \eta^3)$$
$$(5.26) \qquad + 4\pi \hat{R}_{\epsilon\mu}(|\mathbf{k}|\sqrt{2 - 2\eta})\eta^2\}d\eta.$$

Hence

$$(5.27) \qquad W^{(1)} = \frac{v}{\lambda(|\mathbf{k}|) - \Sigma(|\mathbf{k}|)}(\hat{\mathbf{k}} \cdot \nabla_{\mathbf{x}}\phi)I.$$

The second order term $W^{(2)}$ satisfies the equation

$$(5.28) \qquad \frac{\partial W^{(0)}}{\partial t} + v\hat{\mathbf{k}} \cdot \nabla_{\mathbf{x}}W^{(1)} = \mathcal{A}W^{(2)} - \Sigma(|\mathbf{k}|)W^{(2)}$$

which is solvable only if the left side of (5.28) is orthogonal to functions of the form (5.24). Integrating (5.28) with respect to $\hat{\mathbf{k}}$ and taking the trace we find that ϕ satisfies the diffusion equation

$$(5.29) \qquad \frac{\partial \phi}{\partial t} = \nabla_{\mathbf{x}} \cdot [D^{em}(|\mathbf{k}|)\nabla_{\mathbf{x}}\phi].$$

The diffusion coefficient is

$$D^{em} = \frac{vl^*_{em}}{3},$$

where the diffusion mean free path l^*_{em} is defined by

$$(5.30) \qquad \begin{aligned} l^*_{em} = \frac{2}{\pi^2|\mathbf{k}|^4}\Bigg(&\int_{-1}^{1}[(\,\hat{R}\,_{\epsilon\epsilon}(|\mathbf{k}|\sqrt{2-2\eta}) + \hat{R}_{\mu\mu}(|\mathbf{k}|\sqrt{2-2\eta}))(1 + \eta^2 - \eta - \eta^3) \\ &+4\,\hat{R}\,_{\epsilon\mu}(|\mathbf{k}|\sqrt{2-2\eta})(\eta - \eta^2)]d\eta\Bigg)^{-1}. \end{aligned}$$

The initial condition for the diffusion equation (5.29) is determined as in the scalar case. The initial condition for the initial layer solution must be

$$(5.31) \qquad W^{il}(0,\mathbf{x},\mathbf{k}) = W_0(\mathbf{x},\mathbf{k}) - \frac{1}{8\pi}\{\int \mathrm{Tr}W_0(\mathbf{x},\mathbf{k})d\Omega(\hat{\mathbf{k}})\}I,$$

so as to make it decay exponentially in time. Then the initial condition for (5.29) is

$$(5.32) \qquad \phi(0,\mathbf{x},|\mathbf{k}|) = \frac{1}{8\pi}\int \mathrm{Tr}W_0(\mathbf{x},\mathbf{k})d\Omega(\hat{\mathbf{k}}).$$

5.3. Diffusion approximation for elastic waves

We shall now determine the diffusion approximation for the elastic transport equations (4.69) and (4.70). We shall show that in the diffusion regime the S waves are depolarized and energy is "equipartitioned" between S and P waves (equation (5.37)).

We rescale space and time variables $(t, \mathbf{x})$ by $t \to \varepsilon^2 t, \mathbf{x} \to \varepsilon\mathbf{x}$ and rewrite the transport equations (4.69) and (4.70) for elastic waves in the scaled form

$$\varepsilon^2\frac{\partial a^P}{\partial t} + \varepsilon v_P\hat{\mathbf{k}} \cdot \nabla_{\mathbf{x}}a^P = \mathcal{A}_{PP}[a^P] + \mathcal{A}_{PS}[W^S] - (\Sigma^{PP} + \Sigma^{PS})a^P$$

$$(5.33) \quad \varepsilon^2\frac{\partial W^S}{\partial t} + \varepsilon v_S\hat{\mathbf{k}} \cdot \nabla_{\mathbf{x}}W^S = \mathcal{A}_{SS}[W^S] + \mathcal{A}_{SP}[a^P] - (\Sigma^{SS} + \Sigma^{SP})W^S.$$

The integral operators $\mathcal{A}_{ij}$ are defined by comparing (5.33) to (4.69) and (4.70). We expand the solution of (5.33) as

$$(5.34) \qquad \begin{aligned} a^P &= a^{(0)} + \varepsilon a^{(1)} + \varepsilon^2 a^{(2)} + \ldots \\ W^S &= W^{(0)} + \varepsilon W^{(1)} + \varepsilon^2 W^{(2)} + \ldots. \end{aligned}$$

By using (5.34) in (5.33) we find that the principal terms $a^{(0)}$ and $W^{(0)}$ must satisfy the equations

$$\mathcal{A}_{PP}[a^{(0)}] + \mathcal{A}_{PS}[W^{(0)}] = (\Sigma^{PP} + \Sigma^{PS})a^{(0)}$$

(5.35)
$$\mathcal{A}_{SS}[W^{(0)}] + \mathcal{A}_{SP}[a^{(0)}] = (\Sigma^{SS} + \Sigma^{SP})W^{(0)}.$$

This is a pair of coupled equations of the form (5.4) and (5.23). The general theory of positive operators is applicable again and implies that the solutions of (5.35) are of the form

(5.36)
$$a^{(0)}(t,\mathbf{x},\mathbf{k}) = \phi(t,\mathbf{x},|\mathbf{k}|)$$
$$W^{(0)}(t,\mathbf{x},\mathbf{k}) = \phi(t,\mathbf{x},\frac{v_S}{v_P}|\mathbf{k}|)I,$$

where $\phi(t,\mathbf{x},|\mathbf{k}|)$ is a scalar function to be determined. It follows that

(5.37)
$$a^{(0)}(t,\mathbf{x},\mathbf{k})I = W^{(0)}(t,\mathbf{x},\frac{v_P}{v_S}\mathbf{k}).$$

Equation (5.36) implies that in the diffusion regime the S wave is completely depolarized. Equation (5.37) shows that the energy in the wave number shell of interaction in phase space is partitioned between the P waves and each polarization of the S waves.

In physical space the local energy densities

(5.38)
$$\mathcal{E}_P(t,\mathbf{x}) = \int_{R^3} a^P(t,\mathbf{x},\mathbf{k})d\mathbf{k}$$

and

(5.39)
$$\mathcal{E}_S(t,\mathbf{x}) = \int_{R^3} \mathrm{Tr}W^S(t,\mathbf{x},\mathbf{k})d\mathbf{k}$$

are related by

(5.40)
$$\mathcal{E}_P(t,\mathbf{x}) = \frac{v_S^3}{2v_P^3}\mathcal{E}_S(t,\mathbf{x}).$$

This provides an effective criterion for determining the range of validity of the diffusion regime in the analysis of seismic data. Unless the energy densities of the P and S waves, which can be obtained from measurements, satisfy relation (5.40) the diffusion approximation is not valid. This formula shows that in the diffusion regime most of the energy is in the S waves, no matter how it was distributed initially.

The first order terms satisfy the system of equations

(5.41)
$$v_P\hat{\mathbf{k}}\cdot\nabla_{\mathbf{x}}a^{(0)} = \mathcal{A}_{PP}[a^{(1)}] + \mathcal{A}_{PS}[W^{(1)}] - (\Sigma^{PP} + \Sigma^{PS})a^{(1)}$$
$$v_S\hat{\mathbf{k}}\cdot\nabla_{\mathbf{x}}W^{(0)} = \mathcal{A}_{SS}[W^{(1)}] + \mathcal{A}_{SP}[a^{(1)}] - (\Sigma^{SS} + \Sigma^{SP})W^{(1)}.$$

As in sections 5.1 and 5.2, the function $u = \hat{\mathbf{k}}\cdot\nabla_{\mathbf{x}}\phi$ is an eigenfunction of all the operators $\mathcal{A}_{PP}$, $\mathcal{A}_{PS}$, $\mathcal{A}_{SP}$ and $\mathcal{A}_{SS}$. Let the corresponding eigenvalues be λ_{pp}, λ_{ps}, λ_{sp} and λ_{ss}, respectively. This implies that if $W^{(1)} = -l_s\hat{\mathbf{k}}\cdot\nabla_{\mathbf{x}}\phi I$ and $a^{(1)} = -l_p\hat{\mathbf{k}}\cdot\nabla_{\mathbf{x}}\phi$, then (5.41) is satisfied provided the constants l_p and l_s solve the system of two linear equations

(5.42)
$$-v_P = \lambda_{pp}l_p + \lambda_{ps}l_s - (\Sigma_{pp} + \Sigma_{ps})l_p$$
$$-v_S = \lambda_{ss}l_s + \lambda_{sp}l_p - (\Sigma_{ss} + \Sigma_{sp})l_s.$$

Both constants l_s and l_p have the dimension of length and can be considered as diffusion mean free paths for S and P waves, respectively.

The second-order terms in ε satisfy the system of equations

$$(5.43) \qquad \frac{\partial a^{(0)}}{\partial t} + v_P \hat{\mathbf{k}} \cdot \nabla_{\mathbf{x}} a^{(1)} = \mathcal{A}_{PP}[a^{(2)}] + \mathcal{A}_{PS}[W^{(2)}] - (\Sigma^{PP} + \Sigma^{PS}) a^{(2)}$$

$$(5.44) \qquad \frac{\partial W^{(0)}}{\partial t} + v_S \hat{\mathbf{k}} \cdot \nabla_{\mathbf{x}} W^{(1)} = \mathcal{A}_{SS}[W^{(2)}] + \mathcal{A}_{SP}[a^{(2)}] - (\Sigma^{SS} + \Sigma^{SP}) W^{(2)}.$$

This system has a solution when the sum of the integrals with respect to $\hat{\mathbf{k}}$ of the left side of (5.43) multiplied by v_S^3/v_P^3 and of the trace of the left side of (5.44) vanishes. This implies that the function ϕ must satisfy the diffusion equation

$$(5.45) \qquad \frac{\partial \phi}{\partial t} = \nabla_{\mathbf{x}} \cdot [D^{el}(|\mathbf{k}|) \nabla_{\mathbf{x}} \phi]$$

with the diffusion coefficient

$$(5.46) \qquad D^{el} = \frac{1}{\frac{2}{v_S^3} + \frac{1}{v_P^3}} \left(\frac{l_p^* v_P}{3v_P^3} + \frac{2l_s^* v_S}{3v_S^3} \right).$$

Thus D^{el} is the weighted mean of "partial diffusion coefficients" for P waves and for each polarization of S waves, where l_s and l_p satisfy (5.42).

In the special case when the power spectral densities are flat (constant) over the wave numbers of interest and there are no density fluctuations, the mean free paths l_p and l_s that satisfy (5.42) are

$$(5.47) \qquad l_p(|\mathbf{k}|) = \frac{(2\mu + \lambda)^2}{\pi^2 |\mathbf{k}|^4} \frac{1}{2\lambda^2 \hat{R}_{\lambda\lambda} + \frac{8}{3} \lambda\mu \hat{R}_{\lambda\mu} + \frac{8}{5} \mu^2 \hat{R}_{\mu\mu} + \frac{4v_P^5}{15v_S^5} \mu^2 \hat{R}_{\mu\mu}}$$

and

$$(5.48) \qquad l_s(|\mathbf{k}|) = \frac{15\rho v_S}{\pi^2 |\mathbf{k}|^4 \mu^2 \hat{R}_{\mu\mu}} \frac{1}{\frac{8}{v_P(2\mu+\lambda)} + \frac{26v_P^2}{v_S^3 \mu}},$$

with all spectral densities $\hat{R}_{ij}$ constant.

The initial condition for the diffusion equation (5.45) is obtained as in the acoustic and electromagnetic cases, and is

$$(5.49) \qquad \phi_0(\mathbf{x}, |\mathbf{k}|) = \frac{1}{12\pi} \int \mathrm{Tr} W_0^S(\mathbf{x}, \mathbf{k}) d\Omega(\hat{\mathbf{k}}) + \frac{1}{12\pi} \int a_0^P(\mathbf{x}, \mathbf{k}) d\Omega(\hat{\mathbf{k}}).$$

Here W_0^S and a_0^P are the initial values for W^S and a^P.

LECTURE 6
The Geophysical Applications

6.1. Introduction

The propagation of elastic wave energy in the presence of inhomogeneities can be described by radiative transport equations. This description provides a good approximation when (i) typical wavelengths are short compared to the propagation distance, (ii) correlation lengths are comparable to wavelengths so that the inhomogeneities have appreciable effect and (iii) the fluctuations are weak. The relevant transport equations were derived in the previous lecture starting from the elastic wave equations in an unbounded medium. They are a coupled system for the angularly resolved P wave energy density $a^P(t, \mathbf{x}, \mathbf{k})$ and the S wave coherence matrix $W^S(t, \mathbf{x}, \mathbf{k})$. Here $\mathbf{k}$ is the wave vector, t is time and $\mathbf{x}$ is position. They account for the S wave polarization and the P to S and S to P wave energy conversion.

The transport equations explain the dominance of S waves in the deep coda, observed in seismograms and discussed in [1]. The reason for this phenomenon is that in the diffusive regime, over distances long compared to the transport mean free path and over time long compared to the transport mean free time, the P to S energy conversion by the random inhomogeneities equilibrates in a universal way, **independent of the details of the scattering**. There is an equipartition of energy that leads to the relation (5.37)

$$(6.1) \qquad\qquad \mathcal{E}_P(t, \mathbf{x}) = \frac{v_S^3}{2v_P^3} \mathcal{E}_S(t, \mathbf{x}).$$

Here $\mathcal{E}_P$ and $\mathcal{E}_S$ are the P and S spatial energy densities, and v_P and v_S are the P and S wave speeds, respectively. For typical values of the P and S speeds this relation becomes $\mathcal{E}_S \sim 10\mathcal{E}_P$, which is in general agreement with observations.

This equipartition law is derived when P to S mode conversion is generated by volume scattering. It was observed in [29] that the P/Lg ratio stabilizes for elastic waves in the crustal region. This stabilization may be similar to that described in (6.1) and could follow from a radiative transport theory which takes into account the free surface and the crustal waveguide structure.

367

6.2. Radiative transport equations

The theory of radiative transport was originally developed to describe how light energy propagates through a turbulent atmosphere. It is based upon a linear transport equation for the angularly resolved energy density and was first derived phenomenologically at the beginning of this century [**16, 33**]. We show in the previous lectures how this theory can be derived from the governing equations for light and for other waves of any type, in a randomly inhomogeneous medium. Our results take into account nonuniformity of the background medium on the scale large compared to the wavelength, scattering by random inhomogeneities on a scale of the wavelength, the effect of polarization, the coupling of different types of waves, etc. The main new application is to elastic waves, in which shear waves exhibit polarization effects while the compressional waves do not, and the two types of waves are coupled. We also analyze solutions of the transport equations at long times and long distances and show that they have diffusive behavior.

Transport equations arise because a wave with wave vector $\mathbf{k}'$ at a point $\mathbf{x}$ in a randomly inhomogeneous medium may be scattered into any direction $\hat{\mathbf{k}}$ with wave vector $\mathbf{k}$. Therefore one must consider the angularly resolved, wave vector dependent, scalar energy density $a(t, \mathbf{x}, \mathbf{k})$ defined for all $\mathbf{k}$ at each point $\mathbf{x}$ and time t. Energy conservation is expressed by the transport equation

$$\frac{\partial a(t, \mathbf{x}, \mathbf{k})}{\partial t} + \nabla_{\mathbf{k}}\omega(\mathbf{x}, \mathbf{k}) \cdot \nabla_{\mathbf{x}} a(t, \mathbf{x}, \mathbf{k}) - \nabla_{\mathbf{x}}\omega(\mathbf{x}, \mathbf{k}) \cdot \nabla_{\mathbf{k}} a(t, \mathbf{x}, \mathbf{k})$$

$$(6.2) \qquad = \int_{R^3} \sigma(\mathbf{x}, \mathbf{k}, \mathbf{k}') a(t, \mathbf{x}, \mathbf{k}') d\mathbf{k}' - \Sigma(\mathbf{x}, \mathbf{k}) a(t, \mathbf{x}, \mathbf{k}).$$

Here $\omega(\mathbf{x}, \mathbf{k})$ is the frequency at $\mathbf{x}$ of the wave with wave vector $\mathbf{k}$, $\sigma(\mathbf{x}, \mathbf{k}, \mathbf{k}')$ is the differential scattering cross-section, the rate at which energy with wave vector $\mathbf{k}'$ is converted to wave energy with wave vector $\mathbf{k}$ at position $\mathbf{x}$, and

$$(6.3) \qquad \Sigma(\mathbf{x}, \mathbf{k}) = \int \sigma(\mathbf{x}, \mathbf{k}', \mathbf{k}) d\mathbf{k}'$$

is the total scattering cross-section. Both σ and Σ are nonnegative and σ is usually symmetric in $\mathbf{k}$ and $\mathbf{k}'$. For an acoustic wave the differential scattering cross-section is given by

$$\sigma(\mathbf{x}, \mathbf{k}, \mathbf{k}') = \left((\hat{\mathbf{k}} \cdot \hat{\mathbf{k}}')^2 \hat{R}_{\rho\rho}(\mathbf{k} - \mathbf{k}') + 2(\hat{\mathbf{k}} \cdot \hat{\mathbf{k}}')\hat{R}_{\rho\kappa}(\mathbf{k} - \mathbf{k}') + \hat{R}_{\kappa\kappa}(\mathbf{k} - \mathbf{k}') \right)$$

$$(6.4) \qquad \cdot \frac{\pi v^2(\mathbf{x})|\mathbf{k}|^2}{2} \delta(v(\mathbf{x})|\mathbf{k}| - v(\mathbf{x})|\mathbf{k}'|),$$

where $\hat{R}_{\rho\rho}$, $\hat{R}_{\rho\kappa}$ and $\hat{R}_{\kappa\kappa}$ are the power spectra of the fluctuations of the density ρ and compressibility κ defined in Lecture 4. The left side of (6.2) is the total time derivative of $a(t, \mathbf{x}, \mathbf{k})$ at a point moving along a ray in phase space $(\mathbf{x}, \mathbf{k})$, which means that the frequency of the ray is adjusting to the appropriate local value. The right side of (6.2) represents the effects of scattering.

The transport equation (6.2) is conservative because

$$\iint a(t, \mathbf{x}, \mathbf{k}) d\mathbf{x} d\mathbf{k} = \text{const}$$

when the total scattering cross-section is given by (6.3). For simplicity we assumed no intrinsic attenuation in these lecture notes. However, attenuation is easily accounted for by letting the total scattering cross-section be the sum of two terms

$$(6.5) \qquad \Sigma(\mathbf{x}, \mathbf{k}) = \Sigma_{sc}(\mathbf{x}, \mathbf{k}) + \Sigma_{ab}(\mathbf{x}, \mathbf{k})$$

where $\Sigma_{sc}(\mathbf{x}, \mathbf{k})$ is the total cross-section due to scattering and is given by (6.3) and $\Sigma_{ab}(\mathbf{x}, \mathbf{k})$ is the attenuation rate.

The reason that power spectral densities of the inhomogeneities determine the scattering cross-section (6.4) is seen most easily from a Born expansion of the wave equations when the inhomogeneities are weak. This is because the single scattering approximation of (6.2) and the second moments of the single scattering approximation for the underlying wave equations (the Born expansion) must be the same. The latter are determined by the power spectra of the inhomogeneities. In the same manner we can explain the appearance of the delta function in the cross-section (6.4) when the random inhomogeneities do not depend on time and therefore the frequencies are unchanged by the scattering. The transport equation (6.2) arises also when the waves are scattered by discrete scatterers that are randomly distributed in the medium. In this case the scattering cross-section (6.4) is the same as the cross-section of a single scatterer times the density of scatterers. We deal only with continuous random media here.

Equation (6.2) has been derived from equations governing particular wave motions by various authors, such as [**54, 60, 58, 59, 40, 7, 10, 32, 34, 11**] with a recent survey presented in [**8**]. These derivations also determine the functions $\omega(\mathbf{x}, \mathbf{k})$ and $\sigma(\mathbf{x}, \mathbf{k}, \mathbf{k})$ and show how a is related to the wave field. In these notes we derive (6.2) and these functions as a special case of a more general theory.

We expect that radiative transport equations will provide a good description of wave energy transport when, as mentioned in the Introduction, (i) typical wavelengths are short compared to macroscopic features of the medium (high frequency approximation), (ii) correlation lengths of the inhomogeneities are comparable to wavelengths and (iii) the fluctuations of the inhomogeneities are weak. In general, we do not know the correlation lengths of the inhomogeneities. Condition (ii) is, therefore, important because it allows strong interaction between the waves and the inhomogeneities, which is the most interesting and difficult case to analyze. In addition to these three conditions, the inhomogeneities must not be too anisotropic because it is well known that in layered random media, for example, we have wave localization even with weak fluctuations, which is quite different from wave transport phenomena [**4**]. When the fluctuations are strong we can have wave localization even when the inhomogeneities are isotropic [**24, 52**].

In a homogeneous background medium where $\omega = v|\mathbf{k}|$ and v is the uniform speed of the wave and absorption is weak, the solutions of (6.2) exhibit diffusive behavior at times and distances that are long compared to a typical transport mean free time $1/\Sigma$ and a typical transport mean free path $|\nabla_\mathbf{k}\omega|/\Sigma$, respectively. The differential scattering cross-section is always rotationally invariant so that

$$(6.6) \qquad \sigma(\mathbf{k}, \mathbf{k}') = \tilde{\sigma}(|\mathbf{k}|, \hat{\mathbf{k}} \cdot \hat{\mathbf{k}}')\delta(|\mathbf{k}| - |\mathbf{k}'|)$$

which means that scattering from a wave with wave vector $\mathbf{k}'$ into a wave with wave vector $\mathbf{k}$ depends only on the angle between the two wave vectors. In the diffusion regime the phase space energy density $a(t, \mathbf{x}, \mathbf{k})$ is approximately independent of the direction of the wave vector $\mathbf{k}$, $a(t, \mathbf{x}, \mathbf{k}) \sim \bar{a}(t, \mathbf{x}, |\mathbf{k}|)$. In the simplest case of a uniform background medium $\bar{a}$ satisfies the diffusion equation

$$(6.7) \qquad \frac{\partial \bar{a}}{\partial t} = \nabla_{\mathbf{x}} \cdot (D\nabla_{\mathbf{x}}\bar{a}) - \nu\bar{a}.$$

The constant diffusion coefficient is

$$(6.8) \qquad D(|\mathbf{k}|) = \frac{vl^*}{3},$$

where the diffusion mean free path $l^*(|\mathbf{k}|)$ is

$$(6.9) \qquad l^* = v\left(2\pi \int_{-1}^{1} \tilde{\sigma}(|\mathbf{k}|, \mu)(1 - \mu)d\mu\right)^{-1}$$

with the variable of integration μ equal to the cosine of the angle between incident and scattered directions. Note that the diffusion mean free path l^* is in general longer than the transport mean free path l, defined by

$$l = \frac{v}{\Sigma} = v\left(2\pi \int \tilde{\sigma}(|\mathbf{k}|, \mu)d\mu\right)^{-1}.$$

The two are equal in the case of isotropic scattering. In (2.6) the absorption coefficient

$$(6.10) \qquad \nu(|\mathbf{k}|) = \frac{1}{4\pi}\int \Sigma_{ab}(\mathbf{k})d\Omega(\hat{\mathbf{k}})$$

is the attenuation rate $\Sigma_{ab}(\mathbf{k})$ averaged over the angular directions $\hat{\mathbf{k}}$.

The diffusion approximation comes about as follows. The phase space energy density becomes approximately independent of the direction $\hat{\mathbf{k}}$ in the far field regime, or, equivalently, for deep or late coda waves. Mathematically this is because the state independent of $\hat{\mathbf{k}}$ is the only equilibrium state for the scattering operator on the right side of equation (6.2). We integrate equation (6.2) with respect to the directions $\hat{\mathbf{k}}$ and use (6.3) to obtain

$$(6.11) \qquad \frac{\partial \bar{a}}{\partial t} + \nabla_{\mathbf{x}} \cdot \psi = -\nu(|\mathbf{k}|)\bar{a},$$

where ν is given by (2.9),

$$(6.12) \qquad \bar{a}(t, \mathbf{x}, |\mathbf{k}|) = \frac{1}{4\pi}\int a(t, \mathbf{x}, \mathbf{k})d\Omega(\hat{\mathbf{k}})$$

is the average energy density and

$$(6.13) \qquad \psi(t, \mathbf{x}, |\mathbf{k}|) = \frac{v}{4\pi}\int \hat{\mathbf{k}}a(t, \mathbf{x}, \mathbf{k})d\Omega(\hat{\mathbf{k}})$$

is the flux. Next we multiply (6.2) by $\hat{\mathbf{k}}$ and integrate with respect to $\hat{\mathbf{k}}$

$$(6.14) \quad \frac{1}{v}\frac{\partial \psi_i}{\partial t} + \frac{v}{4\pi}\int \hat{k}_i\hat{k}_j\frac{\partial a}{\partial x^j}d\Omega(\hat{\mathbf{k}}) = -\frac{1}{l^*}\psi_i + \frac{1}{4\pi}\int \hat{k}_i\Sigma_{ab}(\mathbf{k})a(\mathbf{k})d\Omega(\hat{\mathbf{k}}).$$

Close to equilibrium a varies slowly in time so we can neglect the term $\partial\psi_i/\partial t$ in (6.14), which is comparable to $\bar{a}_{tt}$. Absorption is weak and so we can also drop the absorption term in (6.14). Note finally that when a is approximately independent of direction $\hat{\mathbf{k}}$, $a \sim \bar{a}$ and the integral on the left of (6.14) is just $(v/3)\nabla_{\mathbf{x}}a$ so (6.14) becomes

$$(6.15) \qquad \psi = -\frac{vl^*}{3}\nabla_{\mathbf{x}}a,$$

where l^* is given by (6.9). We insert (6.15) into (6.11), replace $\bar{a}$ by a and obtain the diffusion equation (6.7) with the diffusion coefficient (6.8). A more systematic derivation of the diffusion approximation for transport equations is given in Lecture 4.

Diffusion approximations for scalar transport equations are very well known [15], including their behavior near boundaries LK,BLP. We show that diffusion approximations are also valid for the more general transport equations that arise for elastic waves.

Before going to elastic waves we note that there is another way to obtain the diffusive behavior. We introduce

$$(6.16) \qquad \mathcal{J}(t,\mathbf{x},|\mathbf{k}|) = \int a(t,\mathbf{x},\mathbf{k})d\Omega(\hat{\mathbf{k}}),$$

the energy density integrated over the angles. When the scattering is isotropic and independent of $\mathbf{x}$ the differential scattering cross-section is given by

$$(6.17) \qquad \sigma(\mathbf{k},\mathbf{k}') = \tilde{\sigma}(|\mathbf{k}|)\delta(|\mathbf{k}| - |\mathbf{k}'|).$$

Consider, for example, the time-independent version of the transport equation (6.2) with a source. Then $\mathcal{J}$ satisfies the time-independent integral equation

$$(6.18) \qquad \mathcal{J}(\mathbf{x},|\mathbf{k}|) = \int_{R^3}[\eta_{sc}\mathcal{J}(\mathbf{x}',|\mathbf{k}|) + \frac{1}{v}f(\mathbf{x}')]\frac{e^{-\eta_{tot}|\mathbf{x}-\mathbf{x}'|}}{4\pi|\mathbf{x}-\mathbf{x}'|^2}d\mathbf{x}',$$

where $f(\mathbf{x})$ is the source energy density function, $\eta_{sc} = \tilde{\sigma}/v$ is the inverse of the scattering mean free path, and $\eta_{tot} = \Sigma/v$, where Σ is the total scattering cross-section (6.5). This equation can be solved explicitly and in the case of a δ-function source the solution in the absence of absorption behaves like $1/r$, which is diffusive behavior. The absorption term Σ_{ab} modifies this solution, making it decay exponentially with a decay rate which vanishes when $\Sigma_{ab} = 0$. This is analogous to the effect of the νa term in the diffusion equation (6.7).

When the differential scattering cross-section is not of the form (6.17) and scattering is not isotropic then instead of a single integral equation (6.18) we get a system of equations for higher angular moments of the intensity $a(\mathbf{x},\mathbf{k})$ [16]. The resulting system is harder to analyze, its explicit solution is complicated and the diffusion approximation is not so transparent. This intermediate step of obtaining equations for integrated quantities is fortunately unnecessary since the solutions of (6.2) converge rapidly to the diffusion approximation for any rotationally invariant differential scattering cross-section of the form (6.6). Therefore in the deep coda regime we need only consider solutions of the diffusion equation.

6.3. Transport theory for elastic waves

Diffusion theory was used in seismology by Wesley [63], Nakamura [46] and others. Dainty and Toksöz [19] derived the diffusion equation for acoustics and suggested the form of the diffusion coefficient for elastic waves. R.S. Wu [66] used extensively the stationary version of the transport equation (6.2), with a source and with isotropic differential scattering cross-section to model propagation of S waves with multiple scattering. Neither the polarization of S waves nor the P to S wave conversion was taken into account. Transport theory was used to effectively separate scattering from absorption attenuation. The diffusive regime for elastic waves was discussed in Wu and Aki [2] where the seismic data for Hindu-Kush region was analyzed and it was found that absorption dominates over scattering in this region. Multiple scattering in the earth's crust was considered in Toksöz et.al. [57] and scalar transport theory was used to separate scattering from absorption effects for Rg waves and other waves. Mayeda et.al. [44] compared data from southern California to the predictions of radiative transport theory for the dependence of the total energy on distance and found very good agreement between the two. McSweeney et.al. [45] found the same results for southcentral Alaska. Fehler et.al. [23] applied multiple lapse-time window analysis to the Kanto-Tokai region and compared the data with predictions from transport theory. Multiple scattering in the time domain was considered by Hoshiba [31] using a Monte Carlo method to model seismic wave propagation in a random medium. The time dependent, scalar transport equation (6.18) was considered by Zeng [68]. Zeng [69] constructed approximate, hybrid single-scattering–diffusive solutions to the time-dependent transport equation and showed that it was a good approximation to the full numerical solution.

In all these papers the vector nature of the underlying elastic wave motion was not taken into consideration. Mode conversion for surface waves was considered in a phenomenological way by Chen and Aki [17] and general mode conversion between longitudinal compressional or P waves and transverse shear or S waves was considered by Sato [49] and by Zeng [70]. However, the transport equations proposed phenomenologically in these papers do not account for polarization of the shear waves. Sato [49] and Zeng [70] incorporated the P to S conversion in the integral equation (6.18) and not in its angularly resolved form (6.2). Starting from the elastic wave equations in a random medium we derive in Lecture 4 a system of transport equations that accounts correctly for P to S mode conversion and for polarization effects.

Longitudinal P waves propagate with local speed

$$v_P(\mathbf{x}) = \sqrt{(2\mu(\mathbf{x}) + \lambda(\mathbf{x}))/\rho(\mathbf{x})}$$

and transverse shear or S waves that can be polarized propagate with local speed

$$v_S(\mathbf{x}) = \sqrt{\mu(\mathbf{x})/\rho(\mathbf{x})}.$$

The corresponding dispersion relations are $\omega_P = v_P|\mathbf{k}|$ and $\omega_S = v_S|\mathbf{k}|$, respectively. The P and S wave modes interact in an inhomogeneous medium because a P wave with a wavenumber $|\mathbf{k}|$, when scattered can generate an S wave with wavenumber $|\mathbf{p}|$ with the same frequency; that is, $v_P(\mathbf{x})|\mathbf{k}| = v_S(\mathbf{x})|\mathbf{p}|$, and vice versa. These scattering processes conserve energy. Therefore the transport equations for P and S waves must be coupled. The transport equation for the P wave energy density should be a scalar equation similar to (6.2) with an additional term that accounts for S to P energy conversion. The S waves can be polarized and are similar in this respect to electromagnetic waves. The phenomenological radiative transport theory for the latter was developed by Chandrasekhar [16]. An important element in that theory is that the transport of energy of polarized waves is described not by a scalar density but by a 2×2 coherence matrix. The coherence matrix has information not only about the total energy of the wave but also about its distribution between the two polarizations and about the cross polarization. This allows for the consideration of polarization effects in the framework of transport theory. In the case of elastic waves, the transport equation for the S wave coherence matrix should be like Chandrasekhar's equation ([16], equation (212)) with an additional term that accounts for P to S energy conversion. We show in these notes that this is in fact the case and we determine explicitly the form of the scattering cross-sections in terms of the power spectral densities of the material inhomogeneities. We show also how the quantities satisfying the transport equations are related to the underlying field properties.

The coupled transport equations for the P wave energy density $a^P(t, \mathbf{x}, \mathbf{k})$ and the 2×2 coherence matrix $W^S(t, \mathbf{x}, \mathbf{k})$ for the S waves have the forms

$$
\frac{\partial a^P}{\partial t} + \nabla_{\mathbf{k}}\omega^P \cdot \nabla_{\mathbf{x}}a^P - \nabla_{\mathbf{x}}\omega^P \cdot \nabla_{\mathbf{k}}a^P
$$

$$
(6.19) \qquad = \int \sigma^{PP}(\mathbf{k}, \mathbf{k}')a^P(\mathbf{k}')d\mathbf{k}' - \Sigma^{PP}(\mathbf{k})a^P(\mathbf{k})
$$

$$
+ \int \sigma^{PS}(\mathbf{k}, \mathbf{k}')[W^S(\mathbf{k}')]d\mathbf{k}' - \Sigma^{PS}(\mathbf{k})a^P(\mathbf{k})
$$

and

$$
\frac{\partial W^S}{\partial t} + \nabla_{\mathbf{k}}\omega^S \cdot \nabla_{\mathbf{x}}W^S - \nabla_{\mathbf{x}}\omega^S \cdot \nabla_{\mathbf{k}}W^S + WN - NW
$$

$$
(6.20) \qquad = \int \sigma^{SS}(\mathbf{k}, \mathbf{k}')[W^S(\mathbf{k}')]d\mathbf{k}' - \Sigma^{SS}(\mathbf{k})W^S(\mathbf{k})
$$

$$
+ \int \sigma^{SP}(\mathbf{k}, \mathbf{k}')[a^P(\mathbf{k}')]d\mathbf{k}' - \Sigma^{SP}(\mathbf{k})W^S(\mathbf{k}).
$$

The differential scattering cross-section $\sigma^{PP}(\mathbf{k}, \mathbf{k}')$ for P to P scattering is similar to (6.4) for scattering of scalar waves and the differential scattering tensor $\sigma^{SS}(\mathbf{k}, \mathbf{k}')$ is similar to Chandrasekhar's tensor ([16]). They have the forms

$$
(6.21) \qquad \sigma^{PP}(\mathbf{k}, \mathbf{k}') = \sigma_{pp}(\mathbf{k}, \mathbf{k}')\delta(v_P|\mathbf{k}| - v_P|\mathbf{k}'|)
$$

and

$$
\sigma^{SS}(\mathbf{k}, \mathbf{k}')[W(\mathbf{k}')] = \{ \ \sigma_{ss}^{TT}T(\mathbf{k}, \mathbf{k}')W(\mathbf{k}')T(\mathbf{k}', \mathbf{k}) + \sigma_{ss}^{\Gamma\Gamma}\Gamma(\mathbf{k}, \mathbf{k}')W(\mathbf{k}')\Gamma(\mathbf{k}', \mathbf{k})
$$

$$
+ \ \sigma_{ss}^{\Gamma T}[T(\mathbf{k}, \mathbf{k}')W(\mathbf{k}')\Gamma(\mathbf{k}', \mathbf{k}) + \Gamma(\mathbf{k}, \mathbf{k}')W(\mathbf{k}')T(\mathbf{k}', \mathbf{k})]\}
$$

$$
(6.22) \qquad \cdot \ \delta(v_S|\mathbf{k}| - v_S|\mathbf{k}'|).
$$

The 2×2 matrices $T(\mathbf{k}, \mathbf{k}')$ and $\Gamma(\mathbf{k}, \mathbf{k}')$ are defined by

$$
(6.23) \qquad T_{ij}(\mathbf{k}, \mathbf{k}') = \mathbf{z}^{(i)}(\mathbf{k}) \cdot \mathbf{z}^{(j)}(\mathbf{k}')
$$

and

$$
(6.24) \quad \Gamma_{ij}(\mathbf{k}, \mathbf{k}') = (\hat{\mathbf{k}} \cdot \hat{\mathbf{k}}')(\mathbf{z}^{(i)}(\mathbf{k}) \cdot \mathbf{z}^{(j)}(\mathbf{k}')) + (\hat{\mathbf{k}} \cdot \mathbf{z}^{(j)}(\mathbf{k}'))(\hat{\mathbf{k}}' \cdot \mathbf{z}^{(i)}(\mathbf{k})).
$$

Here $(\hat{\mathbf{k}}, \mathbf{z}^{(1)}(\mathbf{k}), \mathbf{z}^{(2)}(\mathbf{k}))$ is the orthonormal propagation triple consisting of the direction of propagation $\hat{\mathbf{k}}$ and two transverse unit vectors $\mathbf{z}^{(1)}(\mathbf{k}), \mathbf{z}^{(2)}(\mathbf{k})$, which in polar coordinates are

$$\hat{\mathbf{k}} = \begin{pmatrix} \sin\theta\cos\phi \\ \sin\theta\sin\phi \\ \cos\theta \end{pmatrix}, \ \mathbf{z}^{(1)}(\mathbf{k}) = \begin{pmatrix} \cos\theta\cos\phi \\ \cos\theta\sin\phi \\ -\sin\theta \end{pmatrix}, \ \mathbf{z}^{(2)}(\mathbf{k}) = \begin{pmatrix} -\sin\phi \\ \cos\phi \\ 0 \end{pmatrix}.$$

The scalar functions σ_{pp} and σ_{ss} are given in terms of power spectral densities of the inhomogeneities in Lecture 5. The total scattering cross-sections Σ^{PP} and Σ^{SS} are the integrals of the corresponding differential scattering cross-sections, as in (6.3), since we assume that there is no intrinsic dissipation.

Note that the geometric structure of the matrices T and Γ is built into the transport equations and is independent of the random inhomogeneities. This means that the phenomenological models of isotropic S to S scattering suggested by Zeng [70] and Sato [49] are not valid and polarization of S waves is important.

The coupling matrix N, which arises due to variations of the background on a large scale and vanishes in the case of a uniform medium, is given by

$$(6.25) \qquad N(\mathbf{x}, \mathbf{k}) = \sum_{i=1}^{3} \frac{\partial v(\mathbf{x})}{\partial x^i} |\mathbf{k}| \mathbf{z}^{(1)}(\mathbf{k}) \cdot \frac{\partial \mathbf{z}^{(2)}(\mathbf{k})}{\partial k_i} \begin{pmatrix} 0 & 1 \\ -1 & 0 \end{pmatrix}.$$

The scattering cross-sections for the S to P and P to S coupling terms, σ^{PS} and σ^{SP}, respectively, have the forms

$$(6.26) \quad \sigma^{PS}(\mathbf{k}, \mathbf{k}')[W^S(\mathbf{k}')] = \mathrm{Tr}[\sigma_{ps}(\mathbf{k}, \mathbf{k}')G(\mathbf{k}, \mathbf{k}')W^S(\mathbf{k}')]\delta(v_P|\mathbf{k}| - v_S|\mathbf{k}'|)$$

$$(6.27) \quad \sigma^{SP}(\mathbf{k}, \mathbf{k}')[a^P(\mathbf{k}')] = \sigma_{ps}(\mathbf{k}', \mathbf{k})G(\mathbf{k}', \mathbf{k})a^P(\mathbf{k}')\delta(v_S|\mathbf{k}| - v_P|\mathbf{k}'|)$$

with the 2×2 matrix G given by

$$(6.28) \qquad G_{ij}(\mathbf{k}, \mathbf{k}') = (\hat{\mathbf{k}} \cdot \mathbf{z}^{(i)}(\mathbf{k}'))(\hat{\mathbf{k}} \cdot \mathbf{z}^{(j)}(\mathbf{k}'))$$

and Tr denoting the trace of a matrix. The geometric nature of G shows that S to P scattering is not isotropic, in general, independently of the nature of the inhomogeneities. The scalar function σ_{ps} is given explicitly in terms of power spectral densities of the inhomogeneities in Lecture 4. The delta function in (6.26) and (6.27) appears because P waves with wave number $|\mathbf{k}|$ when scattered generate S waves with wave number $v_P|\mathbf{k}|/v_S$ and vice versa, since the frequency of the waves is not changed by scattering.

The geometrical meaning of the 2×2 matrices T, Γ and G that appear in the differential scattering cross-sections (6.22) and (6.26) and (6.27) is similar to that of T which appears in Chandrasekhar's equations [16]. They arise from a single scattering event of P and S waves with wave vector $\mathbf{k}'$ that scatter to P and S waves with wave vector $\mathbf{k}$ and from the fact that the transport equations deal with quadratic field quantities.

Equations (6.19,6.20) have an important reciprocity property between P to S and S to P scattering. It is expressed by

$$(6.29) \qquad \sigma^{PS}(\mathbf{k}, \mathbf{k}') = \mathrm{Tr}\sigma^{SP}(\mathbf{k}', \mathbf{k}).$$

This property is important because it is responsible for the form of the equilibrium in scattering in the far field regime. That is, it makes the P and S wave energies equilibrate in the ratio (6.1) as we explain below in detail. Aki [1] used reciprocity of single scattering, and the asymptotic forms $1/v_P^2 r$ and $1/v_S^2 r$ of point source P and S wave solutions of the elastic equations in a uniform medium, to show that

$$(6.30) \qquad \frac{g_{PS}}{g_{SP}} = \frac{2v_P^4}{v_S^4}.$$

Here g_{PS} and g_{SP} are the P to S and S to P energy conversion coefficients *for a single scattering*, respectively (the factor of 2 omitted in [1] was corrected by Korneev and Johnson in [39]. The main implication of (6.30) is that S waves will dominate after many scatterings because typically $g_{PS}/g_{SP} \sim 18$. The dominance of S waves is also observed in the seismological data [1]. Zeng [70] phenomenologically incorporated (6.30) (without the factor 2) in the radiative transport theory and considered a system of coupled equations (6.18) with scattering coefficients chosen according to (6.30). He observed the dominance of S wave energy in the numerical solutions but, of course, this is expected since it was put into the equations by adopting (6.30).

The passage from single scattering to the multiple scattering phenomena is not so direct. For example, the far field energy density for single scattering behaves like r^{-2}, but with multiple scattering it behaves like r^{-1}. The analog of (6.30) with multiple scattering is the ratio of the transport mean free paths

$$(6.31) \qquad l_{PS} = \frac{v_P}{\Sigma^{PS}}, \ l_{SP} = \frac{v_S}{\Sigma^{SP}}$$

where the total scattering cross-section Σ^{PS} and Σ^{SP} are given by

$$\Sigma^{PS}(\mathbf{k}) = \int \mathrm{Tr}\sigma^{SP}(\mathbf{k}',\mathbf{k})[I]d\mathbf{k}'$$
$$\Sigma^{SP}(\mathbf{k}) = \int \sigma^{PS}(\mathbf{k}',\mathbf{k})[I]d\mathbf{k}'.$$

We have that

$$(6.32) \qquad \frac{l_{SP}(v_P\mathbf{k}/v_S)}{l_{PS}(\mathbf{k})} = \frac{2v_P^2}{v_S^2},$$

where the wave vectors of the P and S waves are chosen so that they could be scattered into each other. This relation holds in general and is a consequence of the reciprocity relation (6.29), as is (6.30). However, neither (6.31) nor (6.32) implies the equipartition law (6.1). This law holds in the diffusive regime and it is discussed in detail below, (6.37). Note that the diffusive approximation is established there without the intermediate step of deriving a system of integral equations analogous to (6.18). We are unable to derive a closed system of equations like those suggested by Zeng [70] and Sato [49] for the energy densities integrated over angles. The complications that arise are the same as in the case of the scalar transport equation with non-isotropic scattering. The main difference in the case of the elastic transport equations is that because of the polarization of the S waves the S to S and P to S scattering cross-sections are intrinsically non-isotropic. Any phenomenological transport theory similar to the one suggested in [49] and [70] must therefore take into account this anisotropy. This means that the resulting system of integral equations for the angular moments will have many more unknowns than the energies. However, this system of equations is not important for analysis of the deep coda behavior since in this regime the solution of the radiative transport equations (6.19) and (6.20) converges rapidly to the solution of the diffusion equation and so the latter is more relevant than the elastic analog of (6.18).

As in the case of the scalar transport equation (6.2) the elastic transport equations (6.19) and (6.20) simplify considerably in the regime where the diffusion approximation is valid; that is, when the transport mean free path is small compared to the propagation distance. In this regime the solution of the transport equations (6.19) and (6.20) is close to equilibrium. At equilibrium the P wave energy density $a^P(t, \mathbf{x}, \mathbf{k})$ is independent of the direction of the wave vector $\mathbf{k}$, so it depends only upon $|\mathbf{k}|$. Therefore when the field is near equilibrium we can write

$$(6.33) \qquad a^P(t, \mathbf{x}, \mathbf{k}) \sim \phi(t, \mathbf{x}, |\mathbf{k}|).$$

Similarly, at equilibrium the S wave coherence matrix $W^S(t, \mathbf{x}, \mathbf{k})$ is independent of the direction of $\mathbf{k}$ and it is proportional to the identity matrix. Therefore near equilibrium

$$(6.34) \qquad W^S(t, \mathbf{x}, \mathbf{k}) \sim w(t, \mathbf{x}, |\mathbf{k}|)I.$$

Furthermore, at equilibrium the scalar function w in (6.34) is related to the scalar function ϕ in (6.33) by

$$(6.35) \qquad w(t, \mathbf{x}, |\mathbf{k}|) = \phi(t, \mathbf{x}, |\mathbf{k}'|)$$

with the wavenumbers $|\mathbf{k}|$ and $|\mathbf{k}'|$ corresponding to the same frequency

$$(6.36) \qquad v_P|\mathbf{k}'| = v_S|\mathbf{k}|.$$

This expresses the equidistribution of wave energy density between P and S waves with the same frequency.

Integrating (6.35) over $\mathbf{k}$ yields

$$(6.37) \qquad \mathcal{E}_P(t, \mathbf{x}) = \frac{v_S^3}{2v_P^3}\mathcal{E}_S(t, \mathbf{x})$$

where $\mathcal{E}_P$ and $\mathcal{E}_S$ are the P and S wave spatial energy densities. They are related to a^P and W^S by

$$\mathcal{E}_P(t, \mathbf{x}) = \int a^P(t, \mathbf{x}, \mathbf{k})d\mathbf{k}$$

and

$$\mathcal{E}_S(t, \mathbf{x}) = \int \mathrm{Tr}W^S(t, \mathbf{x}, \mathbf{k})d\mathbf{k}.$$

The factor of 2 in (6.37) is due to the polarization of the S waves and shows that the vector nature of the S waves cannot be ignored. Relation (6.37) shows that the S waves dominate in the far field and $\mathcal{E}_S/\mathcal{E}_P \sim 10$. Accidentally this value is close to the value 9 given by the right side of (6.30) without the factor of 2, which is the form given in [1]. However, the mechanism that produces this dominance is through the stabilization or equilibration of P to S mode conversion by multiple scattering. This stabilization, which is derived here from first principles, is reminiscent of the important empirical observation of Hansen et.al. [29] regarding the stabilization of the Lg wave energy. The ratio of S to P energy depends only on the speed of the waves, while the stabilization of the Lg wave energy is presumably due to the fact that the Lg waves are slower than the P waves and, in addition, some analog of (6.1) holds.

To determine the scalar function $\phi(t, \mathbf{x}, |\mathbf{k}|)$ in (3.15a) and (3.16a), we first integrate the transport equations (6.19) and the trace of (6.20) with respect to the direction $\hat{\mathbf{k}}$. Then we add the equations together. The right side of the resulting equation vanishes because energy is conserved by the totality of the scattering processes. Next we multiply the transport equations by $\hat{\mathbf{k}}$, integrate over angles and add them. As in the case of the scalar transport equation (equations (6.11) and (6.14)), after neglecting second order time derivatives we find that ϕ satisfies the diffusion equation (6.7). A systematic derivation of the diffusion approximation for elastic waves is given in Lecture 5. The diffusion coefficient $D(|\mathbf{k}|)$ is given by

$$(6.38) \qquad D(|\mathbf{k}|) = \frac{1}{\frac{2}{v_S^3} + \frac{1}{v_P^3}} \left(\frac{l_p^* v_P}{3v_P^3} + \frac{2l_s^* v_S}{3v_S^3} \right).$$

It can be interpreted as the weighted mean of the "individual" diffusion coefficients of P and S waves. The weights are in the ratio $v_S^3/2v_P^3$ as in (6.1). Here l_p^* and l_s^* are diffusion mean free paths for P and S waves determined in the previous chapter. In general the diffusion mean free paths are longer than the transport mean free paths.

The general form of the diffusion coefficient (6.38) for elastic waves was predicted by Dainty and Toksöz (1977) from physical considerations. Specifically, they correctly argued that D should be a weighted mean of the P and S wave diffusion coefficients. They suggested that the weights should be equal to the roughly one-to-ten observed far field P to S energy ratio as is the case from (6.38). This is based on the equipartition law (6.35), which, along with the one-to-ten ratio of energies (6.37), does not depend on the details of the scattering. Dainty and Toksöz [19] (pp.379-380) had expected that it would depend on the details. After our work was completed we became aware of Weaver's papers ([61, 62] where the equipartition law (6.1) and transport theory for elastic waves are derived by a different method.

6.4. Summary

The main results of ([48]) presented here with seismological applications, are (i) the derivation from first principles of the correct radiative transport equations for elastic wave motion in unbounded media, (ii) the demonstration that polarization of shear waves is important and must be taken into consideration and (iii) the demonstration that in the diffusive regime there is a universal P to S wave energy stabilization. This energy equipartition phenomenon, although intuitively clear, was not known before and its precise form (1.1) is not easy to guess. It is perhaps the simplest instance of many different energy equipartition laws that are valid in other complex situations, such as the ones encountered in crustal wave propagation, that have not been discovered yet.

We thank M. Campillo, L.Johnson, H. Sato and R.S Wu for many detailed discussions on the use of transport theory in seismology.

References

1. K.Aki, Scattering conversions P to S versus S to P Bull. Seism. Soc. Am., **82**, 1969-1972, 1992.
2. K.Aki and R.S.Wu, Scattering and Attenuation of Seismic Waves Parts I, II, III, Birkhäuser, Boston, 1988.
3. E.Akkermans, P.E.Wolf, R.Maynard and G.Maret, Theoretical study of the coherent backscattering of light by disordered media, J. Phys. France, **49**, 1988, 77.
4. M.Asch, W.Kohler, G.Papanicolaou, M.Postel and B.White, Frequency content of randomly scattered signals, SIAM Review, **33**, 1991, 519-625.
5. F.Bailly, J.F.Clouet and J.P.Fouque: Parabolic and white noise approximation for waves in random media, SIAM Journal on Applied Math, **56**, No. 5, 1996, 1445-1470.
6. F.Bailly, PhD Thesis, Orsay, 1996.
7. Yu.Barabanenkov, A.Vinogradov, Yu.Kravtsov and V.Tatarskii, Application of the theory of multiple scattering of waves to the derivation of the radiative transfer equation for a statistically inhomogeneous medium, Radiofizika, **15** 1972, 1852-1860. English translation pp. 1420-1425.
8. Yu.Barabanenkov, Yu.Kravtsov, V.Ozrin and A.Saichev, Enhanced backscattering in optics, Progress in Optics **29**, 1991, 67-190.
9. A.Bensoussan, J.L.Lions and G.Papanicolaou, Boundary Layers and homogenization of transport processes, Publ. RIMS, **15**, 1979, 53-157.
10. I.M.Besieris and F.D.Tappert, Propagation of frequency modulated pulses in a randomly stratified plasma, Jour. Math. Phys. **14**, 1973, 704-707.
11. I.M.Besieris, W.Kohler and H.Freese, A transport-theoretic analysis of pulse propagation through ocean sediments, J. Acoust. Soc. Am., **72**, 1982, 937-946.
12. M.Born and E.Wolf *Principles of optics*, Pergamon Press, Oxford, 1986.
13. R.Burridge *Some Mathematical topics in seismology*, Courant Inst. of Math. Sciences, New York, 1976.
14. R.Burridge and G.Papanicolaou, Transport equations for Stokes' parameters from Maxwell's equations in a random medium, Jour. Math. Phys., **16**, 1975, 2074-2085.
15. K.Case and P.Zweifel *Linear transport theory*, Addison-Wesley Pub. Co, 1967.
16. S.Chandrasekhar, *Radiative transfer*, Dover, New York, 1960.

17. X.Chen and K.Aki, Energy transfer theory of seismic surface waves in a random scattering and absorption in half space-medium, Proc. of 15th annual seismic research symposium, 1993, 58-64.

18. R.Courant and D.Hilbert, *Methods of mathematical physics*, vol. II, Wiley Publ. Co., 1962.

19. A.M.Dainty and M.N.Toksöz, Elastic wave propagation in a highly scattering medium, Journal of Geophysics **43**, 1977, 375-388.

20. G.Dell'Antonio, Large time small coupling behavior of a quantum particle in a random potential, Annals of Inst. H. Poincare, section A, **39**, 1983, 339-384.

21. A.Fannjiang, unpublished notes, 1991.

22. A.Fannjiang and T.Komorowski, Turbulent diffusion in Markovian flows, submitted to Annals of Applied Probablity, 1997.

23. M.Fehler, M.Hoshiba, H.Sato and K.Obara, Separation of scattering and intrinsic attenuation for the Kanto-Tokai region, Japan, Geophys. J. Int., **108**, 1992, 787-800.

24. J.Froelich and T.Spencer, Absence of diffusion in the Anderson tight binding model for large disorder or low energy, Comm. Math. Phys. **88**, 1983, 151-184.

25. P.Garabedian, Partial Differential Equations, Wiley, NY, 1965.

26. P.Gérard, Microlocal defect measures, Comm. PDEs, **16**, 1991, 1761-1794.

27. P.Gérard, P.Markovich, N.Mauser and F.Poupaud, Homogenization limits and Wigner transforms, Comm.Pure Appl. Math., **50**, 1997, 323-380.

28. I.Gihman and A.Skorohod, *The Theory of Stochastic Processes*, vol. 1, Springer-Verlag, New York, 1974.

29. R.A.Hansen, F.Ringdal and P.Richards, The stability of RMS Lg measurements and their potential for accurate estimation of the yields of Soviet underground nuclear explosions, Bull. Seism. Soc. Am., **80**, 1990, 2106-2126.

30. T.G.Ho, L.G.Landau and A.J.Wilkins, On the weak coupling limit for a Fermi gas in a random potential, Rev. Math. Phys., **5**, 1993, 209-298.

31. M.Hoshiba, Simulation of multiple scattered coda wave excitation adopting energy conservation law, Phys. Earth Planet Inter. **67**, 1991, 123-126.

32. M.S.Howe, On the kinetic theory of wave propagation in random media, Phil. Trans. Roy. Soc. Lond. **274**, 1973,523-549.

33. H.C.van de Hulst, *Multiple Light Scattering*, Volumes 1 and 2, Academic Press, NY, 1980.

34. A.Ishimaru, *Wave propagation and scattering in random media*, vol. II, Academic Press, New York, 1978.

35. S.Karlin *Total positivity*, Stanford University Press, 1968.

36. J.B.Keller, The geometrical theory of diffraction, Jour. Opt. Soc. Am., **12**, 1962, 116-132.

37. J.B.Keller and R.Lewis, Asymptotic methods for partial differential equations: The reduced wave equation and Maxwell's equations, in Surveys in applied mathematics, eds. J.B.Keller, D.McLaughlin and G Papanicolaou, Plenum Press, New York, 1995.

38. W.Kohler, G.Papanicolaou and B.White, Localization and mode conversion for elastic waves in randomly layered media I-II, Wave Motion, **23**, 1-22, and 181-201, 1996.

39. V.Korneev and L.Johnson, Elastic scattering by a spherical inclusion III, Preprint, 1993.

40. C.W.Law and K.Watson, Radiation transport along curved ray paths, Jour. Math. Phys., **11**, 1970, 3125-3137.

41. E.Larsen and J.B.Keller, Asymptotic solution of neutron transport problems for small mean free paths, J. Math. Phys., **15**, 1974, 75-81.

42. R.Lewis, Geometrical optics and polarization, I.E.E.E. Trans. on Antennas and Propagation AP-14, 1966, 100-101.

43. P.Martin and G.Emch, A rigorous model sustaining van Hove's phenomenon, Helv. Phys. Acta, **48**, 1975, 59-.

44. K.Mayeda, F.Su and K.Aki, Seismic albedo from the total energy dependence on hypocentral distance in southern California, Phys. Earth Planet. Int., **67**, 1991, 104-114.

45. T.McSweeney, N.Biswas, K.Mayeda and K.Aki, Scattering and anelastic attenuation of seismic energy in central and southcentral Alaska, Phys. Earth Planet. Int., **67**, 1991, 115-122.

46. Y.Nakamura, Seismic energy transmission in an intensively scattering environment, Journal of Geophysics, **43**, 1977, 389-399.

47. G.Papanicolaou, L.Ryzhik and J.Keller, On the stability of the P to S energy ratio in the diffusive regime, Bulletin of the Seismological Society of America, **86**, 1996, 1107-1115.

48. L.Ryzhik, G.Papanicolaou and J.Keller, Transport equations for elastic and other waves in random media, Wave Motion, **24**, 1996, 327-370.

49. H.Sato, Multiple isotropic scattering model including P-S conversions for the seismogram envelope formation, Geophys. J. Int., **117**, 1994, 487-494.

50. L.Schwartz, Théorie des Distributions, Herman, Paris, 1966.

51. T.L.Shang and L.S.Gao, Transportation theory of multiple scattering and its application to seismic coda waves of impulse source, Scientia Sinica, Ser. B. **31**, 1988, 1503-1514.

52. P.Sheng, *Introduction to wave scattering, localization, and mesoscopic phenomena*, Academic Press, San Diego, 1995.

53. H.Spohn, Derivation of the transport equation for electrons moving through random impurities, J. Stat. Phys., **17**, 1977, 385-412.

54. P.Stott, A transport equation for the multiple scattering of electromagnetic waves by a turbulent plasma, Jour. Phys. A, **1**, 1968, 675-689.

55. L.Tartar, H-measures, a new approach for studying homogenization, oscillations and concentration effects in partial differential equations, Proc. Roy. Soc. Edinburgh, **115A**, 1990, 193-230.

56. B.A.van Tiggelen and A.Langendijk, Rigorous treatment of the speed of diffusing classical waves, Europhys. Letters, **23**, 1993, 311.

57. M.N.Toksöz, A.Dainty, E.Reiter and R.S.Wu, A model for attenuation and scattering in earth's crust, PAGEOPH, 1988 **128**, 81-100.

58. K.Watson, Multiple scattering of electromagnetic waves in an underdense plasma, Jour. Math. Phys., **10**, 1969, 688-702.

59. K.Watson, Electromagnetic wave scattering within a plasma in the transport approximation, Physics of Fluids, **13**, 1970, 2514-2523.

60. K.Watson and J.L.Peacher, Doppler shift in frequency in the transport of electromagnetic waves in an underdense plasma, Jour. Math. Phys11, 1970, 1496-1504.

61. R.Weaver, On diffuse waves in solid media, J. Acoust. Soc. Am., **71**, 1982, 1608-1609.

62. R.Weaver, Diffusivity of ultrasound in polycrystals, J. Mech. and Phys. of Solids, **38**, 1990, 55-86.

63. J.P.Wesley, Diffusion of seismic energy in the near range, Journal of Geophysical Research **70**, 1965, 5099-5106.

64. B.White, P.Sheng and B.Nair, Localization and backscattering spectrum in stratified lithology, Geophysics, **55**, 1990, 1158-1165

65. E.Wigner, On the quantum correction for thermodynamic equilibrium, Physical Rev., **40**, 1932, 749-759.

66. R.S.Wu, Multiple scattering and energy transfer of seismic waves–separation of scattering effect from intrinsic attenuation – I. Theoretical modeling, Geophys. J. R. Astr. Soc., **82**, 1985, 57-80.

67. R.S.Wu and K.Aki, Multiple scattering and energy transfer of seismic waves – separation of scattering effect from intrinsic attenuation –II. Application of the theory to Hindu Kush region, Seismic wave scattering and attenuation, vol. I, eds. R.S.Wu and K.Aki,1988, 49-80.

68. Y.Zeng, F.Su and K.Aki, Scattering wave energy propagation in a medium with randomly distributed isotropic scatterers, Jour. Geophys. R., **96**, 1991, 607-619.

69. Y.Zeng, Compact solutions of multiple scattering wave energy in the time domain, Bull. Seism. Soc. Am. **81**, 1991 1022-1029.

70. Y.Zeng, Theory of scattered P-wave and S- wave energy in a random isotropic scattering medium, Bulletin of Seism. Soc. Amer., **83**, 1993, 1264-1276.

Lectures on Geometric Optics

Jeffrey Rauch
with the assistance of Markus Keel

IAS/Park City Mathematics Series
Volume 5, 1999

Geometric Optics

Jeffrey Rauch
with the assistance of Markus Keel

LECTURE 1
Introduction

These notes represent about twice as much material as presented in a series of eight lectures given at the Institute for Advanced Study Park City Mathematics Institute in July 1995. The lectures were forty five minutes long followed by what were very interesting question sessions lasting from fifteen to thirty minutes. I have made an attempt to preserve some of the informality of the lectures in the notes. This is in part achieved by often describing the main ideas of an argument or special cases. The details or general cases are left as exercises. This also preserves the format of a minicourse. I have included many of the topics raised in the discussion sessions and hope that an occasional reader will enjoy seeing the discussion of their questions. My research in this area is joint work with Jean-Luc Joly and Guy Métivier from the Universitées de Bordeaux and Rennes respectively. It has been a wonderful collaboration and I gratefully acknowledge their contribution to all that might be good in these notes.

The lectures and the notes are aimed at the level of graduate students who have studied one hard course in partial differential equations. However, the reader is not assumed to have any familiarity with geometric optics. As a result about two thirds of the notes are devoted to developing background about hyperbolic equations which prepares the way for nonlinear geometric optics. In particular, linear geometric optics is presented first.

The subject of geometric optics begins with the earliest understanding of the propagation of light. Simple observation of sun beams streaming through a partial break in clouds, or a flashlight beam in a dusty room gives the impression that light travels in straight lines. At mirrors the lines reflect with the usual law of equal angles of incidence and reflection. Passing from air to water the lines are bent.

---

[1] Department of Mathematics, University of Michigan, Ann Arbor, MI 48109

[2] Department of Mathematics, University of California at Los Angeles, Los Angeles, CA 90024

All three phenomena are explained by Fermat's Principal of Least Time. The rays are locally paths of least time. Refraction at an interface is explained by positing that light travels at different speeds in the two media. This description is purely geometrical involving only broken rays and times of transit. The appearance of a minimum principal had important philosophical impact, since it was consistent with a world view holding that nature acts in a best possible way. Fermat's principal was enunciated twenty years before Römer demonstrated the finiteness of the speed of light.

Today light is understood as an electromagnetic phenomenon, so is described by the time evolution of electromagnetic fields which are solutions of a system of partial differential equations. When quantum effects are important, this theory must be quantized. A mathematically solid foundation for the quantization of the electromagnetic field in 1+3 dimensional space time has not been found.

The reason that a field theory involving partial differential equations can be replaced by a geometric theory involving rays and speeds is that visible light has very short wavelength compared to the size of human sensory organs and common physical objects. Thus, much observational data involving light occurs in an asymptotic regime of very short wavelength. Happily, the short wavelength asymptotic study of systems of partial differential equations often involves significant simplifications. In particular there are good descriptions involving rays. We will use the phrase *geometric optics* to be synonymous with the study of the short wavelength asymptotic analysis of solutions of systems of partial differential equations.

In optical phenomena, not only is the wavelength short but the wave trains are long. The study of structures which have short wavelength and are in addition very short, say a short pulse, also yields a geometric theory. Long wavetrains have a longer time to allow nonlinear interactions which makes nonlinear effects more important. Long propagation distances also increase the importance of nonlinear effects. An extreme example is the propagation of light across the ocean in optical fibers. The nonlinear effects are very weak, but over 5000 kilometers, the cumulative effects are enormous. To control signal degradation in such fibers the signal is treated about every 30 kilometers. Still, there is free propagation for 30 kilometers which needs to be understood. This poses serious analytic and numerical challenges.

A second way to bring nonlinear effects to the fore is to increase the amplitude of disturbances. It was only with the advent of the laser that sufficiently intense optical fields were produced so that nonlinear effects were routinely observed. The conclusion is that for nonlinearity to be important, either the fields or the propagation distances must be large. For the latter, dissipative losses must be small.

The ray description as a simplification of the Maxwell equations is analogous to the fact that classical mechanics gives a good approximation to solutions of the Schrödinger equation of quantum mechanics. The associated method is called the quasiclassical approximation. The role of rays in optics is played by the paths of classical mechanics. There is an important difference in the two cases. The Schrödinger equation has a small parameter, Planck's constant, which is a natural constant. The quasiclassical approximation is an approximation valid for small Plank's constant. The mathematical theory involves the limit as this constant tends to zero. Maxwell's equations apparently have a small parameter too, the inverse of the speed of light. One might guess that rays occur in a theory where this speed tends to infinity. This is not the case. The upshot is that for Maxwell's equations in vacuum the small parameter which is the wavelength is introduced via the initial data, it is not in the equation. For Schrödinger's equation, the small parameter is already in the equation and the semiclassical limit involves the study of initial data tuned to this wavelength.

It is important to recognize that short wavelength phenomena cannot simply be studied by numerical simulations. If one were to discretize a cubic meter of space with mesh size 10^{-5} cm. so as to have five mesh points per wavelength, there would be 10^{21} data points in each time slice. Since this is nearly as large as the number of atoms per cubic centimeter, there is no chance for the memory of a computer to be sufficient to store enough data, let alone make calculations. Such brute force approaches are doomed to fail. A much more intelligent approach would be to use radical local mesh refinement so that the fine mesh was used only when needed. Still this falls far outside the bounds of present computing power. Happily the asymptotic analysis offers an alternative approach which is not only powerful but is mathematically elegant.

There are several very important omissions from these notes. There is no discussion of modeling and of practical applications. There was no time to treat special cases so as to see the interesting qualitative information which can flow from the expansions of geometric optics. There is no discussion of the phenomena of resonance and focusing which are two of the the most interesting parts of the subject. The effects of resonance are now well understood and information about the joint effects of focusing and nonlinearity is accumulating, though the hardest questions remain open. There is no discussion of the geometric optics approach to shocks. Here the mathematical theory is still very young. To ease the introduction only semilinear problems have been discussed. All that is presented here has quasilinear analogues whose proofs surmount interesting technical barriers. In these senses the lectures represent a first step aimed at a large and rich subject and I hope that some readers and some auditors are sufficiently attracted to probe further. Suggestions for further reading are scattered throughout the text.

Acknowledgments. The research of J. Rauch was partially supported by the National Science Foundation and the Office of Naval Research under grants NSF-DMS-9203413 and OD-G-N0014-92-J-1245 respectively. The authors would also like to thank the Institute for Mathematics and its Applications in Minneapolis for the hospitality offered to M.Keel during August 1995.

LECTURE 2
Basic Linear Existence Theorems

2.1. Energy estimates for symmetric hyperbolic systems

The two classic examples of hyperbolic equations are the Maxwell's equations of electrodynamics and the equations of inviscid compressible fluid flow. The Maxwell equations for the electric and magnetic field strengths $E(t,x), B(t,x)$ include two dynamic equations

$$E_t = c\,\mathrm{curl}\,B - 4\pi\mathbf{j}\,, \qquad B_t = -c\,\mathrm{curl}\,E\,, \qquad c = 3\times 10^{10}\,\mathrm{cm./sec.} \tag{2.1}$$

The vector field $\mathbf{j}(t,x)$ is the current density measuring the flow of charge. These equations determine E, B from their initial data once $\mathbf{j}$ is known. Not all initial data are physically relevant. The physical solutions are a subset of the dynamics defined by the additional Maxwell equations

$$\mathrm{div}\,E = 4\pi\rho, \quad \text{and} \quad \mathrm{div}\,B = 0\,, \tag{2.2}$$

where $\rho(t,x)$ is the charge density.

Taking the divergence of the first equation in (2.1) and the time derivative of the first equation in (2.2) shows that the *continuity equation,*

$$\partial_t\rho = -\,\mathrm{div}\,\mathbf{j}\,, \tag{2.3}$$

follows from the Maxwell system. This equation expresses the *conservation of charge.*

Taking the divergence of (2.1) and using (2.3) yields

$$\partial_t\,\mathrm{div}\,B = \partial_t\,\mathrm{div}(E - 4\pi\rho) = 0\,.$$

Thus, when the continuity equation is satisfied, the constraint equations (2.2) hold as soon as they are satisfied at time $t = 0$.

The system (2.1) is a symmetric hyperbolic system in the following sense. Introduce the $\mathbb{R}^6$ valued unknown $u := (E, B)$. Then equation (2.1) has the form

$$\frac{\partial u}{\partial t} + \sum_{j=1}^{3} A_j\frac{\partial u}{\partial x_j} = f\,, \qquad f := (\mathbf{j}, 0)\,, \tag{2.4}$$

with constant 6×6 real matrices A_j.

389

Exercise. Compute the matrices A_j. In particular, verify that they are symmetric.

The importance of symmetry is that it leads to simple L^2 and more generally H^s estimates which are often related to physical quantities like energy or entropy. As a simple example consider the Maxwell system without charges and currents. Since the system has constant coefficients it is efficiently analyzed using the Fourier transform in x. Thus

$$u(x) = (2\pi)^{-d/2} \int_{\mathbb{R}^d} e^{ix.\xi}\, \hat{u}(\xi)\, d\xi\,,$$

where

$$\hat{u}(\xi) := (2\pi)^{-d/2} \int_{\mathbb{R}^d} e^{-ix.\xi}\, u(\xi)\, d\xi\,.$$

Taking the Fourier transform in x of the dynamic Maxwell equations yields

$$\partial_t \hat{u}(t,\xi) - \sum i\, A_j\, \xi_j\, \hat{u}(t,\xi) = 0\,.$$

Integrating yields

$$\hat{u}(t,\xi) = e^{it \sum A_j \xi_j}\, \hat{u}(0,\xi)\,.$$

The symmetry implies that $\exp(it \sum A_j \xi_j)$ is a unitary matrix-valued function of t, ξ. Thus for all t, ξ

$$\|\hat{u}(t,\xi)\|^2 = \|\hat{u}(0,\xi)\|^2\,.$$

Written out this asserts that

$$|\hat{E}(t,\xi)|^2 + |\hat{B}(t,\xi)|^2 = \ \text{independent of time}$$

which expresses the conservation of energy at every frequency. Integrating $d\xi$ and using the Plancherel Theorem implies that the L^2 norm is conserved, that is for all t

$$\|u(t)\|_{L^2(\mathbb{R}^d)} = \|u(0)\|_{L^2(\mathbb{R}^d)}\,.$$

For the fields this asserts that

$$\int_{\mathbb{R}^d} |E|^2 + |B|^2\, dx = \ \text{independent of time}$$

which is the physical law of *conservation of energy*. More generally, the Sobolev H^s norms defined for $s \in \mathbb{R}$ by

$$\|v\|_{H^s(\mathbb{R}^d)}^2 := \int_{\mathbb{R}^d} (1 + |\xi|^2)^s\, |\hat{v}(\xi)|^2\, d\xi$$

are conserved. These norms are without physical interpretation but are crucially important in the mathematical analysis. Similar estimates and corresponding global existence theorems are valid for variable coefficient operators (for example, Maxwell's equations in a nonhomogeneous dielectric) satisfying a simple symmetry criterion. The introduction of this class of operators and the observation that it is ubiquitous in mathematical physics is due to K.O. Friedrichs.

Definition. In $\mathbb{R}^{1+d}$ introduce coordinates $y = y_0, y_1, \cdots, y_d := t, x_1, \cdots, x_d$. A partial differential operator

$$L(y, \partial) = \sum_{\mu=0}^{d} A_\mu(y) \, \frac{\partial}{\partial y_\mu} + B(y) \tag{2.5}$$

is called *symmetric hyperbolic* if the coefficient matrices A_μ are smooth hermitian symmetric valued functions on $\mathbb{R}^{1+d}$ such that for all α and μ

$$\sup_y \|\partial_y^\alpha A_\mu(y)\| < \infty, \tag{2.6}$$

and, there is a $c > 0$ so that for all y,

$$A_0(y) \geq c\, I \tag{2.7}$$

The first key observation is that the basic L^2 estimate proved above using the Fourier transform has a generalization to such variable coefficient problems. The basic idea is functional analytic. Suppose that $G(t)$ is for each t a linear operator on a Hilbert space such that $G + G^*$ is bounded with a bound independent of t, that is

$$\| G(t) + G(t)^* \| \leq 2C. \tag{2.8}$$

In this sense the operator G is nearly antiselfadjoint. If $u(t)$ satisfies

$$u'(t) - G(t)\, u(t) = f(t) \tag{2.9}$$

then reasoning formally yields the estimate

$$\frac{d}{dt} \|u(t)\|^2 = (u, u') + (u', u) = (u, Gu) + (Gu, u) + 2\Re(u, f). \tag{2.10}$$

The Cauchy-Schwartz inequality shows that

$$2\Re(u, f) \leq 2 \|u(t)\| \, \|f(t)\|, \tag{2.11}$$

and the near antisymmetry implies that

$$(u, Gu) + (Gu, u) = \big(u, (G + G^*)u\big) \leq C\|u\|^2. \tag{2.12}$$

Where $u(t) \neq 0$, dividing (2.10) by the norm of u yields

$$\frac{d \, \|u(t)\|}{dt} \leq C\|u(t)\| + \|f(t)\|. \tag{2.13}$$

Thus, if u does not vanish on $[0, t]$, integrating (2.13) yields

$$\|u(t)\| \leq e^{Ct} \|u(0)\| + \int_0^t e^{C(t-\sigma)} \|f(\sigma)\| \, d\sigma. \tag{2.14}$$

If $u(t) \neq 0$ but u vanishes for some earlier times, let $t_1 \in [0, t]$ be the largest value of t such that $u(t)$ vanishes. Integrating (2.13) from $t_1 + \epsilon$ to t yields

$$\|u(t)\| \leq e^{C(t-t_1+\epsilon)} \|u(t_1 + \epsilon)\| + \int_{t_1+\epsilon}^t e^{C(t-s)} \|f(s)\| \, ds. \tag{2.15}$$

Letting $\epsilon \to 0$ shows that (2.14) is valid in this case too. In fact a stronger estimate without the first term in (2.14) is valid in this case.

Finally, if $u(t) \equiv 0$ then (2.14) is trivially true. Thus, in all cases (2.14) holds.

Equation (2.10) shows that the estimate (2.14) is proved by taking the real part of the scalar product (u, Lu). This argument generalizes to the case of an equation of the form

$$A_0(t)\frac{du}{dt} + G\,u = f\,, \tag{2.16}$$

where A_0 is strictly positive with $\|dA_0/dt\| \leq C'$. The starting point is either

$$\frac{d}{dt}\left(u(t), A_0(t)\,u(t)\right), \tag{2.17}$$

or equivalently

$$0 = \Re\left(u\,, A_0(t)\frac{du}{dt} - G\,u - f\right).$$

One finds that

$$\|u(t)\| \leq C\,e^{Ct}\left(\|u(0)\| + \int_0^t e^{-C\sigma}\,\|f(\sigma)\|\,d\sigma\right). \tag{2.18}$$

Exercise. Carry out the two formal derivations of the estimate (2.14) in this case. What is the change in the definition of C that is required?

The application to symmetric hyperbolic systems is then immediate. The operator

$$G(t) = \sum_{1 \leq j \leq d} A_j(y)\,\partial_j + B(y)$$

has adjoint differential operator given by

$$G^* = -\sum_{1 \leq j \leq d} A_j\partial_j - \sum(\partial_j A_j) + B^*\,.$$

This operator is defined by the identity

$$(G(t)\,\phi, \psi)_{\mathbb{R}^d} = (\phi, G^*(t)\psi)_{\mathbb{R}^d} \tag{2.19}$$

valid for ϕ and ψ belonging to $C_0^\infty(\mathbb{R}^d)$. In particular

$$G(t) + G(t)^* = B(y) + B^*(y) - \sum_{j=1}^{3}(\partial_j A_j(y))$$

is multiplication by a uniformly bounded matrix. In this sense, G is nearly antisymmetric.

Proposition 2.1. *For every $s \in \mathbb{R}$ there is a constant C so that for all smooth u on space time such that* $\operatorname{supp} u \cap ([0, t] \times \mathbb{R}^d)$ *is compact for all t,*

$$\|u(t)\|_{H^s(\mathbb{R}^d)} \leq C\,e^{Ct}\|u(0)\|_{H^s(\mathbb{R}^d)} + \int_0^t C\,e^{C(t-\sigma)}\|(Lu)(\sigma)\|_{H^s(\mathbb{R}^d)}\,d\sigma\,. \tag{2.20}$$

Proof. The proof for integer $s \geq 0$ is as follows. For any $\alpha \in \mathbb{N}^d$ with $|\alpha| \leq s$, the basic L^2 estimate (2.18) implies that

$$\|\partial_x^\alpha u(t)\| \leq C\, e^{Ct}\, \|\partial_x^\alpha u(0)\| + \int_0^t C\, e^{C(t-\sigma)}\, \|L\, \partial_x^\alpha u\, (\sigma)\|\, d\sigma\,. \tag{2.21}$$

Using the product rule for differentiation expresses

$$L\, \partial_x^\alpha u = \partial_x^\alpha L\, u + \sum_{|\beta| \leq s} C_{\alpha,\beta}(y)\, \partial_x^\beta u$$

with smooth bounded matrix valued functions $C_{\alpha,\beta}$. The estimate follows upon summing (2.21) over all $|\alpha| \leq s$ and then applying Gronwall's inequality. $\qquad\square$

Exercise. Fill in the details.

2.2. Existence theorems for symmetric hyperbolic systems

As is the case with many good estimates, the corresponding existence theorem lingers not far behind.

Theorem 2.2 [Friedrichs]. *If* $g \in H^s(\mathbb{R}^d)$ *and* $f \in L^1_{\mathrm{loc}}(\mathbb{R}; H^s(\mathbb{R}^d))$ *for some* $s \in \mathbb{R}$, *then there is one and only one solution* $u \in C(\mathbb{R}; H^s(\mathbb{R}^d))$ *to the initial value problem*

$$Lu = f\,, \qquad u|_{t=0} = g\,. \tag{2.22}$$

In addition, there is a constant $C = C(L,s)$ *independent of* f, g *so that for all* $t > 0$

$$\|u(t)\|_{H^s(\mathbb{R}^d)} \leq C\, e^{Ct}\|u(0)\|_{H^s(\mathbb{R}^d)} + \int_0^t C\, e^{C(t-\sigma)}\|f(\sigma)\|_{H^s(\mathbb{R}^d)}\, d\sigma\,, \tag{2.23}$$

with a similar estimate for $t < 0$.

Remark. If $u \in L^2_{\mathrm{loc}}(\mathbb{R}; H^s(\mathbb{R}^d))$ satisfies (22), then $\partial u/\partial t \in L^1_{\mathrm{loc}}(\mathbb{R}; H^{s-1}(\mathbb{R}^d))$. Therefore u is continuous on $\mathbb{R}$ with values in $H^{s-1}(\mathbb{R}^d)$ so the initial condition in (2.22) makes sense. This remark has nothing to do with hyperbolicity. The improved continuity in the second part of the next theorem uses hyperbolicity in an essential way.

The solution u is constructed as the limit of approximate solutions u^h. The u^h are solutions of a differential-difference equation obtained by replacing x derivatives by centered difference quotients. As a warm up consider the simple initial value problem

$$\partial_t u + \partial_x u = 0\,, \qquad u(0,x) = g(x)\,,$$

with $x \in \mathbb{R}^1$. Define the centered difference operator by

$$\delta^h \phi(x) = \frac{\phi(x+h) - \phi(x-h)}{2h}\,.$$

Approximate solutions are defined as solutions of the evolution equations with bounded generators

$$\partial_t u^h + \delta^h u^h = 0\,, \qquad u^h(0,x) = g(x)\,.$$

Note that as operators on H^s, the norms of the generators diverge to infinity as $h \to 0$. This corresponds to the fact that the difference operators δ^h converge to the unbounded operator ∂_x.

Exercise. Use the Fourier Transform to show that for any $s \in \mathbb{R}$ and $g \in H^s(\mathbb{R})$ this recipe determines a sequence of approximate solutions which as $h \to 0$, converge in $C(]-\infty, \infty[\, ; H^s(\mathbb{R}))$ to the exact solution.

One does not have similar good behavior for the finite difference approximations

$$\partial_t u^h + i\, \delta^h u^h = 0\,, \qquad u^h(0, x) = g(x)\,,$$

to the nonhyperbolic initial value problem

$$\partial_t u + i\, \partial_x u = 0\,, \qquad u(0, x) = g(x)\,.$$

For this initial value problem and generic g there is nonexistence (see chapter 3 of my book).

Exercise. For the approximations to the nonhyperbolic initial value problem prove that

$$\liminf_{h \to 0} \|u^h(t)\|^2_{L^2(\mathbb{R})} \geq \int_{\mathbb{R}} e^{2t\xi}\, |\hat{g}(\xi)|^2\, d\xi\,.$$

In particular, if the right hand side is infinite, $u^h(t)$ does not converge in $L^2(\mathbb{R})$ as h tends to zero. The right hand side is infinite for generic $g \in C_0^\infty(\mathbb{R})$. In the same way, prove that for generic smooth g, $u^h(t)$ is unbounded in $H^s(\mathbb{R})$ for all $t \neq 0$ and $s < 0$.

Exercise. Prove Friedrich's Theorem. Here is an outline. Define approximate solutions u^h as solutions of

$$L^h u^h = f\,, \qquad u^h(0) = g\,,$$

where for $h > 0$, L^h comes from L upon replacing the unbounded antiselfadjoint operators ∂_j with $j \geq 1$ by the bounded antiselfadjoint finite difference operators

$$\delta_j^h \phi := \frac{\phi(x_1, \cdots, x_j + h, \cdots, x_d) - \phi(x_1, \cdots, x_j - h, \cdots x_d)}{2h}\,.$$

Show that estimates like (2.18) hold with constants independent of h.

Pass to the limit to find a solution in $L_{\mathrm{loc}}^\infty(\mathbb{R}\,; L^2(\mathbb{R}^d))$ which in addition satisfies an estimate of type (2.18). (Note that we have not yet established continuity in time.)

Perform a similar argument to prove the case of $s \in \mathbb{N}$.

For $s = 1$ the solution is continuous with values in L^2. Prove continuity in time with values in L^2 for the case $s = 0$ by approximating by solutions with H^1 data. Prove continuity in time for general s by a similar argument.

Prove uniqueness on $[0, T] \times \mathbb{R}$ by a duality argument like that in the proof of the Holmgren Uniqueness Theorem using $C^1(\mathbb{R}\,; H^{|s|}(\mathbb{R}^d))$ solutions of the adjoint equation

$$L'v = \psi \in L_{\mathrm{loc}}^1(\mathbb{R}\,; H^{|s|+1}(\mathbb{R}^d))\,, \qquad v|_{t=T} = 0\,.$$

The existence of such solutions has just been demonstrated.

Discussion. The idea of using difference approximations to prove existence goes back to the work of Peano on ordinary differential equations. In the context of partial differential equations, note the seminal paper of Courant Friedrichs and Lewy in 1928.

Theorem 2.3. *If in addition, $\partial_t^k f \in L_{\mathrm{loc}}^1(\mathbb{R}\,;\,H^{s-k}(\mathbb{R}^d))$ for $k = 0, 1, \cdots, N$, then*

$$u \in C^k(\mathbb{R}\,;\,H^{s-k}(\mathbb{R}^d) \qquad for \quad k = 0, 1, \cdots, N\,.$$

Furthermore, there is a constant $C = C(L, s, N)$ so that for $0 \leq k \leq N$

$$\|\partial_t^k u(t)\|_{H^{s-k}(\mathbb{R}^d)} \leq C\,e^{Ct}\|u(0)\|_{H^s(\mathbb{R}^d)} + \int_0^t C\,e^{C(t-\sigma)} \sum_{k \leq N} \|\partial_t^k f(\sigma)\|_{H^{s-k}(\mathbb{R}^d)}\,d\sigma\,,$$

$$(2.24)$$

with a similar estimate for $t < 0$.

2.3. Finite speed of propagation

An extremely important aspect of the L^2 estimates which form the basis of the results above is that they can be localized. In particular, the energy method can be used to prove that there is finite speed of propagation.

The basic identity in the derivation is the energy balance law

$$\partial_t \left\langle A_0(t)\,u(t,x), u(t,x)\right\rangle + \sum_{j=1}^d \partial_j \left\langle A_j(t)\,u(t,x), u(t,x)\right\rangle =$$

$$\left\langle Z(t,x)\,u(t,x), u(t,x)\right\rangle + \Re\left\langle (Lu)(t,x), u(t,x)\right\rangle,$$

$$(2.25)$$

where $\langle\ ,\ \rangle$ is the scalar product in $\mathbb{C}^N$ and Z is the smooth matrix valued function

$$Z(y) := -B(y) - B^*(y) + \sum_{\mu=0}^d \frac{\partial A_\mu(y)}{\partial y_\mu}\,.$$

$$(2.26)$$

The idea is to integrate this identity over the truncated cones

$$|x - a| \leq R - ct\,, \qquad t \in]0, R/c[\,.$$

The sides of the cones are moving inward at speed c defined as follows. For $\xi \in \mathbb{R}^d$ satisfying $|\xi| = 1$ define

$$r(y, \xi) := \quad \text{spectral radius of} \quad \sum_{j=1}^d A_0^{-1/2}\,A_j(y)\,\xi_j\,A_0^{-1/2}\,.$$

Then

$$c := \sup\{\,r(y, \xi) : y \in \mathbb{R}^{d+1} \ \text{and} \ |\xi| = 1\}$$

$$(2.27)$$

With these definitions,

$$\eta_0 \geq c\,|\eta_1, \cdots, \eta_d| \quad \Rightarrow \quad \sum_{\mu=0}^d \eta_\mu A_\mu \geq 0\,.$$

$$(2.28)$$

In fact, c is exactly the smallest number with this property.

Exercise. Prove the last assertion.

For $a \in \mathbb{R}^d$ and $R > 0$ define the cones $\Omega = \Omega(a, R)$ by

$$\Omega := \{\, (t, x) : 0 \leq t \leq R/c, \quad \text{and} \quad |x - a| < R - ct \,\} \tag{2.29}$$

Denote their sections by

$$\Omega(t) := \{\, x : (t, x) \in \Omega \,\}, \qquad t \in [0, R/c]. \tag{2.30}$$

Theorem 2.4 [Friedrichs]. *Suppose that u_1 and u_2 are smooth solutions of the semilinear equation*

$$L u + F(y, u) = f(y)$$

whose initial values coincide in the ball $\{|x - a| \leq R\}$. Then $u_1 = u_2$ in $\Omega(a, R)$.

Proof. For $0 \leq t \leq R/c$, let

$$\phi(t) := \int_{\Omega(t)} \langle\, A_0 \left(u_1(t, x) - u_2(t, x)\right), \left(u_1(t, x) - u_2(t, x)\right) \,\rangle \, dx \,.$$

In particular, $\phi(0) = 0$. Apply (2.25) to $u := u_1 - u_2$. Integrate over $\Omega \cap [0, t]$. Use the fundamental theorem of calculus to convert the integral of the derivative terms to boundary terms. From the top and bottom of the truncated cone one gets $\phi(t) - \phi(0)$.

Using the simple estimate

$$\|Lu(y)\| = \|F(y, u_1(y)) - F(y, u_2(y))\| \leq C(F, u_1, u_2) \, \|u_1(y) - u_2(y)\|$$

shows that the integral of the terms on the right of (2.25) is bounded above by

$$C(L, F, u_1, u_2) \int_0^t \phi(\sigma) \, d\sigma \,.$$

The key observation is that the contribution from the sides of the cone is given by

$$\int_{|x-a|=R-ct} \left\langle\, \sum_0^d \eta_\mu A_\mu(y) \, u(y) \,,\, u(y) \,\right\rangle d\Sigma(y) \geq 0 \,,$$

where η is the unit outward normal. The speed c is defined exactly so that η satisfies the condition of (2.28), which implies that the matrix $\sum \eta_\mu A_\mu$ in the brackets is nonnegative.

These estimates imply that

$$\phi(t) \leq C(L, F, u_1, u_2) \int_0^t \phi(\sigma) \, d\sigma$$

and Gronwall's inequality completes the proof. $\qquad\qquad\square$

This theorem shows that no signals travel with speed greater than c given by formula (2.28). This formula for the maximal speed is sharp. It can be refined in the sense that speed may depend on direction and there are sharper anisotropic results. They are described in Lax's 1963 Stanford lecture notes on *Hyperbolic Partial Differential Equations*.

The proof of finite speed yields the following important local L^2 stability estimate.

Theorem 2.5. *There is a constant $C = C(L)$ so that for all a, R, T, u such that $0 \le T \le R/c$ and $u \in C^1(\Omega(a, R) \cap \{0 \le t \le T\})$*

$$\|u(t)\|_{L^2(\Omega(t))} \le C \left(\|u(0)\|_{L^2(\Omega(0))} + \int_0^t \|Lu(\sigma)\|_{L^2(\Omega(\sigma))} \, d\sigma \right). \qquad (2.31)$$

2.4. Plane waves, characteristic variety and finite speed

To see that the formula (2.27) is sharp in the case where L has constant coefficients and no lower order terms it suffices to examine plane wave solutions. Such solutions are also important in motivating high frequency asymptotic expansions. It is not unusual for the analysis of a partial differential equation in science texts to consist only of a calculation of all plane wave solutions. This often gives a lot of information.

Plane waves are solutions which depend only on $y.\eta$ for some $\eta \in \mathbb{R}^{1+d}$. That is,

$$u(y) := a(y.\eta) \qquad a : \mathbb{R} \to \mathbb{C}^N. \qquad (2.32)$$

Computing exactly when $L = L_1(\partial)$ yields

$$L(\partial_y) u = L_1(\partial_y) a(y.\eta) = L_1(\eta) a'(y.\eta). \qquad (2.33)$$

Exercises.

1. For $\eta = (\tau, \xi)$ with $|\xi| = 1$, compute all plane wave solutions of Maxwell's equations in vacuum, that is with $\rho = \mathbf{j} = 0$. **Partial Answer.** There is a two dimensional space of standing waves, that is solutions with $\tau = 0$, which do not satisfy the divergence conditions (2.2) and a four dimensional space of solutions which travel at the speed of light. The direction of motion is parallel to ξ while E, B, ξ form an orthogonal basis for $\mathbb{R}^3$.

2. Compute all plane wave solutions of the wave equation $\Box u = 0$.

For problems with lower order terms, the computations are not as clean.

3. Compute all plane wave solutions of the dissipative wave equation $\Box u + 2u_t = 0$.

4. Compute all plane wave solutions of the telegrapher's equation $u_{tt} - u_{xx} + 2u_t + u = 0$.

The definition of *ellipticity* of a partial differential operator is that for all y the constant coefficient homogeneous operator $L_1(y, \partial_y)$ has no nonconstant plane wave solutions. This is equivalent to the invertibility of $L_1(y, \eta)$ for all real η.

Exercise. Verify the ellipticity of your favorite elliptic operators. This should include at least the Laplacian, and the Cauchy-Riemann system. For the Laplacian L_1 must be replaced by L_2 in the definition of ellipticity. In general ellipticity of an mth order operator is equivalent to the invertibility of $L_m(y, \eta)$ for all real η.

Definition. The characteristic variety of L, denoted $\text{Char} \, L$ is the set of pairs $(y, \eta) \in \mathbb{R}^{1+d} \times \mathbb{R}^{1+d} \setminus 0$ such that

$$\det L_1(y, \eta) = 0. \qquad (2.34)$$

For $(y, \eta) \in \operatorname{Char} L$, let $\pi(y, \eta)$ denote the spectral projection of $L_1(y, \eta)$ on its kernel. That is

$$\pi(y, \eta) := \frac{1}{2\pi i} \oint_{|z|=r} \left(zI - L_1(y, \eta) \right)^{-1} dz$$

where r is chosen so small that 0 is the only eigenvalue of $L(y, \eta)$ in the disc $|z| \leq r$.

If $L = L_1$ has constant coefficients, then a necessary and sufficient condition for (2.32) to define a solution of $Lu = 0$ is that $\pi(\eta) a' = a'$. In particular, there are nontrivial solutions if and only if $\eta \in \operatorname{Char} L$. Except for an additive constant vector, the equation $\pi a' = a'$ is equivalent to

$$\pi(\eta) a = a. \tag{2.35}$$

This polarization for a recurs in all of our formulas from geometric optics.

Next consider the speeds of propagation of plane wave solutions. The solution

$$u = a(\tau t + x.\xi)$$

is a plane wave translating rigidly in the direction ξ with velocity $-\tau/|\xi|$. The characteristic equation asserts that

$$\det \left(A_0 \tau + \sum_1^d \xi_j A_j \right) = 0, \tag{2.36}$$

which is equivalent to

$$\tau \in \operatorname{spectrum} \left(\sum_{j=1}^d A_0^{-1/2} A_j \xi_j A_0^{-1/2} \right). \tag{2.37}$$

This shows that c defined by (2.27) is exactly the speed of the fastest moving plane wave. The Theorem shows the supremum of such speeds over all principal parts at points of space time gives an upper bound for the speeds of all waves.

General solutions of constant coefficient initial value problems can often be expressed as a Fourier superposition of exponential solutions of the form $e^{i(\tau t + x.\xi)}$ with ξ real and τ complex. Since τ may not be real these need not be plane waves. For constant coefficient operators and fixed ξ, the solutions of this form come from the complex roots τ of the equation

$$\det L(i\tau, i\xi) = 0.$$

The roots $\tau_j(\xi)$ where they are nice functions of ξ define the *dispersion relations* of the equation. Of particular importance is the case of conservative systems for which the roots are automatically real.

Exercise. Suppose that $L(\partial_y)$ is symmetric hyperbolic and conservative in the sense that $B = -B^*$. This hold in particular if $B = 0$. For such systems the energy

$$\int_{\mathbb{R}^d} \left\langle A_0 u(t, x), u(t, x) \right\rangle dx$$

is independent of time. Prove that the roots τ must be real in this case.

Exercise. Find the dispersion relations for the wave equation, the Klein-Gordon Equation $\Box u + u = 0$, and the Schrödinger equation $u_t + i\Delta_x u = 0$. You must generalize the notion of dispersion relation beyond the first order case to solve these problems.

For dissipative equations the exponential solutions decay in time which corresponds to roots τ with positive imaginary parts. Hadamard's analysis of wellposedness of initial value problems rests on the observation that one does not have continuous dependence on initial conditions if there exist exponential solutions whose imaginary parts tend to $-\infty$. A systematic use of exponential solutions in the study of initial value problems can be found in Chapter 3 of my book.

2.5. Solutions on cones of determinacy

Combining finite speed and Friedrich's Theorem on $\mathbb{R}^{1+d}$ yields an existence and uniqueness theorem on a domain of determinacy Ω as in (2.29). Introduce the notation

$$\Omega_T := \Omega \cap \{0 \le t \le T\}.$$

Theorem 2.6. *If $g \in C^\infty(\Omega(0))$ and $f \in C^\infty(\Omega \cap \{0 \le t \le T\})$ then there is one and only one solution $u \in C^\infty(\Omega \cap \{0 \le t \le T\})$ of the initial value problem*

$$Lu = f \quad on \quad \Omega_T, \qquad u(0) = g \quad on \quad \Omega(0). \tag{2.38}$$

If $s \in \mathbb{N}$ there is a constant $C = C(L, s)$ such that for all $0 \le t \le T$

$$\sum_{|\alpha| \le s} \|\partial_y^\alpha u(t)\|_{L^2(\Omega(t))} \le C \left(\|u(0)\|_{H^s(\Omega(0))} + \int_0^t \sum_{|\alpha| \le s} \|\partial_y^\alpha f(\sigma, x)\|_{L^2(\Omega(\sigma))} \, d\sigma \right).$$

$$\tag{2.39}$$

Proof. Extend f and g to be smooth compactly supported functions on space time and space respectively. Solving the resulting initial value problem on $\mathbb{R}^{1+d}$ constructs a solution. Uniqueness is proved in Friedrich's Theorem above.

Exercise. Prove the estimate (2.39) by using (2.31), a commutation argument and Gronwall's inequality.

Remark. Solutions with finite regularity can be constructed by an approximation argument using (2.39).

LECTURE 3
Examples of Propagation of Singularities and of Energy

The theme of this workshop is wavelike solutions of partial differential equations. These solutions have spatially localized structures whose evolution in time can be followed. The most common are singularities and modulated high frequency (short wavelength) solutions. Both involve radically different spatial scales. For the second it is the scales of macroscopic objects and the much smaller wavelength. The classic example is light with a wavelength on the order of 5×10^{-5} centimeter. Singularities are often restricted to varieties of lower codimension, hence of width equal to zero which is infinitely small compared to the scales of their other variations. Real world waves modeled by such solutions have the singular behavior spread over very small lengths, not exactly zero.

The path of a localized structure in space time is curvelike, and such curves are often called *rays*. Describing the evolution often leads to transport equations along rays. When phenomena are described by partial differential equations, linking the above ideas with the equation means finding solutions whose salient features are localized and in simple cases are described by transport equations. Reflecting the small parameter implicit in the description above, such results often appear in an asymptotic analysis as the parameter tends to zero.

From the basic existence theory of the last section, there is important information about wave propagation. The energy estimates prove good stability estimates under perturbation, and the invariance of $H^s(\mathbb{R}^d)$ spaces under time evolution is a key property. However, exactly the same features are valid for the Schrödinger equations

$$L(y, \partial_y)\, u \pm i\epsilon \Delta_x u = 0\,.$$

In fact the proofs are identical, since the generator has been perturbed by an exactly antiselfadjoint operator. The Schrödinger equations, however, do not have finite speed of propagation. The conjunction of H^s stability and finite speed characterizes hyperbolic wave equations.

Additional information about propagation for constant coefficient systems comes from the study of plane waves. This is especially informative for conservative systems.

In this section some simple, relatively explicit examples are described with the goal of clarifying the roles of propagation of singularities and propagation of energy. The latter reveals the classical group velocities of applied mathematics.

The former concerns only high frequencies, so the group velocities in the limit of short wavelength are relevant and not the group velocities at finite wavelength. The reason is simple. Up to an error as small as one likes in energy, the data can be replaced by data with compactly supported Fourier transform. On the other hand, up to an error as smooth as one likes the data can be replaced by data with Fourier transform vanishing on $|\xi| \leq R$ with R as large as one likes.

3.1. Examples

1. The general solution of the one dimensional wave equation

$$u_{tt} - u_{xx} = 0 \tag{3.1}$$

is the sum of progressing waves

$$f(x - t) + g(x + t). \tag{3.2}$$

The rays are the integral curves of

$$\partial_t \pm \partial_x. \tag{3.3}$$

Structures are rigidly transported at speeds ± 1. The transport equation expresses constancy on the integral curves of (3.3).

There is a simple energy law for solutions suitably small at infinity,

$$\int_{\mathbb{R}} u_t^2 + u_x^2 \, dx = \text{independent of time.}$$

The fundamental solution which solves (3) together with the initial values

$$u(0, x) = 0, \qquad u_t(0, x) = \delta(x), \tag{3.4}$$

is given by the explicit formula

$$u(t, x) = \frac{\operatorname{sgn} t}{2} \, \chi_{[-t,t]} = \frac{1}{2} \left(h(x - t) - h(x + t) \right), \tag{3.5}$$

where h denotes Heaviside's function. Note the singularities which propagate to the left and the right.

2. Interesting things happen if one adds a lower order term. For example consider the Klein-Gordon equation

$$u_{tt} - u_{xx} + u = 0. \tag{3.6}$$

In sharp contrast with (3.2), there are hardly any undistorted progressing wave solutions.

Exercise 1. Find all solution of (3.6) of the form $f(x - ct)$.

Discussion. The special solutions of the form $e^{i(\tau t - x\xi)}$ are particularly important since the general solution is a Fourier superposition of these special plane waves. The equation $\tau = \tau(\xi)$ defining such solutions is called the *dispersion relation* of (3.6).

There is a nice energy conservation law. The symbol $\mathcal{S}(\mathbb{R}^d)$ denotes the Schwartz space of rapidly decreasing smooth functions. That is, functions such that for all α, β

$$\sup_{x \in \mathbb{R}^d} \left| x^\beta \partial_x^\alpha \psi(x) \right| < \infty.$$

Exercise 2. Prove that if $u \in C^\infty(\mathbb{R} : \mathcal{S}(\mathbb{R}))$ is a real valued solution of the Klein-Gordon equation, then

$$\int u_t^2 + u_x^2 + u^2 \, dx$$

is independent of t. This quantity is called the **energy** and is denoted E.

The fundamental solution, that is the solution with initial data (1.4), is not as simple as in the case of the wave equation. However, the singularities can be easily computed. Introduce

$$h_n(x) := \begin{cases} x^n/n! & \text{for } x \geq 0 \\ 0 & \text{for } x \leq 0 \end{cases} \tag{3.7}$$

Then

$$\frac{d}{dx} h_{n+1} = h_n \qquad \text{for} \quad n \geq 0. \tag{3.8}$$

Exercise 3. Show that there are uniquely determined functions $a_n(t)$ satisfying

$$a_0(0) = 1/2, \quad \text{and} \quad a_n(0) = 0 \quad \text{for} \quad n \geq 1,$$

and so that for all $N \geq 2$,

$$\left(\partial_t^2 - \partial_x^2 + 1 \right) \sum_{n=0}^{N} a_n(t) \, h_n(x - t) \in C^{N-2}(\mathbb{R}^2). \tag{3.9}$$

In this case, we say that the series

$$\sum_{n=0}^{\infty} a_n(t) \, h_n(t - x)$$

is a formal solution of $(\partial_t^2 - \partial_x^2 + 1)u \in C^\infty$.

Exercise 4. Suppose that u is the fundamental solution of the Klein-Gordon equation and $M \geq 0$. Find a distribution w_M such that $u - w_M \in C^M(\mathbb{R}^2)$. Show that the fundamental solution of the wave equation and that of the Klein-Gordon equation differ by a Lipshitz continuous function. Show that the singular supports of the two fundamental solutions are equal.

Hint. Add (3.9) to its spatial reflection.

Exercise 5. Study the fundamental solution for the dissipative wave equation

$$u_{tt} - u_{xx} + 2u_t = 0. \tag{3.10}$$

In particular show that it is no longer a continuous perturbation of that for the wave equation, but nevertheless the singular support agrees with that of the wave equation.

Hint. Seek solutions of $(\partial_t^2 - \partial_x^2 + 2\partial_t)u \in C^\infty$ of the form $\sum_n b_n(t)\, h_n(t - x)$.

The method in the above exercises is called **progressing wave expansions**. It is discussed in more generality in chapter 6 of Courant-Hilbert Vol. 2, and in Lax's *Lecture on Hyperbolic PDE*. The higher dimensional analogue of these solutions are singular along codimension one characteristic hypersurfaces in space time. The singularities propagate satisfying transport equations along rays lying in the hypersurface. The general class goes under the name *conormal solutions*. They are discussed, for example, in M. Beals' book cited in the references. In practice they describe propagating wavefronts.

Returning to the Klein-Gordon Equation, or more generally perturbations of the wave equation by constant coefficient lower order terms, one can solve explicitly using the Fourier Transform. The computation of the singularities of the fundamental solution of the Klein-Gordon equation suggests that the main part of solutions travel with speed equal to 1. One might expect that the energy in a disk growing linearly in time at a speed slower than one would be small. For compactly supported data, such a disk would contain no singularities for large time. Thus it is not unreasonable to guess that for any $\sigma < 1$, and $R > 0$

$$\limsup_{t \to \infty} \int_{|x| < R + \sigma t} u_t^2 + u_x^2 + u^2 \, dx = 0. \tag{3.11}$$

The energy method shows that speeds are no larger than one. The idea about the main part of the solution expressed in (3.11) is dead wrong for the Klein-Gordon equation. The next exercise shows that the main part of the solution travels strictly slower than speed 1, even though singularities travel with speed exactly equal to 1.

Exercise 6. Suppose that $f \in H^1(\mathbb{R})$ and $g \in L^2(\mathbb{R})$ and that u is the unique solution of the Klein-Gordon equation with initial data

$$u(0, x) = f(x) \qquad u_t(0, x) = g(x). \tag{3.12}$$

Prove that for any $\epsilon > 0$ and $R \geq 0$, there is a $\delta > 0$ so that

$$\limsup_{t \to \infty} \int_{|x| > (1 - \delta)t - R} u_t^2 + u_x^2 + u^2/2 \, dx . < \epsilon. \tag{3.13}$$

Hint. Replace $\hat{f}, \hat{g}$ by compactly support smooth functions making an error at most $\epsilon/2$ in energy. Then use the method of nonstationary phase as on pages 149-150 of my book.

Discussion. In the process of the proof you should find that the contribution of the part of the data whose Fourier Transform is supported near wave number ξ influences the solution only near the *group lines* which travel with speeds $\pm\xi/(1 + |\xi|^2)^{1/2}$. These speeds are called the **group velocities** corresponding to the wavenumber ξ. Note that they are strictly less than one. This is the motor which drives the present exercise. Note also that as $\xi \to \infty$ the group velocities approach ± 1. Thus high frequencies will propagate at speeds nearly equal to one. In particular they travel at the same speed. High frequency signals stay together better than low frequency signals. Since singularities of solutions are made of only the high frequencies (modifying the data by an element of $\mathcal{S}$ modifies the solution by such an element and therefore by a smooth term) one expects singularities to propagate at speeds ± 1 which is exactly what is true for the fundamental solution. Once known for the fundamental solution it follows for all solutions by the argument in my book on pages 164-165. The notion of group velocity is discussed at some length in the treatise of Whitham. It occurs in asymptotic solutions resembling those discussed in these lectures (see [Donnat], [Donnat-Rauch]).

The analysis of Exercise 6 does not apply to the fundamental solution since the latter does not have finite energy. However it belongs to $C^j(\mathbf{R} : H^{s-j}(\mathbb{R}))$ for all $j \in \mathbb{N}$ and $s < 1/2$. Thus the next result provides a good replacement of (3.13).

Exercise 7. Suppose that u is the fundamental solution of the Klein-Gordon equation (1.6) and that $s < 1/2$. If $0 \le \chi \in C^\infty(\mathbb{R})$ is a plateau cutoff supported on the positive half line, that is

$$\chi(x) = 0 \quad \text{for} \quad x \le 0 \qquad \text{and} \quad \chi(x) = 1 \quad \text{for} \quad x \ge 1 ,$$

then for all $R \ge 0$,

$$\lim_{t \to \infty} \| \chi(R + |x| - t)\, u(t,x) \|_{H^s(\mathbb{R}_x)} = 0 . \tag{3.14}$$

Hint. Prove that

$$\| \chi\, u(t) \|_{H^s(\mathbb{R})} \le C\Big(\|u(0)\|_{H^s(\mathbb{R})} + \|u_t(0)\|_{H^{s-1}(\mathbb{R})} \Big)$$

with C independent of t and the initial data. Conclude that it suffices to prove (3.14) with initial data $u(0), u_t(0)$ dense in $H^s \times H^{s-1}$. Take the dense set to be data with Fourier Transform in $C_0^\infty(\mathbb{R})$.

Discussion. This shows that though the singularities move at speed ± 1, the energy moving at this speed is negligible in the limit $t \to \infty$.

These examples illustrate the important observation that the propagation of singularities in solutions and the propagation of the majority of the energy may be governed by different rules.

LECTURE 4
Elliptic Geometric Optics

4.1. Constant coefficients and linear phases

The study of oscillatory solutions of elliptic equations is easier than the corresponding hyperbolic theory. The reason is simple. Oscillations propagate in the hyperbolic case, and have only local effects in the elliptic case. Nevertheless, the elliptic case is a good starting point for several reasons. First, the linear elliptic results are needed eventually in the proofs of nonlinear hyperbolic results. Second, it is easier to introduce some of the basic asymptotics in this case.

For the analysis of this section there is no need for symmetry or any other hypothesis of hyperbolicity. Similarly the independent variable y is not split into space and time. The partial differential operator

$$L(y, \partial_y) = \sum A_\mu(y) \frac{\partial}{\partial y_\mu} + B(y)$$

is assumed to have smooth matrix valued coefficients on an open set $\mathcal{O} \subset \mathbb{R}^n$.

Our starting point is an observation from the elementary study of constant coefficient ordinary differential equations. Recall that if $L(d/dt)$ is a constant coefficient linear ordinary differential operator

$$L := a_m \frac{d^m}{dt^m} + \cdots + a_1 \frac{d^1}{dt^1} + a_0, \qquad a_m \neq 0,$$

then a particular solution of the homogeneous equation

$$Lu = b\, e^{i\tau t}$$

is given by

$$u = L(i\tau)^{-1} b\, e^{i\tau t}$$

provided only that

$$L(i\tau) \neq 0. \tag{4.1}$$

Since for $|\tau| \geq 1$,

$$|L(\tau)| \geq |a_m\, \tau^m| - \Big(\sum_0^{m-1} |a_j| \Big) |\tau|^{m-1}$$

it follows that (4.1) is satisfied for all large τ.

407

In the same way, a particular solution of a constant coefficient system of partial differential equations

$$L(\partial_y)\, u = b\, e^{iy\cdot\eta}$$

is given by

$$u = L(i\eta)^{-1}\, b\, e^{iy\cdot\eta}$$

provided only that $\det L(i\eta) \neq 0$.

To study the case of a first order system in the limit of small wavelength, use η/ϵ in place of η and consider $\epsilon \to 0$. Since

$$L(i\eta/\epsilon) = \frac{1}{\epsilon}\left(L_1(i\eta) + \epsilon L_0\right),$$

it follows that if η is not characteristic then $\det L(i\eta/\epsilon)$ is nonzero for ϵ sufficiently small.

Assume that η is noncharacteristic and consider the equation

$$L(\partial_y)\, u = b\, e^{iy\cdot\eta/\epsilon},\quad b \in \mathbb{C}^N. \tag{4.2}$$

An explicit solution is given by

$$u = e^{iy\cdot\eta/\epsilon}\left(L_1(i\eta/\epsilon) + L_0\right)^{-1} b = \epsilon\, e^{iy\cdot\eta/\epsilon}\left(L_1(i\eta) + \epsilon\, L_0\right)^{-1} b.$$

For ϵ small, the inverse is given by a convergent Neumann series so

$$u(y) = \epsilon\, e^{iy\cdot\eta/\epsilon} \sum_{n=0}^{\infty} (-\epsilon)^n \left(L_1(i\eta)^{-1} L_0\right)^n L_1(i\eta)^{-1} b$$

$$= \epsilon\, e^{iy\cdot\eta/\epsilon}\left(L_1(i\eta)^{-1} b + \text{higher order terms}\right).$$

Observe that the form of the solution is a series

$$e^{iy\cdot\eta/\epsilon}\left(\epsilon\, a_1 + \epsilon^2 a_2 + \cdots\right)$$

where the vector coefficients are equal to the value b multiplied by a finite number of matrices.

Multiplying the source and the solution by $e^{ic/\epsilon}$ with real c shows that the computation above works for the affine phase $\phi = c + y.\eta$, in which case $\eta = d\phi$.

A key feature of this solution is that the leading term depends only on the principal symbol L_1. The reason is simple. For the highly oscillatory solutions the derivatives are of order $1/\epsilon$ larger than u. Thus so long as the combination of derivatives represented by $L_1(i\eta)\, u$ is nonzero it will be dominant. The general principle here is that for noncharacteristic short wavelength oscillations, the principal symbol dominates. In contrast when $L_1(i\eta)$ is not invertible, the lower order terms play an important role as we will see in Lecture 5.

4.2. Iterative improvement for variable coefficients and nonlinear phases

The next step is a key insight. Suppose that one considers source terms which are rapidly oscillating with possibly nonlinear phase, and a differential operator with possibly variable coefficients

$$L(y, \partial_y)\, u = b(y)\, e^{i\phi(y)/\epsilon}. \tag{4.3}$$

The phase ϕ is a smooth real valued function whose gradient is assumed to be nonvanishing on the support of $b(y)$. The coefficients of L are assumed to be smooth on a neighborhood of this support.

Here is the key intuition, which leads to an approximate solution. Imagine an observer who probes u near a point $\underline{y}$. Suppose that he observes on a region which is large compared to ϵ but small compared to the scale on which b, the coefficients of L, and $d\phi$ vary. Thus to such an observer these quantities appear constant and the differential equation looks like

$$L(\underline{y}, \partial_y)u = b(\underline{y})\, e^{i(\phi(\underline{y}) + d\phi(\underline{y})\cdot(y-\underline{y}))/\epsilon}\,. \tag{4.4}$$

If $(\underline{y}, d\phi(\underline{y}))$ is not in the characteristic variety of L, the previous analysis shows that for ϵ small an approximate solution on this region is given by

$$\begin{aligned} u_{\mathrm{approx}} &\sim \epsilon\, e^{i(\phi(\underline{y}) + d\phi(\underline{y})\cdot(y-\underline{y}))/\epsilon}\, L_1(\underline{y}, id\phi(\underline{y}))^{-1}\, b(\underline{y}) \\ &\approx \epsilon\, e^{i\phi(y)/\epsilon}\, L_1(\underline{y}, id\phi(\underline{y}))^{-1}\, b(\underline{y})\,. \end{aligned}$$

These computations suggest that

$$u(y) \approx \epsilon\, e^{i\phi(y)/\epsilon}\, a_1(y)\,, \qquad a_1(y) := L_1(y, id\phi(y))^{-1}\, b(y)\,. \tag{4.5}$$

The idea leading to this guess was that in the limit of very small wavelength the problem can be replaced by an approximate problem with constant coefficients, a source with constant amplitude, and, an affine phase. To see how successful it is, take u as defined in (4.5) and apply L to find

$$L(y, \partial_y)\left(\epsilon\, e^{i\phi(y)/\epsilon}\, a_1(y)\right) = e^{i\phi(y)/\epsilon}\left(b(y) + \epsilon b_1(y)\right), \tag{4.6}$$

where

$$b_1(y) := L(y, \partial_y)\, a_1(y)\,. \tag{4.7}$$

This is excellent news, since the previous computation tells us a correction term. Let

$$u = e^{i\phi(y)/\epsilon}\left(\epsilon\, a_1(y) + \epsilon^2\, a_2(y)\right), \qquad a_2(y) := L_1(y, id\phi(y))^{-1}\, b_1(y)\,, \tag{4.8}$$

to find

$$L(y, \partial_y)\, u = e^{i\phi(y)/\epsilon}\left(b(y) + \epsilon^2 b_2(y)\right), \tag{4.9}$$

where

$$b_2(y) := L(y, \partial_y)\, a_2(y)\,. \tag{4.10}$$

This process, by induction on m, then proves the following theorem.

Theorem 4.1. *Suppose that $m \geq 1$ is an integer, $\Omega \subset \mathbb{R}^n$ is a bounded open set, $b(y)$ is a smooth amplitude on Ω and ϕ is a smooth real valued phase such that for all $y \in \Omega$, $d\phi(y) \neq 0$ and $(y, d\phi(y)) \notin \operatorname{Char} L$. Then, there are uniquely determined smooth amplitudes a_j on Ω so that*

$$u := e^{i\phi(y)/\epsilon}\left(\epsilon\, a_1(y) + \epsilon^2\, a_2(y) + \cdots + \epsilon^m\, a_m(y)\right) \tag{4.11}$$

satisfies

$$Lu = e^{i\phi(y)/\epsilon}\, b(y) + \epsilon^m\, e^{i\phi(y)/\epsilon}\, r(y)\,, \tag{4.12}$$

with r smooth on Ω. The principal amplitude is given by $a_1 = L(y, id\phi(y))^{-1}\, b(y)$.

4.3. Formal asymptotics approach

Once the form of the expansion of this theorem is known, the exact coefficients can be easily and quickly computed without going through the above recursion. Consider the following more general situation. Given b_j, find a_j so that

$$L(y, \partial_y) \left(e^{i\phi(y)/\epsilon} \left(\epsilon\, a_1(y) + \epsilon^2 a_2(y) + \cdots \right) \right) \sim e^{i\phi(y)/\epsilon} \left(b_0(y) + \epsilon^1 b_1(y) + \cdots \right).$$
$$(4.13)$$

Note the difference of 1 in the exponents of ϵ. Computing the left side explicitly yields

$$e^{i\phi(y)/\epsilon} \left(L_1(y, id\phi(y)) a_1(y) + \sum_{j=1}^{\infty} \epsilon^j \left(L_1(y, id\phi(y)) a_{j+1}(y) + L(y, \partial_y) a_j(y) \right) \right)$$
$$(4.14)$$

with the convention that $a_0 = 0$. The equations determining the a_j are then read off to be

$$L_1(y, id\phi(y))\, a_1(y) = b_0(y),$$
$$(4.15)$$

and for $j > 1$,

$$L_1(y, id\phi(y))\, a_j(y) = -L(y, \partial_y)\, a_{j-1}(y) + b_{j-1}.$$
$$(4.16)$$

The equation (4.13) was left purposely vague to show that the formal computations are really straightforward. To put meat on the bones of the formal series one has to make something of the sums in (4.13). These sums do not usually converge but represent asymptotic expansions as $\epsilon \to 0$. The interpretation is exactly analogous to that of Taylor expansions.

Definition. If $\mathcal{O}$ is an open set in $\mathbb{R}^n$, $b_j(y)$ is a sequence of smooth functions on $\mathcal{O}$, and $b \in C^\infty(]0, 1[\times\mathcal{O} \, ; \, \mathbb{C}^N)$, the asymptotic relation

$$b(\epsilon, y) \sim b_0(y) + \epsilon\, b_1(y) + \epsilon^2\, b_2(y) + \cdots$$
$$(4.17)$$

means that for every integer $m \geq 0$, every multiindex $\alpha \in \mathbb{N}^n$ and every compact subset $K \subset \mathcal{O}$

$$\sup_K \left| \partial_y^\alpha \left(b - \sum_{j=0}^{m} \epsilon^j\, b_j(y) \right) \right| = o(\epsilon^m)$$
$$(4.18)$$

as $\epsilon \to 0$.

Similarly if $c \in C^\infty(]0, 1[\times\mathcal{O})$, the asymptotic relation

$$b(\epsilon, y) \sim c(\epsilon, y)$$

means that $b - c \sim \sum \epsilon^j\, 0$.

Remarks. 1. Instead of $\sim \sum \epsilon^j\, 0$ we will write ~ 0.

2. For smooth functions on a bounded $\overline{\mathcal{O}}$ a similar definition *with control at the boundary* can be given by replacing the supremum over compact subsets K by the supremum over $\overline{\mathcal{O}}$. In this way one defines asymptotic expansions in $C^\infty(\overline{\mathcal{O}})$ in contrast to the expansions in $C^\infty(\mathcal{O})$.

3. If b is smooth on a neighborhood of $\epsilon = 0$, then Taylor's Theorem shows that (4.18) is equivalent to

$$b_j(y) = \frac{1}{j!} \frac{\partial^j b(0,y)}{\partial \epsilon^j}.$$

The definition still leaves the interpretation of (4.13) up in the air since the construction went from b_j to a_j with no mention of functions of $a(\epsilon, y)$ and $b(\epsilon, y)$. The key link is Borel's Theorem.

Borel's Theorem 4.2. *Given a sequence b_j of smooth functions on the open set $\mathcal{O} \subset \mathbb{R}^n$ there is a smooth function $b(\epsilon, y)$ defined on $\mathbb{R} \times \mathcal{O}$ so that*

$$b(\epsilon, y) \sim b_0(y) + \epsilon b_1(y) + \epsilon^2 b_2(y) + \cdots.$$

Remark. Returning to the discussion of the Definition, Borel's Theorem implies that one can choose $c(\epsilon, y) \in C^\infty(\mathbb{R} \times \mathcal{O})$ with $c \sim \sum \epsilon^j b_j(y)$. Then $b \sim c$ and $j! \, b_j = \partial^j c(0,y)/\partial \epsilon^j$. This shows that the smooth in ϵ case of Remark 3 is the general case.

The proof of Borel's Theorem is a direct generalization of the proof of the following seemingly much more special result.

Borel's Theorem 4.3. *Given a sequence b_j $0 \leq j < \infty$ of complex numbers there is a smooth function $b(\epsilon)$ on $\mathbb{R}$ whose Taylor series at the origin is $\sum \epsilon^j b_j$.*

Proof. The idea is to set $b(\epsilon) = \sum \epsilon^j b_j$. However, this series has no reason at all to converge since the b_j may grow arbitrarily rapidly. The clever idea is to cut off the summands so that they live only for $|\epsilon|$ small where the ϵ^j can compensate the b_j.

Choose a function $\chi \in C_0^\infty(]-1, 1[)$ such that $\chi(\epsilon) = 1$ for $|\epsilon| \leq 1/2$.

The summand $\epsilon^j b_j$ is replaced by $\epsilon^j \chi(M_j \epsilon) b_j$ where the sequence of positive numbers $M_0 < M_1 < M_2 < \cdots$ is chosen as follows.

Choose $M_0 \geq 1$ so that $|b_0/M_0| < 1$.

For $j > 0$ choose $M_j > M_{j-1}$ so that for $m = 0, 1, 2, \ldots, j-1$ and all $\epsilon \in \mathbb{R}$,

$$\left| \frac{d^m}{d\epsilon^m} \left(\chi(M_j \epsilon) \, \epsilon^j \, b_j \right) \right| \leq \frac{1}{2^j}. \tag{4.19}$$

The reason that this is possible is that when the derivatives are expanded there are a finite number of terms. Each is a bounded function of ϵ times

$$b_j \, \epsilon^{j-l} \, M_j^k \, \frac{d^k \chi}{d\epsilon^k}(M_j \epsilon), \qquad k + l = m \leq j - 1. \tag{4.20}$$

In the support of the χ term, $\epsilon < 1/M_j$. Thus the term (4.20) is bounded by $c(\chi, j)|b_j|/M_j^{j-k-l}$, so (4.19) can be achieved by choosing M_j sufficiently large.

Then

$$\sum \epsilon^j \, \chi(M_j \epsilon) \, b_j$$

converges uniformly with all of its derivatives to a function $b(\epsilon)$. That it satisfies the conditions of the theorem is immediately verified by termwise differentiation.$\quad\square$

Exercise. Prove the first of the Borel Theorems.

Given the language of asymptotic expansions in ϵ the computations prove the following result.

Theorem 4.4. *Suppose that $\Omega \subset \mathbb{R}^n$ is an open set,*

$$b(\epsilon, y) \sim \sum_{j=0}^{\infty} \epsilon^j \, b_j(y)$$

is a smooth amplitude on Ω and ϕ is a smooth real valued phase such that for all $y \in \Omega$ $d\phi(y) \neq 0$ and $(y, d\phi(y)) \notin \operatorname{Char} L$. Then there is a smooth

$$a(\epsilon, y) \sim \sum_{j=1}^{\infty} \epsilon^j \, a_j(y)$$

such that

$$L(y, \partial_y) \left(e^{i\phi(y)/\epsilon} \, a(\epsilon, y) \right) - b(\epsilon, y) e^{i\phi(y)/\epsilon} \sim 0 \,.$$

The amplitude a is unique in the sense that if $\tilde{a}(\epsilon, y)$ is a second solution then

$$a(\epsilon, y) - \tilde{a}(\epsilon, y) \sim 0 \,.$$

In particular the $a_j(y)$ are uniquely determined and the principal amplitude is given by (4.5).

Remark. It is reasonable to take both $a(\epsilon, y)$ and $b(\epsilon, y)$ as smooth functions on $[0, \infty[\, \times \Omega$. However, neither the source $e^{i\phi(y)/\epsilon} \, b(\epsilon, y)$ nor the response $e^{i\phi(y)/\epsilon} \, a(\epsilon, y)$ is smooth up to $\epsilon = 0$. This follows from

$$\frac{d^k}{d\epsilon^k} \, e^{i\phi(y)/\epsilon} \, b(\epsilon, y) = \frac{(-1)^{k-1}(k-1)!}{\epsilon^k} \, \phi(y) \, e^{i\phi(y)/\epsilon} \, b_0(y) + O(\epsilon^{1-k}) \,,$$

and a similar expression differing by a power of ϵ for the response.

Exercise. Compute two terms of an asymptotic solution of

$$\frac{\partial^2 u}{\partial x_1^2} + \frac{\partial^2 u}{\partial x_2^2} + \frac{\partial u}{\partial x_1} = e^{ix.\xi/\epsilon} \,.$$

It is interesting to ask whether the asymptotic solution $u(\epsilon, y) := a(\epsilon, y) e^{i\phi(y)/\epsilon}$ can be corrected by a term $c(\epsilon, y) \sim 0$ so that

$$L(u + c) = 0 \,.$$

Define the residual by

$$r(\epsilon, y) := L(y, \partial_y) \, u(\epsilon, y) \sim 0 \,.$$

One needs to solve

$$L(y, \partial_y) \, c(\epsilon, y) = -r(\epsilon, y) \,, \qquad c \sim 0 \,.$$

Under a variety of conditions this is possible. For example, if L is symmetric hyperbolic it suffices to supplement the equation for c with the initial condition

$c|_{t=0} = 0$. A similar argument works for parabolic equations determining c in $t \geq 0$.

If L is elliptic, then inhomogeneous equations like that for c are solvable on sufficiently small neighborhoods of arbitrary points. Thus the asymptotic expansion can be locally corrected. A simple argument with a partition of unity allows one to perform the correction on any compact set.

If L has constant coefficients, one can choose a fundamental solution E and a plateau cutoff χ and set $c := E * (\chi r)$. This too works on compact subsets of space time.

However, Levy showed in 1957 that linear partial differential equations with variable coefficients, even with polynomial coefficients, need not be locally solvable. In such cases the equation for c need not have solutions. Then, the construction of an asymptotic solution is the best that one can do.

4.4. Perturbation approach

The fundamental equations, (4.15), (4.16) have now been derived two different ways, one inductive and one by plugging in the formal *ansatz*. Here is a third derivation which has more the feel of perturbation theory. In that sense it resembles the Neumann series computation at the beginning of the section. It is useful to have a variety of approaches for at least three reasons. First one sees that they are all versions of the same thing. In much of the mathematical and scientific literature these different computations are confused as fundamentally different things. Second, in extending these ideas sometimes one or the other point of view is more easily adaptable. Finally, different arguments appeal to different people and you can choose your favorite.

Suppose that as $\epsilon \to 0$,

$$b(\epsilon, y) \sim b_0(y) + \epsilon b_1(y) + \epsilon^2 b_2(y) + \cdots . \tag{4.21}$$

Seek $u(\epsilon, y)$ solving

$$L(y, \partial_y) u \sim e^{i\phi(y)/\epsilon} b . \tag{4.22}$$

Motivated by the case of constant coefficient ordinary differential equations try

$$u = e^{i\phi(y)/\epsilon} a(\epsilon, y) . \tag{4.23}$$

Compute

$$L(y, \partial_y) \left(e^{i\phi(y)/\epsilon} a \right) = e^{i\phi(y)/\epsilon} \left(\frac{1}{\epsilon} L_1(y, id\phi(y)) + L(y, \partial_y) \right) a . \tag{4.24}$$

Thus (4.22) holds if and only if

$$\left(L_1(y, id\phi(y)) + \epsilon L(y, \partial_y) \right) a \sim \epsilon b . \tag{4.25}$$

If there is a solution a which has derivatives which are $O(1)$ as $\epsilon \to 0$, then there are two terms on the left, one of order 1 and the other of order ϵ. Neglecting the latter yields a first approximation which is identical to (4.5). There are at least two natural ways to proceed from here. One is to seek a as an asymptotic (a.k.a. Taylor series) in ϵ

$$a(\epsilon, y) \sim \epsilon a_1(y) + \epsilon^2 a_2(y) + \cdots . \tag{4.26}$$

Plugging this into (4.25) leads immediately to equations (4.15) and (4.16).

Exercise. Verify.

An alternative is to do fixed point iteration on the equation (4.25) generating a sequence of approximations by solving

$$a^\nu(y) = L(y, id\phi(y))^{-1} \left(b - \epsilon L(y, \partial_y) a^{\nu-1} \right). \tag{4.27}$$

The first approximation is found by dropping the ϵL term from the right hand side to find

$$a^1 = \epsilon L(y, id\phi(y))^{-1} b. \tag{4.28}$$

The same choice is generated by choosing $a^0 = 0$. The iteration implies that

$$a^{\nu+1} - a^\nu = -\epsilon L(y, id\phi(y))^{-1} L(y, \partial_y) \left(a^\nu - a^{\nu-1} \right). \tag{4.29}$$

This implies that

$$a^\nu = \epsilon a_1 + \epsilon^2 a_2 + \cdots + \epsilon^\nu a_\nu + O(\epsilon^{\nu+1}), \tag{4.30}$$

with the a_j from (4.15), (4.16).

Having given three distinct approaches to solving (4.25) which all lead to the same answer you may have the misimpression that the solution of this equation is trivial. In fact, that is not the case. The differential operator on the left of (4.25) is a first order operator with the property that the differentiation terms have a coefficients of size ϵ. The derivative terms which are normally the main terms have a small coefficient and so end up playing the role of corrections. One consequence is that the successive correction terms are generated by applying operators of high order in $\epsilon\partial_y$. Normally corrections in differential equations involve successive integration. This is all to say that the approximation which is just produced is subtle, and also that convergence is unlikely except when the operators and sources satisfy real analyticity hypotheses. Such hypotheses are physically unnatural since they imply that knowledge of sources in one place determines them everywhere.

4.5. Elliptic Regularity Theorem

A striking application of the Theorem above is the following proof of the interior elliptic regularity theory. The proof is easily modified to give the microlocal version. The interested reader is referred to Theorem 8.3.1 of the first volume of Hörmander's treatise for that refinement.

Elliptic Regularity Theorem 4.6. *Suppose that on a neighborhood of $\underline{y}$, $L(y, \partial_y)$ is an elliptic operator of order 1, and, Lu belongs to H^s. Then u is H^{s+1} on a neighborhood of $\underline{y}$.*

Proof. For this proof denote by n the dimension of y space. Let

$$f := Lu \tag{4.31}$$

which is defined and H^s on a neighborhood of $\underline{y}$. Choose a smooth χ, compactly supported in this neighborhood and identically equal to 1 on a smaller neighborhood of $\underline{y}$ so that L is elliptic on a neighborhood of the support of χ. Then

$$b(y) := \chi(y)f(y) \in H^s(\mathbb{R}^n). \tag{4.32}$$

Denote by ω the points of the unit sphere, $S^{n-1} \subset \mathbb{R}^n$.

Since L is elliptic, every ω is noncharacteristic. The same is therefore true for the transposed operator $L(y, \partial_y)'$ since the principal symbols of these operators differ by a nonvanishing scalar factor. Thus we have constructed, for each ω, asymptotic solutions $v(\epsilon, \omega, y)$ to

$$L(y, \partial_y)' \, v(\epsilon, \omega, y) \sim \chi(y) e^{-iy.\omega/\epsilon} \,. \tag{4.33}$$

The construction is uniform in the following sense. We construct $a_j(\omega, y)$ smooth in y *and* ω. Then one can choose

$$a(\epsilon, \omega, y) \sim \sum_{j=1}^{\infty} \epsilon^j \, a_j(\omega, y) \quad \text{in} \quad C^{\infty}(S^{n-1} \times \mathbb{R}^n) \,. \tag{4.34}$$

The construction being local the a_j and a can be chosen to vanish outside the support of χ. Setting

$$v(\epsilon, \omega, y) := a(\epsilon, \omega, y) \, e^{iy.\omega/\epsilon} \,, \tag{4.35}$$

(4.33) holds in $C^{\infty}(S^{n-1} \times \mathbb{R}^n)$.

The strategy is to prove that $\chi u \in H^{s+1}$ by studying the Fourier transform $\widehat{\chi u}(\eta)$ for $k = |\eta| \to \infty$. For $k := 1/\epsilon \to \infty$ and $\omega := \eta/|\eta|$, compute

$$\widehat{\chi u}(\eta) = \widehat{\chi u}(k\omega) = \langle \chi u, e^{-ik\omega.y} \rangle = \langle u, e^{-i\omega.y/\epsilon}\chi(y) \rangle = \langle u, L'v \rangle + O(k^{-M}) \,. \tag{4.36}$$

The differential equation asserts that $\langle u, L'v \rangle = \langle Lu, v \rangle$ so the first term on the right is equal to

$$\langle u, L'v \rangle = \langle Lu, v \rangle = \langle f, v \rangle = \left\langle f, \sum_{j=1}^{M-1} k^{-j} v_j e^{-ik\omega.y} \right\rangle + O(k^{-M}) \,. \tag{4.37}$$

Exercise. Are you convinced that (4.37) is correct?

Choose $M \in \mathbb{N}$ so that

$$\int_{|\eta|>1} \frac{<\eta>^{2(s+1)}}{|\eta|^{2M}} \, d\eta < \infty \,. \tag{4.38}$$

The summands in (4.37) are equal to

$$k^{-j} \, \widehat{v_j f}(k\omega) \,, \qquad 1 \leq j \leq M - 1 \,.$$

By construction, the v_j are supported in the domain where f is H^s and are uniformly smooth in ω. Choose a cutoff function $\psi(y)$ which is equal to one on the support of v and so that ψf is H^s. It follows that

$$\int_{k>1} |\widehat{v_j f}(k\omega)|^2 \, k^{2s} \, dk d\omega \leq C(j, \psi) \, \|\psi f\|_{H^s(\mathbb{R}^n)} \,. \tag{4.39}$$

This estimate is proved by the same simple argument as Proposition 2.6.4 in my book.

Since the sum is over $j \geq 1$, estimates (4.36), (4.37), (4.38), and (4.39) imply that

$$\int_{|\eta|>1} |\eta|^{2(s+1)} |\widehat{\chi u}(\eta)|^2 \, d\eta < \infty \tag{4.40}$$

which is the desired result.   □

Corollary 4.7. *If $L(y, \partial_y)$ is elliptic on the open set Ω and $u \in \mathcal{D}'(\Omega)$ satisfies $Lu \in C^\infty(\Omega)$, then $u \in C^\infty(\Omega)$.*

Proof. If $\Omega_1 \subset\subset \Omega$ then u is $H^s_{loc}(\Omega_1)$ for some possibly very negative s. The Theorem implies that $u \in H^{s+1}_{loc}(\Omega_1)$. A second application implies that $u \in H^{s+2}_{loc}(\Omega_1)$. An induction shows that u belongs to $H^{s+m}_{loc}(\Omega_1)$ for all integers m. Sobolev's embedding theorem implies that $u \in C^\infty(\Omega_1)$. Since Ω_1 is arbitrary the proof is complete.   □

LECTURE 5

Linear Hyperbolic Geometric Optics

The mathematical subject of geometric optics is devoted to the construction and analysis of asymptotic solutions of partial differential equations which are accurate when wavelengths are small compared to other natural lengths in the problem. Since the wavelength of visible light is of the order of magnitude 5×10^{-5}cm a great deal of what one sees falls into this category. The description of optical phenomena was and is a great impetus to study short wavelength problems.

The key feature of geometric optics, propagation along rays, is not present in the elliptic case of the last section. Rays occur for hyperbolic problems, and in the same spirit, for singular elliptic problems which arise when looking for high frequency time periodic solutions of a hyperbolic equation. An example of the latter is that solutions of the form

$$u(t,x) = e^{i\tau t}\, w(x)\,, \qquad \tau \gg 1 \tag{5.1}$$

to D'Alembert's equation

$$\Box\, u = 0 \tag{5.2}$$

must satisfy

$$\tau^2\, w + \Delta\, w = 0\,. \tag{5.3}$$

The key dichotomy is that in the (nonsingular) elliptic case, rapid oscillations have only local effects. The values of the coefficients a_j in the neighborhood of a point are determined by the values of the b_j in the same neighborhood. For hyperbolic equations (and their related singular elliptic problems like (5.3)) the oscillations may and usually do propagate to distant parts of space time. This is why they are the main carriers of information in both the communications industry and in the universe.

In this and later chapters, the underlying operator $L(y, \partial_y)$ is assumed to be symmetric hyperbolic. The nonhyperbolic detour of the last Lecture is finished.

5.1. Constant coefficients and linear phases

The starting point for the construction of the asymptotic solutions of geometric optics is the observation that if L has constant coefficients and no lower order terms

417

then there are many exact solutions in the form of plane waves

$$u(y) := a(y.\eta), \qquad y.\eta := \sum_{\mu=0}^{d} y_\mu \eta_\mu.$$

In 2.4 we computed that for homogeneous constant coefficient operators $L(y, \partial_y) = L_1(\partial_y)$, u satisfies $Lu = 0$ if and only if $\eta \in \operatorname{Char} L$ and

$$a'(\sigma) \in \ker L_1(\eta) \tag{5.4}$$

for all σ. Thus, up to an additive constant the plane wave solutions are given by an arbitrary function on $\mathbb{R}$ with values in $\ker L(\eta)$.

Introduce the notation

$$\eta = (\eta_0, \eta_1 \cdots, \eta_d) = (\tau, \xi_1, \cdots, \xi_d). \tag{5.5}$$

The solutions above have the form

$$u = a(\tau t + x.\xi) \tag{5.6}$$

so represent plane waves traveling rigidly in the direction $-\xi$ with velocity $\tau/|\xi|$.

Exercise. Find all such plane wave solutions for the following homogeneous partial differential operators.

    1. $u_{tt} = c^2 \Delta u$.
    2. $\Pi_{j=1}^{m} \left(\frac{\partial}{\partial t} + \lambda_j \frac{\partial}{\partial x} \right)$ where x is one dimensional and the λ are distinct reals.
    3. $d = 1$, $A_0 = I$, and $A_1 = diag\,(\lambda_1, \cdots, \lambda_N)$.

Exercise. For some of the examples above show that if you add a constant coefficient lower term then the solutions of the form (5.3) become much rarer. Instead of being parametrized by an arbitrary function of one variable, the solutions become a family depending on a finite number of parameters.

This exercise clearly poses a challenge for us: find an appropriate analogue of the plane wave solutions of constant coefficient homogeneous equations which probe the propagation properties of systems with variable coefficients and lower order terms.

Among the plane wave solutions described above are those with high frequency sinusoidal oscillatory behavior

$$u = e^{iy.\eta/\epsilon}\, a(y.\eta), \tag{5.7}$$

where a is smooth and ϵ is small compared to the typical distances on which a varies. Equation (5.7) represents plane waves oscillating with wavelength ϵ and wave envelope a which are rigidly transported at speed $\tau/|\xi|$. The existence of such waves relies crucially on there being points in the characteristic variety. In the case of elliptic operators, plane waves are most definitely not present. This is easily seen by noticing that elliptic operators are characterized by estimates of the following form. For relatively compact open subsets $\omega \subset \Omega \subset \mathbb{R}^n$

$$\|u\|_{H^s(\omega)} \le c(L, \omega, \Omega, s) \left(\|Lu\|_{H^{s-1}(\Omega)} + \|u\|_{H^{s-1}(\Omega)} \right). \tag{5.8}$$

If there were high frequency solutions of the above form then the left hand side would grow like ϵ^{-s} and the right hand side would grow like ϵ^{s-1} so that no inequality like (5.8) could be satisfied.

The solutions (5.7) have wavelength of order $\epsilon \ll 1$. Such functions have derivatives of order $1/\epsilon$ which are therefore large. When a partial differential operator with lower order term is applied to such an expression the lower order terms tend to be less important. Also the variation of the coefficients from point to point, when on a scale long compared to ϵ, are also less important. This suggests that there should be analogues of the plane waves beyond the constant coefficient homogeneous case.

5.2. Scalar constant coefficient operators and linear phases

The case of scalar operators is easier than systems and we begin with the example of a second order scalar constant coefficient operator

$$L := \sum_{\mu,\nu=0}^{d} a_{\mu\nu}\partial_\mu\partial_\nu + \sum_{\mu=0}^{d} b_\mu\partial_\mu \, . \tag{5.9}$$

We suppose that the principal coefficients $a_{\mu\nu}$ are real and symmetric in μ, ν and do not depend on y. Otherwise linear phases would be unrealistic, since for variable coefficient operators surfaces of constant phase are typically not planes even if they start that way at time zero. Viewed another way, linear functions $y.\eta$ will almost never satisfy a variable coefficient eikonal equation. Taking a hint from the asymptotic derivation of 4.4, compute

$$L\left(e^{iy.\eta/\epsilon}\,a(y)\right)$$
$$= e^{iy.\eta/\epsilon}\left(-\sum a_{\mu,\nu}\,\eta_\mu\eta_\nu/\epsilon^2 + i\left(2\sum_{\mu,\nu} a_{\mu,\nu}\,\eta_\nu\,\partial_\mu + b.\eta\right)/\epsilon + \sum_\mu b_\mu\partial_\mu\right)a \, .$$
$$\tag{5.10}$$

In order for this to be zero for nonzero a it is necessary that

$$\sum a_{\mu\nu}\,\eta_\mu\eta_\nu = 0 \, , \tag{5.11}$$

which is equivalent to $\eta \in \operatorname{Char} L$. Recall that for an mth order differential operator, the characteristic variety is defined by the equation $\det L_m(y,\eta) = 0$. Given (5.11), one must solve

$$\left(i\left(2\sum_{\mu,\nu} a_{\mu\nu}\eta_\nu\partial_\mu + b.\eta\right) + \epsilon\sum_\mu b_\mu\partial_\mu\right)a = 0 \, . \tag{5.12}$$

A first approximation, $a_0(y)$, to a solution of (5.12) is the solution of the transport equation corresponding to $\epsilon = 0$,

$$\left(2\sum_{\mu,\nu} a_{\mu\nu}\eta_\nu\partial_\mu + b.\eta\right)a_0 = 0 \, . \tag{5.13}$$

Either of two strategies can then be followed. One can posit an asymptotic expansion

$$a(y) \sim a_0(y) + \epsilon\,a_1(y) + \epsilon^2\,a_2(y) + \cdots \, . \tag{5.14}$$

and plug into (5.12). Alternatively, one can take a_0 as first approximation and iterate according to

$$i\left(2\sum_{\mu,\nu} a_{\mu\nu}\eta_\nu\partial_\mu + b.\eta\right) a^{m+1} = -\epsilon\sum b_\mu\partial_\mu\, a^m\,. \tag{5.15}$$

Both approaches lead to (5.14) with coefficients a_j recursively determined by the transport equations

$$i\left(2\sum_{\mu,\nu} a_{\mu\nu}\eta_\nu\partial_\mu + b.\eta\right) a_{j+1} = -\sum b_\mu\partial_\mu\, a_j\,. \tag{5.16}$$

The operator on the left of (5.16) corresponds to the vector field $\sum a_{\mu\nu}\eta_\nu\partial_\mu$. Initial conditions for the a_j must be prescribed on a surface transverse to this vector field.

Exercise. Suppose that $L_2 = \partial_t^2 - c^2\partial_1^2 - \partial_2^2$. Compute the characteristic variety and the direction and speed of transport for every $\eta \in \mathrm{Char}\,L$. Can you find a geometric picture involving ellipses which gives the direction and/or speed?

Exercise. Prove that if L in (5.9) is strictly hyperbolic then each spacelike hyperplane is transverse to every transport vector field. Recall that the strictly hyperbolic operators have characteristic variety given by a two sided light cone and the spacelike hyperplanes are those whose conormals lie inside the conical surfaces.

Theorem 5.1. *Suppose that L from (5.9) is strictly hyperbolic in (t,x) with t timelike, η is characteristic, and that $f_j(x) \in C^\infty(\mathbb{R}^d)$ for $j = 0, 1, \cdots$. Then there are smooth $a_j(t,x)$ uniquely determined by the transport equations (5.16) together with the initial conditions*

$$a_j(0,x) = f_j(x)\,.$$

If a is smooth and satisfies

$$a(\epsilon,y) \sim \sum_{j=0}^{\infty} \epsilon^j\, a_j(y)$$

then

$$L\left(a(\epsilon,y)\, e^{iy.\eta/\epsilon}\right) \sim 0\,. \tag{5.17}$$

It is reasonable to try the analogous *ansatz*

$$u^\epsilon \sim e^{iy.\eta/\epsilon}\left(a_0(y) + \epsilon a_1(y) + \epsilon^2 a_2(y) + \cdots\right) \tag{5.18}$$

with smooth vector valued coefficients a_j in the case of systems. We skip directly to operators with variable coefficients, and nonlinear phases as they are not essentially any more difficult and the *ansatz* for them is as natural as (5.18).

5.3. Variable coefficient systems and nonlinear phases

With the background of the elliptic case from Lecture 4, and the scalar constant coefficient hyperbolic case in 5.2, it is quite natural to seek solutions

$$L(y,\partial_y)\left(a(\epsilon,y)\, e^{i\phi(y)/\epsilon}\right) \sim 0\,. \tag{5.19}$$

with vector valued

$$a(\epsilon, y) \sim \sum_{j=0}^{\infty} \epsilon^j \, a_j(y) \, . \tag{5.20}$$

Here ϕ is assumed to be a smooth real valued function with $d\phi \neq 0$ on the domain of interest. This guarantees that the solutions are rapidly oscillating as $\epsilon \to 0$. Note that the surfaces of constant phase in space time have conormal vectors given by $d\phi$.

Computing formally one has

$$L(y, \partial_y) \left(e^{i\phi(y)/\epsilon} \sum_{j=0}^{\infty} \epsilon^j \, a_j(y) \right)$$

$$\sim e^{i\phi(y)/\epsilon} \left(\frac{1}{\epsilon} L_1(y, id\phi(y)) \, a_0 + \sum_{j=0}^{\infty} \epsilon^j \left[L_1(y, id\phi(y)) \, a_{j+1}(y) + L(y, \partial_y) \, a_j(y) \right] \right)$$

$$\tag{5.21}$$

The desired relation (5.19) holds if and only if

$$L_1(y, i \, d\phi(y)) \, a_0 = 0 \, , \tag{5.22}$$

and

$$L_1(y, id\phi(y)) \, a_j(y) + L(y, \partial_y) \, a_{j-1}(y) = 0 \, , \qquad \text{for} \quad j = 1, 2, \cdots . \tag{5.23}$$

Proposition 5.2. *If*

$$a(\epsilon, y) \sim \sum_{j=0}^{\infty} \epsilon^j \, a_j(y) \quad in \quad C^{\infty}(\Omega)$$

then

$$L(y, \partial_y) \left(a(\epsilon, y) \, e^{i\phi(y)/\epsilon} \right) \sim 0$$

if an only if the coefficients $a_j(y)$ satisfy the system of equations (5.22) and (5.23).

In order for (5.22) to admit nonzero solutions, the matrix valued function $L_1(y, id\phi) = iL_1(y, d\phi)$ must be singular. That is, $a_0(y)$ can only be nonzero where

$$\det L_1(y, d\phi(y)) = 0. \tag{5.24}$$

This is a nonlinear first order partial differential equation for the real valued phase ϕ. It is called the *eikonal equation*. It asserts that the graph of $d\phi$ belongs to the characteristic variety of L. For example, if L is the Maxwell System the equation reads

$$\partial_t \phi \left(|\partial_t \phi|^2 - c^2 |\nabla_x \phi|^2 \right) = 0 \, .$$

The physically relevant solutions are those which satisfy

$$|\partial_t \phi|^2 = c^2 |\nabla_x \phi|^2 \, . \tag{5.25}$$

In addition to plane wave solutions $\phi = t\tau + x.\xi$ with $\tau^2 = c^2|\xi|^2$, classical examples are the spherically symmetric solutions $\phi = t \pm |x|$.

Exercise. Are there other spherically symmetric solutions?

Exercise. Another interesting class of special solutions is those of the form

$$\phi(t,x) := t \pm c\psi(x)\,, \qquad \text{with} \qquad |\nabla \psi| = 1\,.$$

Interpret such functions ψ as distance from x to something.

Exercise. Arrive at equations (5.21), (5.23) beginning with $L\big(e^{i\phi/\epsilon}a(\epsilon, y)\big) \sim 0$ with a phase satisfying the eikonal equation, followed by a perturbative approach to the equation for a as in 4.4.

The eikonal equation (5.24) is a scalar first order nonlinear equation whose solutions can be parametrized by their initial data. The standard theory of such equations implies the following result. As a reference, consult just about any book on basic PDE except mine - for example Courant-Hilbert V.II, chap. II, F. John, chap. I., or Hormander vol.1 Thm 6.4.5.

Theorem 5.3. *Suppose that $g(x)$ is smooth, real valued, with nonvanishing differential on a region $\omega \subset \mathbb{R}^d$, and also that at $\underline{x} \in \omega$, λ is an eigenvalue of*

$$-\sum_{j=1}^{d} A_0(0,\underline{x})^{-1/2}\, A_j(0,\underline{x})\, \frac{\partial g(\underline{x})}{\partial x_j}\, A_0(0,\underline{x})^{-1/2}$$

with one-dimensional eigenspace. Then, there is a unique solution ϕ to (5.24) defined on a neighborhood of $\underline{x}$ and satisfying the initial conditions

$$\phi(0,x) = g(x)\,, \qquad \phi_t(0,\underline{x}) = \lambda.$$

Remark. The idea here is the following. At time 0, the derivatives $\frac{\partial \phi}{\partial x_j}$ are determined by the initial data. The aim is to find ϕ satisfying

$$\det\Big(A_0 \frac{\partial \phi}{\partial t} + \sum_{1}^{d} A_j \frac{\partial \phi}{\partial x_j}\Big) = 0$$

which is asking that $\frac{\partial \phi}{\partial t}(y)$ be an eigenvalue of the symmetric matrix

$$-\sum_{j=1}^{d} A_0^{-1/2}\, A_j(y)\, \frac{\partial \phi}{\partial x_j}\, A_0^{-1/2}\,.$$

The technical hypothesis of the theorem picks one eigenvalue at one point of the initial hyperplane. Since the eigenvalue is simple there is a uniquely determined smooth choice near that point, which specifies ϕ_t on a neighborhood of the point.

Hypothesis. *We assume that on an open connected subset Ω of space time,*

$$\ker L_1(y, d\phi(y))$$

has strictly positive dimension independent of y.

Equation (5.22) then implies that

$$a_0(y) \in \ker L_1(y, d\phi(y))\,.$$

Since $L_1(y, d\phi(y))$ is singular, it is not surjective. Thus the case $j = 1$ of (5.23) has information about a_0, namely

$$L(y, d\phi(y))\, a_0 \in \text{range } L_1(y, d\phi(y))\,.$$

The last two displayed equations are sufficient to determine a_0 from its initial data, though this is by no means clear at first sight.

Definitions. On Ω let $\pi(y)$ denote the orthogonal projection of $\mathbb{C}^N$ onto $\ker L_1(y, d\phi(y))$. Define $Q(y)$ to be the partial inverse of $L_1(y, id\phi(y))$ defined by

$$Q(y)\, \pi(y) = 0\,, \tag{5.26}$$

and for all $v \in \mathbb{C}^N$,

$$Q(y)\, L_1(y, id\phi(y))\, v = (I - \pi(y))\, v\,. \tag{5.27}$$

Then, $\pi(y)$ is a smooth matrix valued function of y, and equation (5.22) is equivalent to

$$\pi(y)\, a_0(y) = a_0(y) \tag{5.28}$$

which is the analogue of equation (5.4) from 5.1. As for equation (5.23), the analysis below models many similar arguments to follow and is therefore quite important.

The condition $L(y, \partial) a_0 \in \text{range } L_1(y, d\phi(y))$ is satisfied if and only if

$$\pi(y)\, L(y, \partial_t, \partial_x)\, a_0 = 0\,. \tag{5.29}$$

Equations (5.28) and (5.29) are our second formulation of the equations which determine a_0. They yield a symmetric hyperbolic system determining a_0 from its initial data. Equations (5.28), (5.29) imply

$$\pi(y) L(y, \partial_y) \pi(y)\, a_0(y) = 0\,. \tag{5.30}$$

Applying $(I - \pi)L$ to $(I - \pi)a_0 = 0$ gives

$$(I - \pi)\, L\, (I - \pi)\, a_0 = 0\,. \tag{5.31}$$

Adding (5.30), (5.31) yields the following system for a_0,

$$\pi\, L\, \pi\, a_0 + (I - \pi)\, L\, (I - \pi)\, a_0 = 0. \tag{5.32}$$

The coefficient matrices of the system,

$$\pi\, A_\mu \pi + (I - \pi)\, A_\mu\, (I - \pi)$$

are symmetric and the coefficient of ∂_t is

$$\pi\, A_0 \pi + (I - \pi)\, A_0\, (I - \pi)$$

which is positive definite since A_0 is. It follows that (5.32) is a symmetric hyperbolic system.

It remains to verify that the solution a_0 of (5.32) satisfies (5.28) and (5.29) as soon as it satisfies (5.28) at $t = 0$. Multiplying (5.32) by $(I - \pi)$ shows that a_0 satisfies

$$(I - \pi)\, L\, (I - \pi)\, a_0 = \pi L \pi\, a_0 = 0\,.$$

It follows that πa_0 satisfies the same symmetric hyperbolic system as a_0, and has the same initial data. It follows from the uniqueness of such solutions that $\pi a_0 = a_0$.

Theorem 5.4. *Suppose that the maximum speed c is defined by (2.28), and that the smooth real valued phase satisfies the eikonal equation and has nonvanishing differential on*

$$\Omega_T := \{\, (t,x) : 0 \le t \le T \le R/c, \quad and, \quad |x - a| \le R - ct \,\} \qquad (5.33)$$

Given smooth functions $g_j(x)$ satisfying $\pi(0,x)g_j = g_j$, there is one and only one sequence $a_j \in C^\infty(\Omega)$ satisfying (22), (23) and the initial conditions

$$\pi(0,x)\,a_j(0,x) = g_j(x) \qquad j = 0, 1, \cdots . \qquad (5.34)$$

As initial data, what is needed is the projections $\pi(y)\,a_j$. For $j = 0$ this is equal to a_0 but for the others represents only part of the values.

This result has a very long history. On one hand it is a generalization to partial differential equations of results with an even longer history in the context of ordinary differential equations. This is clearly presented in the lecture notes of Alinhac cited in the references. On the other hand, one finds such expansions for constant coefficient Cauchy problems with highly oscillatory data by expressing the exact solution as a Fourier integral and then evaluating by the method of stationary phase. A well presented example of this type is given in 12.2 of [H, vol II]. Such methods are crucial, in particular for explaining the behavior beyond caustics, and in particular the double mirror experiment of Guoy. A third line is the closely related WKB method which is much used in quantum theory.

Proof. In order to prove the assertions about the principal term a_0 it suffices to show that the maximal speed of propagation of the symmetric hyperbolic operator $\pi\, L\pi + (I - \pi)\, L\,(I - \pi)$ is no larger than c. In that case the domains Ω are domains of determinacy and the result follows from Theorem 2.6. The assertion about the speeds is a consequence of the fact that if $L_1(y,\eta)$ is a nonnegative matrix, then the same is true of the principal symbol $\pi\, L_1(y,\eta)\pi + (I - \pi)\, L_1(y,\eta)\,(I - \pi)$ for equation (5.32).

We next show how a_1 is determined. The inductive construction of a_j for $j \ge 1$ should then be clear. To simplify notation, let

$$M = M(y) := L_1(y, d\phi(y)) . \qquad (5.35)$$

Write

$$a_1 = \pi\, a_1 + (I - \pi)\, a_1 . \qquad (5.36)$$

The term $(I - \pi)a_1$ is determined by the $(I - \pi)$ projection of the $j = 1$ instance of (5.23),

$$(I - \pi)\left(M a_1 + L(y, \partial_y)a_0 \right) = 0. \qquad (5.37)$$

Recall that Q is a partial inverse of M so multiplying (5.37) by Q yields

$$(I - \pi)a_1 = -Q\, L(y, \partial_y)\, a_0 . \qquad (5.38)$$

Once this is known, πa_1 is determined by constructing a symmetric hyperbolic system as we did for a_0 above. This hyperbolic equation comes from the projection on the kernel of the case $j = 2$ of (5.23),

$$\pi\left(L(y, \partial_y)\, a_1 + M\, a_2 \right) = 0. \qquad (5.39)$$

The pattern is that a_j is determined so that $(I - \pi)$ of the case j and π of the case $j + 1$ are satisfied. Thus the coefficient a_1 must satisfy (5.37) and (5.39).

Denote by C the undetermined part of a_1 so

$$a_1 = C - Q\,L\,a_0\,, \qquad \pi\,C = C\,. \tag{5.40}$$

Use the fact that $\pi M = 0$ and insert the already determined value for $(I - \pi)a_1$ in (39) to find

$$\pi\,L(y, \partial_y)\,C - \pi L(y, \partial_y)\,Q\,L(y, \partial_y)\,a_0 = 0\,. \tag{5.41}$$

Since $C = \pi C$, adding zero to equation (5.41) yields the symmetric hyperbolic system

$$(I - \pi)L(I - \pi)C + \pi L \pi C = \pi\,L(y, \partial_y)\,Q\,L(y, \partial_y)\,a_0 \tag{5.42}$$

with known source term on the right. This uniquely determines C from its initial values. Note that it is the projection of $\pi\,a_1$ at $t = 0$ which is required initial data. The part $(I - \pi)a_1$ at $t = 0$ has already been determined.

As in the case of a_0, multiplying (5.42) by $(I - \pi)$ shows that C satisfies

$$(I - \pi)\,L(y, \partial_y)\,(I - \pi)\,C = 0 = \pi\,L(y, \partial_y)\,\pi\,C - \pi\,L(y, \partial_y)\,Q\,L(y, \partial_y)\,a_0\,. \tag{5.43}$$

Thus, if C satisfies (42) then so does πC. Since $\pi C = C$ at $t = 0$, it follows that πC satisfies the same initial value problem as C so

$$\pi C = C\,. \tag{5.44}$$

It remains to verify that a_1 given by (5.40) indeed satisfies the two equations (5.37) and (5.39). This is immediate since (5.37) follows from (5.38), and (5.39) follows from (5.40), (5.43) and (5.44).

This completes the construction of a_1 at which point the case $j = 1$ and π times the case $j = 2$ of (5.23) are satisfied. The inductive construction of the a_j is left to the reader. $\qquad \square$

Exercise. Construct the inductive proof.

Theorem 5.5 [Lax, Duke J. 1957]. *Suppose that c, Ω_T, ϕ, and a_j are as in the last theorem and that*

$$a(\epsilon, y) \sim \sum_{j=0}^{\infty} \epsilon^j\, a_j(y) \qquad in \quad C^{\infty}(\Omega_T)\,.$$

Suppose that $u(\epsilon, y) \in C^{\infty}(\Omega_T)$ is the exact solution of the initial value problem

$$L(y, \partial_y)\,u(\epsilon, y) = 0\,, \qquad u(\epsilon, 0, x) = a(\epsilon, 0, x)\,e^{i\phi(0, x)/\epsilon}\,. \tag{5.45}$$

Then

$$u(\epsilon, y) \sim e^{i\phi(y)/\epsilon}\, a(\epsilon, y) \qquad in \quad C^{\infty}(\Omega_T)\,. \tag{5.46}$$

Remark. As in Theorem 4.4, neither the exact solution u^ϵ nor the approximation $e^{i\phi(y)/\epsilon}\,a(\epsilon, y)$ is smooth at $\epsilon = 0$.

Proof. For any $m, s \in \mathbb{N}$ there is a constant C so that

$$\|L(y, \partial_y)\left(u(\epsilon, y) - e^{i\phi(y)/\epsilon}\, a(\epsilon, y)\right)\|_{H^s(\Omega_T)} \leq C\, \epsilon^m \,,$$

and

$$\| u(\epsilon, 0, x) - e^{i\phi(0,x)/\epsilon}\, a(0, x) \|_{H^s(|x-a|\leq R)} \leq C\, \epsilon^m \,.$$

The basic linear H^s energy estimate from Lecture 2 implies that

$$\|u(\epsilon, y) - e^{i\phi(y)/\epsilon}\, a(\epsilon, y)\|_{H^s(\Omega_T)} \leq C'(T, m, s)\, \epsilon^m \,.$$

Since this is true for all m, s the result follows from Sobolev's Embedding Theorem. $\qquad\qquad\qquad\qquad\qquad\qquad\qquad\qquad\qquad\qquad\qquad\qquad\qquad\square$

Exercise. Verify that $\phi := t - |x|$ satisfies the eikonal equation for the wave operator $\square$. Determine the transport equation that a_0 must satisfy in order that it be the principal term in an asymptotic expansion

$$u^\epsilon \sim e^{i(t-|x|)/\epsilon}\left(a_0(y) + \epsilon a_1(y) + \cdots\right)$$

of $\square u^\epsilon \sim 0$. Observe that the amplitudes decay more rapidly as d increases. Do you understand why? If not, return to this question after finishing the section. Perform the same computation for each of the three operators

$$\square + \frac{\partial}{\partial t}\,, \qquad \square + \frac{\partial}{\partial x_1}\,, \qquad \square + 1\,.$$

Compare the decay of amplitudes for the four equations.

For the study of nonlinear equations, it is important to understand the effect of oscillatory source terms. The case of nowhere characteristic phase is treated in Lecture 4. The case of an everywhere characteristic phase is analyzed exactly as above. The result is the following.

Theorem 5.6 [Lax]. *Suppose that c, Ω_T and the real phase ϕ satisfying the eikonal equation are as above. Given smooth functions $b_j \in C^\infty(\overline{\Omega})$ there are uniquely determined amplitudes $\underline{a}_j \in C^\infty(\overline{\Omega})$ satisfying $\pi(y)\underline{a}_j(0, x) = 0$ and $\underline{a}_0(0, x) = 0$ and so that if*

$$\underline{a}(\epsilon, y) \sim \sum_{j=0}^{\infty} \epsilon^j\, \underline{a}_j(y), \quad \text{and} \quad b(\epsilon, y) \sim e^{i\phi(y)/\epsilon} \sum \epsilon^j b_j(y)$$

then

$$L(y, \partial_y)\left(\sum \epsilon^j\, \underline{a}_j(y)\, e^{i\phi(y)/\epsilon}\right) \sim b(\epsilon, y)\, e^{i\phi(y)/\epsilon} \quad in \quad C^\infty(\Omega_T).$$

The principal amplitude is determined by the pair of equations

$$\pi(y)\, \underline{a}_0 = \underline{a}_0\,, \qquad \pi(y)\, L(y, \partial_y)\, \pi(y)\, \underline{a}_0 = b_0\,, \tag{5.47}$$

with the initial condition $\underline{a}_0(0, x) = 0$.

This result shows that a source of size one with characteristic phase yields waves of size one. This contrasts with the case of noncharacteristic phases in Lecture 4 where the response is order ϵ.

5.4 Rays and transport

One of the key ideas in geometric optics is transport along rays. The way that the equations for a_0 are presented above, this property is not evident. The advantage of the previous analysis is that some additional phenomena, like conical refraction, are included. In that case the kernel has dimension two and the symmetric hyperbolic system determining a_0 has propagation properties like the 1+2 dimensional wave equation. In particular the propagation is not simply transport. See [Ludwig], or [JMR, Ann. Inst. Fourier] for a discussion of conical refraction. In this section the role of rays is brought to the fore.

It has already been supposed that $\ker L_1(y, d\phi(y))$ has dimension independent of y. For the remainder of this section we suppose that *this dimension is equal to one*. Then $\pi(y)$ is a projector of rank 1, and the polarization (5.28) determines a_0 up to a scalar multiple.

Using the product rule for the derivative $\partial_\mu(\pi a_j)$ in (5.30) yields

$$\pi A_\mu \pi \partial_\mu a_0 + \pi (A_\mu(\partial_\mu \pi) + B)\pi a_0 = 0 \tag{5.48}$$

Each matrix $\pi(y)A_\mu(y)\pi(y)$ defines a linear transformation from $\ker L_1(y, d\phi(y))$ to itself. Since this is one dimensional, there are uniquely determined scalars $v_\mu(y)$ such that

$$\pi(y)A_\mu(y)\pi(y) = v_\mu(y)\,\pi(y)\,.$$

Similarly there is a unique scalar valued $\gamma(y)$ such that

$$\pi(y)\big(A_\mu(\partial_\mu \pi) + B\big)\pi(y) = \gamma(y)\,\pi(y)\,.$$

So, if we define the vector field $V(\partial_y)$ by

$$V(\partial_y) := \sum_{\mu=0}^{3} v_\mu(y)\,\frac{\partial}{\partial y_\mu} \tag{5.49}$$

then equation (5.30) for a_0 is equivalent to the transport equation

$$V(\partial_y)\, a_0 + \gamma(y)\, a_0 = 0\,. \tag{5.50}$$

This is a simple ordinary differential equation along the integral curves of V.

Definition. The integral curves of V are called *rays*. They are determined by L and the phase ϕ. Equation (50) is called the *transport equation for the amplitude* a_0.

The rays are defined using both the principal symbol $L_1(y, \eta)$ of the differential equation and the phase ϕ which satisfies the eikonal equation. Recall that the solutions of the eikonal equation are constructed as follows. The graph of $d\phi$ that is the set of points $(y, d\phi(y))$ is generated from the initial points over $t = 0$ by flowing along the Hamilton vector field of

$$p(y, \eta) := \det L_1(y, \eta)$$

given by the classical expression

$$\sum_{\mu=0}^{d} \left(\frac{\partial p}{\partial \eta_\mu}\frac{\partial}{\partial y_\mu} - \frac{\partial p}{\partial y_\mu}\frac{\partial}{\partial \eta_\mu}\right).$$

The function p is constant on integral curves. The integral curves along which $p = 0$ are curves in y, η space called *bicharacteristics* or *null bicharacteristics*. Thus, the graph of $d\phi$ is foliated by a family of bicharacteristics.

Theorem 5.7. *The projection on spacetime of the bicharacteristics foliating the graph of $d\phi$ are exactly the rays defined above.*

In other words

$$\sum \frac{\partial p}{\partial \eta_\mu}\Big|_{(y,d\phi(y))} \frac{\partial}{\partial y_\mu} \quad \text{and} \quad V(\partial_y)$$

are parallel.

Proof. By hypothesis,

$$L_1(y, d\phi(y)) = \sum A_\mu(y) \frac{\partial \phi(y)}{\partial y_\mu}$$

has kernel of dimension 1. Fix y. Choose an orthonormal basis for $\mathbb{C}^N$ whose first basis vector belongs to this kernel. In this basis, the symmetric matrix $L_1(y, d\phi(y))$ has first row and first column equal to zero. The lower $(N-1) \times (N-1)$ block is nonsingular,

$$L_1(y, d\phi(y)) = \begin{pmatrix} 0 & 0 \\ 0 & M_{(N-1)\times(N-1)}, \end{pmatrix} \quad \det M(y) \neq 0.$$

Denote by $A_{1,1}(y)$ the upper left matrix element of the matrix A. A direct evaluation shows that

$$\frac{\partial p}{\partial \eta_\mu}(y, d\phi(y)) = \left(A_\mu(y)\right)_{1,1} \det M.$$

Exercise. Verify this.

The matrix $\pi(y)$ in these coordinates is given by

$$\pi(y)_{i,j} = \begin{cases} 1 & \text{if } i = j = 1 \\ 0 & \text{otherwise} . \end{cases}$$

Thus, the definition of V yields,

$$V(\partial_y) = \sum \left(A_\mu(y)\right)_{1,1} \frac{\partial}{\partial y_\mu}.$$

Thus, the projection of the Hamilton field is parallel to V. It follows that the projection of the integral curves of the Hamilton field are integral curves of V. $\square$

Remark. The operator $\pi(y) L_1(\partial) \pi(y)$ is a simple transport operator under the more general hypothesis that on a neighborhood of $(y, d\phi(y))$ the characteristic variety of L is a smooth embedded hypersurface (see [JMR 1995-1996]).

Exercise. Let

$$L := \partial_t + \begin{pmatrix} 1 & 0 \\ 0 & -1 \end{pmatrix} \frac{\partial}{\partial x_1} + \begin{pmatrix} 0 & 1 \\ 1 & 0 \end{pmatrix} \frac{\partial}{\partial x_2}, \tag{5.51}$$

and

$$\phi(y) := t - |x| \tag{5.52}$$

on $x \neq 0$. Compute the characteristic variety of L. Show that ϕ satisfies the eikonal equation. Compute the rays associated to L and ϕ.

Exercise. With the same L and for those linear phases $y.\eta$ which satisfy the eikonal equation, find the associated rays.

If $\omega \subset \{t = 0\}$ is a nice bounded open set the family of rays starting in ω is called a *bundle* or *tube of rays*. It is denoted $\mathcal{T}$ and its section at time t is denoted $\omega(t)$. In particular, $\omega = \omega(0)$.

The transport equation (5.50) shows how the amplitude varies along rays. Intuitively, when a bundle narrows the wave is squeezed and the amplitudes grow. If there are dissipative mechanisms, this growth may be suppressed. For conservative systems the intuitive argument just given can be made rigorous as follows.

Recall that the energy identity asserts that when $Lu = 0$,

$$\frac{d}{dt} \int \langle A_0(y)u(t,x), u(t,x) \rangle \, dx + \int \langle (B + B^* - \sum \frac{\partial A_\mu}{\partial y_\mu})u(t,x), u(t,x) \rangle \, dx = 0.$$

Definition. A symmetric hyperbolic system is *conservative* if

$$B + B^* - \sum \frac{\partial A_\mu}{\partial y_\mu} \equiv 0.$$

Proposition. *A system is conservative if and only if the energy $\int_{\mathbb{R}^d} \langle A_0 u(t,x), u(t,x) \rangle \, dx$ is independent of time for all solutions of $Lu = 0$ compactly supported in x.*

Exercise. Sufficiency is clear. Prove necessity.

Theorem 5.8. *Suppose that $\ker L_1(y, d\phi(y))$ is one dimensional for all y and that the system L is conservative. If $\mathcal{T}$ is a tube of rays and $a(\epsilon, y)e^{i\phi(y)/\epsilon}$ is an asymptotic solution of $Lu \sim 0$ as in Lax's theorem, then the energy in the tube is conserved in the sense that*

$$\int_{\omega(t)} \langle A_0(y) \, a_0(t,x), a_0(t,x) \rangle \, dx$$

is independent of t.

Proof. For $0 < \delta << 1$, choose a cutoff function $0 \leq \chi_\delta(x) \leq 1$ such that χ is equal to one on ω_0 and $\chi(x) = 0$ when $\mathrm{dist}\,(x, \omega_0) > \delta$. Construct a Lax solution $\tilde{a}(\epsilon, y)e^{i\phi(y)/\epsilon}$ with

$$\tilde{a}_0(0, x) = \chi_\delta(x) \, a_0(x). \tag{5.53}$$

Then Lax's Theorem together with conservation of energy implies that for all m and t

$$\int_{\mathbb{R}^d} \langle A_0 \, \tilde{a}(\epsilon, t, x), \tilde{a}(\epsilon, t, x) \rangle \, dx - \int_{\mathbb{R}^d} \langle A_0 \, \tilde{a}(0, x), \tilde{a}(0, x) \rangle \, dx = O(\epsilon^m).$$

In addition the quantity on the left is controlled by its principal term, so

$$\int_{\mathbb{R}^d} \big\langle A_0 \, \tilde{a}_0(t,x), \tilde{a}_0(t,x) \big\rangle \, dx - \int_{\mathbb{R}^d} \big\langle A_0 \, \tilde{a}_0(0,x), \tilde{a}_0(0,x) \big\rangle \, dx = O(\epsilon^2) \, . \qquad (5.54)$$

Since the left hand side is independent of ϵ it must vanish identically.

By construction, for all t

$$\int_{\mathbb{R}^d} \big\langle A_0 \, \tilde{a}_0(t,x), \tilde{a}_0(t,x) \big\rangle \, dx - \int_{\omega(t)} \big\langle A_0 \, a_0(t,x), a_0(t,x) \big\rangle \, dx = O(\delta) \, . \qquad (5.55)$$

The result follows from (5.54) and (5.55) by letting δ tend to zero. $\qquad\square$

Consider $\omega(0)$ shrinking to a point $\underline{x}$ so the tube converges to the ray through $\underline{x}$. Then

$$\frac{\mathrm{vol}\,(\omega(t))}{\mathrm{vol}\,(\omega(0))} \to J(t,\underline{x})$$

where $J(t,\underline{x})$ is the Jacobean $\det \frac{d\Psi}{dx}$ of the flow $\Psi(t,x)$ generated by the vector field V.

The law of conservation of energy applied to a tube of diameter δ implies that

$$\mathrm{vol}\,(\omega(t)) \big\langle A_0(\psi(t,\underline{x}))\, a_0(\psi(t,\underline{x})) \,,\, a_0(\psi(t,\underline{x})) \big\rangle =$$
$$\mathrm{vol}\,(\omega(0)) \big\langle A_0(\psi(0,\underline{x}))\, a_0(\psi(0,\underline{x})) \,,\, a_0(\psi(0,\underline{x})) \big\rangle (1 + O(\delta)) \, .$$

Dividing by $\mathrm{vol}\,(\omega(0))$ and passing to the limit $\delta \to 0$ implies that the quantity

$$\big\langle A_0(\psi(t,\underline{x}))\, a_0(\psi(t,\underline{x})) \,,\, a_0(\psi(t,\underline{x})) \big\rangle J(t,\underline{x}) \qquad (5.56)$$

is independent of t.

Proposition 5.9. *If $a(\epsilon,y)\, e^{i\phi(y)/\epsilon}$ is a Lax asymptotic solution of a conservative system then the quantity*

$$\big\langle A_0(y)\, a_0(y) \,,\, a_0(y) \big\rangle J(y)$$

is constant on the rays associated to ϕ.

This expresses precisely the relationship between the amplitude a_0 and the compression or expansion of the rays. The compensation of amplitude and spread of rays arranges exactly for the conservation of energy. An analogous situation occurs for the incompressible flow of fluid in a pipe or channel where conservation of mass implies that flow is faster where the channel narrows and slower where the channel is wide. Wide passages in rivers are slow moving and narrows are rapid.

LECTURE 6
Basic Nonlinear Existence Theorems

6.1. Introduction

Nonlinear equations are classified according to the strength of the nonlinearity. The key criterion is what order terms in the equation are nonlinear and a secondary condition is the growth of the nonlinear terms at infinity. When solutions are uniformly bounded in absolute value, the behavior at infinity is not important.

Among the nonlinear equations in applications two sorts are most common. *Semilinear equations* are linear in their principal part. First order semilinear symmetric hyperbolic systems take the form

$$L(y, \partial_y)\, u + F(y, u) = f(y)\,, \qquad F(y, 0) \equiv 0 \tag{6.1}$$

where L is a symmetric hyperbolic operator, and the nonlinear function is a smooth map from $\mathbb{R}^{1+d} \times \mathbb{C}^N \to \mathbb{C}^N$ whose partial derivatives of all orders are uniformly bounded on sets of the form $\mathbb{R}^{1+d} \times K$, with compact $K \subset \mathbb{C}^N$. The derivatives are standard partial derivatives and not derivatives in the sense of complex analysis.

More strongly nonlinear, and typical of compressible inviscid fluid dynamics, are the quasilinear systems,

$$L(y, u, \partial_y)\, u + F(y, u) = f(y)\,, \tag{6.2}$$

where

$$L(y, u, \partial_y) = \sum_{j=0}^{d} A_j(y, u)\, \partial_j \tag{6.3}$$

has coefficients which are smooth hermitian symmetric matrix valued functions with derivatives bounded on $\mathbb{R}^{1+d} \times K$ as above. A_0 is assumed uniformly positive on such sets.

For semilinear equations there is a natural local existence theorem requiring data in $H^s(\mathbb{R}^d)$ for some $s > d/2$. The theorem gives solutions which are continuous functions of time with values in $H^s(\mathbb{R}^d)$. This shows that the spaces $H^s(\mathbb{R}^d)$ with $s > d/2$, are natural configuration spaces for the dynamics. Once a solution belongs to such a space, it is bounded and continuous so that $F(y, u)$ is well defined and in fact is bounded and continuous.

431

For quasilinear equations, the local existence theorem requires an extra derivative, that is initial data in $H^s(\mathbb{R}^d)$ with $s > 1 + d/2$. Again the solution is a continuous functions of time with values in $H^s(\mathbb{R}^d)$.

We give the details of the the semilinear result. The key step in the proof uses Schauder's Lemma, which bounds the $H^s(\mathbb{R}^d)$ norm of the composition $F(y, u)$ in terms of the $H^s(\mathbb{R}^d)$ norm of u.

6.2. Schauder's Lemma and Sobolev Embedding

Theorem 6.1 Schauder's Lemma. *Suppose that $G(x, u)$ is a smooth function $\mathbb{R}^d \times \mathbb{C}^N \to \mathbb{C}^N$ such that $G(x, 0) = 0$ and every partial derivative of order $\leq s+1$ is uniformly bounded on sets of the form $\mathbb{R}^d \times K$. Then the map $w \mapsto G(x, w)$ sends $H^s(\mathbb{R}^d)$ to itself provided $s > d/2$. The map is uniformly lipshitzian on bounded subsets of $H^s(\mathbb{R}^d)$.*

Proof of Schauder's Lemma for integer s. For the sake of exposition, assume that $G = G(w)$ depends only on w. The key step is to estimate the H^s norm of $G(w)$ assuming that $w \in \mathcal{S}$. We prove that there is a constant $C = C(R)$ so that

$$w \in \mathcal{S}(\mathbb{R}^d) \quad \text{and} \quad \|w\|_{H^s(\mathbb{R}^d)} \leq R \quad \text{imply} \quad \|G(w)\|_{H^s(\mathbb{R}^d)} \leq C(R).$$

An approximation argument then proves the first assertion of the theorem.

For $w \in \mathcal{S}$, consider a derivative $\partial_x^\alpha G(w(x))$ with $|\alpha| \leq s$. Leibniz' rule implies that it is a finite sum of terms of the form

$$G^{(\gamma)}(w) \, \Pi_{j=1}^J \, \partial_x^{\alpha_j} w \tag{6.4}$$

where $|\gamma| = J \leq s$, and $\alpha_1 + \cdots + \alpha_J = \alpha$.

The goal is to estimate (6.4) in L^2. Following [Rauch 1983] we work entirely in L^2 using the Fourier transform to show that

$$\|G^{(\gamma)}(\omega) \, \Pi_{j=1}^J \, \partial_x^{\alpha_j} w\|_{L^2} \leq C(R). \tag{6.5}$$

Since $s > d/2$, $H^s(\mathbb{R}^d) \subset L^\infty(\mathbb{R}^d)$ and

$$\|w\|_{L^\infty(\mathbb{R}^d)} \leq C \, \|w\|_{H^s(\mathbb{R}^d)}. \tag{6.6}$$

Inequality (6.6) for elements of the Schwartz space $\mathcal{S}(\mathbb{R}^d)$ is an immediate consequence of the Fourier Inversion Formula. Write

$$w(x) = (2\pi)^{-d/2} \int_{\mathbb{R}^d} e^{-ix.\xi} \, \hat{w}(\xi) \, d\xi = (2\pi)^{-d/2} \int_{\mathbb{R}^d} \frac{e^{-ix.\xi}}{<\xi>^s} <\xi>^s \, \hat{w}(\xi) \, d\xi.$$

The Schwarz inequality yields

$$|w(x)| \leq \left\| \frac{1}{<\xi>^s} \right\|_{L^2(\mathbb{R}^d)} \|w\|_{H^s(\mathbb{R}^d)}.$$

The first factor on the right is finite if and only if $s > d/2$. An approximation argument finishes the proof of (6.6).

Thus

$$\|G^{(\gamma)}(w)\|_{L^\infty(\mathbb{R}^d)} \leq C(R).$$

The key is the following estimate which is applied to the product in (6.5).

Lemma 6.2. *If $s > d/2$ there is a constant $C = C(s,d)$ so that for all $w_j \in \mathcal{S}(\mathbb{R}^d)$ and all multiindices α_j with $s' := \sum |\alpha_j| \leq s$,*

$$\| \Pi_{j=1}^J \partial_x^{\alpha_j} w_j \|_{L^2(\mathbb{R}^d)} \leq C \, \Pi_{j=1}^J \| w_j \|_{H^s(\mathbb{R}^d)} \, .$$

Proof. Thanks to Plancherel's theorem, it suffices to estimate the L^2 norm of the Fourier transform of the product. Set

$$g_i := < \xi >^{s-|\alpha_i|} \widehat{\partial_x^{\alpha_i} w_i} \, ,$$

where $< \xi > \equiv (1 + |\xi|^2)^{1/2}$, so $\|g_i\|_{L^2} \leq c\|w_i\|_{H^s}$. Denoting by $\mathcal{F}$, the Fourier transform,

$$\mathcal{F}\big(\Pi_{j=1}^J \partial_x^{\alpha_j} w_{l_j}\big)(\xi_1) = \frac{g_1}{< \xi >^{s-|\alpha_1|}} * \frac{g_2}{< \xi >^{s-|\alpha_2|}} * \cdots * \frac{g_J}{< \xi >^{s-|\alpha_J|}}(\xi_1)$$

$$= \int_{\mathbb{R}^{d(J-1)}} \frac{g_1(\xi_1 - \xi_2)}{< \xi_1 - \xi_2 >^{s-|\alpha_1|}} \cdots \frac{g_J(\xi_J)}{< \xi_j >^{s-|\alpha_J|}} \, d\xi_2 \ldots d\xi_J \, .$$

$$(6.7)$$

For each ξ, at least one of the J numbers $< \xi_1-\xi_2 >, \ldots , < \xi_{J-1}-\xi_J >, < \xi_J >$ is maximal. Suppose it's the bth number $< \xi_b - \xi_{b+1} >$ with the convention that $\xi_{J+1} \equiv 0$. Then since $\sum |\alpha_i| \leq s$,

$$\langle \xi_b - \xi_{b+1} \rangle^{s-|\alpha_b|} \geq \langle \xi_b - \xi_{b+1} \rangle^{\sum_{j \neq b} |\alpha_j|} \geq \Pi_{j \neq b} \langle \xi_j - \xi_{j+1} \rangle^{|\alpha_j|}$$

which implies that

$$\Pi_{j=1}^J \langle \xi_j - \xi_{j+1} \rangle^{s-|\alpha_j|} = \langle \xi_b - \xi_{b+1} \rangle^{s-|\alpha_b|} \Pi_{j \neq b} \langle \xi_j - \xi_{j+1} \rangle^{s-|\alpha_j|} \geq \Pi_{j \neq b} \langle \xi_j - \xi_{j+1} \rangle^s \, .$$

Thus the integrand on the right side of (6.7) is dominated by

$$\left| g_b(\xi_b - \xi_{b+1}) \, \Pi_{j \neq b} \frac{g_j(\xi_j - \xi_{j+1})}{\langle \xi_j - \xi_{j+1} \rangle^s} \right| , \qquad (6.8)$$

so for any $\xi_1 \in \mathbb{R}^d$, the integrand in (6.7) is dominated by the sum over b of the terms (6.8). Hence

$$\|\mathcal{F}\big(\Pi_{j=1}^J \partial_x^{\alpha_j} w_{l_j}\big)\|_{L^2(\mathbb{R}^d)} \leq \| \sum_{b=1}^J \frac{|g_1|}{< \xi >^s} * \cdots * |g_b| * \frac{|g_{b+1}|}{< \xi >^s} * \cdots * \frac{|g_J|}{< \xi >^s} \|_{L^2}$$

$$\leq \sum_{b=1}^J \left\| \frac{g_1}{< \xi >^s} \right\|_{L^1} \cdots \left\| \frac{g_{b-1}}{< \xi >^s} \right\|_{L^1} \|g_b\|_{L^2} \left\| \frac{g_{b+1}}{< \xi >^s} \right\|_{L^1} \cdots \left\| \frac{g_J}{< \xi >^s} \right\|_{L^1}$$

where the last step uses Young's inequality.

Since $s > d/2$, $< \xi >^{-s} \in L^2(\mathbb{R}^d)$ so the Schwarz inequality yields

$$\| \frac{g_j}{< \xi >^s} \|_{L^1} \leq c\|g_j\|_{L^2} \leq C\|w_j\|_{H^s} \, .$$

Plugging this in the previous estimate proves the lemma. $\square$

To prove the Lipshitz continuity asserted in Schauder's Lemma it suffices to show that for all R there is a constant $C(R)$ so that

$$w_j \in \mathcal{S}(\mathbb{R}^d) \quad \text{for } j = 1, 2 \qquad \text{and} \qquad \|w_j\|_{H^s(\mathbb{R}^d)} \leq R$$

imply

$$\|G(w_1) - G(w_2)\|_{H^s(\mathbb{R}^d)} \leq C \|w_1 - w_2\|_{H^s(\mathbb{R}^d)} \,.$$

Toward that end write

$$G(w_1) - G(w_2) = \int_0^1 G'(w_2 + \theta(w_1 - w_2))\, d\theta \, (w_1 - w_2) \,.$$

The estimates of the first part show that the family of functions $G'(w_2 + \theta(w_1 - w_2))$ parametrized by θ is bounded in $H^s(\mathbb{R}^d)$. Thus

$$\left\| \int_0^1 G'(w_2 + \theta(w_1 - w_2))\, d\theta \right\|_{H^s(\mathbb{R}^d)} \leq C(R) \,.$$

Applying the Lemma to the expression for $G(w_1) - G(w_2)$ as a product of two terms yields the desired estimate. $\qquad\square$

The standard proof of Schauder's Lemma for integer s uses the L^p version of the Sobolev Embedding Theorem. The simplest such result, which we have already used, is that for $s > d/2$, $H^s(\mathbb{R}^d) \subset L^\infty(\mathbb{R}^d)$ and (6.6) holds. To see how L^p Sobolev estimates are used consider the proof that u^2 belongs to $H^2(\mathbb{R}^2)$ as soon as u does. One must show that u^2, $\partial(u^2)$, and, $\partial^2(u^2)$ are square integrable.

For the first one needs to show that $u \in L^4$. Estimate (6.6) implies that $u \in L^\infty$. Since $u \in H^2$ it follows that $u \in L^2$. It follows that $u \in L^p$ for all $p \geq 2$.

For the first derivative write $\partial(u^2) = 2u\partial u$, which is the product of a bounded function and a square integrable function and so is in L^2.

The second derivative is more interesting. Write

$$\partial^2(u^2) = u\partial^2 u + 2(\partial u)^2 \,.$$

The first is a product $L^\infty \times L^2$ so is L^2. For the second, one needs to know that $\partial u \in L^4$.

The general result of this sort is the following. Proofs can be found in the treatises of Hormander and Taylor for example. In contrast to the special case of estimate (6.6), most proofs proceed by expressing u in terms of its derivatives with the aid of the Fundamental Theorem of Calculus.

Sobolev Embedding Theorem 6.3. *If $1 \leq s \in \mathbb{R}$ and $\alpha \in \mathbb{N}^d$ is a multiindex with $0 < s - |\alpha| < d/2$, there is a constant $C = C(\alpha, s, d)$ independent of $u \in H^s(\mathbb{R}^d)$ so that*

$$\| \partial_y^\alpha u \|_{L^{p(\alpha)}} \leq C \, \| \, |\xi|^s \, \hat{u}(\xi) \, \|_{L^2(\mathbb{R}^d)} \,, \tag{6.9}$$

where

$$p(\alpha) := \frac{2d}{d - 2s + 2|\alpha|} \,. \tag{6.10}$$

For $s - |\alpha| > d/2$, $\partial_y^\alpha u$ is bounded and continuous and

$$\|\partial_y^\alpha u\|_{L^\infty} \leq C \|u\|_{H^s(\mathbb{R}^d)} \,.$$

For $s - |\alpha| = d/2$, one has

$$\|\partial_y^\alpha u\|_{L^p(\mathbb{R}^d)} \leq C(p, s, \alpha) \, \|u\|_{H^s(\mathbb{R}^d)}$$

for all $2 \leq p < \infty$.

The formula for $p(\alpha)$ is forced by dimensional analysis. For a fixed nonzero $\psi \in C_0^\infty$, consider $u_\lambda(x) := \psi(\lambda x)$. The left hand side of (9) then is of the form $c\lambda^a$ for some a. Similarly the right hand side is of the form $c'\lambda^b$ for some b. In order for the inequality to hold, one must have $\lambda^a \leq c''\lambda^b$ for all positive λ so it is necessary that $a = b$. A simple computation shows that this holds exactly when p is given by (6.10).

Exercise. Verify this scaling argument.

Another way to look at the scaling argument is that for dimensionless u the left hand side of (6.9) has dimensions $length^{(d-p|\alpha|)/p}$ while the right hand side has dimensions $length^{(d-2s)/2}$. The formula for p results from equating these two expressions.

Returning to second derivatives $\partial^2(u^2)$, take $s = d = 2$, and $|\alpha| = 1$ to show that $\partial u \in L^p$ for all p. This completes the proof that $u^2 \in H^2$.

Standard proof of Schauder's Lemma. The usual proof uses the Sobolev estimates. together with Hölder's inequality. Hölder's inequality yields

$$\sum_{k=1}^{J} \frac{1}{p_k} = \frac{1}{2} \quad \Rightarrow \quad \| \partial_x^{\alpha_1} w_{l_1} \cdots \partial_x^{\alpha_J} w_{l_J} \|_{L^2} \leq \Pi_{k=1}^{J} \|\partial_x^{\alpha_k} w_{l_k}\|_{L^{p_k}}.$$

Since each factor $\partial_x^{\alpha_k} w_{l_k}$ belongs to L^2 it suffices to find q_k so that

$$\partial_x^{\alpha_k} w_{l_k} \in L^{q_k} \quad \text{and} \quad \sum \frac{1}{q_k} \leq \frac{1}{2}.$$

Let $\mathcal{B}$ denote the set of $k \in \{1,\dots,J\}$ so that $s - |\alpha_k| > d/2$. For these indices the factor in our product is bounded, and so for $k \in \mathcal{B}$ set $q_k := \infty$.

Let $\mathcal{A} \subset \{1,\dots,J\}$ denote those indices i for which $s - |\alpha_i| < \frac{d}{2}$. For $k \in \mathcal{A}$, q_k is chosen as in Sobolev's Theorem,

$$q_k := \frac{2d}{d - 2s + 2|\alpha_k|}.$$

If $s - |\alpha_k| = \frac{d}{2}$, the factor in the product belongs to L^p for all $2 \leq p < \infty$ and the choice of q_k in this range is postponed.

With these choices, the Sobolev embedding theorem estimates

$$\|\partial_x^{\alpha_k} w_{l_k}\|_{L^{q_k}} \leq C\|w\|_{H^s(\mathbb{R}^d)}.$$

Then since $\sum |\alpha_i| \leq s$, and $s > d/2$,

$$\sum_{i \in \mathcal{A} \cup \mathcal{B}} \frac{1}{q_i} = \sum_{i \in \mathcal{A}} \frac{1}{q_i} = \sum_{i \in \mathcal{A}} \frac{d - 2s + 2|\alpha_i|}{2d} \leq \frac{Jd - 2Js + 2s}{2d}$$

$$= \frac{Jd - 2(J-1)s}{2d} < \frac{Jd - (J-1)d}{2d} = \frac{1}{2}.$$

This shows there is room to pick large q_k corresponding to the case $s - |\alpha_k| = d/2$ so that $\sum 1/q_k < 1/2$, and the proof is complete. $\qquad\square$

Another nice proof of Schauder's Lemma can be found in [Beals, pp 11-12]. Other arguments can be built on Littlewood-Paley decomposition of $G(w)$ as in the work of Bony and Meyer, or, on the representation

$$G(u) = \int \hat{G}(\xi)\,(e^{iu\xi} - 1)\,d\xi\,.$$

The latter requires that one prove a bound on the norm of $e^{iu\xi} - 1$ (see [Rauch-Reed 1982]) which grows at most polynomially in ξ. The last two arguments have the advantage of working when s is not an integer.

6.3. Basic existence theorem

It is now straightforward to prove the following local existence theorem.

Theorem 6.4 [Schauder]. *If $s > d/2$ and $f \in L^1_{loc}([0,\infty[\,;\mathbb{R}^d)$, then there is a $T > 0$ and a unique solution $u \in C([0,T]\,;H^s(\mathbb{R}^d))$ to the semilinear initial value problem defined by the partial differential equation (6.2) together with the initial condition*

$$u(0,x) = g(x) \in H^s(\mathbb{R}^d). \tag{6.11}$$

The time T can be chosen uniformly for f and g belonging to bounded subsets of the spaces to which they belong. As a result there is a $T^* \in\,]0,\infty]$ and a maximal solution in $C([0,T^*[\,;H^s(\mathbb{R}^d))$. If $T^* < \infty$ then

$$\liminf_{t \to T^*} \|u(t)\|_{H^s(\mathbb{R}^d)} = \infty\,. \tag{6.12}$$

Schauder proved a quasilinear second order scalar version, but his argument, which is recalled in section 10 of chapter 6 in [Courant], works without essential modification once you add the linear energy inequalities of Friedrichs. The following existence proof is a blend of the proof in Lax's notes and of Picard's original argument for ordinary differential equations. Note how Picard's elegant bounds (6.18) replace the usual contraction argument which is less precise.

Proof. The solution is constructed as the limit of a sequence of Picard iterates. The first approximation is not really important. Set

$$u^1(t,x) := g(x)$$

for all t, x. For $\nu > 1$, the basic linear existence theorem implies that the Picard iterates defined as solutions of the linear initial value problems

$$L(y, \partial_y)\,u^{\nu+1} + F(y, u^\nu) = f(y)\,, \qquad u^{\nu+1}(0) = g$$

are well defined elements of $C([0,\infty[\,;H^s(\mathbb{R}^d))$.

Let C denote the constant in the linear energy estimate (2.23). Choose a real number

$$R > 2C\,\|g\|_{H^s(\mathbb{R}^d)}\,. \tag{6.13}$$

Schauder's lemma implies that there is a constant $B(R) > 0$ so that

$$\|w(t,\cdot)\|_{H^s(\mathbb{R}^d)} \le R \Rightarrow \|F(t,\cdot,w)\|_{H^s(\mathbb{R}^d)} \le B\,.$$

Thanks to (6.13) one can choose $T > 0$ so that

$$C\left(e^{CT}\|g\|_{H^s(\mathbb{R}^d)} + \int_0^T e^{C(T-\sigma)}\big(B + \|f(\sigma)\|_{H^s(\mathbb{R}^d)}\big)\,d\sigma\right) \le R. \tag{6.14}$$

A straightforward application of (2.23) then shows that for all $\nu \ge 1$ and all $0 \le t \le T$

$$\|u^\nu(t)\|_{H^s(\mathbb{R}^d)} \le R. \tag{6.15}$$

Schauder's Lemma implies that there is a constant Λ so that for all t,

$$\|w_j\|_{H^s(\mathbb{R}^d)} \le R \Rightarrow \|F(t,x,w_1(x)) - F(t,x,w_2(x))\|_{H^s(\mathbb{R}^d_x)} \le \Lambda\|w_1 - w_2\|_{H^s(\mathbb{R}^d_x)}. \tag{6.16}$$

Then for $\nu \ge 2$, the basic linear estimate applied to the difference $u^{\nu+1} - u^\nu$ implies that

$$\|u^{\nu+1}(t) - u^\nu(t)\|_{H^s(\mathbb{R}^d)} \le C\,\Lambda \int_0^t e^{C(t-\sigma)} \|u^\nu(\sigma) - u^{\nu-1}(\sigma)\|_{H^s(\mathbb{R}^d)}\,d\sigma. \tag{6.17}$$

Let

$$M_1 := \sup_{0 \le t \le T} \|u^1(t) - u^2(t)\|_{H^s(\mathbb{R}^d)}$$

and

$$M_2 := C\,\Lambda\,e^{CT}.$$

Then a simple induction on ν using (6.17) shows that for all $\nu \ge 2$

$$\|u^{\nu+1}(t) - u^\nu(t)\|_{H^s(\mathbb{R}^d)} \le M_1\frac{(M_2 t)^{\nu-1}}{(\nu-1)!}. \tag{6.18}$$

Exercise. Prove (6.18).

Estimate (6.18) shows that u^ν is a Cauchy sequence in the Banach space $C([0,T]\,;\,H^s(\mathbb{R}^d))$. Let u be the limit.

Exercise. Prove u satisfies the initial value problem (6.1), (6.11).

This completes the proof of existence.

Uniqueness is a consequence of the integral inequality

$$\|u_1(t) - u_2(t)\| \le C_1 \int_0^t e^{C(t-\sigma)} \|u_1(\sigma) - u_2(\sigma)\|_{H^s(\mathbb{R}^d)}\,d\sigma, \tag{6.19}$$

which is proved exactly as (6.17). Gronwall's inequality implies that $\|u_1 - u_2\|$ vanishes identically. $\qquad\square$

Remarks.

1. Similar estimates show that there is continuous dependence of the solutions when the data f and g converge in $L^1_{loc}(\mathbb{R}\,;\,H^s(\mathbb{R}^d))$ and $H^s(\mathbb{R}^d)$ respectively.

2. Approximating the data by smooth data, and therefore the solutions by smooth solutions of approximating problems, the finite speed of propagation from 2.3 extends to the solutions just constructed.

Exercise. Prove these two assertions.

6.4. Moser's inequality and the nature of the breakdown

The breakdown (6.12) could in principal occur in a variety of ways. For example, the function might stay bounded and become more and more rapidly oscillatory. In fact this does not occur. Where the domain of existence ends the maximal amplitude of the solution must diverge to infinity. To prove this requires more refined inequalities than those of Sobolev and Schauder.

The Schauder Lemma implies that

$$\|G(y,w)\|_{H^s(\mathbb{R}^d_x)} \le h(\|w\|_{H^s(\mathbb{R}^d_x)})$$

for a nonlinear function h which depends on G.

Theorem 6.5. Moser's Inequality. *With the same hypotheses as Schauder's Lemma, there is a smooth function $h : [0,\infty[\to [0,\infty[$ so that for all $w \in H^s(\mathbb{R}^d)$ and t,*

$$\|G(x,w)\|_{H^s(\mathbb{R}^d_x)} \le h(\|w\|_{L^\infty(\mathbb{R}^d_x)})\,\|w\|_{H^s(\mathbb{R}^d_x)}\,. \tag{6.20}$$

This is proved by using Leibniz' rule and Holder's inequality as in the standard proof of Schauder's Lemma. However in place of the Sobolev inequalities one uses the Galiardo-Nirenberg interpolation inequalities which we now recall.

Theorem 6.6. Gagliardo-Nirenberg Inequalities. *If $w \in H^s(\mathbb{R}^d) \cap L^\infty(\mathbb{R}^d)$ and $0 < |\alpha| < s$ then*

$$\partial_x^\alpha w \in L^{2s/|\alpha|}(\mathbb{R}^d)$$

In addition, there is a constant $C = C(|\alpha|, s, d)$ so that

$$\|\partial^\alpha w\|_{L^{2s/|\alpha|}(\mathbb{R}^d)} \le C\,\|w\|_{L^\infty(\mathbb{R}^d)}^{1-|\alpha|/s} \Big(\sum_{|\beta|=s} \|\partial^\beta w\|_{L^2(\mathbb{R}^d)} \Big)^{|\alpha|/s} \tag{6.21}$$

Remarks. 1. The second factor on the right in (6.21) is equivalent to the L^2 norm of the operator $|\partial_x|^s$ applied to u where $|\partial_x|^s$ is defined to be the Fourier multiplier by $|\xi|^s$. This gives the correct extension to non integer s.

2. The indices in (6.21) are nearly forced. Consider which inequalities

$$\|\partial^\alpha w\|_{L^p(\mathbb{R}^d)} \le C\,\|w\|_{L^\infty(\mathbb{R}^d)}^{1-\theta} \Big(\sum_{|\beta|=s} \|\partial^\beta w\|_{L^2(\mathbb{R}^d)} \Big)^{\theta}$$

homogeneous of degree one in w might be true. The test functions $w = e^{ix\cdot\xi/\epsilon}\psi(x)$ with $\epsilon \to 0$ show that a necessary condition is $|\alpha| \le s\theta$. The idea is to use the L^∞ norm as much as possible and the s-norm as little as possible, which yields $|\alpha| = s\theta$. Considering $w = \psi(\epsilon x)$, or equivalently comparing the dimensions of the two sides forces $p = 2s/\alpha$.

Proof. The following paragraphs are meant to lead you through the main steps in a proof. The first thing is to realize that the motor is a clever use of integration by parts. To illustrate that we begin by proving the special case $s = 2$, $p = 4$. Precisely we show that for real valued $u \in C_0^\infty(\mathbb{R}^d)$,

$$\|Du\|_{L^4} \le C\|u\|_{L^\infty}^{1/2}\,\|D^2u\|_{L^2}^{1/2}\,.$$

The centerpiece of the proof is the following case of the product rule for derivatives

$$D\left(u\,(Du)^3\right) = (Du)^4 + 3\,(u)\,(Du)^2\,(D^2u)\,.$$

Since u is compactly supported,

$$0 = \int_{\mathbb{R}^d} D\left(u\,(Du)\,(Du)^2\right)\,dx\,.$$

Combining these two observations yields

$$\int_{\mathbb{R}^d} |Du|^4\,dx = -3\int_{\mathbb{R}^d} |u\,(D^2u)\,(Du)^2|\,dx\,.$$

Applying Hölder's inequality to estimate the right hand side yields

$$\int_{\mathbb{R}^d} |Du|^4\,dx \le 3\|u\|_{L^\infty}\left(\int_{\mathbb{R}^d} |D^2u|^2\,dx\right)^{1/2}\left(\int_{\mathbb{R}^d} |Du|^4\,dx\right)^{1/2}\,.$$

If $u \ne 0$ the last factor is nonzero so dividing both sides by this term yields,

$$\left(\int_{\mathbb{R}^d} |Du|^4\,dx\right)^{1/2} \le 3\|u\|_{L^\infty}\left(\int_{\mathbb{R}^d} |D^2u|^2\,dx\right)^{1/2}\,.$$

Equivalently,

$$\|Du\|_{L^4} \le \sqrt{3}\,\|u\|_{L^\infty}^{1/2}\,\|D^2u\|_{L^2}^{1/2}\,,$$

which is the desired estimate.

Exercise. Prove the general $s = 2$, $|\alpha| = 1$ estimate by a similar argument. **Hint** Use the primitive $c_p t|t|^{p-1}$ of t^p.

To see how to go to higher derivatives, consider the case $s = 3$ in which case (6.21) asserts that $D^2u \in L^3$ and $Du \in L^6$. This is proved by appealing twice to Gagliardo-Nirenberg estimates with $s = 2$. Namely, one estimates

$$\|Du\|_{L^6} \le C\|u\|_{L^\infty}^{1/2}\,\|D^2u\|_{L^3}^{1/2} \qquad \text{and} \qquad \|D^2u\|_{L^3} \le C\|Du\|_{L^6}^{1/2}\,\|D^3u\|_{L^2}^{1/2}\,.$$

Using the second in the first yields the desired estimate for Du. Using that in the second, yields the desired estimate for D^2u. Note that the second inequality is *not* a special case of (6.21). The bottom line, is that it suffices to prove the following more general inequality than (6.21) for the case $s = 2$,

$$\|Du\|_{L^p} \le C\,\|u\|_{L^q}^{1/2}\,\|D^2u\|_{L^r}^{1/2}\,, \qquad \text{where} \qquad \frac{2}{p} := \frac{1}{q} + \frac{1}{r}\,.$$

Exercise. Prove this Gagliardo-Nirenberg estimate then derive (6.21).

Proof of Moser's Inequality. For $w \in \mathcal{S}(\mathbb{R}^d)$ and $\sigma := |\alpha| \le s$, the quantity $\partial_x^\alpha(G(w))$ is a sum of terms of the form

$$G^{(\gamma)}(w)\,\Pi_{j=1}^{J}\,\partial_x^{\alpha_j} w \tag{6.22}$$

where $|\gamma| = J$, and $\alpha_1 + \cdots + \alpha_J = \alpha$. The first factor in (6.22) is bounded with L^∞ norm bounded by a nonlinear function of the L^∞ norm of w.

For the second factor, Hölder's inequality yields

$$\|\,\partial_x^{\alpha_1} w \cdots \partial_x^{\alpha_J} w\,\|_{L^2} \le \Pi_{k=1}^{J}\,\|\partial_x^{\alpha_k} w\|_{L^{2/\lambda_k}}$$

provided the nonnegative λ_k satisfy $\sum \lambda_k = 1$.

The Gagliardo-Nirenberg inequalities yield

$$\|\partial_x^{\alpha_k} w\|_{L^{2\sigma/|\alpha_k|}} \leq \|w\|_{L^\infty}^{(\sigma-|\alpha_k|)/\sigma} \|w\|_{H^\sigma}^{|\alpha_k|/\sigma}.$$

With these choices

$$\sum \lambda_k = \sum \frac{|\alpha_k|}{\sigma} = 1$$

as needed. $\qquad\square$

Theorem 6.7. *If $T^* < \infty$ in the basic existence theorem of 6.3, then*

$$\limsup_{t \to T^*} \|u(t)\|_{L^\infty} = \infty. \tag{6.23}$$

Proof. It suffices to show that it is impossible to have $T^* < \infty$ and $|u| \leq R < \infty$ on $[0, T^*[\times\mathbb{R}^d$. The strategy is to show that if $|u(t,x)| \leq R < \infty$ on $[0, T^*[\times\mathbb{R}^d$, then (6.12) is violated.

Use the linear inequality for $0 \leq t < T$,

$$\|u(t)\|_{H^s(\mathbb{R}^d)} \leq C \left(\|u(0)\|_{H^s(\mathbb{R}^d)} + \int_0^t \|(Lu)(\sigma)\|_{H^s(\mathbb{R}^d)} \, d\sigma \right). \tag{6.24}$$

Then use Moser's inequality to give

$$\|(Lu)(\sigma)\|_{H^s(\mathbb{R}^d)} = \|F(\sigma, x, u(\sigma, x)) - f(\sigma, x)\|_{H^s(\mathbb{R}^d)} \leq C(R) \left(\|u(\sigma)\|_{H^s(\mathbb{R}^d)} + 1 \right). \tag{6.25}$$

Using (6.25) in (6.24) yields

$$\|u(t)\|_{H^s(\mathbb{R}^d)} \leq C \left(\|u(0)\|_{H^s(\mathbb{R}^d)} + \int_0^t \left(\|u(\sigma)\|_{H^s(\mathbb{R}^d)} + 1 \right) \, d\sigma \right). \tag{6.26}$$

Gronwall's inequality shows that there is a constant $C'' < \infty$ so that for $t \in [0, T^*[$

$$\|u(t)\|_{H^s(\mathbb{R}^d)} \leq C''. \tag{6.27}$$

This violates (6.12), and the proof is complete. $\qquad\square$

A mild sharpening of this argument (due to Yudovich) shows that weaker norms than L^∞, for example the BMO norm, must also blow up at T^*.

LECTURE 7
One Phase Nonlinear Geometric Optics

The goal is to find analogues of the asymptotic solutions of Lax in the nonlinear context. There are two important nonlinear effects which must be understood in order to arrive at the appropriate *ansatz*.

7.1. Amplitudes and harmonics

For linear equations, any solution may be multiplied by a constant to yield another solution. This is not the case for nonlinear equations. If one studies short wavelength oscillatory solutions, the propagation and interactions depend crucially on the amplitudes. The easiest case to understand, and therefore a natural starting point, is small oscillations.

Consider the semilinear case (6.1) with nonlinear function satisfying

$$F(y, 0) = 0 \qquad D_u F(y, 0) = 0. \tag{7.1}$$

When u is complex valued the expression $D_u F = 0$ is a shorthand for $D_{\Re u} F = 0 = D_{\Im u} F$. Similarly the expression $D_u F\, u$ is a shorthand for $D_{\Re u} F\, \Re u + D_{\Im u} F\, \Im u$. Hypothesis (7.1) is reasonable since the Taylor polynomial,

$$F(y, 0) + D_u F(y, 0)\, u$$

can be absorbed as source term f and zeroth order term in $L(y, \partial_y)$. Nevertheless, for simplicity of exposition we suppose that $f = 0$ in the following paragraphs.

Suppose that $a(\epsilon, y)\, e^{i\phi(y)/\epsilon}$ is a Lax solution as in 5.4 and that a has compact support for each t. Consider the semilinear initial value problem with the initial data

$$u(\epsilon, 0, x) = \epsilon^m\, a(\epsilon, 0, x)\, e^{i\phi(0,x)/\epsilon}. \tag{7.2}$$

The power m scales the amplitude as a function of the wavelength. The larger is m the smaller is the data. The initial data is bounded in $H^s(\mathbb{R}^d)$ if and only if $s \le m$.

If $m > d/2$ then the data converges to zero in $H^s(\mathbb{R}^d)$ for all $s \in\,]d/2, m[$. It follows from the basic nonlinear existence theorem that solutions exist on an ϵ independent neighborhood and are given by a Taylor series

$$u(\epsilon, y) \sim \sum_{j=1}^{\infty} M_j(u(\epsilon, 0, x)) = \sum_{j=1}^{\infty} \epsilon^{jm}\, M_j(a(\epsilon, 0, x)e^{i\phi(0,x)/\epsilon}), \tag{7.3}$$

441

where

$$M_j \; : \; H^s(\mathbb{R}^d) \to C([0,T] \, ; \, H^s(\mathbb{R}^d)) \tag{7.4}$$

is a continuous symmetric j-linear form. Taylor's Theorem implies that the expansion (7.3) is an asymptotic expansion in $H^s(\mathbb{R}^d)$. Truncating at the ϵ^l term, the error is $O(\epsilon^{l+1})$ in $H^s(\mathbb{R}^d)$. Such approximations derived directly from Taylor's Theorem go under the name *regular perturbation theory*. Along with linear algebra it is the most common university level technique employed in the sciences.

Before computing the series (7.3), note in passing that the local existence theorem yields existence on a domain independent of ϵ. For $m \leq d/2$, the existence theorem guarantees existence only on a domain which shrinks with ϵ because the H^s norm of the data grows to ∞ for all $s > d/2$. We will see that for $m \geq 0$, there is, nevertheless, existence on an ϵ independent domain. The simple explicitly solvable example

$$\partial_t u(\epsilon, y) = u(\epsilon, y)^2 \,, \qquad u(\epsilon, 0, x) = \epsilon^m \, e^{\frac{ix \cdot \xi}{\epsilon}}$$

shows that the domain may shrink to zero for $m < 0$.

Exercise. Verify.

It is easier to compute the terms of the expansion in (7.3) directly rather than to use the definition of Taylor coefficients in infinite dimensions. The idea is to follow the standard science book strategy of considering initial data $\delta u(\epsilon, 0, x)$ and then computing a Taylor series expansion in δ for the solution. One then sets $\delta = 1$ to obtain the Taylor series (7.3).

The series in δ is computed by setting powers of δ equal to zero in the following relation

$$L(y, \partial_y) \left(\sum_{j=1}^{\infty} \delta^j \, u_j(\epsilon, y) \right) + F\left(y, \sum_{j=1}^{\infty} \delta^j \, u_j(\epsilon, y)\right) \sim 0 \,. \tag{7.5}$$

Expand

$$L(y, \partial_y) \left(\sum_{j=1}^{\infty} \delta^j \, u_j(\epsilon, y) \right) \sim \sum_{j=1}^{\infty} \delta^j \, L(y, \partial_y) \, u_j(\epsilon, y) \,, \tag{7.6}$$

and

$$F\left(y, \sum_{j=1}^{\infty} \delta^j \, u_j(\epsilon, y)\right) \sim \delta^2 \, D_u^2 F(y, 0)\left(u_1, u_1\right) + \text{h.o.t.} \tag{7.7}$$

Setting the sum of these two expressions equal to zero yields the differential equations for the leading two terms

$$L \, u_1 = 0 \,, \qquad L \, u_2 + D_u^2 F(y, 0)\left(u_1, u_1\right) = 0 \,. \tag{7.8}$$

The initial conditions are

$$u_1(\epsilon, 0, x) = \epsilon^m \, a(\epsilon, 0, x) \, e^{i\phi(0,x)/\epsilon} \,, \qquad u_2(0, x) = 0 \,. \tag{7.9}$$

The computation of equations for the higher order terms is left to the interested reader. They share with the above equations the form that the source terms for the function u_j involves only the earlier terms $u_1, \cdots, u_{j-1}$.

Equations (7.8) and (7.9) show that as $\epsilon \to 0$, u_1 is given asymptotically by the Lax solution $\epsilon^m \, a(\epsilon, y) \, e^{i\phi(y)/\epsilon}$. Once u_1 is known the next term, u_2 can be found.

Exercise. Find an initial value problem which determines u_3 once u_1 and u_2 are known.

To see the form of u_2, it is crucial to consider the source term $D^2F(u_1, u_1)$ on the right hand side of (7.8). It is a quadratic expression in u_1 and u_1 oscillates with phase $\phi(y)/\epsilon$. Squaring such a term yields a source oscillating with phase $2\phi(y)/\epsilon$. The square of the complex conjugate, which is a second example of a smooth quadratic expression, yields a phase $-2\phi(y)/\epsilon$. Finally an expression in the product of u with its conjugate yields a nonoscillatory source. The source term has the form

$$\epsilon^{2m}\left(c_{-2}(y)\,e^{-i2\phi(y)/\epsilon} + c_0(y) + c_{+2}(y)e^{i2\phi(y)/\epsilon}\right). \tag{7.10}$$

From Lax's Theorem with oscillatory source, the oscillatory parts of this source yields terms of the form

$$\epsilon^{2m}\left(a_{\pm 2}(y) + O(\epsilon)\right)e^{\pm i2\phi(y)/\epsilon}$$

in the solution $u_2(\epsilon, y)$.

The key observation is that the Taylor expansion begins with a term linear in the initial data, and, which is equal to a Lax solution of order ϵ^m. The next term, quadratic in the initial data, is of order ϵ^{2m} and has terms oscillating with the new phases $\pm 2\phi(y)/\epsilon$. It may also have nonoscillating terms. The cubic and higher order terms in the Taylor expansion are of order ϵ^{jm} for integer j and will have terms oscillating with phases including higher integer multiples of $\pm\phi(y)/\epsilon$.

This generation and interaction of harmonics is one of the key signatures of nonlinear problems. Note that the wavelength of the jth harmonic is $1/j$ times the original wavelength. Thus the interaction also is an interaction between different length scales. A classical experiment from the early sixties involved passing monochromatic red laser light through glass and observing the blue harmonic in the output. This was the birth of the new science of Nonlinear Optics.

Though this computation is so far only justified for $m > d/2$, it is an interesting indication that something better is true. Formally, the expansion seems to work provided only that $m > 0$, in which case the supposedly higher order corrections are indeed higher order in ϵ. In fact, using local existence results tailored to oscillatory data as in Lecture 8, the expansions can be justified in this range for a domain of time independent of ϵ. On the other hand, for $m < 0$ we know that the domain of existence may shrink.

A fundamental lesson to be learned is that for $m > 0$ linear phenomena are accompanied by *creation of harmonics at higher order in ϵ*. This leads to correction terms in the solution which have amplitudes with higher powers of ϵ and phases which are integer multiples of $\phi(y)/\epsilon$. The higher is m, the smaller is the initial data and the greater is the gap between the amplitudes of the principal term and the harmonics. Viewed another way, the smaller is m, the larger are the data, and the more important are the nonlinear effects.

There is another important lesson. The leading nonlinear term is of order ϵ^{2m} while the Lax solution enters at order ϵ^m. As $m \to 0$, these orders become equal. This leads the courageous to suspect that there may be something interesting occurring when $m = 0$ in which case the harmonics should appear in the principal term. This in fact is the case as we will show. For $m = 0$ oscillations can be described on an ϵ independent domain, and the leading term in the expansion involves a nonlinear interaction among oscillations with phases $j\phi(y)/\epsilon$ for all $j \in \mathbb{Z}$. This critical scaling of the amplitudes is called *nonlinear geometric optics*, or *weakly nonlinear geometric optics* depending on the author.

Nonlinear geometric optics described here is more complicated than but descendant from earlier work on pulses of width ϵ and height one. We have wave trains with a large number of wavelengths, not just one pulse. A description of the pulses and the relation to wavetrains can be found in [Hunter-Majda-Rosales, Studies in Applied Math, 75(1986)] and in the survey article of [Majda]. Wavetrains are blessed with interesting additional nonlinear interactions which go under the name of *resonance*. These too are described in the articles just cited. They would have been the next topic of this series of lectures. I recommend the articles of [Joly-Metivier-Rauch, Ann. Inst. Fourier, College de France] for an introduction to our point of view. Two main advances are that rigorous proofs have been constructed and the formal arguments have been buttressed to include the possibility of the simultaneous existence of an infinite number of distinct basic phases ϕ_k.

7.2. More on the generation of harmonics

Here are three simple ordinary differential equation calculations aimed at making you more familiar with the creation of harmonics.

Exercise. Consider the solution $x(\epsilon, t)$ of the nonlinear initial value problem

$$\frac{d^2x}{dt^2} + \omega^2 x + x^2 = 0\,, \qquad x|_{t=0} = \epsilon\,, \qquad \frac{dx}{dt}\Big|_{t=0} = 0\,.$$

Then x is an analytic function of ϵ, t on its domain of existence. Compute the first three terms in the Taylor expansion

$$x(\epsilon, t) = a_0(t) + \epsilon a_1(t) + \epsilon^2 a_2(t) + \cdots\,.$$

Note the presence of harmonics when they appear, and the amplitude of the harmonics.

In the last exercise, the harmonics appeared in a regular perturbation expansion of small solutions to a nonlinear equation. An entirely equivalent problem is the expansion of solutions of fixed amplitude with a weak nonlinearity.

Exercise. Consider the solution $x(\epsilon, t)$ of the weakly nonlinear initial value problem

$$\frac{d^2x}{dt^2} + \omega^2 x + \epsilon x^2 = 0\,, \qquad x|_{t=0} = 1\,, \qquad \frac{dx}{dt}\Big|_{t=0} = 0\,.$$

Then x is an analytic function of ϵ, t on its domain of existence. Compute the first two terms in the Taylor expansion

$$x(\epsilon, t) = a_0(t) + \epsilon a_1(t) + \epsilon^2 a_2(t) + \cdots\,.$$

Note the presence of harmonics when they appear, and the amplitude of the harmonics.

Finally, here is an example of the generation of harmonics for forced oscillations.

Exercise. Consider the solution $x(\epsilon, t)$ of the nonlinear initial value problem

$$\frac{d^2 x}{dt^2} + x + x^2 = \epsilon \cos \beta t\,, \qquad x|_{t=0} = 0\,, \qquad \frac{dx}{dt}\bigg|_{t=0} = 0\,,$$

where $\beta \neq 0, \pm 1$. Then x is an analytic function of ϵ, t on its domain of existence. Compute the first three terms in the Taylor expansion

$$x(\epsilon, t) = a_0(t) + \epsilon a_1(t) + \epsilon^2 a_2(t) + \cdots \,.$$

Note the presence of harmonics when they appear, and the amplitude of the harmonics.

7.3. Formulating the ansatz

Summarizing the results of 7.1, we are lead to consider semilinear initial value problems with initial data of the form $a(\epsilon, 0, x)\, e^{i\phi(0,x)/\epsilon}$ which are initial data of a Lax solution in the linear case. The key fact is that the amplitude is of order ϵ^0. For this amplitude one expects harmonics to be present in the leading ϵ^0 term and for these harmonics to interact. Our aim is describe these phenomena.

The computations suggest that the solution will have oscillations with all the phases $n\phi(y)/\epsilon$. Thus the principal term is expected to be at least as complicated as an infinite sum from the leading terms of the nth harmonics. The amplitude of the nth harmonic is denoted

$$a_0(n, \epsilon, y) \sim a_0(n, y) + \epsilon\, a_1(n, y) + \cdots \,. \tag{7.11}$$

At this stage it seems that the natural thing to do is to derive dynamic equations for the infinite set of amplitudes $a_0(n, y)$ which must include both the linear hyperbolic propagation properties given by rays and transport equations for each $a_0(n, y)$ and also nonlinear interaction terms which express at least the idea that if one starts with $a_0(1, y) \neq 0$ and all others vanishing then the other modes will tend to be illuminated.

There is a very effective method for managing this infinity of unknowns. Introduce the generating function

$$U_0(y, \theta) := \sum_{n=-\infty}^{\infty} a_0(n, y)\, e^{in\theta}\,. \tag{7.12}$$

Then U_0 is periodic in θ and the amplitudes $a_0(n, y)$ are the Fourier coefficients of U. Knowing U is equivalent to knowing the $a_0(n, y)$ for all $n \in \mathbb{Z}$.

Adding correctors we search for asymptotic solutions of first order semilinear symmetric hyperbolic systems in the form

$$u(\epsilon, y) = U(\epsilon, y, \phi(y)/\epsilon) \tag{7.13}$$

where $U(\epsilon, y, \theta)$ is periodic in θ and is given by an asymptotic series

$$U(\epsilon, y, \theta) \sim \sum_{j=0}^{\infty} \epsilon^j\, U_j(y, \theta)\,. \tag{7.14}$$

The leading term, $U_0(y, \phi(y)/\epsilon)$ clearly presents two scales. If $U_0(y, \theta)$ and $\phi(y)$ vary on the length scale 1, then the leading term varies on the scale 1, and the scale

ϵ. The expansion (7.14) is called a two scale or multiscale expansion. In the case of ordinary differential equations, where there is only one independent variable, often called time, such expansions are often called two timing, after the presence of two time scales.

Using an *ansatz* containing a principal term as in (7.14) is a classical procedure in applied mathematics. Our approach here can be viewed as originating in the articles of Choquet-Bruhat and Hunter-Keller. Neither treats the more subtle issue of resonance. However, it is exactly because there is no resonance, that correctors as in (7.14) can be found. The theory with resonance must often content itself with leading order asymptotics only.

7.4. Equations for the profiles

We proceed to see whether it is possible to find profiles $U_j(y, \theta)$ so that

$$L(y, \partial_y)\, U(\epsilon, y, \phi(y)/\epsilon) + F(y, U(\epsilon, y, \phi(y)/\epsilon))\ \sim\ 0\,. \tag{7.15}$$

The left hand side of (7.15) is an expression of the form $W(\epsilon, y, \phi(y)/\epsilon)$ where $W(\epsilon, y, \theta)$ is periodic in θ. Plugging (7.14) into the first term of (7.15) yields

$$L(y, \partial_y)\, U \sim \sum \epsilon^j\, L(y, \partial_y)\, (U_j(y, \phi(y)/\epsilon))\,.$$

Expanding the second term from (7.15) in a Taylor series about U_0 yields

$$F(y, U_0 + \epsilon U_1 + \cdots) \sim F(y, U_0) + \epsilon\Big(F_u(y, U_0)\, U_1 + F_{\overline{u}}(y, U_0)\, \overline{U}_1\Big) + \text{ h.o.t.} \tag{7.16}$$

Note that the linear terms in U_1 are real linear and not necessarily complex linear, since F is assumed to be smooth but not necessarily holomorphic. These two expansions show that

$$W(\epsilon, y, \theta) \sim \sum_{j=-1}^{\infty} \epsilon^j\, W_j(y, \theta) = \epsilon^{-1} W_{-1}(y, \theta) + W_0(y, \theta) + \epsilon^1\, W_1(y, \theta) + \cdots\,. \tag{7.17}$$

The strategy is to choose the U_j so that all the W_j vanish identically. In the linear case we did the same thing, except that all the terms were simple exponentials in their dependence on θ.

The leading, ϵ^{-1} term in (7.17) comes from the terms in (7.15) where the y derivatives fall on the $\phi(y)/\epsilon$ part. That yields

$$W_{-1}(y, \theta) = \sum_{\mu=0}^{d} A_\mu(y)\, \frac{\partial \phi(y)}{\partial y_\mu}\, \frac{\partial U_0}{\partial \theta} = L_1(y, d\phi(y))\, \frac{\partial U_0}{\partial \theta}\,. \tag{7.18}$$

In order for there to be nontrivial oscillations, one must have $\partial_\theta U_0 \neq 0$ so the first constraint we place on the expansion is that the matrix $L_1(y, d\phi(y))$ have nontrivial kernel. Equivalently, ϕ must satisfy the familiar eikonal equation

$$\det L_1(y, d\phi(y)) = 0\,. \tag{7.19}$$

As usual we suppose that not only is (7.19) satisfied, but the dimension of the kernel of $L_1(y, d\phi(y))$ does not depend on y and denote by $\pi(y)$ orthogonal projection of $\mathbb{C}^N$ onto the kernel.

Setting $W_{-1} = 0$ then yields the equation

$$U_0 \in \ker L_1(y, d\phi(y)) \frac{\partial}{\partial \theta} \, . \tag{7.20}$$

The ϵ^0 coefficient W_0 has two contributions. The first when the differential operator acts on the y variables in U_0 and the second when y derivatives fall on the $\phi(y)/\epsilon$ part of U_1. One finds

$$W_0(y, \theta) = L(y, \partial_y) U_0(y, \theta) + F(y, U_0(y, \theta)) + L_1(y, d\phi(y)) \frac{\partial}{\partial \theta} U_1(y, \theta) \, . \tag{7.21}$$

Setting $W_0 = 0$ yields an equation which mixes U_0 and U_1. As in the linear case, information about U_0 is contained in the assertion

$$L(y, \partial_y) U_0(y, \theta) + F(y, U_0(y, \theta)) \in \operatorname{range}\left(L_1(y, d\phi(y)) \frac{\partial}{\partial \theta} \right). \tag{7.22}$$

Equations (7.20) and (7.22) are our first form of the profile equations of nonlinear geometric optics. Written this way, it is not at all clear that they determine U_0 from its initial data. They are open invitations to study the action of the operator $L_1(y, d\phi(y))\partial_\theta$ on periodic functions of θ.

The action of this operator is revealed by expanding in Fourier series in θ. For

$$V(y, \theta) = \sum_{n=-\infty}^{\infty} V_n(y) \, e^{in\theta} \, , \tag{7.23}$$

$V_0(y)$ is the nonoscillating contribution, and, $V - V_0$ the rapidly oscillating part. Direct computation yields

$$L_1(y, d\phi(y)) \frac{\partial}{\partial \theta} V = i \sum L_1(y, d\phi(y)) \, n \, V_n(y) \, e^{in\theta} \, . \tag{7.24}$$

The kernel consists of functions such that for $n \neq 0$, V_n takes values in the kernel of $L_1(y, d\phi(y))$. Equivalently,

$$\pi(y) \, V_n(y) = V_n(y) \quad \text{for } n \neq 0 \, .$$

Define a projection operator $\mathbf{E}$ on Fourier series by

$$\mathbf{E} \sum_{n=-\infty}^{\infty} V_n(y) \, e^{in\theta} := V_0 + \sum_{n \neq 0} \pi(y) \, V_n(y) \, e^{in\theta} \, . \tag{7.25}$$

For each y, $\mathbf{E}$ acts as an orthogonal projection in $L^2(S^1)$. Equation (7.20) is equivalent to

$$\mathbf{E} \, U_0 = U_0 \, , \tag{7.26}$$

that is, U_0 belongs to the subspace onto which $\mathbf{E}$ projects. This equation shows that the oscillating part of U_0 satisfies the familiar polarization from Lecture 5.

Equation (7.22) can be analyzed similarly. The image of $L_1(y, d\phi(y))\partial_\theta$ consists of those Fourier series whose constant term vanishes, and whose other coefficients lie in the image of $L_1(y, d\phi(y))$. Equivalently, those coefficients are annihilated by $\pi(y)$. Thus equation (7.22) is equivalent to

$$\mathbf{E} \Big(L(y, \partial_y) U_0(y, \theta) + F(y, U_0(y, \theta)) \Big) = 0 \, . \tag{7.27}$$

The pair of equations (7.26), (7.27) is analogous in form to the pair of equations (5.28) and (5.29) which determined a_0. Note that equations (7.26-27) hold if and only if

$$W_{-1} = 0, \qquad \text{and,} \qquad \mathbf{E}\,W_0 = 0. \tag{7.28}$$

A note about our strategy here. The equations for the profiles U_j will be found by induction. Suppose that $j \geq 1$ and that $U_0, \ldots, U_{j-1}$ have been determined so that

$$W_{-1} = \cdots = W_{j-1} = 0, \qquad \text{and} \qquad \mathbf{E}W_j = 0. \tag{7.29}$$

The equations determining U_j are then equivalent to

$$(I - \mathbf{E})\,W_{j-1} = 0, \qquad \text{and} \qquad \mathbf{E}\,W_j = 0. \tag{7.30}$$

In this way, solving the system of equations for the profiles is equivalent to solving $W_j = 0$ for all j. Thus the profiles U_j satisfy the profile equations if and only if any $U(\epsilon, y, \theta)$ satisfying

$$U(\epsilon, y, \theta) \sim \sum_{j=0}^{\infty} \epsilon^j \, U_j(y, \theta)$$

satisfies

$$L(y, \partial_y)\, U(\epsilon, y, \phi(y)/\epsilon) + F(y, U((\epsilon, y, \phi(y)/\epsilon))) \sim 0.$$

We next find the profile equations for U_1. Equation (7.21) shows that $(I - \mathbf{E})W_0 = 0$ if and only if

$$(I - \mathbf{E})\, L_1(y, d\phi(y))\, \frac{\partial}{\partial \theta}\, U_1 = -(I - \mathbf{E}) \left(L(y, \partial_y)\, U_0 + F(u, U_0) \right) := F_0(y, U_0), \tag{7.31}$$

where the right hand side, denoted F_0 is a function of the profile U_0 and its derivatives which are assumed known. The dependence on the derivatives is not indicated in the notation, since for the sequel it is not important just how many derivatives occur in the terms F_j. This equation determines U_1 modulo the kernel of the operator $(I - \mathbf{E})\, L_1(y, d\phi(y))\partial_\theta$, which from the definition of $\mathbf{E}$, is the same as the kernel of $L_1(y, d\phi(y))\partial_\theta$.

Define the partial inverse $\mathbf{Q}$ of the operator $(I - \mathbf{E})\, L_1(y, d\phi(y))\partial_\theta$, by

$$\mathbf{Q}\left(\sum V_n(y)\, e^{in\theta} \right) := \sum_{n \neq 0} \frac{1}{in}\, Q(y)\, V_n(y)\, e^{in\theta}, \tag{7.32}$$

where $Q(y)$ is the partial inverse of $L_1(y, d\phi(y))$ defined in (5.26), (5.27). Then

$$\mathbf{Q}\left((I - \mathbf{E})\, L_1(y, d\phi(y))\partial_\theta \right) V = (I - \mathbf{E})\, V, \tag{7.33}$$

for all trigonometric series V. Then, (7.31) is equivalent to the equation

$$(I - \mathbf{E})\, U_1 = -\mathbf{Q}\, F_0(y, U_0). \tag{7.34}$$

As indicated before, the determination of $\mathbf{E}\,U_1$ comes from setting $\mathbf{E}\,W_1 = 0$. The expansions at the beginning of this section yield

$$W_1(y,\theta) = L(y,\partial_y)\,U_1(y,\theta) + F_u(y,U_0)\,U_1(y,\theta)$$
$$+ F_{\overline{u}}(y,U_0)\,\overline{U}_1(y,\theta) + L_1(y,d\phi(y))\,\frac{\partial}{\partial\theta}\,U_2(y,\theta)\,.$$
$$(7.35)$$

Multiplying by $\mathbf{E}$ eliminates the last term and one gets the second equation for the profile U_1,

$$\mathbf{E}\left(L(y,\partial_y)\,U_1(y,\theta) + F_u(y,U_0)\,U_1(y,\theta) + F_{\overline{u}}(y,U_0)\,\overline{U}_1(y,\theta) \right) = 0\,. \qquad (7.36)$$

The pattern is now established. Setting $(I - \mathbf{E})W_1 = 0$ yields

$$(I - \mathbf{E})\,U_2 = F_1(y,U_0,U_1)\,, \qquad (7.37)$$

and the equation $\mathbf{E}W_2 = 0$ yields

$$\mathbf{E}\left(L(y,\partial_y)\,U_2(y,\theta) + F_u(y,U_0)\,U_2(y,\theta) + F_{\overline{u}}(y,U_0)\,\overline{U}_2(y,\theta) + q(U_1,\overline{U}_1) \right) = 0\,. \qquad (7.38)$$

The q is a quadratic expression in U_1 and its complex conjugate which comes from the second order terms in the Taylor expansion (7.16).

Continuing in this fashion yields for all $j \geq 1$ a pair of equations

$$(I - \mathbf{E})\,U_j = F_{j-1}(y,U_0,U_1,\ldots,U_{j-1})\,, \qquad (7.39)$$

and

$$\mathbf{E}\left(L(y,\partial_y)\,U_j(y,\theta) + F_u(y,U_0)\,U_j(y,\theta) \right.$$
$$\left. + F_{\overline{u}}(y,U_0)\,\overline{U}_j(y,\theta) + G_j(y,U_0,\ldots,U_{j-1}) \right) = 0 \qquad (7.40)$$

which are equivalent to (7.30).

On first regard, it is not at all obvious how to attack these equations. Our approach is to find analogues of the arguments in section 5.3.

Important observation. The equation for the principal profile U_0 is nonlinear in U_0 whereas the equations for the higher profiles U_j with $j \geq 1$, are $\mathbb{R}$-linear in U_j.

7.5. Solving the profile equations

This subsection shows that the equations derived above determine the profiles U_j from suitable initial data. Once this is done, the asymptotic expansion is constructed and one needs to show that the asymptotic series so computed is asymptotic to the exact solution of the original nonlinear equation which has the same initial data. Part of this is to show that the solution of the nonlinear equation exists on a domain independent of ϵ. These tasks are carried out in Lecture 8.

To see that U_0 is determined from its initial data, first note that (7.26) and (7.27) together imply that

$$\mathbf{E}\,L(y,\partial_y)\,\mathbf{E}U_0 + \mathbf{E}\,F(y,\mathbf{E}U_0(y,\theta)) = 0\,. \qquad (7.41)$$

Applying $(I - \mathbf{E})\, L(y, \partial_y)$ to (26) yields

$$(I - \mathbf{E})\, L(y, \partial_y)\, (I - \mathbf{E})\, U_0 = 0 \,. \tag{7.42}$$

Adding these two equations yields

$$(I - \mathbf{E})\, L(y, \partial_y)\, (I - \mathbf{E})\, U_0 + \mathbf{E}\, L(y, \partial_y)\, \mathbf{E}\, U_0 + \mathbf{E}\, F(y, \mathbf{E}\, U_0(y, \theta)) = 0 \,, \tag{7.43}$$

an analogue of (5.32).

Define a linear operator

$$\mathbf{L} := \sum_{\mu=0}^{d} \mathbf{A}_\mu\, \partial_\mu + \mathbf{B}\,, \tag{7.44}$$

where the coefficients are the operators

$$\mathbf{A}_\mu := (I - \mathbf{E})\, A_\mu(y)\, (I - \mathbf{E}) + \mathbf{E}\, A_\mu(y)\, \mathbf{E}\,, \qquad \mathbf{B} := (I - \mathbf{E})\, B(y)\, (I - \mathbf{E}) + \mathbf{E}\, B(y)\, \mathbf{E}\,. \tag{7.45}$$

Denote by Ω a domain of determinacy as in (5.33) in the Theorems of Lecture 5. The unknown U_0 is sought as a $\mathbb{C}^N$ valued function on $\Omega \times S^1$. The notation was chosen so that $\mathbf{L}$ looks like a differential operator. Some care must be taken since the coefficient operators are not simple matrix multiplications. However, the idea behind the basic energy estimate for symmetric hyperbolic operators extends nearly immediately to $\mathbf{L}$. For example writing

$$\mathbf{L} := \mathbf{A}_0 \partial_t + \mathbf{G}(t)\,, \tag{7.46}$$

a straight forward integration by parts shows that for U rapidly decaying as $y \to \infty$ one has

$$\frac{d}{dt}\, (U(t), \mathbf{A}_0\, U(t))/2 = \Re\, (U(t), \mathbf{L}\, U(t)) + (U(t)\,, \mathbf{Z}\, U(t))\,, \tag{7.47}$$

where $(\ ,\ )$ denotes the scalar product in $L^2(\mathbb{R}^d \times S^1)$ and

$$\mathbf{Z} := (I - \mathbf{E})\, Z(y)\, (I - \mathbf{E}) + \mathbf{E}\, Z(y)\, \mathbf{E}\,, \tag{7.48}$$

with $Z(y)$ defined in (2.26). The measure $d\theta$ on the unit circle is normalized to be of total mass equal to one, that is

$$\int_{S^1} 1\, d\theta := 1\,. \tag{7.49}$$

Since $\mathbf{A}_0$ is a strictly positive operator, (47) can be made the centerpiece of existence theorems like those in Lecture 2.

In order to treat phases which need not be globally defined, this argument needs to be localized to domains of the form $\Omega \times S^1$ where Ω denotes a domain of determinacy as in (5.33). That too is straightforward starting from the energy law

$$\sum_{\mu=0}^{d} \partial_\mu\, e_\mu = \Re \int_{S^1} \langle U, \mathbf{L} U \rangle \, d\theta + \int_{S^1} \langle U, \mathbf{Z} U \rangle \, d\theta\,, \tag{7.50}$$

where

$$e_\mu(y) := \int_{S^1} \langle U, \mathbf{A}_\mu U \rangle \, d\theta\,. \tag{7.51}$$

The key observation here is that if one integrates by parts in the domain $\Omega \times S^1 \cap \{0 \le t \le \underline{t}\}$, the contribution from the sides of the cones is nonnegative and this leads to the localized energy law for smooth U

$$\|U(t)\|_{L^2(\Omega(t)\times S^1)} \le C(L,\phi)\left(\|U(0)\|_{L^2(\Omega(0)\times S^1)} + \int_0^t \|(\mathbf{L}\,U(\sigma))\|_{L^2(\Omega(\sigma)\times S^1)}\, d\sigma \right).$$
$$(7.52)$$

A commutation argument like that in 2.1 yields the more general estimate for $s \in \mathbb{N}$

$$\|U(t)\|_{H^s(\Omega(t)\times S^1)}$$
$$\le C(s,L,\phi)\left(\|U(0)\|_{H^s(\Omega(0)\times S^1)} + \int_0^t \|(\mathbf{L}\,U(\sigma))\|_{H^s(\Omega(\sigma)\times S^1)}\, d\sigma \right). \qquad (7.53)$$

Replacing derivatives by difference quotients leads then to a convergent sequence of approximating equations which can be used to prove the following linear existence theorem.

Theorem 7.1.   *If $s \in \mathbb{N}$, $g \in H^s(\Omega(0) \times S^1)$, and $f \in L^1(\Omega \times S^1)$ satisfies*

$$\int_0^T \left(\int_{\Omega(t)\times S^1} \sum_{|\alpha|\le s} |\partial_{x,\theta}^\alpha f(t,x,\theta)|^2 \, dx d\theta \right)^{1/2} dt < \infty, \qquad (7.54)$$

then there is a unique $U \in L^2(\Omega \times S^1)$ satisfying

$$\mathbf{L}\,U = f, \qquad \text{and} \qquad U|_{t=0} = g. \qquad (7.55)$$

In addition, the solution satisfies the estimate (7.53). If $g \in C^\infty(\Omega(0) \times S^1)$ and $f \in C^\infty(\Omega \times S^1)$, then $U \in C^\infty(\Omega \times S^1)$.

To treat nonlinear problems as in Lecture 6, note that for $s > (d+1)/2$, Schauder's Lemma implies that the map

$$U(t) \mapsto \mathbf{E}\,F(y, \mathbf{E}\,U(t))$$

is a locally Lipshitzean map of $H^s(\Omega(t) \times S^1)$ to itself, uniformly for $0 \le t \le T$. Standard Picard iteration,

$$\mathbf{L}\,U^{\nu+1} + \mathbf{E}\,F(y, \mathbf{E}\,U^\nu) = 0, \qquad U^{\nu+1}|_{t=0} = g \qquad (7.56)$$

as in Lecture 6 leads to the basic nonlinear local existence theorem. Existence is proved on $\Omega_T \times S^1$ where $\Omega_T := \Omega \cap \{0 \le t \le T\}$.

Local Solvability of the Principal Profile Equation 7.2.   *If $(d+1)/2 < s \in \mathbb{N}$ and $g_0 \in H^s(\Omega(0) \times S^1)$, then there is a $0 < T < R/c$ and unique $U_0 \in C(\Omega_T \times S^1)$ satisfying (7.43) together with the initial condition $U_0(0,.) = g$. If $g_0 \in C^\infty(\Omega(0) \times S^1)$ then $U_0 \in C^\infty(\Omega_T \times S^1)$.*

It is important to note that what is solved here is equation (7.43) which follows from the desired equations (7.26) and (7.27). Thus we have shown that the latter equations determine uniquely U_0 but we have not yet shown that there exists a U_0 satisfying (7.26) and (7.27). Clearly if the initial data g do not satisfy (7.26), then there is no chance for that equation.

Lemma 7.3. *If in addition to the hypotheses of the last theorem, $\mathbf{E}\,g = g$, then the resulting solution U_0 satisfies (7.26) and (7.27).*

Proof. As in the analysis of 5.4 an important first step is to observe that the left hand side of (7.43) is the sum of two orthogonal parts so that equation (7.43) implies that both vanish. Equivalently, multiplying (7.43) by $\mathbf{E}$ shows that U_0 satisfies the pair of equations

$$(I - \mathbf{E})\,L(y, \partial_y)\,(I - \mathbf{E})\,U_0 = 0\,, \quad \text{and} \quad \mathbf{E}\,L(y, \partial_y)\,\mathbf{E}\,U_0 + \mathbf{E}\,f(y, \mathbf{E}\,U_0(y, \theta)) = 0\,. \tag{7.57}$$

It follows that $\mathbf{E}\,U_0$ also satisfies both of these equations, and therefore equation (7.43). Since $\mathbf{E}\,U_0$ has the same initial data as U_0, it follows by uniqueness of solutions of the initial value problem for (7.43) that $\mathbf{E}\,U_0 = U_0$, which is equation (7.26).

Finally, (7.26) and the second equation of (7.57) imply (7.27). $\qquad\square$

Fix $0 < T < R/c$ as in the Existence Theorem 7.2. Then the higher order profiles can be found on $\Omega_t \times S^1$ so as to satisfy (7.39) and (7.40). The argument is as follows. In (7.40) write

$$U_j = \mathbf{E}\,U_j + (I - \mathbf{E})\,U_j = \mathbf{E}\,U_j + F_{j-1}$$

where (7.39) is used in the last equality. This yields an equation of the form

$$\mathbf{E}\,L(y, \partial_y)\,\mathbf{E}\,U_j + \text{ linear in } \mathbf{E}\,U_j = \text{ known}\,.$$

To this equation add $(I - \mathbf{E})\,L(y, \partial_y)$ applied to (7.39) to find an equation for

$$C := \mathbf{E}U_j \tag{7.58}$$

of the form

$$\mathbf{L}\,C + \text{ linear in } C = \text{ known}\,. \tag{7.59}$$

This linear equation determines C from its initial data. Imitating arguments which by now should be familiar one shows that if C satisfies $\mathbf{E}\,C = C$ at $t = 0$ then it does so throughout $\Omega \times S^1$ and that $U_j := C + \mathbf{Q}\,F_{j-1}$ satisfies the two profile equations (7.39) and (7.40).

Theorem 7.4 [Joly-Rauch]. *Suppose that $g_j = \mathbf{E}\,g_j \in C^\infty(\Omega(0) \times S^1)$, and that $T > 0$ is chosen as in the local existence theorem for the principal profile. Then there are uniquely determined profiles $U_j(y, \theta) \in C^\infty(\Omega_T \times S^1)$ with*

$$\mathbf{E}\,U_j\big|_{t=0} = g_j \quad on \quad \Omega(0) \times S^1 \tag{7.60}$$

so that if

$$U(\epsilon, y, \theta) \sim \sum_{j=0}^{\infty} \epsilon^j\,U_j(y, \theta) \quad in \quad C^\infty(\Omega_T \times S^1) \tag{7.61}$$

and

$$u(\epsilon, y) := U(\epsilon, y, \phi(y)/\epsilon)\,, \tag{7.62}$$

then

$$L(y, \partial_y)\,u + f(y, u) \sim 0 \quad in \quad C^\infty(\Omega_T) \tag{7.63}$$

Exercise. Carry out the argument evoked before the statement.

This completes the construction of an infinitely accurate approximate solution u. One point of view toward this, and that expressed in most science texts, is that the partial differential equations involve parameters which are only known approximately so an infinitely accurate approximation is for all practical purposes as good as an exact solution.

Hadamard offered a deeper appreciation of this remark. He observes that since there are uncertainties in the equations and data, in order for the equations to lead to well defined predictions, it is crucial that the predictions be unchanged or only very slightly changed when the equations and data are changed within the limits of the uncertainties. This lead to his notion of well posed problems.

In our case, the point of view of Hadamard leads to the question of showing that a pair of infinitely accurate approximate solutions with infinitely close initial data are in fact close. This does not follow from the basic existence theorem of section 6, because the approximate solutions tend to infinity in the configuration space H^s with $s > d/2$ and the sensitivity of the equation to perturbations grows for large data. One approach to circumventing this is to find a different configuration space in which a good existence theory is available and in which the approximate solutions do not grow. In the case $d = 1$, L^∞ does the trick. For higher dimensions, the space of bounded stratified solutions introduced by Rauch and Reed works and is the heart of the proof in [Joly-Rauch]. In Lecture 8 we give a different proof borrowing ideas from O. Gues and P. Donnat.

7.6. Rays and nonlinear transport

The formal expansion having been constructed, one can ask what has been gained. Put another way, can one learn anything interesting from the expansions? I will briefly assess the state of affairs.

First, since an infinitely accurate approximate solution exists on an interval $[0, T]$ of time, this strongly suggests that there is smooth existence for the nonlinear problem on such a domain. This is not apparent from the basic existence theorem and even given the approximate solution, is not really easy to prove as you will see in Lecture 8.

If you want to see how the solutions behave, you must solve the equations for the principal profile. These are an integro differential system which is basically a hyperbolic problem with one more space variable, namely θ. In this sense to find the principal profile is a little harder than to solve a single hyperbolic Cauchy problem. The payoff is not the solution of a single initial value problem, but the solution (asymptotically) of a one parameter family of such problems which have short wavelength oscillations. As pointed out in the introduction, these small structures make such a family particularly difficult to resolve, for example by numerical methods. It is worth noting that though the number of space variables is increased, there are no derivatives with respect to θ. Also if $rank\,\pi(y) = k$ then the unknown has essentially k components. The number of unknown functions is reduced from N to k. Thus the equation for the profile is often actually simpler that solving a single initial value problem for the original problem.

It is crucial to note that in most cases the equation for the profile is much simpler indeed. It is usually a nonlinear transport equation along rays. The notable exception to this rule is conical refraction which is briefly discussed in [JMR, Ann. Inst. Fourier] and the references in that paper.

In the remainder of this section we derive a typical transport equation. Toward that end we suppose as in 5.4, that $rank\, \pi(y) = 1$ for all y and introduce the vector field V from equation (5.49).

In addition, we suppose that the nonlinearity is odd, that is

$$F(y, -u) = -F(y, u)$$

for all y, u. This implies that if U is a solution of the profile equations, then so is $-U$. In addition, there is always reflection symmetry in θ, that is $U(y, -\theta)$ is a solution as soon as U is. Combining these two it follows that $-U(y, -\theta)$ is a solution as soon as U is.

From the uniqueness of solutions to the profile equation, it follows that U_0 will be odd in θ as soon as this is true at $t = 0$. Thus there is an interesting subclass of profiles which are odd in θ and therefore have no nonoscillatory Fourier component. Consider such profiles, namely those which satisfy

$$U_0(y, -\theta) = -U_0(y, \theta)\,.$$

For such profiles, the operator $\mathbf{E}$ acts simply by multiplication by π. The definition of V then shows that equation (41) is equivalent to the nonlinear transport system

$$V(y, \partial_y)\, U_0(y, \theta) + \pi(y)\, F(y, U_0(y, \theta)) = 0\,, \qquad \pi(y)\, U_0 = U_0\,.$$

For each fixed θ this is a semilinear ordinary differential equation for U_0 along the integral curves of V. Solving such a family of equations is radically simpler than solving a multidimensional hyperbolic system.

In addition, in special cases there are explicit solutions which give insight into the underlying dynamics defined by the nonlinear hyperbolic equation. It is in this way that the subject was developed and used in the applied mathematics community. The reader is encouraged to browse the references given in the bibliography to find interesting applications both mathematical and physical. In the applied literature the method often goes under the name *slowly varying envelope approximation.*

The equations (7.41) simplify even without the oddness assumption. In the more general case, equation (7.41) has both an oscillatory and nonoscillatory part. Denote with an underline, the average value of a periodic function of θ

$$\underline{g(\theta)} := \frac{1}{2\pi} \int_0^{2\pi} g(\theta)\, d\theta\,.$$

The oscillatory part is denoted with an asterisk,

$$g^*(\theta) := g - \underline{g}\,.$$

Equation (7.41) is equivalent to the coupled system

$$L(y, \partial_y)\, \underline{U_0} + \underline{F(y, U_0)} = 0\,,$$

$$V(y, \partial_y)\, (U_0)^*(y, \theta) + \pi(y)\, F(y, U_0(y, \theta))^* = 0\,, \qquad \pi(y)\, (U_0)^* = (U_0)^*\,.$$

Note that neither the mean field $\underline{U}_0$ nor the oscillatory part can be found by itself. They interact. The equation for the mean field is as complicated as a family of hyperbolic systems on $\mathbb{R}^{1+d}$ parametrized by θ. The equation for the oscillatory part is as complicated as a family of nonlinear transport equations parametrized by θ. Both equation are integro-differential. They are differential in t, x and integral in θ.

LECTURE 8
Justification of One Phase Nonlinear Geometric Optics

In the last section profiles $U_j(y, \theta)$, periodic in θ were constructed so that if

$$U(\epsilon, y, \theta) \sim \sum_{j=0}^{\infty} \epsilon^j \, U_j(y, \theta) \tag{8.1}$$

then

$$u^\epsilon(y) := U(\epsilon, y, \phi(y)/\epsilon) \tag{8.2}$$

satisfies

$$L(y, \partial) \, u^\epsilon + F(y, u^\epsilon) \sim 0 \,. \tag{8.3}$$

Denote by $v^\epsilon(y)$ the exact solution of

$$L(y, \partial) \, v^\epsilon + F(y, v^\epsilon) = 0\,, \qquad v^\epsilon\big|_{t=0} = u^\epsilon\big|_{t=0} \,. \tag{8.4}$$

To show that the asymptotic expansion is correct amounts to showing that

$$u^\epsilon(y) \sim v^\epsilon(y) \tag{8.5}$$

The proof is contained in this section. An important part of the proof is that the exact solution v^ϵ exists on an ϵ independent time interval. Since the $H^s(\mathbb{R}^d)$ norm of the initial data grows infinitely large for any $s > 0$ this is not at all obvious. It is about as hard to prove this existence as to prove the asymptotic equality (8.5).

8.1. The spaces $H_\epsilon^s(\mathbb{R}^d)$

A key to the proof is the introduction of ϵ dependent Sobolev norms. The asymptotic solution has the form (8.2) so that though the derivatives grow as ϵ decreases, the operator $\epsilon \partial$ applied to the asymptotic solution is bounded independent of ϵ. This suggests that one looks for estimates for these operators applied to the exact solution. This strategy was used by O. Gues [1993] to study the quasilinear version of the one phase theorems. In that context, he later showed [Gues 1992] that the results can be refined using other norms. The chronological order just recounted is correct. The reversed publication dates represent delays of publication. The spaces $H_\epsilon^s(\mathbb{R}^d)$ are essential in the more general setting of multiphase nonlinear geometric optics [Joly, Metivier, Rauch, Duke J.], and offer a simple approach in the one phase case.

457

Definition. For $s \in \mathbb{Z}$, $0 < \epsilon \le 1$, and $w \in H^s(\mathbb{R}^d)$ define the $H^s_\epsilon(\mathbb{R}^d)$ norm by

$$\|w\|^2_{H^s_\epsilon(\mathbb{R}^d)} := \sum_{|\alpha| \le s} \|\,(\epsilon \partial_x)^\alpha \, w\,\|^2_{L^2(\mathbb{R}^d)} \,. \tag{8.6}$$

A family w^ϵ is *bounded in* $H^s_\epsilon(\mathbb{R}^d)$ when

$$\sup_{0 < \epsilon \le 1} \|w^\epsilon\|_{H^s_\epsilon(\mathbb{R}^d)} \;<\; \infty \,.$$

Example. For t fixed, the family $u^\epsilon(y)$ defined in (2) is bounded in $H^s_\epsilon(\mathbb{R}^d)$ provided that the support of $U(\epsilon, t, x, \theta)$ is bounded in x and contained in the domain of definition of ϕ.

The norm in $H^s_\epsilon(\mathbb{R}^d)$ is equivalent to the norm whose square is equal to

$$\int_{\mathbb{R}^d} (1 + |\epsilon\,\xi|^2)^s \, |\hat{u}(\xi)|^2 \, d\xi, \tag{8.7}$$

which shows how the definition can be generalized to noninteger s.

The analogues of the Sobolev inequalities with these norms are immediate consequences of the following scaling identity,

$$v(x) := w(\epsilon x) \quad \Rightarrow \quad \partial_x v = (\epsilon \partial_x w)(\epsilon x) \,. \tag{8.8}$$

Thus,

$$\partial_x^\alpha v = \big((\epsilon \partial_x)^\alpha w\big) \,(\epsilon x) \,, \tag{8.9}$$

so

$$\|\partial_x^\alpha v\|^2_{L^2(\mathbb{R}^d)} = \int |(\epsilon \partial_x^\alpha) w\,(\epsilon x)|^2 \, dx \,.$$

The change of variable $X := \epsilon x$ shows that this is equal to

$$\int |\,(\epsilon \partial_x)^\alpha w(X)\,|^2 \, \epsilon^{-d} \, dX = \epsilon^{-d} \,\|\,(\epsilon \partial_x)^\alpha \, w\,\|^2_{L^2(\mathbb{R}^d)} \,.$$

Summing shows that

$$\epsilon^{d/2} \,\|v\|_{H^s(\mathbb{R}^d)} = \|w\|_{H^s_\epsilon(\mathbb{R}^d)} \,. \tag{8.10}$$

Using this one finds the following embedding of $H^s_\epsilon(\mathbb{R}^d)$ in L^∞. For $s > d/2$,

$$\|w\|_{L^\infty(\mathbb{R}^d)} = \|v\|_{L^\infty(\mathbb{R}^d)} \le C(s,d) \,\|v\|_{H^s(\mathbb{R}^d)} = \epsilon^{-d/2} \, C(s,d) \,\|w\|_{H^s_\epsilon(\mathbb{R}^d)} \,. \tag{8.11}$$

Similarly for smooth $F(w)$ which vanish for $w = 0$ one has the ϵ version of Schauder's Lemma for $s > d/2$,

$$\begin{aligned}
\|F(w)\|_{H^s_\epsilon(\mathbb{R}^d)} &= \epsilon^{d/2} \,\|F(v)\|_{H^s(\mathbb{R}^d)} \\
&\le \epsilon^{d/2} \, G(\|v\|_{H^s(\mathbb{R}^d)}) = \epsilon^{d/2} \, G(\epsilon^{-d/2} \,\|w\|_{H^s_\epsilon(\mathbb{R}^d)}) \,.
\end{aligned} \tag{8.12}$$

Note that if one estimates the L^∞ norm of a function using (8.11) there is a large factor $\epsilon^{-d/2}$ intervening. Similarly, in the Schauder estimate (8.12), the negative power of ϵ in the the argument of G is a potential source of trouble. The Moser estimate is better behaved,

$$\begin{aligned}
\|F(w)\|_{H^s_\epsilon(\mathbb{R}^d)} &= \epsilon^{d/2} \,\|F(v)\|_{H^s(\mathbb{R}^d)} \\
&\le \epsilon^{d/2} \, G(\|v\|_{L^\infty}) \,\|v\|_{H^s(\mathbb{R}^d)} = G(\|w\|_{L^\infty}) \,\|w\|_{H^s_\epsilon(\mathbb{R}^d)} \,.
\end{aligned} \tag{8.13}$$

The fortuitous cancellation of powers of $\epsilon^{\pm d/2}$ in the last inequality shows that the Moser inequality for $H^s_\epsilon(\mathbb{R}^d)$ has exactly the same form, with the same nonlinear function G.

In the justification of the asymptotic expansion, it is crucial to estimate the difference $F(u) - F(v)$ when u and v are close. In our application, the function u is our approximate solution and v is the exact solution. In this way, more is known of u than of v. There is sup norm control of the operators $\epsilon\partial$ applied to u which will be used to get L^2 control on these operators applied to v.

Lemma 8.1.   *For any $R > 0$ and s there is a constant $C = C(F, R, s)$ so that if u satisfies*

$$\|(\epsilon\partial)^\alpha u\|_{L^\infty(\mathbb{R}^d)} \leq R \qquad \text{for all } |\alpha| \leq s \,,$$

and w satisfies

$$\|w\|_{L^\infty(\mathbb{R}^d)} \leq R \,,$$

then

$$\|F(y, u + w) - F(y, u)\|_{H^s_\epsilon(\mathbb{R}^d)} \leq C(R) \|w\|_{H^s_\epsilon(\mathbb{R}^d)} \,.$$

Proof. To simplify the exposition suppose that F does not depend on y. Then it suffices to prove the assertion with $\epsilon = 1$ since both sides of the inequality scale as $\epsilon^{d/2}$.

To prove the assertion for $\epsilon = 1$ write

$$F(u + w) - F(w) = w \int_0^1 F'(u + \lambda w) \, d\lambda := w \, \mathcal{G}(u, w) \,.$$

Expanding $\partial^\nu (w\mathcal{G}(u, w))$ using Leibniz' rule yields a finite number of terms of the form

$$H_{\alpha, \beta}(u, w) \, (\partial^{\alpha_1} u) \cdots (\partial^{\alpha_m} u) \, (\partial^{\beta_1} w) \cdots (\partial^{\beta_n} w)$$

with $\sum \alpha_k + \sum \beta_l = \nu$. The product of the first $m+1$ factors has sup norm bounded by $C(R)$. The Gagliardo-Nirenberg estimates imply that the product of the last n has $L^2(\mathbb{R}^d)$ norm bounded by $C(R)\|w\|_{H^s(\mathbb{R}^d)}$. This completes the proof. $\square$

Exercise. Prove the Lemma for F which depend on y.

For phases which are only locally defined, spaces $H^s_\epsilon(B_r)$ are needed where B_r denotes a ball of radius r in $\mathbb{R}^d$. The definition is exactly as in (8.6) with the change that $\mathbb{R}^d$ is replaced by B_r. The scaling $v = w(rx)$ shows that for balls centered at the origin and $0 < r_0 \leq r \leq r_1 < \infty$, there is a constant C so that

$$C^{-1} \|w\|_{H^s_\epsilon(B_r)} \leq \|v\|_{H^s_\epsilon(B_1)} \leq C \|w\|_{H^s_\epsilon(B_r)} \,.$$

The usual reflection operators construct a bounded linear extension operator $v \to Ev$ from $H^s_\epsilon(B_1)$ to $H^s_\epsilon(\mathbb{R}^d)$ so that $Ev = v$ on B_1.

Exercise.   For $s = 0$ extending v to vanish outside B works. For $s = 1$ and $|x| > 1$ denote by $R(x) := x/|x|^2$ the reflected point in the unit sphere. Choose a smooth function χ which is equal to 1 on a neighborhood of 1 and vanishes outside $]1/2, 3/2[$. Show that setting $Ev(x) := \chi(|x|) \, v(R(x))$ for $|x| > 1$ works for $s = 1$. For larger s construct an appropriate extension operator by setting

$$Ev := \sum_{j=1}^s c_j \, v(R(2^j x)) \,.$$

The key is the choice of the constants c_j so that $s - 1$ derivatives match at the boundary of the ball. This elegant idea is called Lions reflection after J.L. Lions.

Proposition. *Lemma 8.1 holds with $\mathbb{R}^d$ replaced by B_r with uniform constant provided that r is bounded away from 0 and ∞.*

Exercise. Write out the details of the proof by combining the above remarks.

8.2. H_ϵ^s estimates for linear symmetric hyperbolic systems

In addition to the estimates of the last section, the proof relies on the fact that linear hyperbolic systems propagate the $H_\epsilon^s(\mathbb{R}^d)$ norms. This fact depends on the basic linear estimate and commutation identities between the operator L and the operators $(\epsilon\partial)^\alpha$. The argument is entirely analogous to the proof of the proposition in §2.1. Square brackets are used to denote the commutator.

Introducing the new variable $\underline{u} := A_0^{-1/2} u$ and multiplying the resulting system for $\underline{u}$ by $A_0^{-1/2}$ yields a semilinear equation of the same form as before with new coefficient matrices $\underline{A}_\mu := A_0^{-1/2} A_\mu A_0^{-1/2}$. In particular, the coefficient of the time derivative is equal to the identity matrix. Thus, without loss of generality we may suppose that $A_0 = I$.

Lemma 8.2. *If $A_0 = I$, then for any $\alpha \in \mathbb{N}^d$ there are matrix valued functions $C_{\alpha\beta}(\epsilon, y)$ with uniformly bounded derivatives on $]0, 1] \times \mathbb{R}^{1+d}$ so that*

$$[L(y, \partial_y), (\epsilon\partial_x)^\alpha] = \sum_{|\beta| \leq |\alpha|} C_{\alpha\beta}(\epsilon, y)(\epsilon\partial_x)^\beta.$$

Proof. The proof is by induction on $|\alpha|$. For $|\alpha| = 1$ compute

$$[L(y, \partial_y), \epsilon\partial_j] = -\sum_k (\partial_j A_k)\epsilon\partial_k + \epsilon(\partial_j B)$$

Suppose next that $m \geq 1$, and the result is true for derivatives of length less than or equal to m. A differention of length $m + 1$ is of the form $\epsilon\partial_j(\epsilon\partial_x)^\alpha$ with $|\alpha| = m$. Then

$$L\,\epsilon\partial_j\,(\epsilon\partial_x)^\alpha - \epsilon\partial_j\,(\epsilon\partial_x)^\alpha\,L = [L, \epsilon\partial_j]\,(\epsilon\partial_x)^\alpha + \epsilon\partial_j\,L\,(\epsilon\partial_x)^\alpha - \epsilon\partial_j\,(\epsilon\partial_x)^\alpha\,L$$

$$= [L, \epsilon\partial_j]\,(\epsilon\partial_x)^\alpha + \epsilon\partial_j\,[L, (\epsilon\partial_x)^\alpha].$$

Using the inductive hypothesis to express the commutators, the result follows. $\square$

Theorem 8.3. *If $A_0 = I$ then for any $s \in \mathbb{N}$ and $T \in]0, \infty[$ there is a constant $C = C(s, T, L)$ so that for all $0 \leq t \leq T$, and $u \in C([0, t]; H^s(\mathbb{R}^d))$ with $Lu \in L^1([0, t]; H^s(\mathbb{R}^d))$,*

$$\|u(t)\|_{H_\epsilon^s(\mathbb{R}^d)} \leq C\left(\|u(0)\|_{H_\epsilon^s(\mathbb{R}^d)} + \int_0^t \|(Lu)(\sigma)\|_{H_\epsilon^s(\mathbb{R}^d)}\,d\sigma\right). \qquad (8.14)$$

Proof. For $|\alpha| \leq s$ use the commutation lemma to write

$$L\left(\epsilon\partial_x\right)^\alpha u = \left(\epsilon\partial_x\right)^\alpha Lu + \sum C_{\alpha\beta}(\epsilon, y)\left(\epsilon\partial_x\right)^\beta u. \tag{8.15}$$

The basic linear estimate (2.20) then implies that for any $0 \leq t \leq \underline{t}$,

$$
\begin{aligned}
\left\|(\epsilon\partial_x)^\alpha u(t)\right\|_{L^2(\mathbb{R}^d)} \leq\; &C\left(\left\|(\epsilon\partial_x)^\alpha u(0)\right\|_{L^2(\mathbb{R}^d)}\right.\\
&\left.+ \int_0^t \left\{\left\|(\epsilon\partial_x^\alpha)(Lu)(\sigma)\right\|_{L^2(\mathbb{R}^d)} + \|u(\sigma)\|_{H_\epsilon^s(\mathbb{R}^d)}\right\} d\sigma\right).
\end{aligned}
\tag{8.16}
$$

Summing over all $|\alpha| \leq s$ yields with a new constant

$$\|u(t)\|_{H_\epsilon^s(\mathbb{R}^d)} \leq C\left(\|u(0)\|_{H_\epsilon^s(\mathbb{R}^d)} + \int_0^t \left\{\|(Lu)(\sigma)\|_{H_\epsilon^s(\mathbb{R}^d)} + \|u(\sigma)\|_{H_\epsilon^s(\mathbb{R}^d)}\right\} d\sigma\right). \tag{8.17}$$

Gronwall's inequality completes the proof. $\qquad\square$

For applications where the phase is only locally defined one must have local versions of the above estimates. Denote by Ω the usual domain of determinacy

$$\Omega_T := [0,T] \cap \left\{|x - a| \leq R - ct\right\}, \qquad cT < R. \tag{8.18}$$

Denote by $\Omega(t)$ the cross section which is a solid ball in $\mathbb{R}^d$,

$$\Omega(t) := \left\{x \;:\; (t,x) \in \Omega\right\}. \tag{8.19}$$

The $H_\epsilon^s(\Omega(t))$ norm is equivalent to the $H^s(\Omega(t))$ norm and is defined exactly as in (8.6) with $\mathbb{R}^d$ replaced by $\Omega(t)$. The following local estimate is sufficient for our needs. The proof is exactly like the proof of the estimate in $H^s(\mathbb{R}^d)$.

Theorem 8.4. *If Ω_T is defined by (8.18) and $s \in \mathbb{N}$ there is a constant $C = C(s, L, \Omega)$ so that for all $u \in C^\infty(\Omega_T)$ and all $t \in [0,T]$*

$$\|u(t)\|_{H_\epsilon^s(\Omega(t))} \leq C\left(\|u(0)\|_{H_\epsilon^s(\Omega(0))} + \int_0^t \|Lu(\sigma)\|_{H_\epsilon^s(\Omega(\sigma))}\, d\sigma\right). \tag{8.20}$$

8.3. Justification of the nonlinear asymptotics

Theorem 8.5 [Joly-Rauch, 1992]. *Suppose that the phase ϕ and smooth profiles $U_j(y, \theta)$ satisfy the profile equations on the domain of determinacy Ω_T in (8.18), and the approximate solution u^ϵ is defined by (8.2) with $U(\epsilon, y, \theta) \sim \sum \epsilon^j U_j(y, \theta)$ in $C^\infty(\Omega_T \times S^1)$. Then for ϵ small the exact solution v^ϵ defined in (8.4) exists and is smooth on Ω_T and*

$$v^\epsilon \sim u^\epsilon \qquad in \quad C^\infty(\Omega_T). \tag{8.21}$$

Proof. We present the proof of Gues 1993 with a simplification of Donnat 1994. Fix $d/2 < s \in \mathbb{N}$. The local existence theorem implies either the existence of a smooth solution v^ϵ on Ω_T or the existence of a $T^*(\epsilon) \leq T$ so that v^ϵ is smooth on the half open cone

$$\Omega_* := \Omega \cap \{0 \leq t < T^*\} \tag{8.22}$$

and v^ϵ blows up at T^* in the sense that

$$\lim_{t \to T^*(\epsilon)} \|v^\epsilon(t)\|_{H^s_\epsilon(\Omega(t))} = \infty. \tag{8.23}$$

We show that for ϵ sufficiently small, the second alternative does not occur and that (8.21) holds.

Choose $R > 0$ so that for $0 < \epsilon \leq 1$ and $|\alpha| \leq s$

$$\sup_{0 \leq t \leq T} \left(\|(\epsilon\partial_x)^\alpha u^\epsilon\|_{L^\infty(\Omega(t))} \right) \leq R/2. \tag{8.24}$$

Denote by $r^\epsilon(y)$ the residual in the equation for the approximate solution,

$$L(y, \partial_y) u^\epsilon + F(y, u^\epsilon) = r^\epsilon. \tag{8.25}$$

By construction

$$r^\epsilon(y) \sim 0 \quad \text{in} \quad C^\infty(\Omega_T). \tag{8.26}$$

Define the error

$$w^\epsilon := v^\epsilon - u^\epsilon. \tag{8.27}$$

An initial value problem for the error is derived by subtracting (8.25) from (8.4). Suppressing the y dependence of F this yields

$$L w^\epsilon + F(u^\epsilon + w^\epsilon) - F(u^\epsilon) = -r^\epsilon, \quad \text{on} \quad \Omega_*, \tag{8.28}$$

$$w^\epsilon\big|_{t=0} = 0. \tag{8.29}$$

The linear H^s_ϵ estimate gives a $C = C(s, L, \Omega)$ independent of ϵ and of t so that for $t < T^*$,

$$\|w^\epsilon(t)\|_{H^s_\epsilon(\Omega(t))} \leq C\left(\int_0^t \|F(u^\epsilon + w^\epsilon) - F(u^\epsilon)\|_{H^s_\epsilon(\Omega(\sigma))} \, d\sigma + \int_0^T \|r^\epsilon\|_{H^s_\epsilon(\Omega(\sigma))} \, d\sigma \right). \tag{8.30}$$

So long as

$$\sup_{0 \leq \sigma \leq t} \left(\|w^\epsilon\|_{L^\infty(\Omega(\sigma))} \right) \leq R/2, \tag{8.31}$$

the Lemma yields with new constant $C = C(s, L, \Omega, F, R)$

$$\|w^\epsilon(t)\|_{H^s_\epsilon(\Omega(t))} \leq C\left(\int_0^t \|w^\epsilon\|_{H^s_\epsilon(\Omega(\sigma))} \, d\sigma + \int_0^T \|r^\epsilon\|_{H^s_\epsilon(\Omega(\sigma))} \, d\sigma \right). \tag{8.32}$$

The first application of this estimate is to show that v^ϵ is smooth on Ω_T for ϵ sufficiently small. If v^ϵ is not smooth on Ω_T then since $w^\epsilon(0) = 0$ and w^ϵ explodes as $t \nearrow T^*$, there is a smallest $\underline{t} \in]0, T[$ so that

$$\|w^\epsilon\|_{H^s_\epsilon(\Omega(\underline{t}))} + \|w^\epsilon\|_{L^\infty(\Omega(\underline{t}))} = R/2, \tag{8.33}$$

Then, (8.32) holds for $0 \leq t \leq \underline{t}$ and Gronwall's inequality implies that

$$\sup_{0 \leq t < \underline{t}} \|w^\epsilon(t)\|_{H^s_\epsilon(\Omega(t))} \leq C \, e^{CT} \int_0^T \|r^\epsilon\|_{H^s_\epsilon(\Omega(\sigma))} \, d\sigma \,. \tag{8.34}$$

Estimate (8.34) is used to recover norms without epsilons. The key inequalities in this direction are

$$\|w\|_{L^\infty(\Omega(t))} \leq C(s, \Omega) \, \|w\|_{H^s(\Omega(t))} \,,$$

and

$$\|w\|_{H^s(\Omega(t))} \leq \epsilon^{-s} \|w\|_{H^s_\epsilon(\Omega(t))} \quad \text{for} \quad 0 < \epsilon \leq 1 \,. \tag{8.35}$$

Using the facts that $\epsilon \leq 1$ and T is fixed, it follows that for $0 \leq t \leq \underline{t}$,

$$\sup_{0 \leq t < \underline{t}} \left(\|w^\epsilon\|_{H^s(\Omega(t))} + \|w^\epsilon\|_{L^\infty(\Omega(t))} \right) \leq \frac{C}{\epsilon^s} \int_0^T \|r^\epsilon\|_{H^s_\epsilon(\Omega(\sigma))} \, d\sigma \,. \tag{8.36}$$

Since $r^\epsilon \sim 0$, it follows that there is a $0 < \epsilon_0 \leq 1$ such that

$$\epsilon \in]0, \epsilon_0] \quad \Rightarrow \quad \text{r.h.s. of (8.36)} \leq R/4 \,. \tag{8.37}$$

In particular, if $0 < \epsilon \leq \epsilon_0$,

$$\sup_{0 \leq t \leq \underline{t}} \left(\|w^\epsilon\|_{H^s(\Omega(\underline{t}))} + \|w^\epsilon\|_{L^\infty(\Omega(\underline{t}))} \right) \leq R/4 \,. \tag{8.38}$$

This contradicts the definition of $\underline{t}$. The conclusion is that for $\epsilon \in]0, \epsilon_0]$, v^ϵ is smooth on Ω_T and in addition

$$\sup_{0 \leq \sigma \leq T} \left(\|w^\epsilon\|_{H^s_\epsilon(\Omega(t))} + \|w^\epsilon\|_{L^\infty(\Omega(t))} \right) \leq R/4 \,. \tag{8.39}$$

For these values of ϵ, (8.32) holds for $0 \leq t \leq T$ and Gronwall's inequality implies that

$$\sup_{0 \leq t < T} \left(\|w^\epsilon\|_{H^s(\Omega(t))} + \|w^\epsilon\|_{L^\infty(\Omega(t))} \right) \leq \frac{C}{\epsilon^s} \int_0^T \|r^\epsilon\|_{H^s_\epsilon(\Omega(\sigma))} \, d\sigma \,. \tag{8.40}$$

Since $r^\epsilon \sim 0$, the right hand side is $\leq C_M \epsilon^M$ for every $M > 0$. We have shown that v^ϵ is smooth on Ω_T for $\epsilon \leq \epsilon_0$, and, that for any $s > d/2$ and $M > 0$ there is a $C = C(s, M)$ so that for $\epsilon \leq \epsilon_0$

$$\sup_{0 \leq t < T} \|w^\epsilon\|_{H^s(\Omega(t))} \leq C_M \, \epsilon^M \,.$$

This is exactly the conclusion of the Theorem thanks to Sobolev's inequality. $\square$

Exercise. There is gap here. Sobolev's Theorem shows that $\partial_x^\beta w^\epsilon = O(\epsilon^\infty)$. Thus the space derivatives are infinitely small as required. However, this asserts nothing at all about time derivatives. Show how to control time derivatives and mixed time and space derivatives. **Hint.** It is a straightforward consequence, you do not have to restart the proof.

References

S. Alinhac, *Equations Differentielles: Etude Asymptotique; Application aux Equations aux Dérivées Paritelles*, Lecture Notes, University of Paris Sud, Orsay, 1980.

M. Beals, *Propagation and Interaction of Singularities in Nonlinear Hyperbolic Problems*, Birkhäuser, Boston, 1989.

Y. Choquet-Bruhat, Ondes asymptotiques et approchées pour les systèmes d'équations aux dérivées partielles non linéaires, J. Math. Pures. Appl. 48(1969), 117-158.

R. Courant, *Methods of Mathematical Physics, vol. II.* Interscience Publishers, 1962.

P. Donnat, Quelque contributions mathématiques in optique non linéaire, Ph.D. thesis, Ecole Polytéchnique, Paris, 1994.

P. Donnat and J. Rauch, Modelling the dispersion of light, in *Singularities and Oscillations*, eds J. Rauch and M. Taylor, IMA Volumes in Mathematics and its Applications, Springer Verlag, 1997.

O. Gues, Ondes multidimensionelles epsilon stratifiées et oscillations, Duke Math. J. 1992.

O. Gues, Dévelopment asymptotique de solutions exactes de systèmes hyperboliques quasilinéaires, Asympt. Anal. 6(1993), 241-269.

L. Hörmander, *The Analysis of Linear Partial Differential Operators vols.I,II*, Springer-Verlag, Berlin, 1983.

J. Hunter, and J. Keller, Weakly nonlinear nonlinear waves, Comm. Pure Appl. Math. 36(1983), 547-569.

J. Hunter, A. Majda, and R. Rosales, Resonantly interacting weakly nonlinear hyperbolic waves II, Stud. Appl. Math. 75(1986), 187-226.

F. John, *Partial Differential Equations* 4th ed., Springer-Verlag, New York, 1982.

J.L. Joly and J. Rauch, Justification of multidimensional single phase semilinear geometric optics, Trans. A.M.S. 330(1992), 599-623.

J.L. Joly, G. Métivier and J. Rauch, Resonant one dimensional nonlinear geometric optics, J. Funct. Anal. 114(1993), 106-231.

J.L. Joly, G. Métivier and J. Rauch, Coherent nonlinear waves and the Wiener algebra, Ann. Inst. Fourier, 44(1994), 167-196.

J.L. Joly, G. Métivier and J. Rauch, Generic rigorous asymptotic expansions for weakly nonlinear multidimensional oscillatory waves, Duke Math. J. 70(1993), 373-404.

J.L. Joly, G. Métivier and J. Rauch, A nonlinear instability for 3×3 systems of conservation laws, Comm. Math. Phys. 162(1994), 47-59.

J.L. Joly, G. Métivier and J. Rauch, Dense oscillations for the compressible two dimensional Euler equations, in *Nonlinear Partial Differential Equations and their Applications, College de France Seminar 1992-1993*, eds. H. Brezis and J.L. Lions, Pitman Research Notes in Math. to appear.

J.L. Joly, G. Métivier and J. Rauch, Diffractive nonlinear geometric optics, in Séminaire Equations aux Dérivées Partielles, Ecole Polytéchnique, Palaiseau, 1995-1996.

P.D. Lax, Asymptotic solutions of oscillatory initial value problems, Duke Math. J. 24(1957), 627-646.

P. D. Lax, *Lectures on Hyperbolic Partial Differential Equations*, Stanford University Lecture Notes, 1963.

D. Ludwig, Conical refraction in crystal optics and hydromagnetics, Comm. Pure Appl. Math. XIV(1961), 113-124.

A. Majda, Nonlinear geometric optics for hyperbolic systems of conservation laws, in *Oscillations Theory, Computation, and Methods of Compensated Compactness* eds. C. Dafermos, J. Ericksen, D. Kenderlehrer, and I. Muller, IMA Volumes on Mathematics and its Applications vol.2, pp. 115-165, Springer-Verlag, New York, 1986.

A. Majda and R. Rosales, Resonantly interacting weakly nonlinear hyperbolic waves , Stud. Appl. Math. 71(1986), 149-179.

A. Majda, R. Rosales, and M. Schonbek, A canonical system of integro-differential equations in nonlinear acoustics, Stud. Appl. Math. 79(1988), 205-262.

D. McLaughlin, G. Papanicolaou, and L. Tartar, Weak limits of semilinear hyperbolic systems with oscillating data, pp. 277-298 in *Macroscopic Modeling of Turbulent Flows*, Lecture Notes in Physics vol 230, Springer-Verlag, 1985.

J. Rauch, An L^2 proof that H^s is invariant under nonlinear maps for $s > n/2$, in *Global Analysis - Analysis on Manifolds*, ed. T, Rassias, Teubner Texte zur Math., band 57, Leipzig, 1983.

J. Rauch, *Partial Differential Equations*, Graduate Texts in Math 128, Springer-Verlag, New York, 1991.

J. Rauch and M. Reed, Nonlinear microlocal analysis of semilinear hyperbolic systems in one space dimension, Duke Math. J. 49(1982), 379-475.

S. Schochet, Fast singular limits of hyperbolic equations, J. Diff. Eq. 114(1994), 476-512.

M. Taylor, *Partial Differential Equations, Basic Theory*, Springer-Verlag, 1997.

T. Wagenmaker, Analytic solutions and resonant solutions of hyperbolic partial differential equations, Ph,D Thesis, University of Michigan, 1993.

G. Whitham, *Linear and Nonlinear Waves*, Wiley-Interscience, New York, 1974.